D1687373

BIOLOGICAL AND MEDICAL PHYSICS, BIOMEDICAL ENGINEERING

For further volumes:
http://www.springer.com/series/3740

BIOLOGICAL AND MEDICAL PHYSICS, BIOMEDICAL ENGINEERING

The fields of biological and medical physics and biomedical engineering are broad, multidisciplinary and dynamic. They lie at the crossroads of frontier research in physics, biology, chemistry, and medicine. The Biological and Medical Physics, Biomedical Engineering Series is intended to be comprehensive, covering a broad range of topics important to the study of the physical, chemical and biological sciences. Its goal is to provide scientists and engineers with textbooks, monographs, and reference works to address the growing need for information.

Books in the series emphasize established and emergent areas of science including molecular, membrane, and mathematical biophysics; photosynthetic energy harvesting and conversion; information processing; physical principles of genetics; sensory communications; automata networks, neural networks, and cellular automata. Equally important will be coverage of applied aspects of biological and medical physics and biomedical engineering such as molecular electronic components and devices, biosensors, medicine, imaging, physical principles of renewable energy production, advanced prostheses, and environmental control and engineering.

Editor-in-Chief:
Elias Greenbaum, Oak Ridge National Laboratory, Oak Ridge, Tennessee, USA

Editorial Board:

Masuo Aizawa, Department of Bioengineering,
Tokyo Institute of Technology, Yokohama, Japan

Olaf S. Andersen, Department of Physiology,
Biophysics & Molecular Medicine,
Cornell University, New York, USA

Robert H. Austin, Department of Physics,
Princeton University, Princeton, New Jersey, USA

James Barber, Department of Biochemistry,
Imperial College of Science, Technology
and Medicine, London, England

Howard C. Berg, Department of Molecular
and Cellular Biology, Harvard University,
Cambridge, Massachusetts, USA

Victor Bloomfield, Department of Biochemistry,
University of Minnesota, St. Paul, Minnesota, USA

Robert Callender, Department of Biochemistry,
Albert Einstein College of Medicine,
Bronx, New York, USA

Britton Chance, Department of Biochemistry/
Biophysics, University of Pennsylvania,
Philadelphia, Pennsylvania, USA

Steven Chu, Lawrence Berkeley National
Laboratory, Berkeley, California, USA

Louis J. DeFelice, Department of Pharmacology,
Vanderbilt University, Nashville, Tennessee, USA

Johann Deisenhofer, Howard Hughes Medical
Institute, The University of Texas, Dallas,
Texas, USA

George Feher, Department of Physics,
University of California, San Diego, La Jolla,
California, USA

Hans Frauenfelder,
Los Alamos National Laboratory,
Los Alamos, New Mexico, USA

Ivar Giaever, Rensselaer Polytechnic Institute,
Troy, New York, USA

Sol M. Gruner, Cornell University,
Ithaca, New York, USA

Judith Herzfeld, Department of Chemistry,
Brandeis University, Waltham, Massachusetts, USA

Mark S. Humayun, Doheny Eye Institute,
Los Angeles, California, USA

Pierre Joliot, Institute de Biologie
Physico-Chimique, Fondation Edmond
de Rothschild, Paris, France

Lajos Keszthelyi, Institute of Biophysics, Hungarian
Academy of Sciences, Szeged, Hungary

Robert S. Knox, Department of Physics
and Astronomy, University of Rochester, Rochester,
New York, USA

Aaron Lewis, Department of Applied Physics,
Hebrew University, Jerusalem, Israel

Stuart M. Lindsay, Department of Physics
and Astronomy, Arizona State University,
Tempe, Arizona, USA

David Mauzerall, Rockefeller University,
New York, New York, USA

Eugenie V. Mielczarek, Department of Physics
and Astronomy, George Mason University, Fairfax,
Virginia, USA

Markolf Niemz, Medical Faculty Mannheim,
University of Heidelberg, Mannheim, Germany

V. Adrian Parsegian, Physical Science Laboratory,
National Institutes of Health, Bethesda,
Maryland, USA

Linda S. Powers, University of Arizona,
Tucson, Arizona, USA

Earl W. Prohofsky, Department of Physics,
Purdue University, West Lafayette, Indiana, USA

Andrew Rubin, Department of Biophysics, Moscow
State University, Moscow, Russia

Michael Seibert, National Renewable Energy
Laboratory, Golden, Colorado, USA

David Thomas, Department of Biochemistry,
University of Minnesota Medical School,
Minneapolis, Minnesota, USA

Thomas M. Deserno
Editor

Biomedical Image Processing

With 254 Figures

Springer

Editor
Dr. Thomas M. Deserno
RWTH Aachen
Institut für Medizinische Informatik
Pauwelsstr. 30
52074 Aachen, Germany
E-mail: deserno@ieee.org

Biological and Medical Physics, Biomedical Engineering   ISSN 1618-7210
ISBN 978-3-642-15815-5   e-ISBN 978-3-642-15816-2
DOI 10.1007/978-3-642-15816-2
Springer Heidelberg Dordrecht London New York

Library of Congress Control Number: 2011921516

© Springer-Verlag Berlin Heidelberg 2011
This work is subject to copyright. All rights are reserved, whether the whole or part of the material is concerned, specifically the rights of translation, reprinting, reuse of illustrations, recitation, broadcasting, reproduction on microfilm or in any other way, and storage in data banks. Duplication of this publication or parts thereof is permitted only under the provisions of the German Copyright Law of September 9, 1965, in its current version, and permission for use must always be obtained from Springer. Violations are liable to prosecution under the German Copyright Law.
The use of general descriptive names, registered names, trademarks, etc. in this publication does not imply, even in the absence of a specific statement, that such names are exempt from the relevant protective laws and regulations and therefore free for general use.

Cover design: eStudio Calamar Steinen

Printed on acid-free paper

Springer is part of Springer Science+Business Media (www.springer.com)

To Verena – the beauty and the beauty of images

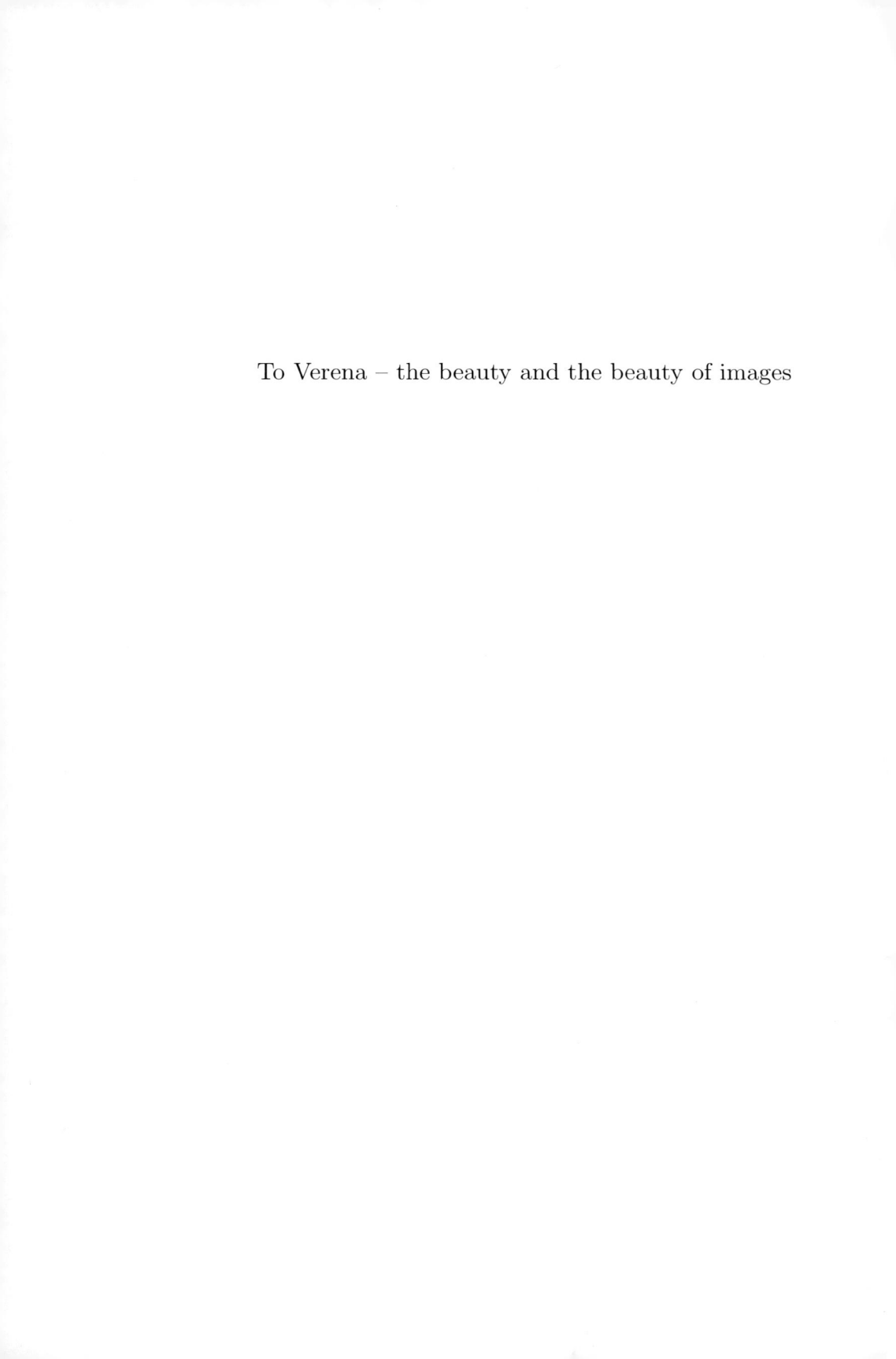

Preface

YATBIP: Yet another textbook on biomedical image processing? – Hopefully not...

Based on the tutorial *SC086 – Fundamentals of Medical Image Processing* regularly offered at the International SPIE Symposium on Medical Imaging, the Springer-Verlag Series Editor of *Biological and Medical Physics, Medical Engineering* invited me in January 2009 to compile this book. Actually, the idea of providing a "suitable" textbook – comprehensive but short, up-to-date but essential, and detailed but illustrative – for novices like experts, and at reasonable costs, is not new. For years, the lack of any such textbook in image processing covering all of the special needs in biology and medicine is evident. In any teaching lecture, tutorial as well as graduate class. I'm always asked by the students to suggest literature but cannot answer satisfyingly, simply because there isn't a "suitable" textbook yet.

So we aimed at compiling a high-quality collection of chapters, written for scientists, researchers, lectures and graduate students as well, covering the recent advantages in the broad field of biomedical imaging and image processing in an exemplary way. In February 2009, several fruitful discussions with colleagues at SPIE Medical Imaging convinced me to face the challenge, and I started recruiting author teams for contributions. Finally, 47 authors from 11 nations all over the world collaborated – all of them leading experts in their field. Intensive efforts were made to direct all authors towards a similar style of presentation and equal degree of technical details. Beside some written guidelines, the overview chapter was provided to the authors as an example before they started writing. All authors first provided a short outline and a detailed table of content, which were distributed between all contributors together with a strictly enforced time line. In October 2009, submission of chapters started, and each manuscript was edited carefully. Editor requests have been processed by the authors improving completeness and clarity of presentation, and finally in June 2010, the manuscript was submitted to the publisher.

Fig. 1. *Eierlegende Wollmilchsau.* Every morning, this special animal provides a cooked egg with chilled fresh milk. Its wool is used for high-quality clothes and the meat for excellent dining. It is the first *all-in-one* approach documented in history (Courtesy of: http://neulehrer.wordpress.com/)

As a result, this book has appeared as uniform monograph with an overview chapter contributed by the editor, followed by some twenty chapters focusing on particular parts selected from biomedical imaging and image processing. Each chapter gives an introduction and overview of recent trends in its field and provides particular case examples, usually taken from the author's own research.

Primarily addressing engineers and system developers in computer sciences, the book covers the entire processing pipeline of biomedical imaging. In particular, the following parts are included, with about three chapters in each of it:

1. Image formation
2. Image enhancement
3. Feature extraction and selection
4. Segmentation
5. Classification and measurements
6. Image data visualization
7. Image management and integration
8. Evaluation and customizing

Many people might object me at this point, because we clearly aimed at reaching the unreachable. In Germany, we have the common phrase "eierlegende Wollmilchsau", a metaphor that directly translates to "egg-providing wool-milk-pig" describing the union of all benefits (Fig. 1).

You as the reader shall judge our success realizing this all-in-one approach: YATBIP or eierlegende Wollmilchsau? Any feedback is deeply welcome and should be directed personally to me as the editor.

Facing now the final manuscript, I want to thank Claus Ascheron for encouraging me to initiate this project, and all contributers for timely delivering their high-quality material and appropriately responding to the editorial remarks and suggestions. Jens Hoffmann was assisting me in LaTeX programming and Max Jung helped in text and image conversion and optimization.

Also, I want to mention Peter Jentsch and Dirk Bartz, who have passed away during the very last iterations of the manuscript, which leaves me behind speechless. We have included the obituaries in the next pages.

Aachen, December 2010   *Thomas M. Deserno, né Lehmann*

Obituaries

Prof. Dr. Peter Jensch died unexpectedly during the period of the proof-reading of this book chapter on April 15, 2010 after a fulfilling life. Peter Jensch was the initiator of the DICOM research activities at the OFFIS - Institute for Information Technology, Oldenburg, Germany, in the early 1990s and was pushing this topic forward for the rest of his life. The most popular result of this engagement is the well-known Open Source DICOM toolkit DCMTK that is hosted and maintained by OFFIS since 1993. Against this background, all members of the DICOM team at OFFIS would like to thank Peter Jensch for establishing this extraordinary project and for being such a likeable, energetic boss, mentor, and colleague to us. Without him, OFFIS would not be the popular name in the world of DICOM it is today and we all would not have such encouraging opportunities and research projects we still enjoy. As Chap. 17 of this book is the last publication Peter Jensch participated in and since the content of this chapter is the very topic that strongly influenced his work, we like to use this opportunity to express our sincere gratitude to Peter Jensch.

Oldenburg, June 2010

Michael Onken
Marco Eichelberg
Jörg Riesmeier

Prof. Dr. Dirk Bartz died unexpectedly on March 28, 2010 while attending the thirtieth Vattenfall Berlin Half Marathon. Running half-marathon in Berlin was one of his favorite activities.

During his academic career, Dirk strongly supported the idea of building a German Interest Group on Medical Visualization and actively took part the whole time giving advice to many students; particularly supporting female researchers was an important issue. Furthermore, Dirk organized many tutorials at Visualization, Eurographics, and Computer-Assisted Radiology and Surgery (CARS).

In 2005, I was very glad that Dirk joined the effort of writing a textbook on "Visualization in Medicine". For an 18 month period, we communicated daily on the various aspects of the book. It was enlightening and a pleasure to discuss with Dirk all the time. He was always perfectly reliable and good-humored even in situations where he had a very high workload.

In the end of 2006, Dirk became appointed as Full Professor for Computer-Assisted Surgery at the International Center for Computer-Assisted Surgery (ICCAS), Leipzig, Germany, and started to build a new research group. He focused on visualization techniques, such as illustrative rendering, perceptual studies (from Dirk I learned the term "psychophysical studies"), and applications in neurosurgery and Ear, Nose and Throat (ENT) surgery.

Dirk belonged to the core team which tried to establish a new workshop series "Visual Computing in Biology and Medicine". It was quite natural that Dirk would host the second event, scheduled to take place in July in Leipzig. Until the very last days of his life, he discussed strategies for this workshop.

Dirk was only 42 years old, leaving behind Heidi, his wife, and his two little sons.

Magedeburg, June 2010   *Berhard Preim*

Contents

1 Fundamentals of Biomedical Image Processing
Thomas M. Deserno.. 1
1.1   Introduction 1
      1.1.1   Steps of Image Processing 2
      1.1.2   Remarks on Terminology 3
      1.1.3   Biomedical Image Processing 4
1.2   Medical Image Formation 4
      1.2.1   Basic Physics 5
      1.2.2   Imaging Modalities 6
      1.2.3   Digitalization 13
1.3   Image Enhancement 16
      1.3.1   Histogram Transforms 16
      1.3.2   Convolution 18
      1.3.3   Mathematical Morphology 18
      1.3.4   Calibration 19
      1.3.5   Registration 20
1.4   Image Data Visualization 22
      1.4.1   Marching Cube Algorithm 23
      1.4.2   Surface Rendering 23
      1.4.3   Volume Rendering 23
1.5   Visual Feature Extraction 25
      1.5.1   Data Level 25
      1.5.2   Pixel Level 25
      1.5.3   Edge Level 25
      1.5.4   Texture Level 26
      1.5.5   Region Level 26
1.6   Segmentation 27
      1.6.1   Pixel-Based Segmentation 27
      1.6.2   Edge-Based Segmentation 30
      1.6.3   Region-Based Segmentation 31

	1.6.4	Over- and Under-Segmentation	32
	1.6.5	Model-Based Segmentation	34
1.7	Classification		37
	1.7.1	Statistic Classifiers	39
	1.7.2	Syntactic Classifiers	39
	1.7.3	Computational Intelligence-Based Classifiers	40
1.8	Quantitative Measurements and Interpretation		41
	1.8.1	Partial Volume Effect	42
	1.8.2	Euclidean Paradigm	42
	1.8.3	Scene Analysis	42
	1.8.4	Examples	43
1.9	Image Management		45
	1.9.1	Archiving	45
	1.9.2	Communication	45
	1.9.3	Retrieval	47
1.10	Conclusion and Outlook		48
References			49

Part I Image Formation

2 Fusion of PET and MRI for Hybrid Imaging
Zang-Hee Cho, Young-Don Son, Young-Bo Kim, and Seung-Schik Yoo 55

2.1	Introduction		55
2.2	Positron Emission Tomography		57
	2.2.1	Basic Principles	57
	2.2.2	Image Reconstruction	59
	2.2.3	Signal Optimization	59
	2.2.4	High-Resolution Research Tomograph	60
2.3	Magnetic Resonance Imaging		62
	2.3.1	Basic Principles	62
	2.3.2	Image Reconstruction	63
	2.3.3	Signal Optimization	64
	2.3.4	High-Field MRI	65
2.4	Hybrid PET Fusion System		67
	2.4.1	PET/CT Systems	68
	2.4.2	PET/MRI Systems	68
	2.4.3	High-Resolution Fusion	70
	2.4.4	PET/MRI Fusion Algorithm	72
2.5	Conclusions		76
References			76

3 Cardiac 4D Ultrasound Imaging
Jan D'hooge 81
3.1 The Role of Ultrasound in Clinical Cardiology 81
3.2 Principles of Ultrasound Image Formation 82
    3.2.1 The Pulse-Echo Measurement 82
    3.2.2 Gray Scale Encoding 83
    3.2.3 Gray Scale Imaging 85
    3.2.4 Phased Array Transducer Technology 85
3.3 Limitations of 2D Cardiac Ultrasound 86
    3.3.1 Complex Anatomy (Congenital Heart Disease) 87
    3.3.2 Geometric Assumptions to Assess Volumes 88
    3.3.3 Out-of-Plane Motion and Foreshortening 89
3.4 Approaches Towards 3D Cardiac Ultrasound 89
    3.4.1 Freehand 3D Ultrasound 90
    3.4.2 Prospective Gating 90
    3.4.3 Retrospective Gating 91
    3.4.4 Two-Dimensional Arrays 92
3.5 Validation of 3D Cardiac Ultrasound Methodologies 95
3.6 Emerging Technologies 96
    3.6.1 Transesophageal 3D Imaging 96
    3.6.2 True Real-Time Volumetric Imaging 97
3.7 Remaining Challenges in 4D Cardiac Ultrasound 98
    3.7.1 Resolution 98
    3.7.2 Image Quality 99
    3.7.3 Data Visualization and Interaction 101
    3.7.4 Segmentation/Automated Analysis 101
References 102

Part II Image Enhancement

4 Morphological Image Processing Applied in Biomedicine
Roberto A. Lotufo, Leticia Rittner, Romaric Audigier,
Rubens C. Machado, and André V. Saúde 107
4.1 Introduction 107
4.2 Binary Morphology 108
    4.2.1 Erosion and Dilation 108
    4.2.2 Opening and Closing 110
    4.2.3 Morphological Reconstruction from Markers 111
    4.2.4 Reconstruction from Opening 112
4.3 Gray-Scale Operations 114
    4.3.1 Erosion and Dilation 115
    4.3.2 Opening and Closing 116
    4.3.3 Component Filters and Morphological Reconstruction 119
    4.3.4 Regional Maxima 121

4.4	Watershed Segmentation		122
	4.4.1	Classical Watershed Transform	122
	4.4.2	Filtering the Minima	123
	4.4.3	Watershed from Markers	124
	4.4.4	Inner and Outer Markers	125
4.5	Segmentation of Diffusion MRI		126
4.6	Conclusions		128
References			128

5 Medical Image Registration
Daniel Rueckert and Julia A. Schnabel 131

5.1	Introduction		131
5.2	Transformation Model		132
	5.2.1	Rigid Transformation	133
	5.2.2	Affine Transformation	133
	5.2.3	Projective Transformation	134
	5.2.4	Non-Rigid Transformation: Parametric Models	134
	5.2.5	Non-Rigid Transformation: Non-Parametric Models	138
5.3	Registration Basis		139
	5.3.1	Feature-Based Registration	140
	5.3.2	Voxel-Based Registration	141
5.4	Optimization		144
5.5	Validation of Registration		144
5.6	Application		146
	5.6.1	Intra-Subject Registration	146
	5.6.2	Inter-Subject Registration	147
5.7	Summary and Conclusions		149
References			150

Part III Feature Extraction and Selection

6 Texture in Biomedical Images
Maria Petrou 157

6.1	Introduction		157
6.2	Characterizing the Texture of Swatches		158
	6.2.1	From Grammars to Markov Random Fields	158
	6.2.2	From Markov Random Fields to Fractals	159
	6.2.3	From Markov Random Fields to Gibbs Distributions	159
	6.2.4	Co-occurrence Matrices	160
	6.2.5	Generalized Co-occurrence Matrices	161
	6.2.6	Orientation Histograms	162
	6.2.7	Textons	163
	6.2.8	Features from the Discrete Fourier Transform	163
6.3	Simultaneous Texture Segmentation and Recognition		165

		6.3.1	From Spatio-Frequency to Spatio-Structural Space....... 166

- 6.3.1 From Spatio-Frequency to Spatio-Structural Space....... 166
- 6.3.2 Statistical Spatio-Structural Space..................... 168
- 6.3.3 Monogenic Signal 169
- 6.3.4 From Monogenic Signal Back to Gabor Functions 170
- 6.3.5 Beyond Spatial Patterns into Gray Value Distributions... 171

6.4 Examples of the Use of Texture Features
in Biomedical Applications 172
- 6.4.1 Mammography.. 172
- 6.4.2 Brain Image Data 173

6.5 Discussion and Conclusions 174

References 175

7 Multi-Scale and Multi-Orientation Medical Image Analysis
Bart M. ter Haar Romeny 177

7.1 Introduction 177
7.2 The Necessity of Scale 178
- 7.2.1 The Optimal Aperture Function....................... 178
- 7.2.2 Derivatives of Sampled, Discrete Data, Such as Images ... 180

7.3 Differential Invariants 181
- 7.3.1 Gauge Coordinates 181
- 7.3.2 Invariants from Tensor Contraction 182

7.4 Second Order Image Structure and Features 183
- 7.4.1 Isophote Curvature 183
- 7.4.2 Flowline Curvature 184
- 7.4.3 Corners 184
- 7.4.4 Principal Curvatures................................. 185
- 7.4.5 The Shape Index 186

7.5 Third Order Image Structure: T-Junctions..................... 187
7.6 Adaptive Blurring and Geometry-Driven Diffusion 187
7.7 Edge Focusing 189
7.8 Orientation Analysis 190
7.9 Applications 192
- 7.9.1 Catheter Detection 192
- 7.9.2 Endocard Contour Detection 193
- 7.9.3 Denoising of Crossing Lines.......................... 193

7.10 Conclusion 194
References 195

8 Feature Extraction and Selection for Decision Making
*Agma J.M. Traina, Caetano Traina Jr., André G.R. Balan,
Marcela X. Ribeiro, Pedro H. Bugatti, Carolina Y.V. Watanabe,
and Paulo M. Azevedo-Marques*..................................... 197

8.1 Introduction 197
8.2 Image Representation 198
- 8.2.1 Medical Image Segmentation and Feature Extraction 199
- 8.2.2 Color Features 201

		8.2.3	Texture Features 203
		8.2.4	Shape Features.................................... 204
8.3	Image Features and Distance Functions 205
		8.3.1	Similarity Search and Metric Spaces 206
		8.3.2	Distance Functions 206
		8.3.3	Case Study: Evaluating Distance Functions for Separating Data 208
8.4	Feature Selection 210
		8.4.1	Curse of Dimensionality 211
		8.4.2	Traditional Algorithm for Feature Selection 211
		8.4.3	Combined Feature Selection and Discretization 213
8.5	Association Rule Mining 215
		8.5.1	Definition 215
		8.5.2	Case Study: Improving Computer-Aided Diagnosis by Association Rule Mining......................... 215
8.6	Conclusions... 220
References 221

Part IV Segmentation

9 Parametric and Non-Parametric Clustering for Segmentation

Hayit Greenspan and Tanveer Syeda-Mahmood 227

9.1	Introduction 227
9.2	Image Modeling and Segmentation 229
	9.2.1 Image Modeling 230
	9.2.2 Segmentation 230
	9.2.3 State of the Art 231
9.3	Probabilistic Modeling of Feature Space 231
	9.3.1 Gaussian Mixture Models 232
	9.3.2 Expectation Maximization........................... 232
	9.3.3 Visualization 233
9.4	Using GMMs for Brain Tissue and Lesion Segmentation 234
	9.4.1 Application Domain 234
	9.4.2 Spatial Constraints 234
	9.4.3 Modeling Spatial Constraints Through GMM 235
	9.4.4 Tissue Segmentation 238
	9.4.5 Lesion Segmentation 238
9.5	Non-Parametric Clustering Approaches to Segmentation......... 240
	9.5.1 Description of the Feature Space 241
	9.5.2 Clustering Intensity, Geometry, and Motion............. 243
9.6	Using Non-Parametric Clustering for Cardiac Ultrasound 245
	9.6.1 Application Domain 245
	9.6.2 Cardiac Motion Estimation 246
	9.6.3 Segmentation of Meaningful Regions 246

| 9.7 | Discussion | 248 |

References 248

**10 Region-Based Segmentation: Fuzzy Connectedness,
Graph Cut and Related Algorithms**
Krzysztof Chris Ciesielski and Jayaram K. Udupa 251
10.1 Introduction and Overview 251
    10.1.1 Digital Image Scene 252
    10.1.2 Topological and Graph-Theoretical Scene
           Representations 253
    10.1.3 Digital Image 253
    10.1.4 Delineated Objects 254
10.2 Threshold-Indicated Fuzzy Connected Objects 254
    10.2.1 Absolute Fuzzy Connectedness Objects 255
    10.2.2 Robustness of Objects 256
    10.2.3 Algorithm for Delineating Objects 256
10.3 Optimization in Foreground-Background Case 257
    10.3.1 Relative Fuzzy Connectedness 258
    10.3.2 Algorithm for Delineating Objects 259
    10.3.3 Graph Cut Delineation 259
10.4 Segmentation of Multiple Objects 262
    10.4.1 Relative Fuzzy Connectedness 262
    10.4.2 Iterative Relative Fuzzy Connectedness 263
    10.4.3 Algorithm for Iterative Relative Fuzzy Connectedness 265
    10.4.4 Variants of IRFC 266
10.5 Scale-Based and Vectorial Fuzzy Connectedness 266
10.6 Affinity Functions in Fuzzy Connectedness 267
    10.6.1 Equivalent Affinities 267
    10.6.2 Essential Parameters in Affinity Functions 269
10.7 Other Delineation Algorithms 270
    10.7.1 Generalized Graph Cut 270
    10.7.2 Level Set vs. Generalized Graph Cut 271
10.8 Medical Image Examples 273
10.9 Concluding Remarks 276
References 276

11 Model-Based Segmentation
Tobias Heimann and Hervé Delingette 279
11.1 Introduction 279
11.2 Deformable Simplex Meshes 281
    11.2.1 Internal Forces on Simplex Meshes 282
    11.2.2 Image Forces 283
    11.2.3 Globally Constrained Deformation 285
    11.2.4 3D+t Deformable Simplex Meshes 286
    11.2.5 Advanced Segmentation Strategies 288
    11.2.6 Geometric Representations
           for Model-Based Segmentation 290

- 11.3 Statistical Models of Shape and Appearance 291
  - 11.3.1 Shape Representation 292
  - 11.3.2 Point Correspondence 292
  - 11.3.3 Construction of Statistical Shape Models 295
  - 11.3.4 Modeling Object Appearance 297
  - 11.3.5 Local Search Algorithms 298
- 11.4 Conclusion 300
- References 301

Part V Classification and Measurements

12 Melanoma Diagnosis
Alexander Horsch 307
- 12.1 The Cutaneous Melanoma 307
  - 12.1.1 Medical Basics 307
  - 12.1.2 Relevance of Early Diagnosis 309
- 12.2 State of the Art in CM Diagnosis 309
  - 12.2.1 Diagnostic Algorithms 309
  - 12.2.2 Imaging Techniques 311
  - 12.2.3 Diagnostic Accuracies 313
- 12.3 Dermoscopy Image Analysis 314
  - 12.3.1 Image Analysis Approaches 314
  - 12.3.2 Segmentation of Skin Lesions 315
  - 12.3.3 Feature Extraction 316
  - 12.3.4 Feature Visualization 317
  - 12.3.5 Classification Methods 319
- 12.4 Commercial Systems 322
  - 12.4.1 System Design Principles 322
  - 12.4.2 Image Capture Devices 323
  - 12.4.3 Dermoscopy Computer Systems 324
- 12.5 Evaluation Issues 324
  - 12.5.1 Case Databases 325
  - 12.5.2 Evaluation Methods 325
- 12.6 Conclusion 325
- References 326

13 CADx Mammography
Lena Costaridou 329
- 13.1 Introduction 329
- 13.2 Basic Terms and Definitions 330
  - 13.2.1 Breast Imaging Modalities 330
  - 13.2.2 Mammographic Lesions 331
  - 13.2.3 CADe Schemes 332
  - 13.2.4 CADx Architectures 333

13.3	CADx Schemes in X-ray Mammography	335
	13.3.1 Morphology Analysis of MC Clusters	335
	13.3.2 Texture Analysis of MC Clusters	338
	13.3.3 Morphology and Texture Analysis of Masses	339
13.4	CADx Schemes in Breast Ultrasound	341
13.5	CADx Schemes in Breast MRI	344
13.6	Application Examples	346
	13.6.1 Segmentation Accuracy on MC Cluster Content	346
	13.6.2 Heterogeneity of Enhancement Kinetics in DCE-MRI	349
13.7	Discussion and Conclusions	351
References		353

14 Quantitative Medical Image Analysis for Clinical Development of Therapeutics

Mostafa Analoui 359

14.1	Introduction	359
14.2	Key Issues in Drug Research and Clinical Development	361
	14.2.1 Biological Marker	361
	14.2.2 Imaging Modality	362
14.3	Quantitative Image Analysis	363
	14.3.1 Assessment of Osteoarthritis	364
	14.3.2 Assessment of Carotid Atherosclerosis	365
	14.3.3 Assessment of Cancer	367
14.4	Managing Variability in Imaging Biomarkers	369
	14.4.1 Technical Validation	370
	14.4.2 Standard Operation Procedures	371
	14.4.3 Regulatory Issues	372
14.5	Future Directions	373
References		374

Part VI Image Data Visualization

15 Visualization and Exploration of Segmented Anatomic Structures

Dirk Bartz and Bernhard Preim 379

15.1	Introduction	379
15.2	Indirect and Direct Volume Rendering	380
	15.2.1 Indirect Volume Rendering	380
	15.2.2 Rendering of Multiple Objects	380
	15.2.3 Direct Volume Rendering	382
	15.2.4 Rendering of Segmented Data	383
	15.2.5 Discussion	384
15.3	Generation of Smooth and Accurate Surface Models	386
	15.3.1 Mesh Smoothing with Fairing	386
	15.3.2 Improving Mesh Quality	388

15.4 Visualization of Vascular Structures 389
      15.4.1 Surface-based Vessel Visualization 390
      15.4.2 Model-based Surface Visualization
            of Vascular Structures 390
      15.4.3 Volume Rendering of Vascular Structures............... 392
15.5 Virtual Endoscopy.. 394
      15.5.1 Graphical Representation 395
      15.5.2 Interaction Model 396
      15.5.3 User Interface.. 396
      15.5.4 Case Study: Virtual Colonoscopy 397
15.6 Conclusions.. 397
References 398

16 Processing and Visualization of Diffusion MRI
James G. Malcolm, Yogesh Rathi, and Carl-Fredrik Westin 403
16.1 Introduction 403
16.2 Modeling 404
      16.2.1 Imaging the Tissue 404
      16.2.2 Parametric Models 405
      16.2.3 Non-parametric Models 405
      16.2.4 Regularization 407
      16.2.5 Characterizing Tissue 407
16.3 Tractography 408
      16.3.1 Deterministic Tractography............................ 408
      16.3.2 Probabilistic Tractography 409
      16.3.3 Global Tractography.................................. 411
      16.3.4 Validation 412
16.4 Applications 413
      16.4.1 Volume Segmentation 413
      16.4.2 Fiber Clustering 414
      16.4.3 Connectivity... 416
      16.4.4 Tissue Analysis 417
16.5 Summary.. 418
References 419

Part VII Image Management and Integration

17 Digital Imaging and Communications in Medicine
Michael Onken, Marco Eichelberg, Jörg Riesmeier, and Peter Jensch ... 427
17.1 DICOM Basics 427
      17.1.1 Introduction and Overview 428
      17.1.2 Information Objects 428
      17.1.3 Display Pipeline...................................... 430
      17.1.4 Network and Media Services 433
      17.1.5 Conformance 437

17.2 Advanced DICOM Services 438
    17.2.1 Advanced Image Display Services 438
    17.2.2 DICOM Structured Reporting 442
    17.2.3 Application Hosting 447
17.3 Conclusions and Outlook 452
References 453

18 PACS-Based Computer-Aided Detection and Diagnosis
*H.K. (Bernie) Huang, Brent J. Liu, Anh HongTu Le,
and Jorge Documet* 455

18.1 Introduction 455
18.2 The Need for CAD-PACS Integration 456
    18.2.1 Approaches of CAD-PACS Integration 457
    18.2.2 CAD Software 459
18.3 DICOM Standard and IHE Workflow Profiles 459
    18.3.1 DICOM Structured Reporting 460
    18.3.2 IHE Profiles 461
18.4 The CAD-PACS^{TM} Toolkit 461
    18.4.1 Concept 462
    18.4.2 Structure, Components, and Editions 462
18.5 Example of CAD-PACS Integration 463
    18.5.1 The Digital Hand Atlas 463
    18.5.2 CAD Evaluation in a Laboratory Setting 464
    18.5.3 CAD Evaluation in a Clinical Environment 465
    18.5.4 CAD-PACS Integration Using DICOM-SR 466
18.6 Conclusion 467
References 469

19 Content-Based Medical Image Retrieval
Henning Müller and Thomas M. Deserno 471

19.1 Introduction 471
    19.1.1 Motivation and History 472
    19.1.2 Query-by-Example(s) Paradigm 472
19.2 General Image Retrieval 473
    19.2.1 Classification vs. Retrieval 473
    19.2.2 System Components and Computation 474
    19.2.3 Features and Signatures 474
    19.2.4 Distance and Similarity Measures 476
19.3 Medical Image Retrieval 476
    19.3.1 Application Fields 477
    19.3.2 Types of Images 477
    19.3.3 Image Preprocessing 478
    19.3.4 Visual and Non-Visual Image Features 478
    19.3.5 Database Architectures 479
    19.3.6 User Interfaces and Interaction 480
    19.3.7 Interfacing with Clinical Information Systems 480

19.4	Evaluation		481
	19.4.1	Available Databases	481
	19.4.2	Tasks and User Models	481
	19.4.3	Ground Truth and Gold Standards	482
	19.4.4	Benchmarks and Events	483
19.5	Examples for Medical CBIR Systems		483
	19.5.1	Medical Gnu Image Finding Tool	484
	19.5.2	Image Retrieval in Medical Applications	484
19.6	Discussion and Conclusions		487
	19.6.1	Strengths and Weaknesses of Current Systems	488
	19.6.2	Gaps of Medical CBIR Systems	488
	19.6.3	Future Developments	488
References			490

Part VIII Evaluation and Customizing

20 Systematic Evaluations and Ground Truth
Jayashree Kalpathy-Cramer and Henning Müller 497

20.1	Introduction		497
20.2	Components for Successful Evaluation Campaigns		498
	20.2.1	Applications and Realistic Tasks	498
	20.2.2	Collections of Images and Ground Truth	499
	20.2.3	Application-Specific Metrics	500
	20.2.4	Organizational Resources and Participants	501
20.3	Evaluation Metrics and Ground Truth		502
	20.3.1	Registration	502
	20.3.2	Segmentation	503
	20.3.3	Retrieval	506
20.4	Examples of Successful Evaluation Campaigns		508
	20.4.1	Registration	508
	20.4.2	Segmentation	509
	20.4.3	Annotation, Classification and Detection	511
	20.4.4	Information Retrieval	512
	20.4.5	Image Retrieval	512
20.5	Lessons Learned		517
20.6	Conclusions		517
References			518

21 Toolkits and Software for Developing Biomedical Image Processing and Analysis Applications
Ivo Wolf 521

21.1	Introduction	521
21.2	Toolkits	522

		21.2.1	The NA-MIC Kit 522
		21.2.2	Insight Segmentation and Registration Toolkit 523
		21.2.3	The Visualization Toolkit 524
		21.2.4	Open Inventor 525
		21.2.5	Medical Imaging Interaction Toolkit 526
		21.2.6	The Image-Guided Surgery Toolkit 527
		21.2.7	The Multimod Application Framework 528
		21.2.8	vtkINRIA3D 529
		21.2.9	OFFIS DICOM ToolKit 529
		21.2.10	Grassroots DICOM Library 530
		21.2.11	The Common Toolkit 530
		21.2.12	Simulation Open Framework Architecture 530
	21.3	Development Environments 531
		21.3.1	SCIRun 532
		21.3.2	OpenXIP 532
		21.3.3	DeVIDE 533
		21.3.4	VisTrails 534
		21.3.5	LONI Pipeline 534
		21.3.6	MeVisLab 535
		21.3.7	MATLAB® 535
		21.3.8	Interactive Data Language 536
	21.4	Extensible Software 537
		21.4.1	3D Slicer 537
		21.4.2	MITK ExtApp and MITK 3M3 538
		21.4.3	Graphical Interface for Medical Image Analysis and Simulation....................................... 539
		21.4.4	OsiriX 539
		21.4.5	ParaView 539
		21.4.6	ImageJ and Fiji 540
		21.4.7	MIPAV 541
		21.4.8	VolView 541
		21.4.9	Analyze 541
		21.4.10	Amira 542
	21.5	Conclusion and Discussion 543
	References 543

22 Image Processing and the Performance Gap
Steven C. Horii and Murray H. Loew 545

22.1	Introduction 545
22.2	Examples of Clinically Useful Image Processing 546
	22.2.1	Windowing and Image Display 546
	22.2.2	Contrast and Edge Enhancememt 546
	22.2.3	Noise Reduction and Color Coding 547
	22.2.4	Registration and Segmentation 547
	22.2.5	Image Compression and Management 548

- 22.3 Why are there Gaps? ... 549
  - 22.3.1 The Conservative Radiologist ... 549
  - 22.3.2 The Busy Radiologist: Digital vs. Analog Workflow ... 549
  - 22.3.3 The Wary Radiologist: Malpractice Concerns ... 550
  - 22.3.4 The Skeptical Radiologist: Evidence-Based Requirements ... 551
  - 22.3.5 Tails, Dogs, and Gaps ... 552
- 22.4 The Goals of Image Processing for Medical Imaging ... 553
  - 22.4.1 Automation of Tasks ... 553
  - 22.4.2 Improvement of Observer Performance ... 555
- 22.5 Closing the Gap ... 561
  - 22.5.1 Education ... 561
  - 22.5.2 Research ... 562
- 22.6 Conclusion ... 563
- References ... 563

Index ... 567

List of Contributors

Mostafa Analoui
Healthcare and Life Sciences
The Livingston Group
New York, NY, USA
analoui@yahoo.com

Romaric Audigier
Laboratoire Vision et Ingénierie
des Contenus, CEA-LIST
romaric.audigier@cea.fr

Paulo M. Azevedo-Marques
Medical School of Ribeirão Preto
University of São Paulo
São Paulo, Brazil
pmarques@fmrp.usp.br

André G.R. Balan
Computer Science, Mathematics
and Cognition Center
Federal University of ABC
São Paulo, Brazil
agrbalan@icmc.usp.br

Dirk Bartz[†]
Innovation Center for Computer
Assisted Surgery (ICCAS)
University of Leipzig
Leipzig, Germany

Pedro H. Bugatti
Computer Science Department
University of São Paulo
São Paulo, Brazil
pbugatti@icmc.usp.br

Zang-Hee Cho
Neuroscience Research Institute
Gachon University of Medicine
and Science, Seoul, Korea
zcho@gachon.ac.kr

Krzysztof Chris Ciesielski
Department of Mathematics
West Virginia University
Morgantown, WV, USA
and
Department of Radiology
University of Pennsylvania
Philadelphia, PA, USA
kcies@math.wvu.edu

Lena Costaridou
Department of Medical Physics
University of Patras
Patras, Greece
costarid@upatras.gr

Hervé Delingette
Asclepios Team, INRIA
Sophia-Antipolis, France
herve.delingette@inria.fr

Thomas M. Deserno
Department of Medical Informatics
RWTH Aachen University
Aachen, Germany
deserno@ieee.org

Jan D'hooge
Department of Cardiovascular
Diseases, Katholieke Universiteit
Leuven, Leuven, Belgium
jan.dhooge@uz.kuleuven.ac.be

Jorge Documet
Image Processing and Informatics
Lab, University of Southern
California, Los Angeles, CA, USA
documet@usc.edu

Marco Eichelberg
OFFIS Institute for Information
Technology, Oldenburg, Germany
eichelberg@offis.de

Hayit Greenspan
Department of Biomedical
Enginering, Tel-Aviv University
Tel-Aviv, Israel
hayit@eng.tau.ac.il

Bart M. ter Haar Romeny
Department of Biomedical
Engineering, Eindhoven
University of Technology
Eindhoven, The Netherlands
b.m.terhaarromeny@tue.nl

Tobias Heimann
French Research Institute
of Computer Science and Automatic
Control, INRIA
Sophia Antipolis Cedex, France
and
German Cancer Research Center
Heidelberg, Germany
t.heimann@dkfz.de

Steven C. Horii
Department of Radiology
University of Pennsylvania
Philadelphia PA, USA
steve.horii@uphs.upenn.edu

Alexander Horsch
Department of Medical Statistics
and Epidemiology, Technische
Universität München
Munich, Germany
and
Computer Science Department
University of Tromsø
Tromsø, Norway
alexander.horsch@tum.de

H.K. (Bernie) Huang
Image Processing and Informatics
Lab, University of Southern
California, Los Angeles, CA, USA
hkhuang@aol.com

Peter Jensch[†]
OFFIS Institute for Information
Technology, Oldenburg, Germany

Jayashree Kalpathy-Cramer
Department of Medical Informatics
and Clinical Epidemiology
Oregon Health & Science University
Portland, OR, USA
kalpathy@ohsu.edu

Young-Bo Kim
Neuroscience Research Institute
Gachon University of Medicine
and Science, Seoul, Korea
neurokim@gachon.ac.kr

Anh HongTu Le
Image Processing and Informatics
Lab, University of Southern
California, Los Angeles, CA, USA
anhhle@usc.edu

Brent J. Liu
Image Processing and Informatics
Lab, University of Southern
California, Los Angeles, CA, USA
brentliu@usc.edu

Murray H. Loew
Biomedical Engineering Program
Department of Electrical
and Computer Engineering
The George Washington University
Washington, DC, USA
loew@gwu.edu

Roberto A. Lotufo
School of Electrical and Computer
Engineering, State University
of Campinas (UNICAMP)
Campinas, Brazil
lotufo@unicamp.br

Rubens C. Machado
Center for Information Technology
Renato Archer (CTI), Ministry
of Science and Technology (MCT)
Campinas, Brazil
rubens.machado@cti.gov.br

James G. Malcolm
Department of Psychiatry, Brigham
and Women's Hospital, Harvard
Medical School, Boston, MA, USA
malcolm@bwh.harvard.edu

Henning Müller
University of Applied Sciences
Western Switzerland (HES-SO)
Sierre, Switzerland

and

University and Hospitals of Geneva
Geneva, Switzerland
henning.mueller@hevs.ch

Michael Onken
OFFIS Institute for Information
Technology, Oldenburg, Germany
onken@offis.de

Maria Petrou
Informatics and Telematics Institute
Centre for Research
and Technology Hellas (CERTH)
Thessaloniki, Greece
petrou@iti.gr

Bernhard Preim
Department of Simulation and
Graphics, University of Magdeburg
Magdeburg, Germany
preim@isg.cs.uni-magdeburg.de

Yogesh Rathi
Department of Psychiatry, Brigham
and Women's Hospital
Harvard Medical School
Boston, MA, USA
yogesh@bwh.harvard.edu

Marcela X. Ribeiro
Computer Science Department
University of São Paulo
São Paulo, Brazil
mxavier@icmc.usp.br

Jörg Riesmeier
ICSMED AG, Oldenburg, Germany
riesmeier@icsmed.de

Leticia Rittner
School of Electrical and Computer
Engineering, State University
of Campinas (UNICAMP)
Campinas, Brazil
lrittner@dca.fee.unicamp.br

Daniel Rueckert
Department of Computing, Imperial
College London, London, UK
d.rueckert@imperial.ac.uk

André V. Saúde
Department of Computer Science
Federal University of Lavras
Lavras, Minas Gerais, Brazil
saude@dcc.ufla.br

Julia A. Schnabel
Institute of Biomedical Engineering
Department of Engineering
Science, University of Oxford
Oxford, UK
julia.schnabel@eng.ox.ac.uk

Young-Don Son
Neuroscience Research Institute
Gachon University of Medicine
and Science, Seoul, Korea
ydson@gachon.ac.kr

Tanveer Syeda-Mahmood
IBM Almaden Research Center
San Jose, CA, USA
stf@almaden.ibm.com

Agma J.M. Traina
Computer Science Department
University of São Paulo
São Paulo, Brazil
agma@icmc.usp.br

Caetano Traina
Computer Science Department
University of São Paulo
São Paulo, Brazil
caetano@icmc.usp.br

Jayaram K. Udupa
Department of Radiology
University of Pennsylvania
Philadelphia, PA, USA
jay@mail.med.upenn.edu

Carolina Y.V. Watanabe
Computer Science Department
University of São Paulo
São Paulo, Brazil
carolina@icmc.usp.br

Carl-Fredrik Westin
Department of Radiology
Brigham and Women's Hospital
Harvard Medical School
Boston, MA, USA
westin@bwh.harvard.edu

Ivo Wolf
Department of Medical Informatics
Mannheim University of Applied
Science, Mannheim, Germany
i.wolf@hs-mannheim.de

Seung-Schik Yoo
Department of Radiology
Brigham and Women's Hospital
Harvard Medical School
Boston, MA, USA
yoo@bwh.harvard.edu

Acronyms

1D	One-Dimensional
2D	Two-Dimensional
3D	Three-Dimensional
4D	Four-Dimensional
AAM	Active Appearance Model
AAPM	American Association of Physicists in Medicine
ABCD	Asymmetry, Border, Color, and Differential structures
ACE	Associative Classifier Engine
ACR	American College of Radiology
ACSE	Association Control Service Element
ADNI	Alzheimer's Disease Neuroimaging Initiative
AE	Application Entity
AFC	Absolute Fuzzy Connectedness
AHA	American Heart Association
AIF	Arterial Input Function
AJAX	Asynchronous Javascript XML
AJCC	American Joint Committee on Cancer
ALM	Acral Lentiginous Melanoma
AMN	Atypical Melanocytic Nevi
ANN	Artificial Neural Network
AOM	Area Overlap Measure
APD	Avalanche Photodiode
API	Application Programming Interface
ASCF	Alternating Sequential Component Filter
ASD	Atrial Septal Defect
ASF	Alternating Sequential Filter
ASM	Active Shape Model
AVD	Absolute Volumetric Difference

BAA	Bone Age Assessment
BDWG	Biomarker Definitions Working Group
BI-RADS	Breast Imaging Reporting and Data System
BIR	Biomedical Imaging Resource
BSD	Berkeley Software Distribution
BSPS	Blending Softcopy Presentation States
CAD	Computer-Aided Diagnosis
CADe	Computer-Assisted Detection
CADx	Computer-Assisted Diagnostics
CARS	Computer-Assisted Radiology and Surgery
CART	Classification And Regression Tree
CAS	Chinese Academy of Sciences; Computer-Assised Surgery
CASH	Color, Architecture, Symmetry, Homogeneity
CAT	Computer-Aided Therapy
CAVE	Cave Automatic Virtual Environment
CBIR	Content-Based Image Retrieval
CBVIR	Content-Based Visual Information Retrieval
CC	Cranio Caudal
CCD	Charge-Coupled Device
CGMM	Constrained GMM
CI	Computational Intelligence; Confidence Interval
CICE	Cumulative Inverse Consistency Error
CIE	Commission Internationale de L'Eclairage
CIMT	Carotid Intima-Media Thickness
CISTIB	Center for Computational Image and Simulation Technologies in Biomedicine
CIT	Center for Information Technology
CLEF	Cross Language Evaluation Forum
CM	Cutaneous Melanoma
CMY	Cyan, Magenta, Yellow
CNMD	Consensus Net Meeting on Dermoscopy
CNR	Contrast to Noise Ratio
CNS	Central Nervous System
CPU	Central Processing Unit
CR	Computed Radiography
CRT	Cathode Ray Tube
CS	Conformance Statement
CSF	Cereborspinal Fluid
CSI	Chemical Shift Imaging
CSPS	Color Softcopy Presentation State
CT	Computed Tomography
CTA	CT Angiography
CTC	CT Colonography

CTE	Cumulative Transitive Error
CTK	Common Toolkit
CTR	Cardio-Thoracic Ratio
CVP	Closest Vessel Projection
DAG	Directed Acyclic Graph
DBM	Deformation-Based Morphometry
DBT	Digital Breast Tomosynthesis
DCE	Dynamic Contrast-Enhanced
DCE-MRI	Dynamic Contrast-Enhanced Magnetic Resonance Imaging
DCMR	DICOM Content Mapping Resource
DCMTK	OFFIS DICOM ToolKit
DDSM	Digital Database for Screening Mammography
DES	Density Emitter Model
DeVIDE	Delft Visualisation and Image Processing Of Development Environment
DFT	Discrete Fourier Transform
DICOM	Digital Imaging and Communications in Medicine
DICOM SR	DICOM Structured Reporting
DIMSE	DICOM Message Service Element
DKFZ	Deutsches Krebsforschungszentrum
dMRI	Diffusion Magnetic Resonance Imaging
DNA	Deoxyribonucleic Acid
DOF	Degree Of Freedom
DP	Detection Performed
DPV	Dermatoscopic Point Value
DR	Digital Radiography
DSA	Digital Subtraction Angiography
DSI	Diffusion Spectrum Imaging
DTI	Diffusion Tensor Imaging
DTM	Decision Tree Method
DVD	Digital Versatile Disc
DWT	Discrete Wavelet Transform
ECG	Electrocardiography
EEG	Electroencephalography
ELM	Epi-Luminescence Microscopy
EM	Expectation Maximization
EN	European Norm
ENT	Ear, Nose, and Throat
EPI	Echo Planar Imaging
EXACT	Extraction of Airways from CT
F-FP-CIT	^{18}FluoroPropyl-CarbomethoxyIodophenyl-norTropane

FA	Fractional Anisotropy
FB	Filtered Backprojection
FC	Fuzzy Connectedness
FDA	Food and Drug Administration
FDG	^{18}F-Fludeoxygloucose
FDI	Fédération Dentaire Internationale
FEM	Finite Element Model
FFD	Free-Form Deformation
FFDM	Full-Field Digital Mammography
FID	Free Induction Decay
Fiji	Fiji Is Just ImageJ
FISH	Fluorescent In-Situ Hybridization
FLT	^{18}F-L-Thymidine
fMRI	Functional MRI
FND	False Negative Dice
FOV	Field-Of-View
FPD	False Positive Dice
FROC	Free-Response Receiver Operating Characteristic
FSC	File Set Creator
FSR	File Set Reader
FSU	File Set Updater
FWHM	Full Width Half Maximum
GA	Genetic Algorithms
GC	Graph Cut
GDCM	Grassroots DICOM Library
GG	Generalized Graph
GIF	Graphics Interchange Format
GIFT	GNU Image Finding Tool
GIMIAS	Graphical Interface for Medical Image Analysis and Simulation
GLCM	Gray-Level Co-occurrence Matrices
GMM	Gaussian Mixture Model
GMP	Good Manufacturing Practice
GNU	GNU's Not Unix
GPA	Generalized Procrustes Analysis
GPU	Graphics Processing Unit
GSPS	Grayscale Softcopy Presentation State
GTC	Generalized Tanimoto Coefficient
GUI	Graphical User Interface
HARAG	Hierarchical Attributed Region Adjacency Graph
HARDI	High Angular Resolution Diffusion Imaging
HD	Hausdorff Distance
HIPAA	Health Insurance Portability and Accountability Act

HIS		Hospital Information System; Hue, Intensity, Saturation
HL7		Health Level Seven
HRRT		High-Resolution Research Tomograph
HSV		Hue-Saturation-Value
HU		Hounsfield Unit
IBSR		Internet Brain Segmentations Repository
ICA		Independent Component Analysis
ICC		International Color Consortium
ICCAS		International Center for Computer-Assisted Surgery
ICP		Iterative Closest Point
ID		Identifier
IDEA		Image Diagnosis Enhancement Through Associations
IDL		Interactive Data Language
IFT		Image Foresting Transform
IGS		Image-Guided Surgery
IGSTK		Image-Guided Surgery Toolkit
IGT		Image-Guided Therapy
IHE		Integrating the Healthcare Enterprise
IHS		Intensity, Hue, Saturation
IOD		Information Object Definition
IP		Internet Protocol
IR		Information Retrieval
IRB		Institutional Review Board
IRFC		Iterative Relative Fuzzy Connectedness
IRMA		Image Retrieval in Medical Applications
ISO		International Organization for Standardization
ITK		Insight Segmentation and Registration Toolkit
JPEG		Joint Photographic Experts Group
JSW		Joint Space Width
k-NN		k-Nearest Neighbor
KIN		Key Image Note
KLT		Karhunen-Loève Transform
LA		Left Atrium
LAC		Los Angeles County Hospital
LCD		Liquid Crystal Display
LDA		Linear Discriminant Analysis
LGPL		Lesser General Public License
LIDC		Lung Image Database Consortium
LMM		Lentigo Maligna Melanoma
LoG		Laplacian Of Gaussian
LONI		Laboratory Of Neuro Imaging
LRA		Logistic Regression Analysis
LS		Level Set
LSA		Lenticulostriate Arterie

LSO	Lutetium Oxyorthosilicate
LUT	Look-Up Table
LV	Left Ventricle
MAF	Multimod Application Framework
MAP	Maximization A Posterior; Mean Average Precision
MATLAB	MATrix LABoratory
MC	Micro-Calcification
MDL	Minimum Description Length
MedGIFT	Medical GIFT
MEDICOM	Medical Image Communication
MeSH	Medical Subject Headings
MHD	Manhattan Distance
MI	Mutual Information
MICCAI	Medical Image Computing and Computer Assisted Intervention
MIP	Maximum Intensity Projection
MIPAV	Medical Image Processing, Analysis, and Visualization
MIT	Massachusetts Institute of Technology
MITK	Medical Imaging Interaction Toolkit
ML	Maximum Likelihood; MeVis Image Processing Library
MLO	Medio-Lateral Oblique
MOD	Magneto Optical Disk
MP	Morphological Processing
MPM	Maximization of the Posterior Marginals
MPPS	Modality Performed Procedure Step
MPU	Multi-Level Partition of Unity
MRA	Magnetic Resonance Angiography
MRF	Markov Random Field
MRI	Magnetic Resonance Imaging
MRM	Magnetic Resonance Mammography
MRML	Medical Reality Markup Language; Multimedia Retrieval Markup Language
MRS	Magnetic Resonance Spectroscopy
MRSI	Magnetic Resonance Spectroscopy Imaging
MS	Multiple Sclerosis
MSW	Multi-Scale Watershed
MTF	Modulation Transfer Function
MV	Mitral Valve
NA-MIC	National Alliance for Medical Image Computing
NASA	National Aeronautics and Space Administration
NCBC	National Centers for Biomedical Computing
NCC	Normalized Cross Correlation

NCI	National Cancer Institute
NEMA	National Electrical Manufacturers Association
NHLBI	National Heart Lung and Blood Institute
NIH	National Institutes of Health
NIREP	Non-Rigid Image Registration Evaluation Project
NLM	National Library of Medicine
NM	Nodular Melanoma
NMF	Non-Negative Matrix Factorization
NMI	Normalised Mutual Information
NMR	Nuclear Magnetic Resonance
NN	Nearest Neighbor
NNT	Number Needed to Treat
NRRD	Nearly Raw Raster Data
NURBS	Non-Uniform Rational B-Spline
OA	Osteo arthritis
OAI	Osteo arthritis Initiative
OCT	Optical Coherence Tomography
ODF	Orientation Distribution Function
OFFIS	Institute for Information Technology
OOI	Object Of Interest; Organ Of Interest
OSA	Obstructive Sleep Apnea
OSGi	Open Services Gateway Initiative
OSI	Open System Interconnection
OTS	Off-The-Shelf
PA	Postero-Anterior
PACS	Picture Archiving and Communication System
PCA	Principal Component Analysis
PCS	Profile Connection Space
PCSP	Pseudo-Color Softcopy Presentation States
PD	Pharmacodynamic
PDE	Partial Differential Equation
PDM	Point Distribution Model
PET	Positron Emission Tomography
PK	Pharmacokinetic
PMT	Photo-Multiplier Tube
POC	Proof Of Concept
POM	Proof Of Mechanism
ppm	Parts Per Million
pQCT	Peripheral Quantitative Computed Tomography
PSF	Point Spread Function
PSL	Pigmented Skin Lesion

PSP	Photostimulable Phosphor
PWF	Post-Processing Work Flow
Q/R	Query/Retrieve
QBE	Query By Example, Query By Image Example
QBI	Q-Ball Imaging
QBIC	Query By Image Content
QDE	Quantum Detection Efficiency
RA	Right Atrium
rCMRGlc	Regional Cerebral Metabolic Rate of Glucose
RF	Radio Frequency
RFC	Relative Fuzzy Connectedness
RGB	Red, Green, and Blue
RIRE	Retrospective Image Registration Evaluation
RIS	Radiology Information System
RNA	Ribonucleic Acid
ROC	Receiver Operating Characteristic
ROI	Region Of Interest
RPC	Rich Client Platform
RREP	Retrospective Registration Evaluation Project
RSNA	Radiological Society of North America
RV	Random Variable, Right Ventricle
SCP	Service Class Provider
SCU	Service Class User
SD	Standard Deviation
SE	Standard Error
SER	Signal Enhancement Ratio
SFM	Screen-Film Mammography
SGLD	Spatial Gray Level Dependence
SIF	Single Image Finding
SIFT	Shift-Invariant Feature Transform
SIM	Scaling Index Method
SINR	Simple Image and Numeric Report
SiPM	Silicon Photomultiplier
SMART	System for the Mechanical Analysis and Retrieval of Text
SNR	Signal-to-Noise Ratio
SOFA	Simulation Open Framework Architecture
SOP	Service Object Pair; Standard Operating Procedure
SOR	Successive Over-Relaxation
SPECT	Single Photon Emission Computed Tomography
SR	Structured Reporting; Super Resolution
SRN	Square Root of the Norm of Coefficients
SSD	Sum of Squared Differences

SSM	Statistical Shape Model, Superficial Spreading Melanoma
STAPLE	Simultaneous Truth and Performance Level Estimation
SVD	Singular Value Decomposition
SVM	Support Vector Machine
TBM	Tensor-Based Morphometry
TCL	Tool Command Language
TCP	Transmission Control Protocol
TEE	Transesophageal Echocardiography
TIFF	Tagged Image File Format
TMG	Tensorial Morphological Gradient
TREC	Text Retrieval Conference
UofU	University of Utah
UCLA	University of California at Los Angeles
UID	Unique Identifier
UK	United Kingdom
UPF	Universitat Pompeu Fabra
US	Ultrasonography, Ultrasound, United States
USB	Universal Serial Bus
USC	University of Southern California
VBM	Voxel-Based Morphometry
VD	Volumetric Difference
VL	Virtual Machine
VME	Virtual Medical Entities
VOI	Volume Of Interest; Value Of Interest
VOLCANO	VOLume Change Analysis of NOdules
VPE	Visual Programming Environment
VR	Value Representation; Virtual Reality
VSD	Ventricular Septum Defect
VSG	Visualization Sciences Group
VTK	Visualization Toolkit
W3C	World Wide Web Consortium
WS	Watershed
WSDL	Web Service Description Language
XA	X-ray Angiography
XIP	eXtensible Imaging Platform
XRF	X-ray Radio-Fluoroscopy
ZIB	Zuse Institute Berlin

1

Fundamentals of Biomedical Image Processing

Thomas M. Deserno

Summary. This chapter gives an introduction to the methods of biomedical image processing. After some fundamental preliminary remarks to the terminology used, medical imaging modalities are introduced (Sect. 1.2). Sections 1.3 and 1.4 deal with low-level image processing and visualization, respectively, as far as necessary to understand the following chapters. Subsequently, the core steps of image analysis, namely: feature extraction, segmentation, classification, quantitative measurements, and interpretation are presented in separate sections. On account of its high relevance, the focus is on segmentation of biomedical images. Special segmentation methods and techniques have been developed in the medical application domain. Section 1.9 provides a brief summary of image communication. The electronic transmission and exchange of medical images will become more important in future for multimedia applications such as electronic patient records in health telematics and integrated care. Section 1.10 completes this chapter with an overview of past, present, and future challenges to biomedical image processing.

1.1 Introduction

By the increasing use of direct digital imaging systems for medical diagnostics, digital image processing becomes more and more important in health care. In addition to originally digital methods, such as Computed Tomography (CT) or Magnetic Resonance Imaging (MRI), initially analogue imaging modalities such as endoscopy or radiography are nowadays equipped with digital sensors. Digital images are composed of individual pixels (this acronym is formed from the words "picture" and "element"), to which discrete brightness or color values are assigned. They can be efficiently processed, objectively evaluated, and made available at many places at the same time by means of appropriate communication networks and protocols, such as Picture Archiving and Communication Systems (PACS) and the Digital Imaging and Communications in Medicine (DICOM) protocol, respectively. Based on digital imaging techniques, the entire spectrum of digital image processing is now applicable in medicine.

T.M. Deserno (ed.), *Biomedical Image Processing*, Biological and Medical Physics,
Biomedical Engineering, DOI: 10.1007/978-3-642-15816-2_1,
© Springer-Verlag Berlin Heidelberg 2011

Fig. 1.1. *Modules of image processing.* In general, image processing covers four main areas: image formation, visualization, analysis, and management. The algorithms of image enhancement can be assigned as pre- and post-processing in all areas

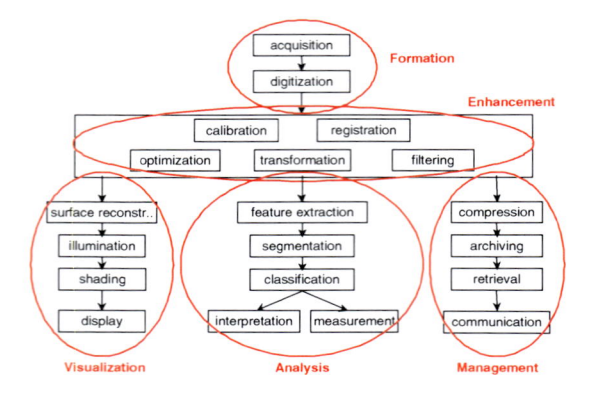

1.1.1 Steps of Image Processing

The commonly used term "biomedical image processing" means the provision of digital image processing for biomedical sciences. In general, digital image processing covers four major areas (Fig. 1.1):

1. *Image formation* includes all the steps from capturing the image to forming a digital image matrix.
2. *Image visualization* refers to all types of manipulation of this matrix, resulting in an optimized output of the image.
3. *Image analysis* includes all the steps of processing, which are used for quantitative measurements as well as abstract interpretations of biomedical images. These steps require a priori knowledge on the nature and content of the images, which must be integrated into the algorithms on a high level of abstraction. Thus, the process of image analysis is very specific, and developed algorithms can be transferred rarely directly into other application domains.
4. *Image management* sums up all techniques that provide the efficient storage, communication, transmission, archiving, and access (retrieval) of image data. Thus, the methods of telemedicine are also a part of the image management.

In contrast to image analysis, which is often also referred to as high-level image processing, low-level processing denotes manual or automatic techniques, which can be realized without a priori knowledge on the specific content of images. This type of algorithms has similar effects regardless of the content of the images. For example, histogram stretching of a radiograph improves the contrast as it does on any holiday photograph. Therefore, low-level processing methods are usually available with programs for image enhancement.

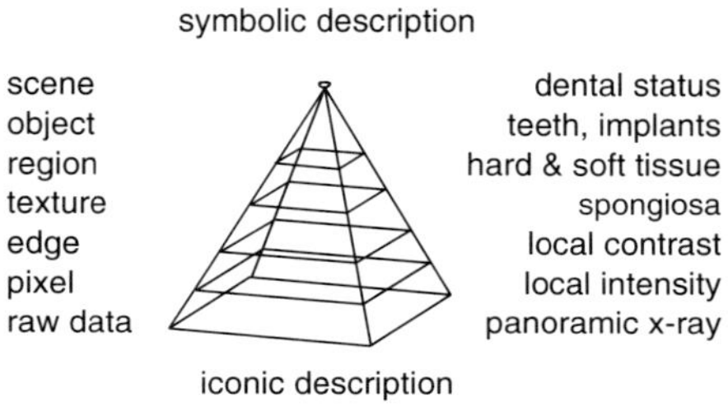

```
              symbolic description
    scene          △         dental status
    object        ╱ ╲        teeth, implants
    region       ╱___╲       hard & soft tissue
    texture     ╱    ╱╲      spongiosa
    edge       ╱____╱__╲     local contrast
    pixel     ╱    ╱    ╲    local intensity
    raw data ╱____╱_____╲   panoramic x-ray
              iconic description
```

Fig. 1.2. *Levels of abstraction.* The general terms (*left*) are exemplified for a panoramic radiograph of upper and lower jaws (*right*). At the pyramid's top, the dental status corresponds to an abstract scene analysis, which only contains standardized information (existence and condition) on the tooth positions

1.1.2 Remarks on Terminology

The complexity of an algorithm, the difficulty of its implementation, or the computation time required for image processing plays a secondary role for the distinction between low-level and high-level processing methods. Rather, the degree of abstraction of the a priori knowledge is important for this meaning. Although the following definitions are not standardized in the literature, they are used consistently within this book (Fig. 1.2):

- The *raw data level* records an image as a whole. Therefore, the totality of all raw data pixels is regarded on this level.
- The *pixel level* refers to discrete individual pixels.
- The *edge level* represents the One-dimensional (1D) structures, which are composed of at least two neighbored pixels.
- The *texture level* refers to Two-Dimensional (2D) or Three-Dimensional (3D) structures. On this level however, the delineation of the area's contour (in three dimensions: the surface of the volume) may be unknown.
- The *region level* describes 2D or 3D structures with a well-defined boundary or surface.
- The *object level* associates textures or regions with a certain meaning or name, i.e., semantics is introduces on this level.
- The *scene level* considers the ensemble of image objects in spatial and/or temporal terms. If 3D structures are imaged over the time, also Four-Dimensional (4D) data is acquired.

From an iconic (concrete) to a symbolic (abstract) description of images, information is gradually reduced. Methods of low-level image processing operate on the raw data as well as on pixel, edge, or texture levels, and thus at a minimally level of abstraction. Methods of high-level image processing include the texture, region, object, and scene levels. The required abstraction can be achieved by increased modeling of a priori knowledge.

1.1.3 Biomedical Image Processing

With these definitions, a particular problem in high-level processing of biomedical images is inherently apparent: resulting from its complex nature, it is difficult to formulate medical a priori knowledge such that it can be integrated directly and easily into automatic algorithms of image processing. In the literature, this is referred to as the *semantic gap*, which means the discrepancy between the cognitive interpretation of a diagnostic image by the physician (high level) and the simple structure of discrete pixels, which is used in computer programs to represent an image (low level). In the medical domain, there are three main aspects hindering bridging this gap:

1. *Heterogeneity of images*: Medical images display living tissue, organs, or body parts. Even if captured with the same modality and following a standardized acquisition protocol, shape, size, and internal structures of these objects may vary remarkably not only from patient to patient (inter-subject variation) but also among different views of a patient and similar views of the same patients at different times (intra-subject variation). In other words, biological structures are subject to both inter- and intra-individual alterability. Thus, universal formulation of a priori knowledge is impossible.
2. *Unknown delineation of objects*: Frequently, biological structures cannot be separated from the background because the diagnostically or therapeutically relevant object is represented by the entire image. Even if definable objects are observed in biomedical images, their segmentation is problematic because the shape or borderline itself is represented fuzzily or only partly. Hence, medically related items often can be abstracted at most on the texture level.
3. *Robustness of algorithms*: In addition to these inherent properties of medical images, which complicate their high-level processing, special requirements of reliability and robustness of medical procedures and, when applied in routine, image processing algorithms are also demanded in the medical area. As a rule, automatic analysis of images in medicine should not provide wrong measurements. That means that images, which cannot be processed correctly, must be automatically classified as such, rejected and withdrawn from further processing. Consequently, all images that have not been rejected must be evaluated correctly. Furthermore, the number of rejected images is not allowed to become large, since most medical imaging procedures are harmful and cannot be repeated just because of image processing errors.

1.2 Medical Image Formation

Since the discovery of X-rays by Wilhelm Conrad Röntgen in 1895, medical images have become a major component of diagnostics, treatment planning and procedures, and follow-up studies. Furthermore, medical images are used

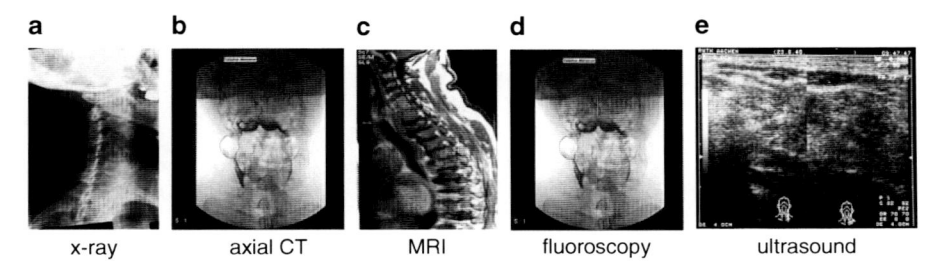

	a	b	c	d	e
	x-ray	axial CT	MRI	fluoroscopy	ultrasound

Fig. 1.3. *Medical imaging modalities.* The body region (here: cervical vertebra) appears completely different when altering the imaging modality

Name	Symbol	Mass	Charge
Proton	p	1 u	+1 e
Neutron	n	1 u	0 e
Alpha particle	α	4 u	+2 e
Electron	β	0 u	−1 e
Positron	β^+	0 u	+1 e
Photon	γ	0 u	0 e

Table 1.1. *Atomic particles.* The given values for mass and charge are only rough estimates. The atomic mass unit $1\,\text{u} = 1.660538782 \cdot 10^{-27}$ kg. The elementary charge $1\,\text{e} = 1.602176487 \cdot 10^{-19}$ C

for education, documentation, and research describing morphology as well as physical and biological functions in 1D, 2D, 3D, and even 4D image data (e.g., cardiac MRI, where up to eight volumes are acquired during a single heart cycle). Today, a large variety of imaging modalities have been established, which are based on transmission, reflection or refraction of light, radiation, temperature, sound, or spin. Figure 1.3 emphasizes the differences in image characteristic with respect to the imaging modality. Obviously, an algorithm for delineation of an individual vertebra shape that works with one imaging modality will not be applicable directly to another modality.

1.2.1 Basic Physics

To understand the different nature of medical images and imaging modalities, we need to recall some basic physics of matter. Roughly, all matter is build from atoms, where a nucleus composed of protons and neutrons is surrounded by a electron shell. Table 1.1 lists charge and mass of nuclear particles.

The number of protons determines the element number. In the equilibrium state, the number of electrons equals the number of protons and there is no external Coulomb field. However, the positions of the particles are not constant. In particular, the electrons orbit the nucleus. According to the Maxwell laws, accelerated (continuously changing its direction) charge induces electromagnetic radiation: the electron would lose energy gradually spiraling inwards and collapsing into the nucleus.

Within the Bohr model of the atom, there are certain shells where an electron can orbit its nucleus without releasing electromagnetic radiation. These

Fig. 1.4. *Bohr model of the atom.* The shells where an electron can orbit the nucleus without releasing electromagnetic radiation are numbered, and there is a maximal number of electrons for each shell. Sometimes, the shells are also referred to by letters k, l, m, etc. The difference of energy between shells is released as radiation when an electron changes its position

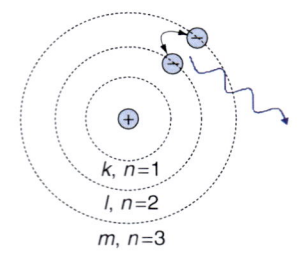

shells are numbered n (Fig. 1.4) and allow for $2 \cdot n^2$ electrons. The energy of an electron $E_n = (-13.6 \text{ eV})\frac{1}{n^2}$ depends on the orbit number n, where inner shells are energetically preferred and ionizing needs higher energy if an electron of an inner shell is removed. The unit Electron Volt (eV) refers to the kinetic energy of an electron after passing the acceleration voltage of 1.0 V.

1.2.2 Imaging Modalities

From the plenty of medical imaging modalities, we will focus on X-ray imaging, CT, MRI, and ultrasound. However, optical modalities such as endoscopy, microscopy, or photography are not less important.

X-Ray Imaging

According to the Bohr model, *X-radiation* – the term was initially introduced by Röntgen – can be generated, for instance, if an electron from a higher shell jumps over into a free position of an inner shell (Fig. 1.4). The discrete difference of energy ΔE is released as a photon (γ particle). ΔE is characteristic to the numbers of shells and the element.

Technically, free positions in inner shells are produced from shooting electrons to the atom. Figure 1.5 schematically shows an X-ray tube. The high voltage between cathode and anode accelerates the electrons that are released from a filament. Passing the acceleration voltage, these electrons are loaded with kinetic energy. Hitting the target material, usually tungsten for skeletal imaging and molybdenum for mammography, two types of interactions may occur, i.e., the accelerated electron interacts with the:

- *Nucleus*, where the electron is slowed down by the Coulomb field of the protons, and a photon is released with an energy equal to the loss of kinetic energy of the electron (*Bremsstrahlung*).
- *Shell*, where the *characteristic radiation* is released as described above.

When X-radiation passes through matter, e.g., the human body we would like to image, the X-ray photons again may interact with the nucleus or the shell resulting in:

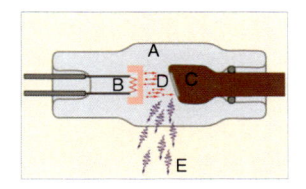

Fig. 1.5. *X-ray tube.* The vacuum tube (**A**) houses cathode (**B**) and anode (**C**). A current heats up the filament, releasing electrons (**D**), which are accelerated towards the anode. Interacting with either the nucleus or the shell of the target material, Bremsstrahlung and characteristic radiation are released (**E**), respectively

- *Absorption*: The photon is completely vanished giving all its energy to the absorbing material. This effect is harmful and causes damage to living cells, but it is required to obtain a contrasted image.
- *Scattering*: A secondary photon is produced, that might be coherent (Thomson effect) or incoherent (Compton effect). Both effects lower the Signal to Noise Ratio (SNR), since the secondary photon usually travels in another direction and contributes to the image at a wrong location, and scatter rasters from lead are used to filter the scattered radiation.

The absorption coefficient μ uniquely describes the material, and is mapped to the gray scale for image display. In plain radiography, high-attenuating material (e.g., bone) is displayed in white (see Fig. 1.3a) while in fluoroscopy, the scale is inverted (see Fig. 1.3d), and the high-absorbing contrast agent is displayed in black.

However, the absorption sums up along the path through the matter. In particular, the absorption is described by an exponential function. In a first approximation, the intensity I of radiation depends on the thickness d of the imaged material $I = I_0 e^{-\mu d}$. However, a human body is not made from constant material $\mu \sim \mu(d)$, and furthermore, the absorption depends on the photon's energy $\mu \sim \mu(E)$. Since X-radiation cannot be obtained mono-energetic (Bremsstrahlung), the absorption equation yields

$$I = \int I_0(E) e^{-\int \mu(z,E) dz} dE \qquad (1.1)$$

The dependence of the absorption on the energy of the photon is obvious. Photons with low energy are more likely absorbed or scattered than high-energetic photons. Consequently, the spectrum of X-radiation, which is released from the X-ray tube, hardens when passing matter. This effect is called beam hardening.

Computed Tomography (CT)

X-ray imaging produces summation images, where all attenuation coefficients along the path are integrated. From a single image, one cannot determine the order of overlapping objects. This is different with CT, where the absorption is determined in 3D for each volume element (voxel). For imaging, a volume is acquired slice by slice, and each slice is reconstructed from several measures in different angulation.

Although back projection or filtered back projection in k-space (spatial frequency domain after Fourier transform) are nowadays used for image reconstruction, we will explain the principle based on the arithmetic reconstruction technique, which in fact was applied to the first CT machines. Suppose a slice being divided into four pixels, and two parallel rays passing it in two different directions (Fig. 1.6). This results in four independent equation (Fig. 1.6B) allowing to compute the four absorption coefficients μ_{ij} (Fig. 1.6E). To obtain more rows and columns, the number of parallel rays and the number of angles must be increased accordingly. Today, fan beam gantries are build (Fig. 1.7), enabling continuous rotation of the imaging fan and continuous longitudinal movement of the patient (spiral CT). Further speedup of acquisition is obtained from scanners, where up to 64 lines of X-ray receptors are mounted. From 1972, the acquisition time per slice has decreased about 10^5 (Table 1.2).

Fig. 1.6. *Arithmetic CT reconstruction.* Two parallel X-rays pass the slice in 90° and the measures are recorded. (**A**) logarithm allowing assignment of absorption coefficients μ; (**B**) four linear equations are obtained; (**C**) iterative solution; (**D**) assignment; (**E**) inverse logarithm

Fig. 1.7. *CT gantry.* Detaching the housing from the CT exposes the X-ray tube and the detector fan (http://wikepedia.org)

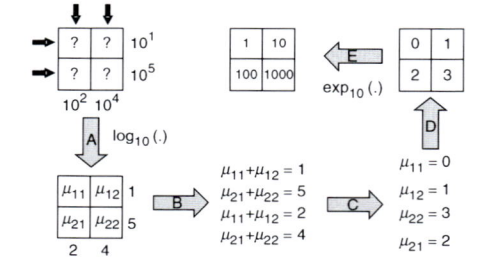

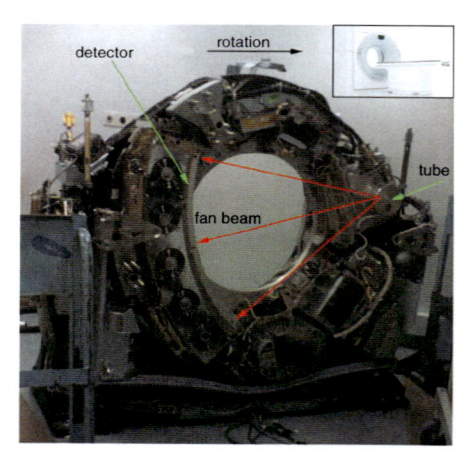

Year	Resolution	Gray scales	Thickness	Time
1974	80 × 80	64 (6 bit)	10 mm	300 s
1984	256 × 256	256 (8 bit)	5 mm	10 s
1994	512 × 512	512 (9 bit)	0.8 mm	0.5 s
2004	1024 × 1024	1024 (10 bit)	0.4 mm	0.005 s

Table 1.2. *CT slice parameters.* Today, a 64 line scanner rotates about three times a second yielding up to 192 slices per second

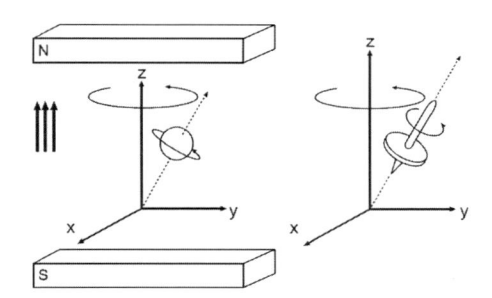

Fig. 1.8. *Precession [1].* A spinning proton in a magnetic field (*left*) moves like a gyroscope in the mass gravitation field of the earth (*right*). According to the Larmor theorem, the precession frequency is determined by the strength of the external magnetic field

Magnetic Resonance Imaging (MRI)

Almost simultaneously to CT, MRI has been introduced to medicine. It is based on electromagnetic effects of the nucleus. Since the human body consists of about 70 % water, we focus on hydrogen. Its nucleus is composed of only one proton. As mentioned before, the particles forming the nucleus are continuously moving. For hydrogen, this movement is a self rotation (spin), which has a magnetic moment. As shown in Fig. 1.8, the magnetic moment is aligned to an outer magnetic field, and precession ν is started.

To understand MRI, we need to regard a probation of tissue, which is composed of billions of hydrogen atoms. In other words, we move from a microscopic to a macroscopic view, where the spins sum up to a macroscopic magnetic moment M. Suppose the external magnetic field B_z is directed along the z-axis, the magnetic moments can align parallel or anti-parallel, where the latter occurs slightly minor (six less in a million). Therefore, $M_z > 0$. In addition, all precession is dephased ($M_{xy} = 0$).

The next component of MRI is a so called Radio Frequency (RF) impulse. Such an impulse can excite the system of spinning protons if the electromagnetic frequency equals the precession frequency. Depending on the amplitude and time of excitement, M can be arbitrarily directed. In Fig. 1.9b, for example, M has been turned into the (x,y)-plane; the corresponding RF impulse is referred to as 90° impulse.

RF excitation is followed by exponential relaxation, where the system is restoring its equilibrium state. The stored energy is released as signal (i.e., the Free Induction Decay (FID) when measured in Fourier domain), which can be detected and transformed to an image. However, the relaxation process is complex, since two independent effects superimpose:

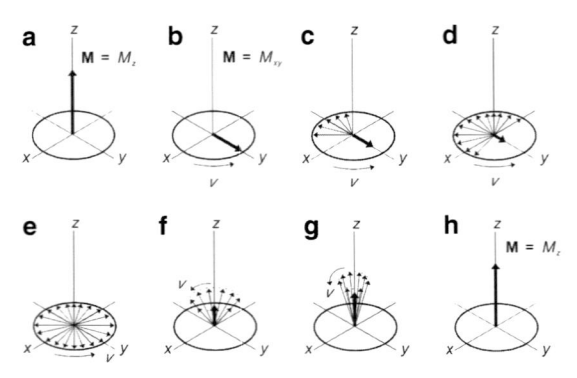

Fig. 1.9. *Excitement and relaxation [1].* Without excitement, $M = M_z$ (**a**); a 90° RF impulse turns $M = M_{xy}$ (**b**); dephasing (spin–spin relaxation) starts first, continuously decreasing M_{xy} (**c,d**); spin–spin relaxation completed (**e**); dealignment (spin–lattice relaxation) starts, steadily increasing M_z (**f,g**); relaxation completed (**h**)

- *spin–spin relaxation* with relaxation time T_2 affects the phase of the spins. For water-based and fat-based tissues, T_2 is in the $40 - 200$ ms and $10 - 100$ ms range, respectively.
- *spin–lattice relaxation* with relaxation time T_1 affects the parallel vs. anti-parallel alignment of spins. For water-based and fat-based tissues, T_1 is in the $0.4 - 1.2$ s and $0.10 - 0.15$ s range, respectively.

Therefore, spin–spin relaxation is almost completed before spin–lattice relaxation is detectable. Relaxation is visualized in Fig. 1.9. After 90° impulse, $M = M_{xy}$ and $M_z = 0$. Note that M_{xy} rotates with precession frequency ν in the (x, y)-plane. When T_2-relaxation is completed (Fig. 1.9e), $M_{xy} = 0$ and $M_z = 0$. The T_1-relaxation is visualized in Fig. 1.9f–h. In Fig. 1.9h, the spins gave back the energy they obtained from the RF pulse to the surrounding lattice.

To obtain a high-quality relaxation signal, spin-echo sequences are applied, where different RF impulses are induced, and readout of T_1 and T_2 is performed in between. Therefore, spin-echo sequences are characterized by the:

- *echo time* T_E determining half of the delay between a 90° and a 180° RF impulse, and the
- *repetition time* T_R denoting the rate of re-applying a 90°/180° sequence.

Figure 1.10 emphasizes the differences in contrast and appearance depending on the echo sequence. In particular, a M_0-, T_1-, or T_2-weighted MRI is obtained if $(T_E \ll T_2$ and $T_R \gg T_1)$, $(T_E \ll T_2$ and $T_R \approx T_1)$, or $(T_E \approx T_2$ and $T_R \gg T_1)$, respectively.

However, the theory we discovered so far does not allow us to obtain such images because we do not have any spatial alignment with the signal yet. This is obtained using gradient fields, which are superimposed to the

a	b	c
M_0 weighted	T_1 weighted	T_2 weighted

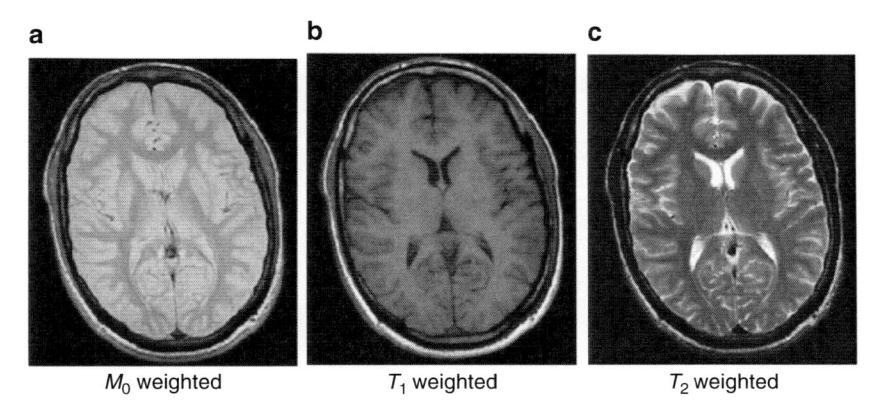

Fig. 1.10. *Forming MRI with spin-echo sequences.* (Courtesy: Tony Stöcker, FZ Jülich)

constant external field B. For instance, let us superimpose a gradient field in x-direction: $B = B_0 + B_G(x)$, $B_G(x_1) < B_G(x_2)$ $\forall x_1 < x_2$. Now, the Larmor frequency of precession $\nu \sim \nu(x)$ is slightly shifted along the x-axis. Since the induced RF impulse covers all frequencies, excitement results in the entire probation, and from the frequency of the FID signal, the according slice can be located. Another gradient, for instance in y-direction, allows for addressing a line rather than a plane. As we have seen with CT reconstruction, capturing signals from different lines through the volume finally allows voxel assignment. Advantageous to CT, gradient fields in MRI can be generated with gradient coils, where the current is adopted, and no mechanical rotation is required. In fact, steering the gradient fields produces the noise of MRI devices, since strong currents need to be turned on and off quickly.

Ultrasound

In contrast to CT and MRI, ultrasound is a medical imaging modality that is based on reflection of sound waves. Depending on the transducer, 1D to 4D data is obtained. We start from the 1D case (signal), where a longitudinal sound wave is traveling through the tissue of the human body. At transitions between different matter (e.g., muscle and fat), the sound wave is partly reflected and transmitted (refracted if the surface is not hit perpendicular). In other words, the echo runtime indicates the distance between transducer and tissue border while the echo strength is related to material properties. More precisely, these sound-relevant properties of matter are described by the speed of sound c_s and the density ρ, yielding the acoustic impedance $Z = c_s \cdot \rho$. Interfacing two materials $Z_1 = Z_0$ and $Z_2 = Z_0 + \Delta Z$, the reflection ratio r and transmission ratio t are given by

Table 1.3. *Speed of sound in matter.* The acoustic impedance $Z = c_s\rho$ computes from density ρ and speed of sound c_s. All numbers refer to body temperature of $37°$ centigrade

Material	c_s in m/s	ρ in kg/m^3	Z in kg/m^{2}s
Bone	3,600	$1.70 \cdot 10^3$	$6.12 \cdot 10^6$
Marrow	1,700	$0.97 \cdot 10^3$	$1.65 \cdot 10^6$
Blood	1,570	$1.02 \cdot 10^3$	$1.61 \cdot 10^6$
Muscle	1,568	$1.04 \cdot 10^3$	$1.63 \cdot 10^6$
Water	1,540	$0.99 \cdot 10^3$	$1.53 \cdot 10^6$
Fat	1,400	$0.97 \cdot 10^3$	$1.36 \cdot 10^6$
Air	340	1.20	$4.08 \cdot 10^2$

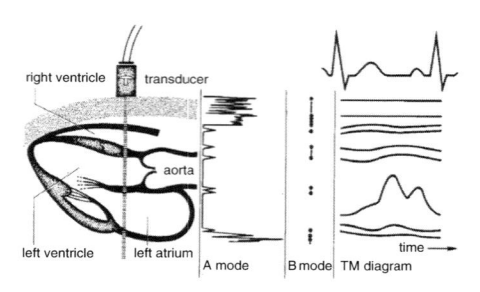

Fig. 1.11. *Ultrasound visualization modes [2].* A section through the heart is drawn schematically (*left*). From a simple sampling line, reflections of sound may be plotted (**i**) according to their amplitude (A-mode), (**ii**) coded as dots with a gray scale mapped to the amplitude (B-mode), which supports 2D images if a array of Piezoelectric crystals is applied, and (**iii**) in their position over the time, to visualize motion (TM diagram)

$$r = \sqrt{\frac{I_0}{I_A}} = \frac{Z_2 - Z_1}{Z_2 + Z_1} = \frac{\Delta Z}{2Z_0 + \Delta Z} \quad \text{and} \quad t = 1 - r \approx \begin{cases} 1, \text{ if } \Delta Z \ll Z_0 \\ 0, \text{ if } \Delta Z \gg Z_0 \end{cases}$$
(1.2)

where I_0 and I_R denote the intensity of the initial and reflected wave, respectively. As we can see from Table 1.3, $t \approx 0$ from air to water and soft tissue to bone, while $t \approx 1$ within the soft tissue. Therefore, a sonographic view behind bony structures or through organs filled with air is almost impossible. Furthermore, water-based gel must be used for air-free coupling the transducer to the human body.

Furthermore, the sound intensity is attenuated from expansion. The attenuation increases linear with the sound frequency but spatial resolution requires high frequency. Therefore, typical diagnostic scanners operate in the frequency range of $2-18$ MHz trading-off spatial resolution and imaging depth.

Technically, a piezoelectric crystal is used to convert an electrical signal into a mechanical movement, and the deformation of the crystal is coupled into the body. Then, the same transducer is used to detect the echos. There are several options to form an image from this pulse-echo signal (Fig. 1.11):

- *A-mode*: In amplitude mode, the echo intensity is plotted on the screen as a function of depth;

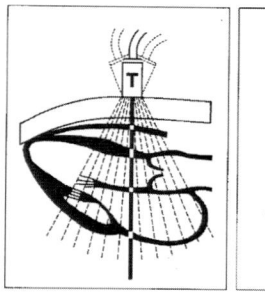

1 Fundamentals of Biomedical Image Processing 13

Fig. 1.12. *B-mode sector scan [2].* An array of transducers is used to scan a sector. In B-mode, reflections at tissue borders are displayed within a fan-shaped aperture, which is typically for medical ultrasound. Turning the transducer array perpendicularly allows for imaging cone-shaped volumes

- *B-mode*: In brightness mode, the echo intensity is coded with gray scales. This allows composing an array of transducers simultaneously scanning a plane through the body (Fig. 1.12). Parallel and sector scanners are available;
- *TM-mode*: Time motion diagrams visualize movements of sound-reflecting tissue borders. This mode offers functional rather than morphological inspection;
- *M-mode*: In motion mode, a sequence of rapidly acquired B-mode scans is displayed as moving picture. This is the most common mode in clinical ultrasound;
- *D-mode*: The doppler mode makes use of the doppler effect (i.e., a shift in frequency that occurs if the source of sound, the receptor, or the reflector is moved) in measuring and visualizing blood flow. Several visualization modes are used:
  - *Color Doppler*: The velocity information is presented as a color-coded overlay on top of a B-mode image;
  - *Continuous Doppler*: Doppler information is sampled along a line through the body, and all velocities detected at each point in time are presented (on a time line);
  - *PW Doppler*: Pulsed-wave Doppler information is sampled from only a small sample volume (defined in the 2D B-mode image), and presented on a time line;
  - *Duplex*: Color and (usually) PW Doppler are simultaneously displayed.

1.2.3 Digitalization

Digital image processing implies a discrete nature of the images. Regardless whether a film-based radiograph is digitized secondarily with a scanner, or the device primarily delivers a digital pixel (voxel) matrix, digitization effects alter the image. Digitization applies to both the definition (sampling) and the value range (quantization).

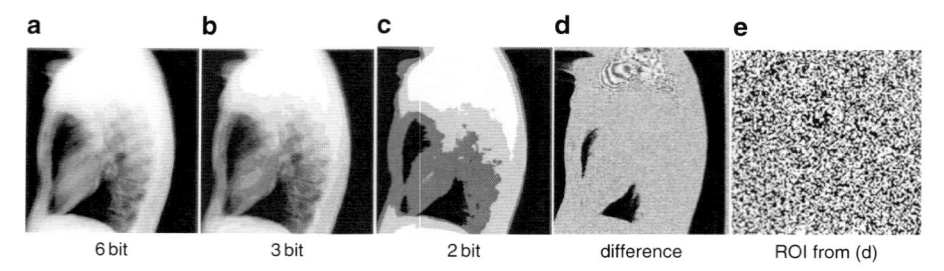

Fig. 1.13. *Quantization.* A lateral chest radiograph is shown with 6 bit, 3 bit, and 2 bit, which equals 64, 8, and 4 different gray scales in panel (**a**), (**b**), and (**c**), respectively. Disregarding saturation effects occurring within bone and air, the difference between 8 bit and 7 bit representation results in white noise (**d**,**e**). Panel (**d**) shows the histogram-optimized difference image, and (**e**) a centered Region of Interest (ROI) of 100 × 100 pixel

Quantization

Quantization refers to the digitization of the value range. We need to determine the maximal number of gray scales for every image. Usually, 8 bit and 24 bit are chosen for gray scale and full color images, respectively, allowing 256 different values in each band. In medicine, radiography or CT usually delivers 12 bit = 4,096 different values. If we assume a continuous brightness, quantization always worsen the image quality. The alteration can be modeled as additive noise, and the SNR of our digital image is improved by an increased number of gray scales.

Quantization noise is visualized in Fig. 1.13. With printing, we do not see any differences between 8 bit, 7 bit, or 6 bit quantization. If the number of gray scales becomes small, artefacts are apparent (Fig. 1.13b,c). Subtracting the 7 bit representation from the original 8 bit image illustrates the quantization noise (Fig. 1.13d,e).

Sampling

Sampling refers to the digitization of the definition range. According to the linear system theory, an analogue signal can be unambiguously represented with a discrete set of samples if the sampling rate exceeds two times the highest frequency occurring in the image (Nyquist theorem). Shannon's popular version of the sampling theorem states [3]:

> If a function $x(t)$ contains no frequencies higher than f_b Hz, it is completely determined by giving its ordinates at a series of points spaced $t = \frac{1}{2f_b}$ seconds apart.

Once the sampling theorem is satisfied, we cannot improve the image quality adding more pixels, which is contrarily to the effects of quantization. In

a	b	c	d	e
256×256	128×128	64×64	32×32	16×16

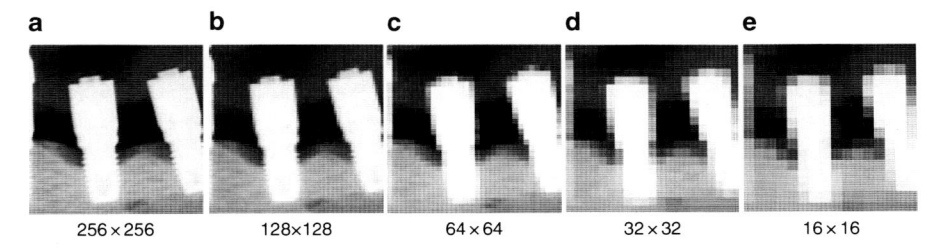

Fig. 1.14. *Sampling.* Artefacts from sub-sampling are visible if the printer resolution is smaller than the pixel scale. In panel (**b**), the first partial pixel effects should be notable

Fig. 1.15. *Moiré pattern.* This pattern is obtained from radiographing a lead grid. The spatial resolution of the entire X-ray imaging chain, disregarding whether it ends analogously or digitally, is measured by the distance from the center to that radius where individual lines can be differentiated. Furthermore, a four leaf clover can be seen in the center although it is neither with the lead lamella nor the squared pixel grid

spatial discretization, increasing the number of samples beyond the Nyquist rate only increases the file size of raw data, but not the information coded in it.

Figure 1.14 emphasizes the loss of information that results from applying an insufficient number of pixels for image acquisition. For instance, the spongious structure of the jaw bone disappears (Fig. 1.14c–e). Furthermore, gray scales are obtained misleadingly indicating a different material. For instance at the border of the implants, pixel values with μ of bone are obtained. In CT imaging, this partial pixel effect is also known as partial volume effect, see Sect. 1.8.1.

Similar to insufficient quantization, subsampling suppresses information in the digital image representation. In contrast to quantization, information is falsely added to an image if the sampling theorem is not fulfilled. Partial effects are one example. More important are aliasing effects. In 2D imaging, Moiré patterns are obtained whenever a regular structure mismatches the grid (Fig. 1.15).

1.3 Image Enhancement

Low-level methods of imaging processing, i.e., procedures and algorithms that are performed without a priori knowledge about the specific content of an image, are mostly applied to pre- or post-processing of medical images (Fig. 1.1). Therefore, the basic methods of histogram transforms, convolution and (morphological) filtering are mostly disregarded unless required for further understanding of this text (see the list of related textbooks on page 49). As a special preprocessing method for medical images, techniques for calibration and registration are briefly introduced.

1.3.1 Histogram Transforms

Point operations (pixel transforms) are based on the histogram of the image. Modifying the pixel values, all pixels are transformed independently from their positions in the image and their immediate neighborhood. Therefore, these type of transform is also referred to as point operation.

Histogram

The histogram shows the frequency distribution of pixel values (e.g., gray scales) disregarding the certain positions where the gray scales occur in the image. Simple pixel transforms can be defined using a histogram. For example, through the stretching of gray scales, the contrast of an image is improved (Fig. 1.16). After determining the histogram, upper and lower bounds are

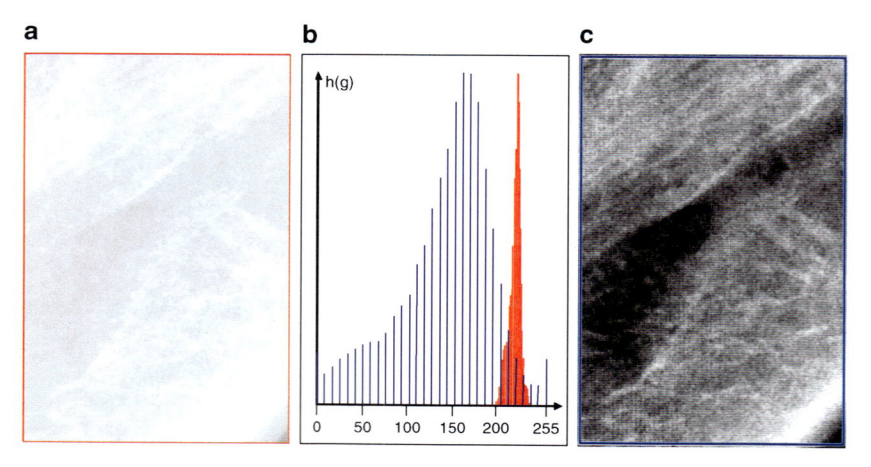

Fig. 1.16. *Histogram stretching.* A ROI is taken in the area of the temporomandibular joint from an intra-oral radiograph (**a**). Resulting from under-exposure, the spongy bone structure is displayed quite poorly. The associated histogram (**b**) is only narrow occupied (*red*). By stretching the histogram, the columns are linearly pulled apart (*blue*) and the contrast of the transformed radiograph is increased (**c**)

Old	New pixel value			Old	New pixel value		
Gray	Red	Green	Blue	Gray	Red	Green	Blue
0	0	0	0	...	...	...	...
1	1	0	2	246	254	244	239
2	2	0	3	247	255	245	243
3	3	0	5	248	255	245	244
4	4	1	7	249	255	246	246
5	5	1	9	250	255	247	248
6	5	2	12	251	255	249	250
7	5	2	14	252	255	251	252
8	5	3	16	253	255	251	253
9	5	4	18	254	255	253	254
...	...	...	...	255	255	255	255

Table 1.4. *Look-up table for pseudo coloring.* For each value in the range of the input image, the lookup table holds a value from the range of the output image. The color palette shown here is used for pseudo coloring keeping the original brightness progression of the input image [4]

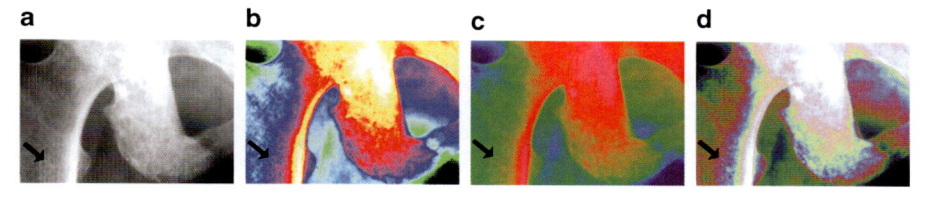

Fig. 1.17. *Pseudo coloring [4].* X-ray image of a pelvic bone metastasis after radiotherapy (**a**); pseudo-colored image of (**a**) using the corner colors of the RGB cube (**b**), colors of constant brightness (**c**), and colors with continuous brightness progression obtained from a spiral around the gray diagonal of the RGB cube (**d**). The arrow indicates local contrast enhancement

located, and a linear transform is applied that maps the lower bound to zero and the upper bound to the maximal gray scale (i.e., 255 for 8 bit images). If the histogram of the initial image does not contain all possible gray scales, the gray scale distance between neighbored pixels is enlarged, which results in an enhanced contrast.

Look-Up Table (LUT)

Technically, computation of histogram transforms is based on a Look-Up Table (LUT). For all pixel values, the lookup table contains a new value, which can also originate from another range of values. The example in Table 1.4 assigns each gray scale with a triple for Red, Green, and Blue (RGB). This transform is called pseudo coloring, and it is frequently used in the biomedical domain to enhance local contrast (Fig. 1.17). Computer graphic boards may limit the number of gray scales to 256 (8 bit), but offer $256^3 = 16,777,216$ colors. Special algorithms are recommended for the pseudo coloring in the medical context. In other words, pseudo coloring allows presentation of data, where the range of values exceeds the length of the RGB cube's edges without reducing the information as it would result from windowing.

a			b			c			d			e			f		
1	1	1	1	2	1	−1	−2	−1	0	−1	0	−1	−2	−1	−2	−1	0
1	1	1	2	4	2	−2	12	−2	−1	5	−1	0	0	0	−1	0	1
1	1	1	1	2	1	−1	−2	−1	0	−1	0	1	2	1	0	1	2

Fig. 1.18. *Convolution templates.* The sliding average (**a**) and the binomial low-pass filter (**b**) cause a smoothing of the image. The binomial high-pass filter (**c**), however, increases contrast and edges, but also the noise in the image. The templates (**a**) to (**c**) must be normalized to make sure that the range domain of values is not exceeded. The contrast filter (**d**) is based on integer pixel values. The convolution with (**d**) is therefore easy to calculate. The anisotropic templates (**e**) and (**f**) belong to the family of Sobel operators. Eight Sobel masks can be generated by rotation and mirroring for direction-selective edge filtering (see Fig. 1.24)

1.3.2 Convolution

In contrast to point operations (histogram transforms), the considered pixels are combined with the values of their neighborhood when discrete filtering is applied. The underlying mathematical operation, i.e., convolution, can be characterized with the help of so-called templates (Fig. 1.18). A *template* is a mostly small, squared mask of usually odd lateral length. This template is mirrored along two axes (hence, the name "convolution" is commonly used) and positioned in one corner of the input image. The image pixels under the mask are named *kernel*[1]. Each pair of corresponding pixel values of template and kernel are multiplied and then summed up. The result is registered at the position of the mask's center pixel in the output image. Then, the template is shifted row by row and column by column to the next positions on the input image, until all the positions have been visited, and thus, the output image has been calculated completely.

The pixel values of the template determine the effect of the filter. If only positive values are used in the template, basically a (weighted) averaging is calculated in the local neighborhood of each pixel (Fig. 1.18a,b). The resulting image is smoothed and appears with reduced noise. However, the sharpness of edges is also reduced. If the template is composed of positive and negative coefficients, the contrast in the image is intensified, and the edges are highlighted (Fig. 1.18c–f). Anisotropic (i.e., not rotationally symmetric) templates also have a preferred direction (Fig. 1.18e,f). Hereby, the contrasts can be direction-selectively strengthened.

1.3.3 Mathematical Morphology

Another approach to filtering is adapted from the mathematical morphology. Although morphologic operators can also be defined for gray scale images,

---

[1] In the literature, "mask", "kernel", and "template" frequently are used as synonyms.

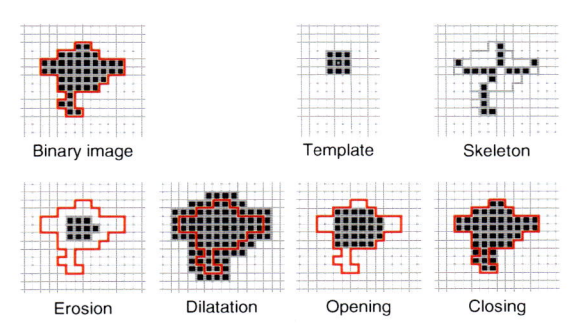

Fig. 1.19. *Binary morphology.* The binary pattern is outlined in red. The small circle marks the center pixel in disk-shaped template with radius 1

morphologic filtering is principally performed on binary input images, i.e., each pixel is assigned either TRUE or FALSE. According to a general convention, the white pixels in the binary image indicate relevant segments and the black pixels indicate the background. For printing, however, this assignment may be inverted. The binary template, which is also referred to as *structural element* (structuring element, structel, strel), is associated to the binary image using logical operations, in particular:

- *erosion* (based on logical AND of structel and binary image),
- *dilation* or dilatation (based on logical OR of structel and binary image),
- *opening* (erosion followed by dilatation using the same structel),
- *closing* (dilation followed by erosion using the same structel), and
- *skeleton* (e.g., by erosion with various structels).

As it can be seen in Fig. 1.19, the erosion reduces the size of a segment, and the dilation leads to its enlargement. The opening removes small details on the outline of segments or the background, without affecting the total size of relevant regions. The closing is able to remove holes in the interior of a region and smooth its contour. Here, the size of the segment is roughly maintained, too. The skeleton is a path with thickness of one pixel, which is located in the middle of the segment.

Binary morphology is applied frequently in medical image processing, for instance to clean up shapes after pixel-based segmentation (see Sect. 1.6.1). Gray scale morphology is simply a generalization from 1 bit (binary) images to images with multiple bits per pixel, where MIN and MAX operations replace the AND and OR operations of binary morphology, respectively.

1.3.4 Calibration

If the physician intents to take quantitative measurements from an image, a careful calibration of the imaging modality is required. Both, geometry (spatial domain) and brightness or color intensity (value domain) must be adapted to the modality. Calibration is device-specific but disregards the biological content captured, and thus, it is part of low-level processing methods. While reading a radiograph, calibration is made unconsciously by the radiologist.

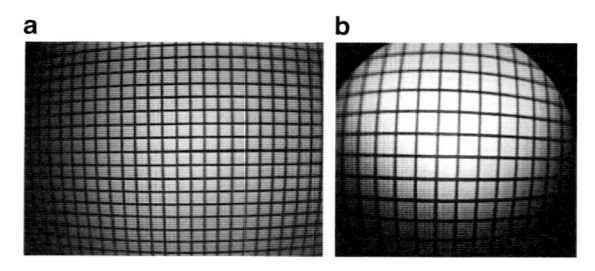

Fig. 1.20. *Geometric distortion and brightness variation [5].* By endoscopic examinations, barrel distortions are often generated, which must be corrected before the image can be analyzed quantitatively. In addition, the boundary areas in the video appear darker and blurred. Image (**a**) is generated with a rigid laryngoscope, which is used for the examination of the larynx. Image (**b**) is taken with a flexible endoscope for nasal laryngoscopy. Both endoscopes are used in clinical routine. Microscopy and other optical methods may produce similar artifacts

However, it must be explicitly implemented for computerized image analysis and measurements.

Geometric aberrations (distortions) have the consequence, that relevant structures of the same size are displayed depending on the position within the image. In the biomedical sciences, the positioning of the imaging device must not affect any measurements. For example in endoscopy, resulting from the optical devices in use, so called barrel distortions are originated (Fig. 1.20). Even in simple planar radiography, the objects, which are far away from the image plane, appear larger than those, which are located close to the imaging device. This must be kept in mind whenever geometric measurements in digital X-rays are taken and displayed to the physicians: point distances in digital images can be converted into length measurements only if a fixed scale is assumed, which is often not fulfilled.

In the same way, the absolute assignment of the pixel values to physical measurements usually is problematic. For example in X-ray imaging, the linear correspondence of brightness values to the accumulated absorption coefficient of the imaged structure is possible, if an aluminum (step) wedge with known X-ray absorption properties is placed beside the object. In digital video recording, white balancing must be performed such that the color values corresponds with reality. However, different illumination of the same scene may still alter the captured colors.

1.3.5 Registration

Often, an absolute calibration of examination procedures is not possible or only limitedly feasible. Then, registration can be used to achieve an approximation of two or more images such that at least a change in measured dimensions can be quantified. For example, an acute inflammation turns tissue

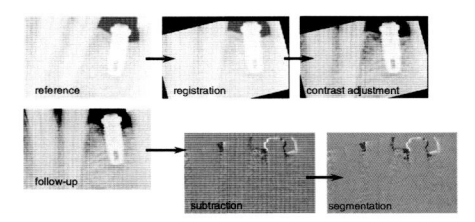

Fig. 1.21. *Unimodal Registration.* In dental implantology, reference and follow-up images are taken at various points of time. Geometric registration with subsequent contrast adjustment enables pixel-by-pixel subtraction. In the subtraction image, bone destruction is clearly emphasized and can be segmented easily on the pixel level of features (*red*)

into a reddish color. Under treatment, the absolute redness of the tissue is less interesting than its relative change as compared to the findings of previous recordings.

Unimodal Registration

This term refers to the relative calibration of images that have been acquired with the same modality. For instance, images that have been taken from the same patient but at different points of time are adjusted in order to quantify the course of the disease. As in the field of calibration, we differ between geometric registration and color or contrast adjustment, if the registration is performed in the spatial domain or the value range, respectively. Figure 1.21 illustrates the diagnostic potential of registration in dental implantology. After registration, the appraisal of the status of peri-implant bone is significantly simplified by the subtraction of recall and follow-up recordings.

Multi-Modal Registration

The images to be compared are captured with different modalities. For example, a 3D rigid registration is illustrated as the movement of the hat on the head. Especially in neurology, these methods have a crucial meaning. Since tumor resection in the brain must be executed very carefully, in order to avoid damage of neighbored brain areas, functional and morphological brain images are registered to plan the procedure. While morphology can be adequately represented in MRI or CT data, function of brain areas is frequently localized using Positron Emission Tomography (PET) or Single Photon Emission Computed Tomography (SPECT). Thus, multi-modal registration of functional and morphological data provides valuable additional information for diagnosis and therapy (Fig. 1.22).

Fig. 1.22. *Multi-modal registration and fusion [6]. 1. Row*: T1-weighted MRI of a 66 year old subject with right parietal glioblastoma; *2. Row*: Corresponding PET layers after multi-modal registration; *3. Row*: Fusion of registered layers to support intervention planning; *4. Row*: The fusion of MRI with PET of the sensorimotor-activated cortex area proves that the relevant area is out of focus

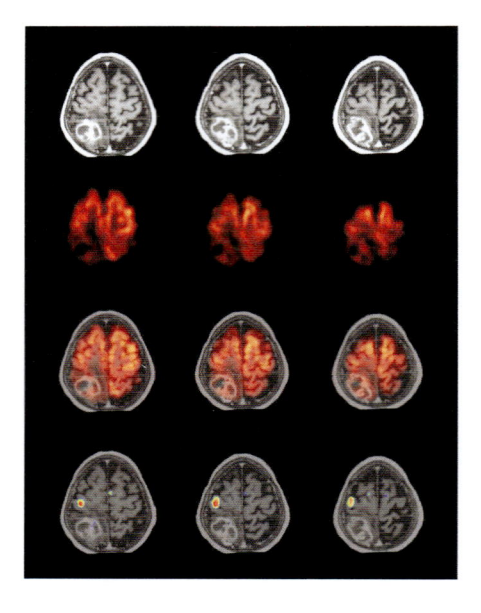

Table 1.5. *Taxonomy of 3D visualization methods.* Triangulation for surface-based rendering is described in textbooks on computer graphics. The marching cube approach is described in the text. As a simple example of surface-based direct volume rendering methods, depth shading visualizes the length of rays passing through the volume until they hit the surface. Integral shading codes the sum of voxel values along the ray as gray scale. It is therefore frequently used to obtain radiograph-like images based on CT data

Concept	Surface-oriented method	Volume-oriented method
Surface reconstruction and rendering	Triangulation	Cuberille approach
		Marching cube
Direct volume rendering	Depth shading	Integral shading
	Depth gradient shading	Transparent shading
	Gray gradient shading	Maximum projection

1.4 Image Data Visualization

Under the concept of image visualization, we had summarized all the transforms which serve the optimized output of the image. In medicine, this includes particularly the realistic visualization of 3D data. Such techniques have found broad applications in medical research, diagnostics, treatment planning and therapy. In contrast to problems from the general area of computer graphics, the displayed objects in medical applications are not given implicitly by formal, mathematical expressions, but as an explicit set of voxel. Consequently, specific methods have been established for medical visualization. These methods are based either on a surface reconstruction or on a direct volume visualization, and lighting and shading are also regarded (Table 1.5).

1.4.1 Marching Cube Algorithm

The marching cube algorithm was specifically developed for surface reconstruction from medical volumes. Here, the voxel is no longer interpreted as a cube of finite edge length but as a point. It is equivalent to a point grid for visualizing volumes. In this volume, a cube is considered with four corners in each of the two adjacent layers. Utilizing symmetry, the complex problem of surface production is reduced to only 15 different topologies, which can be calculated most efficiently since the polygon descriptions that belong to the basic topologies can be stored in a lookup table. Similar to the process of spatial convolution, the cube is positioned successively at all points in the volume dataset (marching). After completion of the marching cube algorithm, a segmented volume is transformed into a triangulated surface. However, the surface is build from a very large number of triangles, which may be reduced significantly by heuristic procedures without any discernible loss of quality. Reducing the number of elements to be visualized supports real-time visualization of the volume.

1.4.2 Surface Rendering

To generate photo-realistic presentations of the volume surface, the lighting is simulated analog to natural scenes. According to the lighting model by Phong, ambient light is created through overlapping of multiple reflections, diffuse scattering on non-shiny surfaces, and direct mirroring on shiny surfaces. While the intensity of the ambient light remains constant in the scene for all surface segments, the intensities of diffuse and speckle reflections depend on the orientation and characteristics of surfaces as well as their distances and directions to the light source and the observing point of viewing.

Without shading, one can recognize the initial triangles. This is a nasty artifact in computer graphics. Therefore, various strategies for shading have been developed to improve significantly the visual impression. For instance, the Gouraud shading results in smooth blunt surfaces, and the Phong shading also provides realistic reflections. In newer applications, transparencies are also modeled to glance at encapsulated objects. Moreover, textures or other bitmaps on the surfaces can be projected to reach a more realistic impression of the scene.

1.4.3 Volume Rendering

Direct volume visualization is abstained from preliminary calculation of the object surface. The visualization is based directly on the voxel data and, therefore, possible without any segmentation. This strategy allows visualization of medical 3D and 4D data by radiologists for interactive localization of pathological areas. The volume is processed either along the data layers (back-to-front or front-to-back) or along an imaginary light ray. Based on the observer position, rays will be pursued through the volume (ray-tracing). Hereby, the recursive follow-up of secondarily reflected rays is also possible (ray-casting). Although quite realistic visualizations can be provided to the observer, problems arising from the discrete nature of pixel topology (see above) have led to a multitude of algorithmic variants.

In general, parameters are extracted from voxel intensity along the rays and applied as gray or color value at the corresponding position in the viewing plane. This procedure is also referred to as shading. By the methods of the surface-based shading, light source and image plane are placed on the same side of the object, while the volume-oriented procedures radiograph the entire object according to X-ray imaging, i.e., the object is located between light sources and the observation (Table 1.5). Combining direct volume with surface-based approaches, amazingly realistic scenes can be created (Fig. 1.23).

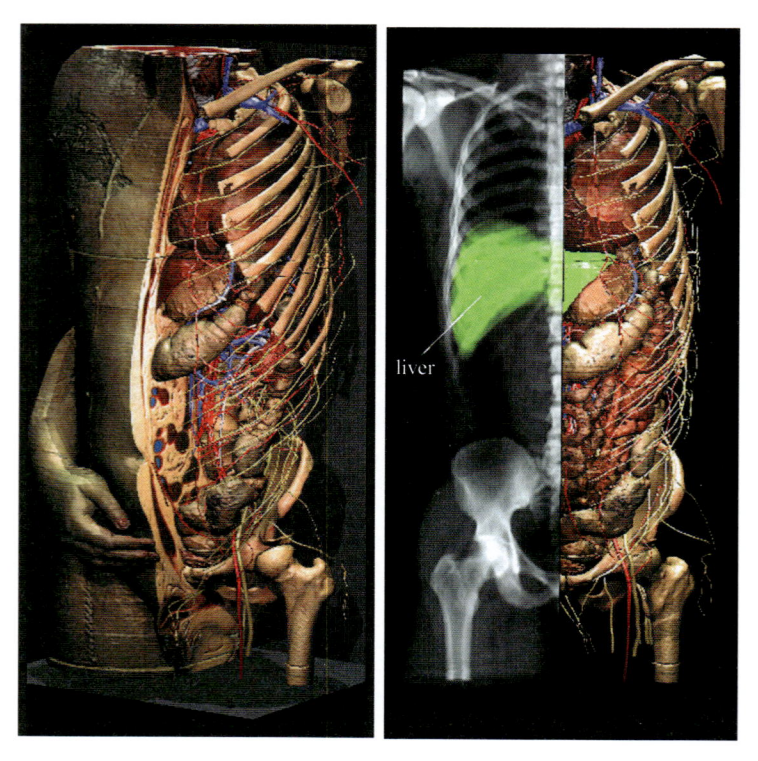

Fig. 1.23. *3D-visualization with Voxel–Man [7].* This 3D model of the internal organs is based on the Visible Human data. The Voxel–Man 3D-Navigator provides unprecedented details and numerous interactive possibilities (*left*). Direct volume rendering and surface-based visualization of segmented objects are combined with integral shading (*right*)

1.5 Visual Feature Extraction

In Fig. 1.1, feature extraction is defined as the first stage of intelligent (high level) image analysis. It is followed by segmentation and classification, which often do not occur in the image itself, i.e., the data or pixel level, but are performed on higher abstraction levels (Fig. 1.2). Therefore, the task of feature extraction is to emphasize image information on the particular level, where subsequent algorithms operate. Consequently, information provided on other levels must be suppressed. Thus, a data reduction to obtain the characteristic properties is executed. The schema in Fig. 1.1 is greatly simplified because many connections between the modules were left out on behalf of readability. So for example, cascades of feature extraction and segmentation at various levels of abstraction can be realized gradually, before classification is eventually performed at a high level of abstraction. Just before classification, a step of feature extraction that is based on the region level is often performed as well.

1.5.1 Data Level

Data-based features depend on the joint information of all pixels. Therefore, all transforms manipulating the whole matrix of an image at once can be regarded for data feature extraction. The most famous example of a data feature transform is the Fourier transform, which describes a 2D image in terms of frequencies, according to their amplitude and phase. Furthermore, the Hough, wavelet or Karhunen-Loève transforms provide possibilities of data feature extraction (see list of textbooks on image processing on page 49). These methods are not in the focus of research in biomedical image processing. In fact, these procedures are rather adapted from technical areas into medical applications.

1.5.2 Pixel Level

Since pixel-based features depend on the values of individual pixels, all point operations that have been defined in Sect. 1.3 can be regarded as feature extraction on the pixel level. Another example was already presented in Fig. 1.21, namely, the arithmetic combination of two images. The subtraction of reference and recall images after appropriate registration in both spatial and value ranges enforce local changes in the images as characteristic pixels.

1.5.3 Edge Level

Edge-based features are defined as local contrast, i.e., a strong difference of (gray scale or color) values of adjacent pixels. Thus, the discrete convolution introduced in Sect. 1.3 can be used with appropriate templates for edge

Fig. 1.24. *Edge extraction using the Sobel operator.* The X-ray image (*center*) was convolved with the eight direction-selective Sobel templates. The strong contrasts on the edges of metallic implants are further strengthened by binarization of the edge images. An isotropic edge image is obtained if, e.g., the maximum at each pixel position is chosen from the eight direction-selective sub-images

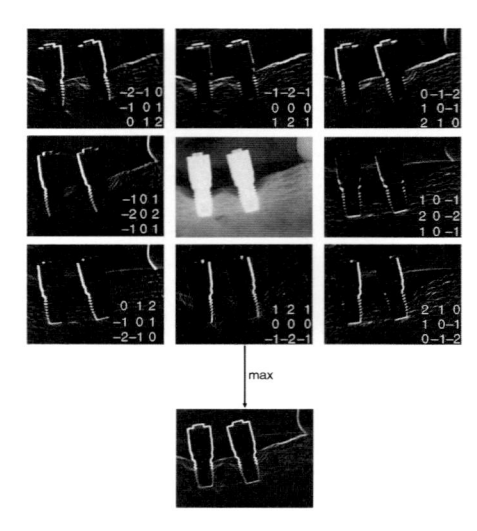

extraction. All masks for high-pass filtering amplify edges in an image. The templates of the so called Sobel operator (Fig. 1.18) are particularly suited for edge extraction. Figure 1.24 exemplarily presents the result of the orientation-selective Sobel masks when applied to a dental radiograph. The edges of the metallic implants are clearly highlighted. An isotropic Sobel-based edge image is achieved, e.g., by a linear or maximum combination of the eight sub-images.

1.5.4 Texture Level

Textural features have been used in medicine for a long time. In textbooks on pathology one can read many metaphors to describe texture, such as a cobblestone-shaped mucosal relief, onion-like stratification of subintima, or honeycomb-structured lung tissue. As intuitive as these metaphors are for people, as difficult is their computational texture processing, and a variety of procedures and approaches have been developed.

Texture analysis attempts to quantify objectively the homogeneity in a heterogeneous but at least subjectively periodic structure (see the spongious bone structure in Fig. 1.18c as an example). In general, we can distinguish:

- *structural approaches* that are based on texture primitives (textone, texture element, texel) and their rules of combinations and
- *statistical approaches* that describe texture by a set of empirical parameters.

1.5.5 Region Level

Regional features are used primarily for object classification and identification. They are normally calculated for each segment after the segmentation process.

The most important parameters to be mentioned here are:

- *localization-descriptive* measurements such as size, position, and orientation of the major axis and
- *delineation-descriptive* measures such as shape, convexity, and length of the border.

Since the degree of abstraction on the region level is rather high as compared to the previous levels, a priori knowledge has already been largely integrated into the image processing chain. Therefore, universal examples cannot be specified. In fact, the definition of regional feature extraction is strongly dependent on the respective application (see Sects. 1.5.5 and 1.6.3).

1.6 Segmentation

Segmentation generally means dividing an image into connected regions. With this definition, the production of regions is emphasized as the pre-stage of classification. Other definitions accentuate the various diagnostically or therapeutically relevant image areas and, thus, focus the most common application of medical imaging, namely, the discrimination between healthy anatomical structures and pathological tissue. By definition, the result of segmentation is always on the regional level of abstraction (cf., Fig. 1.2). Depending on the level of feature extraction as an input to the segmentation, we can methodically classify pixel-, edge-, and texture- or region-oriented procedures. In addition, there are hybrid approaches, which result from combination of single procedures.

1.6.1 Pixel-Based Segmentation

Pixel-based procedures of segmentation only consider the gray scale or color value of current pixels disregarding its surroundings. It should be noted that pixel-based approaches are not segmentation procedures in the strict sense of our definition. Since each pixel is considered only isolated from its neighborhood, it cannot be ensured that actually only connected segments are obtained. For this reason, post-processing is required, e.g., by morphologic filtering (see Sect. 1.3.3). Most pixel-based procedures use thresholds in the histogram of an image and employ more or less complex methods to determine this threshold. Furthermore, statistical methods for pixel clustering are used.

Static Thresholding

If the assignment of pixel intensities is well known and constant for a certain type of tissue, static thresholds are applicable. A static threshold is independent of the individual instance in a set of similar images. For example, bone or soft tissue windows in the CT can be realized (Fig. 1.25) with static thresholds on the Hounsfield Unit (HU).

Fig. 1.25. *Static thresholding [5].* Pixel-based segmentation in CT relies on Hounsfield Units (HU), which allow the definition of windows for different types of tissue: bone $[200\ldots 3{,}000]$, water $[-200\ldots 200]$, fat $[-500\cdots -200]$, or air $[-1{,}000\cdots -500]$

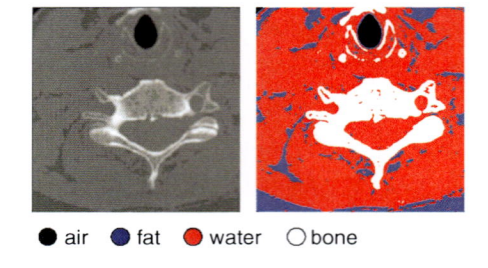

● air ● fat ● water ○ bone

Adaptive Thresholding

Globally adaptive thresholds result from analyzing each individual image entirely. They are exclusively used in this image. The well-known method of Otsu is based on a simple object vs. background model. The threshold in the histogram is determined such that the two resulting classes minimize the intra-class variance of gray scale values, while the inter-class variance is maximized. For example in skeletal radiography, bone, soft tissue and background can be seen, but the actual mean gray scale of this tissue classes may vary with respect to illumination and exposure parameters. By adopting the threshold to the image, the Otsu segmentation is able to balance this variation in imaging.

Using *locally adaptive* thresholds, the threshold is computed not only for each image individually, but also for each region within an image. In the extreme case, an individual threshold is determined for every pixel position (i.e., pixel-adaptive). This is particularly necessary if the simple object to background assumption is globally invalid because of continuous brightness gradients. For example, due to the irregularity of optical illumination, the background in microscopy imaging of cell cultures (Fig. 1.26a) runs from light shades of gray (top right) to dark shades of gray (bottom left), where also the gray scale values of the cells are located. A global threshold determined with the dynamic procedure of Otsu (Fig. 1.26b) does not separate the cells from backgrounds, although the global threshold had been determined image-individually. The locally adaptive segmentation (Fig. 1.26c) leads to a significantly improved result, but isolated block artifacts appear. These artifacts can be avoided only by pixel-adaptive thresholding (Fig. 1.26d).

Clustering

Pixel clustering is another way of pixel-based segmentation. This statistical method is particularly suitable if more than one value is assigned to each pixel and regarded in the segmentation process (e.g., color images). Figure 1.27 illustrates the iso-data clustering algorithm (also referred to as k-means clustering) in a simple 2D case. All pixel values are registered as data points in the 2D feature space. Initialized by the number of segments to be obtained,

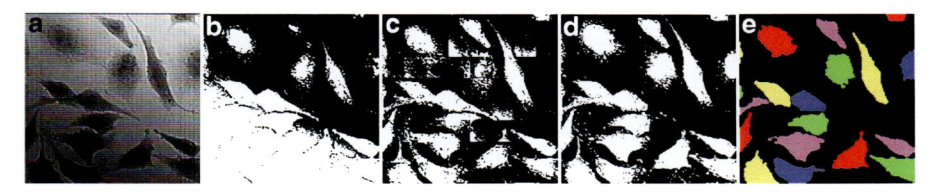

Fig. 1.26. *Dynamic thresholding in microscopy [8].* The microscopy of a cell culture (**a**) was segmented using a global threshold (**b**), locally adaptive (**c**) and pixel-adaptive (**d**). According to morphological post-processing for noise reduction and a connected components analysis, the final segmentation is shown in (**e**)

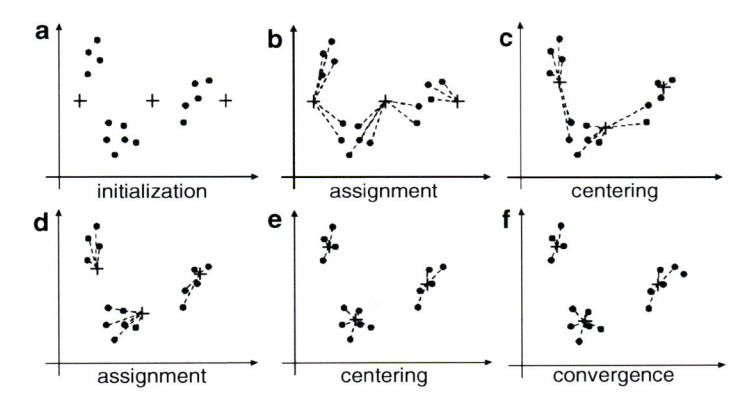

Fig. 1.27. *Iso-data pixel clustering.* The iterative iso-data algorithm for pixel clustering is exemplified in a 2D feature space. The number of clusters is given a priori. After arbitrary initialization, the data points are assigned to the nearest cluster center. Then, the positions of the centers are recalculated and the assignment is updated until the process finally converges. The final location of cluster centers is not affected by their initial position. This may only have impact to the number of iterations

the initial cluster centers are arbitrarily placed by the algorithm. Then, the following two steps are repeated iteratively until the process converges:

1. Each data point is aligned to the closest cluster center.
2. Based on the current assignment, the cluster centers are recalculated.

It can be proven mathematically that the resulting cluster centers are independent of initial positions, which may only impact the number of iterations and hence, the calculation time. However, either a fixed distance metrics (e.g., Euclidean (geometric) distance) or a data-adaptive metrics (e.g., Mahalanobis distance) must be selected, which certainly impacts the clustering result. Also, the predefined number of cluster centers is an important parameter. If the application domain does not allow to determine the number of segments a priori, pixel clustering can be performed for a different number of centers and the residual error of the computed model can be analyzed to determine the appropriate number of centers.

Post-Processing

Segments obtained from pixel-based analysis usually are incoherent and highly noisy (see Fig. 1.25 or Fig. 1.26b). Therefore, post-processing is required. Noisy structures can be effectively reduced with methods of mathematical morphology. While a morphologic opening removes spread parts from the segments, holes are closed by morphologic closing (see Sect. 1.3.3). The connected components algorithm provides each separated segment with a unique reference number. In the segmentation of the cell image (Fig. 1.26a); clustering provides a rough cluster of "cells", which is separated from the "background", although many individual cells are shown separately in Panel 1.26(d). After morphological post-processing and connected components analysis, cells are separated and colored (labeled) differently according to their segment number. Now, they can be further processed as independent objects (Fig. 1.26e).

1.6.2 Edge-Based Segmentation

This type of segmentation is based on the abstract level of edges and tries to capture the objects due to their closed outline in the image. Hence, edge-based segmentation procedures are only used for such problems, where objects are represented as clearly defined boundaries. As described in Sect. 1.1.3, this occurs rather seldom when biological tissue is imaged. One of these special cases is a metallic implant, which is displayed in a radiograph.

In general, the image processing chain for edge-based segmentation is composed of edge extraction and edge completion. Edge extraction is usually obtained by edge-based feature extraction, as described in Sect. 1.5.3, such as generated with the Sobel filter (see Fig. 1.24). The next steps of processing are binarization, to obtain only edge pixels and non-edge pixels, morphological filtering to reduce noise and artifacts, and, finally, a skeleton of the edge is computed. Tracing and closing of binary contours are the main tasks of the edge-based segmentation. Almost exclusively, heuristic methods are used. For example, one can search along differently directed rays to find connecting pieces of a contour. This procedure aims at bridging local gaps on the edge profile.

Livewire Segmentation

In practice, edge-based segmentation is often realized semi-automatically. By the interactive livewire segmentation, the user clicks onto or near by the edge of the Object of Interest (OOI), and the computer determines the exact edge location based on local gradients. Then, the computer calculates a cost function, which again is based on local gradients. For all paths (wire) to the current position of the cursor, the path with the lowest cost is displayed in real time (live) as the cursor is moved manually. Therefore, the metaphor "livewire" is commonly used to refer to this interactive method of segmentation. If the cursor moves far from the object, the contour is lost but if the cursor is placed near to the contour again, the cost function ensures that the wire snaps back to the desired object. Finally, the user must provide only a few supporting points by hand and can directly verify the correctness of segmentation (Fig. 1.28). Application of such procedures can be found at computer-assisted (semi-automatic) segmentations in layers of CT data, e.g., to produce a model for surgical intervention planning. Guided by the cost function, the segmentation result (delineation) is independent of the user placing the supporting point (localization).

1.6.3 Region-Based Segmentation

As an advantage of region-based segmentation, only connected segments are produced, and morphological post-processing is avoided. There are agglomerative (bottom-up) and divisive (top-down) approaches. All approaches are based on a certain distance or similarity measure to guide the assignment of neighbored pixels or regions. Here, plenty of methods are used. Easiest, one can compare the mean gray value but complex texture measures (see Sect. 1.5.4) are often used, too.

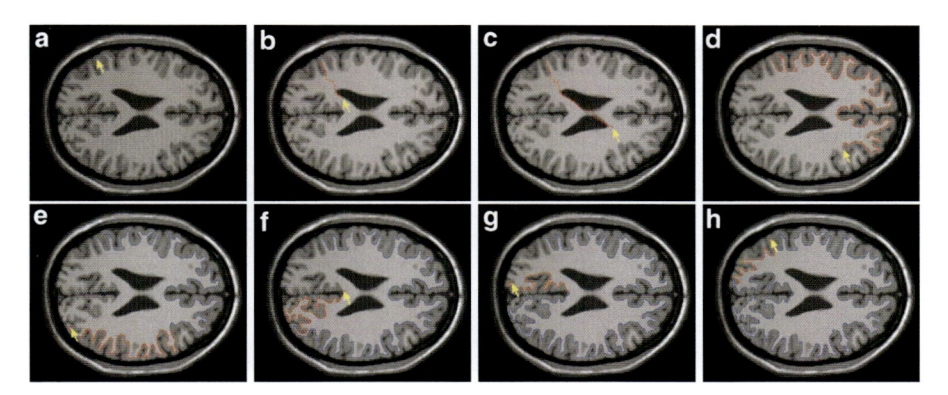

Fig. 1.28. *Edge-based interactive livewire segmentation [9].* The user marks a starting point with the cursor (*yellow*) on the border between white and gray matter (**a**). The connection to the current cursor position is denoted with red, cf. (**b**) to (**e**). Depending on the cursor position, the contour can also jump between very different courses (**d**, **e**). So, the user can interactively place an appropriate fix point. The fixed curve segment is shown in blue, cf. (**e**) to (**g**). In this example, only five points are manually marked to achieve a complete segmentation (**h**)

Agglomerative Algorithm

Region growing, in 3D also referred to as volume growing, is a well known example of an agglomerative procedure. Starting from seed points, which may be placed either automatically or manually, neighbored pixels are iteratively associated to the growing areas if the distance measure is below a certain threshold. This process is iterated until no more merges can be carried out. From this qualitative description, the variety and sensitivity of the parameters of such procedures are already clear. Special influence on the result of agglomerative segmentation has:

- the number and position of seed points,
- the order in which the pixels or voxels are iteratively processed,
- the distance or similarity measure applied, and
- the threshold used to guide merging.

Therefore, agglomerative algorithms for segmentation often are affected by small shifts or rotations of the input image. For instance, if x- and y-axis of the image matrix are transposed, the result of segmentation is different regarding size and shape of OOI, which is an unwanted effect in medical image processing.

Divisive Algorithm

The divisive approach somehow inverts the agglomerative strategy. By splitting, the regions are iteratively subdivided until they are considered sufficiently homogeneous in terms of the chosen similarity measure. As an advantage, seed points are not required anymore, because the first split is performed throughout the whole image. As a drawback, the dividing lines are usually drawn horizontally or vertically, and this arbitrary separation may separate the image objects. Therefore, split is unusually performed as a self standing segmentation procedure, but rather combined with a subsequent merging step (split and merge). Another drawback of divisive segmentation procedures is the resulting wedge-formed boundary of objects, which may require post-processing such as contour smoothing.

1.6.4 Over- and Under-Segmentation

A fundamental problem of pixel- and region-based segmentation is the dualism between over- and under-segmentation. For a definition of these terms, we rely on the general model of the image processing chain (see Fig. 1.1). Here, segmentation is regarded as a pre-stage for classification, in which the extracted image segments are assigned to their semantic meaning. This can take the form of automatically assigning concrete terms for the segments (for example, the organ "heart" or the object "TPS implant screws" or, more abstract, a "defect" or an "artifact").

In any case, the segment should be related directly to an object, if an automatic classification is desired. In this context, under-segmentation occurs if resulting segments are composed from parts of several objects. Analogously, over-segmentation is obtained if a particular object is disintegrated into several segments or parts of segments. The big problem with segmentation of medical images is that over- and under-segmentation usually occur simultaneously.

Hierarchical Algorithm

Hierarchical procedures are one of the concepts to deal with the dualism between over- and under segmentation. Starting on a lower resolution of the image, where it is represented with a small number of pixles only, the chance of splitting objects into more than one segment is decreased. Then, the exact outline of each segment is reconstructed on higher resolutions, where more details are contained (Fig. 1.29).

Hybrid Algorithm

In the practice of medical image processing, hybrid approaches of segmentation have come to the greatest importance. Here, one is trying to combine the advantages of individual (usually edge- and region-based) algorithms without maintaining their disadvantages.

For example, the watershed transform extends an agglomerative, regional segmentation procedure with edge-based aspects of segmentation. Indeed, it is based on the very intuitive analogy of the image with a topographic surface: the gray levels of the pixels correspond to the altitude of the relief. In

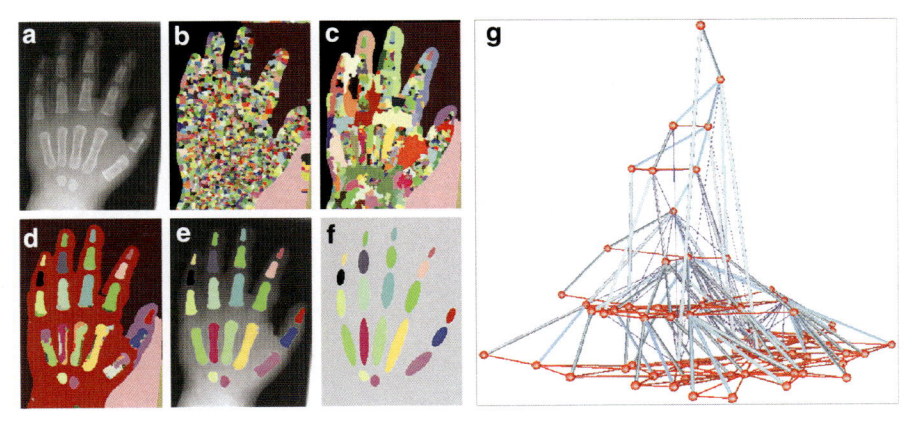

Fig. 1.29. *Hierarchical region merging.* The skeletal radiograph of the hand (**a**) has been segmented at various levels of resolution, cf. (**b**) to (**d**). The initial step is obtained with the watershed transform (see Sect. 1.6.4). Depending on the size of the objects, they can be localized in the appropriate level (**e**), approximated by ellipses (**f**), or visualized as nodes in a graph (**g**)

hydrology, a catchment basin relative to a body of water (e.g., a river or an ocean) is defined as a region where water from rain drains downhill into this reference body, whereas the watershed lines (also known as water parting or divide) separate different drainage basins. Similarly, a catchment basin surrounded with watershed lines can be defined for every regional minimum in the image. In general, the image gradient is taken as the topographic surface, so as the catchment basins to correspond to connected and homogeneous regions (structures of interest), and watershed lines to lie on higher gradient values.

The so-called classical watershed transform takes into account all regional minima of the image to compute a primitive catchment basin for each one. As natural images contain many regional minima, in general, too many basins are created. The image is over-segmented (see, for example, Fig. 1.29b). However, over-segmentation can be reduced by filtering the image and, therefore, decreasing the number of minima.

On the other hand, when applied to segmentation of medical images, the watershed transform especially has the following advantages:

- From the region-based idea of the flooding process, contiguous segments are determined inherently.
- From the edge-based approach of the watersheds, the objects are exactly delineated.
- The problem of under-segmentation is avoided, since the merging of smaller pools is prevented by the watersheds.

1.6.5 Model-Based Segmentation

State of the art methods for model- or knowledge-based segmentation involve active contour models and deformable templates as well as active shape and active appearance models.

Active Contour Model

Active contour models apply edge-based segmentation considering region-based aspects and an object-based model of a priori knowledge. In the medical application domain, so called snake and balloon approaches are applied for segmentation of 2D and 3D image data and the tracing of contours in 2D image and 3D image sequences, i.e., 3D and 4D data, respectively. The contour of the objects, which is usually closely modeled, is presented by individual nodes, which are – in the simplest case – piecewise connected with straight lines forming a closed polygon. For the nodes, a scalar quality measure (e.g., energy) is calculated and optimized in the local environment of the nodes. Alternatively, adjusted forces are determined that directly move the nodes. The iterative segmentation process completes at minimal energy or if an optimum balance of forces was found. Thus, the potential of this approach is kept in the choice of capable quality criteria (e.g., energy) or forces.

Snake

In 1988, Kass et al. have introduced classical snake approach [10]. It models an internal and an external quality criterion, both as undirected energy. The internal energy results from a predefined elasticity and stiffness of the contour, which is high in places of strong bends or on buckling. The external energy is calculated from an edge-filtered image. The external energy is small, if the contour runs along edges. The idea behind this approach is an edge-based segmentation combined with the a priori knowledge that biological objects rarely have sharp-bending boundaries. With an optimal weighting of energy terms, the contour course is primarily determined by the information of edges in the image. However, if the object's contour is partially covered or incompletely captured, the internal energy ensures an appropriate interpolation of the region's shape.

So simple this approach has been formulated verbally, so difficult it is to implement. During the iteration, the number of nodes must be constantly adjusted to the current size of the contour. Furthermore, crossovers and entanglements of the moving contour must be avoided. The classical snake approach also requires an already precisely positioned starting contour, which often must be defined interactively. Then, the two steps of segmentation, i.e., localization and delineation are performed again by man and machine, respectively. This concept was also applied in the first publications of this segmentation method. For a contour tracking of moving objects in image sequences, the segmentation of image at time t serves as initial contour of iteration in image $t + 1$. After a single initialization for the image $t = 0$, the procedure runs automatically. Hence, fluoroscopy and endoscopy are suitable modalities for the application of the snake approach to track the shape of moving objects.

Balloon

Balloons are based on forces rather than energies. Besides the internal and external force, an inner pressure or suction is modeled, which lets the contour continuously expand or shrink. Figure 1.30 shows the inflation movement of a balloon to segment the cell membrane, which is visualized by the synaptic boutons of contacting dendrites in a microscopy of a motoneuron. Although segmentation is done without an accurate initial contour, in the course of iteration the balloon nestles onto the real contour of cell membrane. Another advantage of the balloon model is that this concept is directly transferable into higher dimensions (Fig. 1.31).

Other Variants

In recent developments of active contour models, it is attempted to incorporate further a priori knowledge, e.g., in the form of anatomical models. Prototypes of the expected object shapes are integrated into the algorithm: In each iteration, the distance of the current object shape to a suitable selected

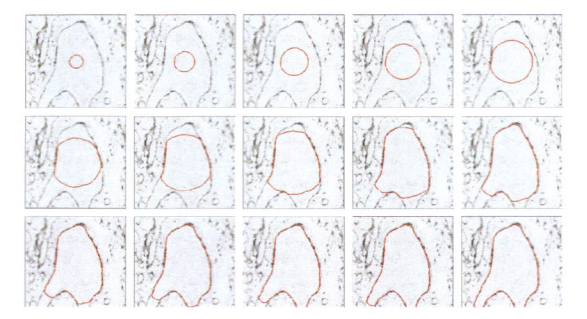

Fig. 1.30. *Balloon segmentation of motoneuron cell membrane [11].* The frames show the balloon at different iterations. By touching the cell membrane, the strong image forces prevent further movement of the active contour. In this application, the internal forces correspond physically to a membrane. This is clearly recognizable at the "adhesion border" of the balloons reaching the dendrites (*bottom left*)

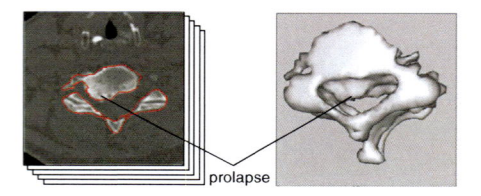

Fig. 1.31. *Segmentation with a 3D balloon model [12].* The CT of a spine (*left*) was segmented with a 3D balloon. In the surface-based rendering after automatic segmentation, the prolapse is clearly visible (*right*). The visualization is based on Phong shading (see Sect. 1.4.2)

prototype is modeled as an additional force on the node. With those extensions, a "break out" of the active contour model is prevented also for long passages of the local object boundary without sufficient edge information.

The complex and time-consuming parameterization of an active contour model for a specific application can be based on manual and also automatic reference segmentations. For the latter approach, different combinations of parameters are determined and the segmentation is performed for all cases. All resulting segmented contours are compared with the appropriate reference contour, a priori defined as the ground truth of the training data. Then, that set of parameters with the best approximation of the reference contour is selected automatically.

Active Shape Model

In the biomedical sciences, OOIs such as bones or organs often have a similar form or projected shape that may vary between individuals or different points of time. Therefore, a probabilistic model may be applied to explain the shape variation. Segmenting an image imposes constraints using this model as a prior. Usually, such a task involves:

1. registration of the training examples to a common pose,
2. probabilistic representation of the variation of the registered samples, and
3. statistical inference between the model and the image.

Introduced by Cootes et al. in 1995, active shapes aim at matching the model to a new image [13]: for probabilistic representation (Step 2), the shapes are constrained by the Point Distribution Model (PDM) allowing variation only in ways that have been seen in the training set of labeled examples. For statistical inference (Step 3), a local neighborhood in the image around each model point is analyzed for a better position. Alternating, the model parameters are updated to best match to these newly determined positions, until convergence is reached.

Similar to active contour models, each training shape is represented by a set of points, where each point corresponds to a certain landmark. To form a feature vector x_i, all landmark coordinates are concatenated. A mean shape $\bar{x}$ and its covariance matrix S from N training sets is obtained by

$$\bar{x} = \frac{1}{N}\sum_{i=0}^{N-1} x_i \quad \text{and} \quad S = \frac{1}{N}\sum_{i=0}^{N-1}(x_i - \bar{x})(x_i - \bar{x})^T \tag{1.3}$$

The Principle Component Analysis (PCA) is applied for dimension reduction computing normalized eigenvectors and eigenvalues of S across all training shapes. The base Φ of eigenvectors ϕ represents the principle modes of variation, and the eigenvalues λ indicate the variance per mode. The prior model is generated from the t largest eigenvalues. Now, any shape x may be approximated by $x \approx \bar{x} + \Phi \nu$, where the weighting vector ν is determined minimizing a distance measure in the image, e.g., the Mahalanobis distance.

Figure 1.32 shows an application of the active shape method for bone age assessment. The BoneXpert® method[2] robustly detects carpal bones and phalanges as well as epiphysis using active shapes.

1.7 Classification

According to the general processing chain (see Fig. 1.1), the task of the classification step is to assign all connected regions, which are obtained from the segmentation, to particularly specified classes of objects. Usually, region-based features that sufficiently abstract the characteristics of the objects are used to guide the classification process. In this case, another feature extraction step is performed between segmentation and classification, which is not visualized

---

[2] Visiana Ltd, Holte, Denmark, http://www.bonexpert.com

Fig. 1.32. *Active shape segmentation of hand radiopraph.* BoneXpert® detects relevant bones and measures distances, geometry and sizes to compute skeletal maturity. Although in this example the first metacarpal bone is misaligned, the automatically suggested bone age, which is computed over several regions, lies within the range of human inter-observer variation

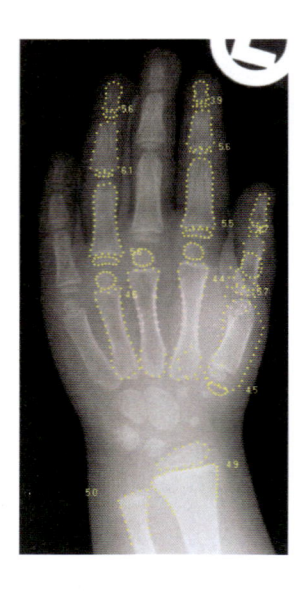

in Fig. 1.1. These features must be sufficiently discriminative and suitably adopted to the application, since they fundamentally impact the resulting quality of the classifier.

For all types of classifiers, we can differ supervised (trained), unsupervised (untrained) and learning classification. For example, pixel clustering, which has been already introduced for pixel-based segmentation, is an unsupervised classification process (see Fig. 1.27). As a goal, individual objects are divided into similar groups. If the classification is used for identification of objects, the general principles or an exemplary reference must be available, from which the ground truth of classification can be created. The features of these samples are then used for parameterization and optimization of the classifier. Through this training, the performance of the classifier can be drastically improved. However, supervised object classification is always problematic, if the patterns that are classified differ remarkably from the trained patterns. In such cases, the training set does not sufficiently reflect the real world. A learning classifier has advantages here, because it changes its parameterization with each performed classification, even after the training phase. In the following, however, we assume a suitable set of features that are sufficiently characteristic and large set of samples.

The classification itself reverts mostly to known numerical (statistical) and non-numerical (syntactic) procedures as well as the newer approaches of Computational Intelligence (CI), such as neural networks, evolutionary algorithms, and fuzzy logic. In general, the individual features, which can be determined by different procedures, are summarized either to numerical feature vectors (also referred to as signature) or abstract strings of symbols. For example, a closed contour object can be described by its Fourier-descriptors as a feature vector, or by means of basic line items such as "straight", "convex", and "concave" forming a symbol chain.

1.7.1 Statistic Classifiers

Statistical classification regards object identification as a problem of the statistical decision theory. Parametric procedures for classification are based on the assumption of distribution functions for the feature specifications of the objects, and the parameters of the distribution functions are determined from the sample. Non-parametric methods, however, waive such model assumptions, which are sometimes unobtainable in biomedical image processing. A common example of such a non-parametric statistical object classifier is the Nearest Neighbor (NN) classifier. All features span the feature space, and each sample is represented by a point in this feature space. Based on the signature of a segment, which has not been included in the training and now is assigned to its nearest neighbor in feature space, the segment is classified to the associated class of the assigned feature vector. The k-Nearest Neighbor (k-NN) classifier assigns the majority class from the k nearest neighbors in feature space (usually, $k = 3$ or $k = 5$). An example of the k-NN classifier is given in Fig. 1.33.

1.7.2 Syntactic Classifiers

In symbol chains, it is neither useful nor possible to define distance measurements or metrics and to evaluate the similarity between two symbol chains, such as used for feature vectors. An exception of this statement is given with the Levenshtein distance, which is defined as the smallest number of modifications such as exchange, erase, or insert, required to transform a symbol chain into another.

The syntactic classification is therefore based on grammars, which can possibly generate an infinite amount of symbol chains with finite symbol formalism. A syntactic classifier can be understood as a knowledge-based classification system (*expert system*), because the classification is based on a formal heuristic, symbolic representation of expert knowledge, which is transferred into image processing systems by means of facts and rules. If the expert system is able to create new rules, a learning classifier is also realizable as a knowledge-based system.

It should be noted that the terms "expert system" or "expert knowledge", however, are not standardized in the literature. Therefore, "primitive" image processing systems, which use simple heuristics as implemented distinction of cases to classification or object identification, are also referred to as "knowledge-based".

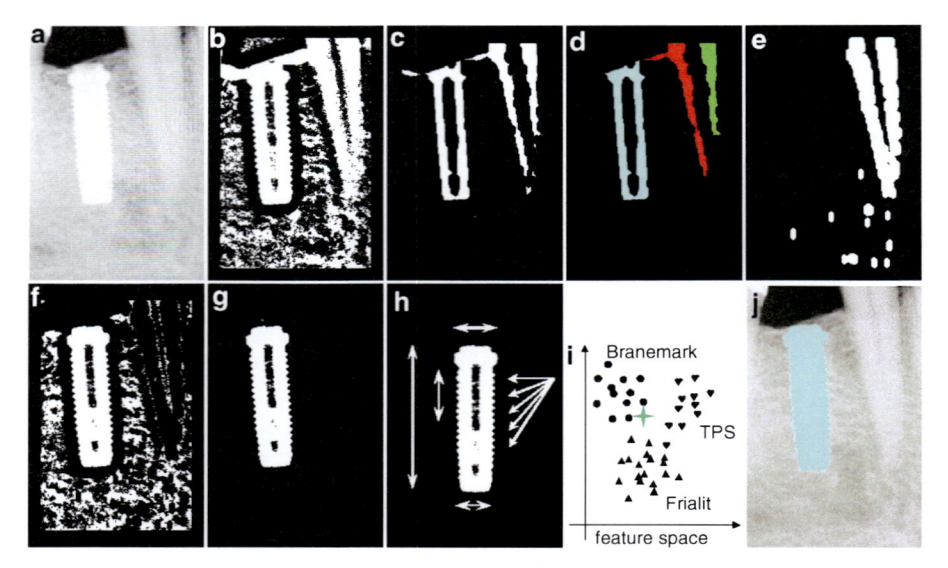

Fig. 1.33. *Identification of dental fixtures [14].* An implant is shown in the intraoral radiograph of the lower jaw (**a**). For feature extraction, the image is binarized with a local adaptive threshold (**b**). The morphological filtering (erosion) separates individual areas (**c**) and eliminates interference. In this example, three regions were segmented (**d**). Further processing is shown for the blue segment. After its fade-out, the gap of morphological erosion is compensated by a subsequent dilation (**e**), and the result is subtracted from the intermediate image (**b**). Any coordinate of blue segment from (**d**) identifies the corresponding region, which can be extracted now (**g**) and aligned into a normal position using the Karhunen-Loève transform. Geometric dimensions are determined as region-based features and stored in a feature vector (signature). As part of the training, the reference measures of different implant types have been recorded in the feature space. The classification in the feature space is done with the statistical k-NN classifier (**i**), which identifies the blue segment reliably as Branemark implant screw (**j**)

1.7.3 Computational Intelligence-Based Classifiers

As part of the artificial intelligence, the methods of CI include neural networks, evolutionary algorithms and fuzzy logic. These methods have their examples in biological information processing. Although they usually require high computational power, they are frequently used in biomedical image processing for classification and object identification. Thereby, all the procedures have a mathematical-based, complex background.

Neural Network

Artificial neural networks simulate the information processing in the human brain. They consist of many simply constructed basic elements (i.e., neurons), which are arranged and linked in several layers. Each neuron calculates the weighted sum of its input excitations, which is mapped over a nonlinear function (i.e., characteristic curve) to the output. The number of layers, the number of neurons per layer, the network's topology, and the characteristic curve of the neurons are predefined within the network dimensioning step. On the one hand, heuristics are usually applied rather than methodological derivations. On the other hand, the individual weights of the excitements are identified numerically during the training of the network. Then, the network remains unchanged and can be used as a classifier.

Evolutionary Algorithm

Evolutionary algorithms are based on the constant repetition of a cycle of mutation and selection following the Darwinian paradigm of the survival of the fittest. Genetic algorithms work on a number of individuals (the population). The crossing of two randomly selected individuals and afterwards the mutation changes the population. A fitness function evaluates the population in terms of their goodness to problem solution. Although the selections are equipped with a random component, fit individuals are frequently selected for reproduction. Evolutionary algorithms can solve complex optimization problems amazingly well, but for object classification, they are less successfully used than other methods.

Fuzzy Algorithm

The idea of fuzzy logic is to extend the binary (**TRUE** or **FALSE**) computer model with some uncertainty or blur, which exists in the real world, too. Many of our sensory impressions are qualitative and imprecise and, therefore, unsuitable for accurate measurements. For example, a pixel is perceived as "dark", "bright" or even "very bright", but not as a pixels with the gray scale value "231". Fuzzy quantities are based mathematically on the fuzzy set theory, in which the belonging of an element to a set of elements is not restricted to the absolute states **TRUE** (1) or **FALSE** (0), but continuously defined within the entire interval $[0..1]$.

Beside classification, applications of fuzzy logic in biomedical image processing can be found also for pre-processing (e.g., contrast enhancement), feature extraction (e.g., edge extraction, skeleton), and segmentation.

1.8 Quantitative Measurements and Interpretation

While the visual appraisal by experts is qualitative and sometimes subject to strong inter- as well as intra-individual fluctuations, in principle, a suitable computer-aided analysis of biomedical images can deliver objective and reproducible results. First of all, this requires a precise calibration of the imaging modality. Furthermore, partial (volume) effects of the imaging system and particularities of the discrete pixel topology must be taken into account and handled accordingly to ensure reliable and reproducible measurements.

Quantitative measurement is focused on automatic detection of objects as well as their properties. Image interpretation steps further towards analyzing the order of individual objects in space and/or time. It may be understood in the sense of analyzing an abstract scene that corresponds to the ambiguous goal of developing a "visual sense for machines", which is as universal and powerful as that of humans.

1.8.1 Partial Volume Effect

The digitalization of the local area or volume of a pixel or voxel, respectively, always yields an averaging of the measured value in the appropriate field. For example in CT, a voxel containing different tissue is assigned a certain Hounsfield value that results from the proportional mean of the individual Hounsfield values of the covered tissue classes. Thus, a voxel containing only bone and air preserves the Hounsfield value of soft tissue and, thus, may distort quantitative measurements. In general, this partial (volume) effect occurs in all modalities and must be accounted appropriately for any automatic measurement (see Fig. 1.14).

1.8.2 Euclidean Paradigm

The common paradigms of the Euclidean geometry do not apply in the discrete pixel domain. For example, the discrete representations of two straight lines may not join in a common pixel although the lines are crossing. Furthermore, different neighborhood concepts of discrete pixel's topology have remarkable impact on the result of automatic image measurements. In particular, the areas identified in region growing may be significantly larger if the 8-neighborhood is applied, i.e., if eight adjacent pixels are analyzed instead of the four direct neighbors (4-neighborhood).

1.8.3 Scene Analysis

The fundamental step of image interpretation is to generate a spatial or temporal scene description on the most abstract level (symbolic image description, see Fig. 1.2). A suitable form of representation is the attributed relational graph (semantic web), which can be analyzed at different hierarchy levels (see Fig. 1.29, right). Therefore, the considered grid matrix of pixels (iconic image description, see Fig. 1.2) so far is inappropriate for image interpretation.

The primitives of the graph (node) and their relationships (edges) must be abstracted from the segmented and identified objects or object parts in the image. So far, only a few algorithms can execute this level of abstraction. Examples for the abstraction of primitives are given by the numerous approaches to shape reconstruction: Shape-from-shading, -texture, -contour, -stereo, etc. Examples for the abstraction of relationships can be found at the depth reconstruction by trigonometric analysis of the projective perspective. Recently, considerable progress has been achieved in symbolic image analysis in the fields of industrial image processing and robotics. Because of the special peculiarities of the biomedical imagery (see Sect. 1.1.3) the transfer of these approaches into health care applications and medical image processing is only sparingly succeeded so far.

1.8.4 Examples

We will now discuss some examples for image measurements. For instance in Fig. 1.33, geometrical features are used for the automatic classification of implant systems. The feature measures are extracted on the abstract level of regions. Frequently, further measures are extracted after object identification, which use the information of the certain object detected, i.e., they operate on the level of objects. In Fig. 1.33i, we can use the knowledge that the blue segment corresponds to a Branemark implant to parameterize a special morphological filter that is adapted to the geometry of Branemark implants and count the number of windings of the screw.

Another example of object-based image measurements is given in Fig. 1.34. The result of balloon segmentation of a cell membrane (see Fig. 1.30) is labeled automatically with local confidence values based on model assumptions (Fig. 1.34a). These values indicate the contour segment belonging to a cell membrane and thus a classification via fuzzy logic (see Sect. 1.7.3).

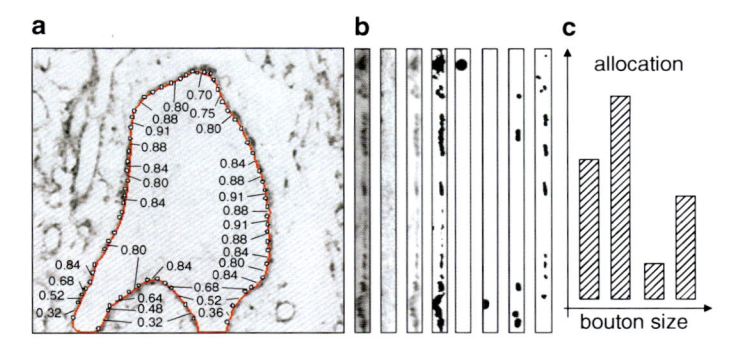

Fig. 1.34. *Quantification of synaptic boutons on a cell membrane [15].* The cell membrane was segmented with a balloon (see Fig. 1.28). Analyzing the impact of internal vs. external forces at a certain vertex, local confidences can be determined to fuzzily classify the affiliation of the contour section to the actual cell membrane (**a**). The cell contour is extracted, linearized, normalized, and binarized before the occupation of the cell membrane with synaptic boutons of different sizes is analyzed by morphological filtering (**b**). The confidence values are considered for averaging the occupation measure along the cell membrane (**c**)

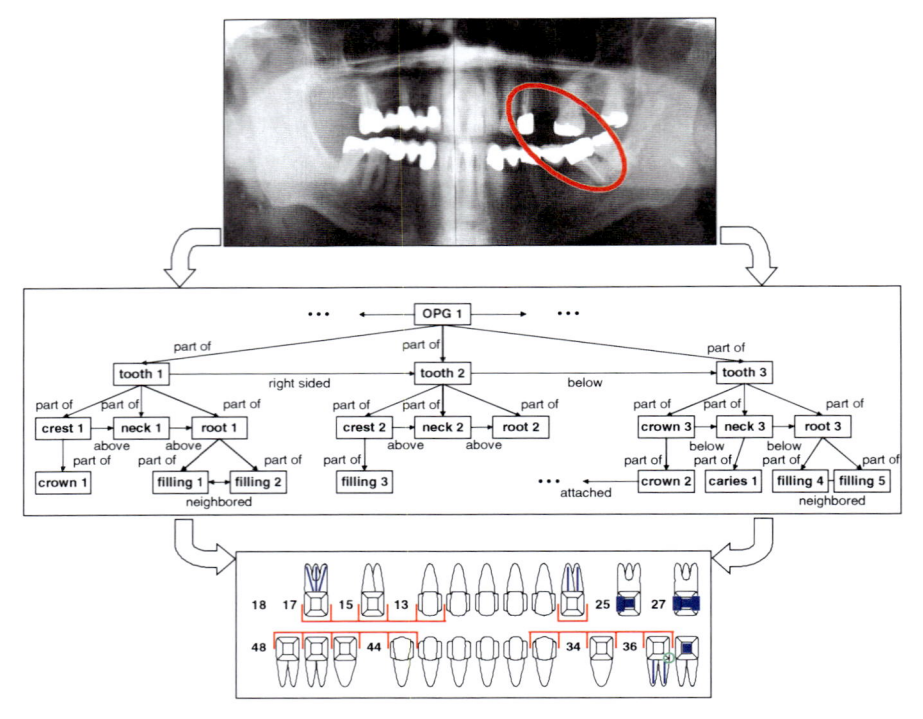

Fig. 1.35. *Scheme of automatic image interpretation.* The panoramic radiograph contains all relevant information of a dental chart. The symbolic description of the scene is obtained with a semantic network. Despite its already considerable complexity, the shown part of the network represents only the marked ROI. In the dental chart, information is coded differently. The teeth are named in accordance with the key of the Fédération Dentaire Internationale (FDI): the leading digit denotes the quadrant clockwise, the second digit refers to the number of the tooth, counting from inside to outside. Existing teeth are represented by templates, in which dental fillings, crowns and bridges are recorded. The green circle at tooth 37 (say: three, seven) indicates a carious process

To increase robustness and reliability of measurements, the confidence values are accounted for an averaging of quantitative measures along the contour, which are extracted, linearized, normalized, and morphologically analyzed (Fig. 1.34b), such that finally a reliable distribution statistics of connecting boutons according to their size is obtained (Fig. 1.34c).

Figure 1.35 displays exemplarily the automatic extraction of a dental chart based on image processing of a panoramic radiograph. It clearly shows the immense difficulties, which have to be faced by the automatic interpretation of biomedical images. Initially, the segmentation and identification of all relevant image objects and object parts must succeed, so that the semantic network can be built. This includes the instances ("tooth 1", "tooth 2", etc.) of the previously identified objects (e.g., "teeth", "crown", "filling"). The interpretation of the scene based on the network must be carried out in a further, not less difficult step of processing. Thus, all teeth must be named according to their position and shape. Then, crowns, bridges, fillings, and carious processes can be registered in the dental chart. However, the automation of this process, which can be accomplished by dentist in a few minutes, is not yet possible automatically with sufficient robustness.

1.9 Image Management

Introductorily, we have summed with the term "image management" all image manipulation techniques, which serve the effective archiving (short and long term), transmission (communication) and the access (retrieval) of data (see Fig. 1.1). For all three points, the specifics in medical applications and health care environments have led to specific solutions, which are briefly introduced in the following sections.

1.9.1 Archiving

Already in the seventies, the invention of CT and its integration with clinical routine has involved the installation of the first Picture Archiving and Communication System (PACS), which main task is the archiving of image data. The core problem of archiving medical images is the immensely large volume of data. A simple radiography with 40×40 cm (e.g., a chest X-ray) with a resolution of five line pairs per millimeter and 10 bit = 1,024 gray levels per pixel already requires a storage capacity of more than 10 MB. Digital mammography, which is captured with high resolution on both breasts in two views results in about 250 MB of raw data for each examination. Ten years ago, radiography, CT, and MRI accumulated in a university hospital to already about 2 TB of image data each year (Table 1.6). This estimate can easily increase tenfold with the resolution-increased novel modalities such as spiral CT and whole-body MRI. For instance in Germany, according to relevant legislations, the data must be kept at least for 30 years. Therefore, efficient storage, retrieval, and communication of medical images have required effective compression techniques and high speed networks. Due to noise in biomedical images, lossless compression usually has a limited effect of compression rates of two or three. Only in recent years, feasible hybrid storage concepts have become available. Storage of and access to medical image data is still of high relevance.

1.9.2 Communication

With increasing digitization of diagnostic imaging, the motto for medical information systems, i.e., to provide "the right information at the right time and

Table 1.6. *Volume of medical image data [5].* The data is taken from the Annual Report 1999 of the University Hospital of RWTH Aachen University, Aachen, Germany (about 1,500 beds). The data is based on the Departments of (i) Diagnostic Radiology, (ii) Neuroradiology, (iii) Nuclear Medicine, and (iv) Dentistry, Oral and Maxillofacial Surgery for a total of 47,199 inpatient and 116,181 outpatient images. Services (such as ultrasound, endoscopic or photographic) from other departments were excluded. For modalities of nuclear medicine, 20 slices per study are assumed. For comparison, the total number of analyses performed in the central laboratory of the Institute for Clinical Chemistry and Pathobiochemistry was estimated with an average of 10 measured values per analysis with highest precision of 64 bit. But still, the annual image data volume is about 10,000 times larger

Modality	Resolution Spatial [pixel]	Range [bit]	Size per image [MB]	Units in year 1999	Total per year [GB]
Chest radiography	4000 × 4000	10	10.73	74,056	775.91
Skeleton radiography	2000 × 2000	10	4.77	82,911	386.09
CT	512 × 512	12	0.38	816,706	299.09
MRI	512 × 512	12	0.38	540,066	197.78
Other radiography	1000 × 1000	10	1.19	69,011	80.34
Panoramic and skull	2000 × 1000	10	2.38	7,599	17.69
Ultrasound	256 × 256	6	0.05	229,528	10.11
Dental radiography	600 × 400	8	0.23	7,542	1.69
PET	128 × 128	12	0.02	65,640	1.50
SPECT	128 × 128	12	0.02	34,720	0.79
Σ					1,770.99
For comparison					
Laboratory tests	1 × 10	64	0.00	4,898,387	0.36

the right place," is projected to the field of medical image processing. Hence, image communication is the core of today's PACS. Image data is not only transferred electronically within a department of radiology or the hospital, but also between widely separated institutions. For this task, simple bitmap formats such as the Tagged Image File Format (TIFF) or the Graphics Interchange Format (GIF) are inadequate, because beside the images, which might have been captured in different dimensions, medical meta information on patients (e.g., Identifier (ID), name, date of birth, ...), the modality (e.g., device, parameters, ...) and organization (e.g., investigation, study, ...) must also be transferred in a standardized way.

Since 1995, the communication is based on the Digital Imaging and Communications in Medicine (DICOM) standard. In its current version, DICOM includes:

- structural information about the contents of the data ("object classes"),
- commands on what should happen to the data ("service classes"), and
- protocols for data transmission.

DICOM is based on the client-server paradigm and allows the coupling of PACS in Radiology Information System (RIS) or Hospital Information Systems (HIS). DICOM incorporates existing standards for communication: the

International Organization for Standardization (ISO) Open System Interconnection (OSI) model, the Transmission Control Protocol (TCP) Internet Protocol (IP), and the Health Level 7 (HL7) standard. Full DICOM compliance for imaging devices and image processing applications is achieved with only a few supported object or service classes, since other DICOM objects, which are not relevant for the current device, simply are handed over to the next system in the DICOM network. The synchronization between the client and server is regularized by conformance claims, which are also specified as part of the DICOM standard. However, the details of implementation of individual services are not specified in the standard, and so in practice, vendor-specific DICOM dialects have been developed, which can lead to incompatibilities when building PACS. In recent years, the Integrating the Healthcare Enterprises (IHE) initiative became important. IHE aims at guiding the use of DICOM and other standards such that complete inter-operability is achieved.

1.9.3 Retrieval

In today's DICOM archives, images can be retrieved systematically, only if the patient name with date of birth or the internal system ID is known. Still, the retrieval is based on alphanumerical attributes, which are stored along the image data. It is obvious that diagnostic performance of PACS is magnified significantly if images would be directly available from similar content of a given example image. To provide the Query by Example (QBE) paradigm is a major task of future systems for Contend-Based Image Retrieval (CBIR). Again, this field of biomedical research requires conceptually different strategies as it is demanded in commercial CBIR systems for other application areas, because of the diverse and complex structure of diagnostic information that is captured in biomedical images.

Figure 1.36 shows the system architecture of the Image Retrieval in Medical Applications (IRMA) framework[3]. This architecture reflects the chain of processing that we have discussed in this chapter, i.e., registration, feature extraction, segmentation, classification of image objects towards the tip of the pyramid (see Fig. 1.2), which is the symbolic interpretation respective scene analysis. In IRMA, the image information that is relevant for retrieval is gradually condensed and abstracted. The image bitmap is symbolically represented by a semantic network (hierarchical tree structure). The nodes contain characteristic information to the represented areas (segments) of the image. Its topology describes the spatial and/or temporal condition of each object. With this technology, radiologists and doctors are supported similarly in patient care, research, and teaching.

---

[3] http://irma-project.org

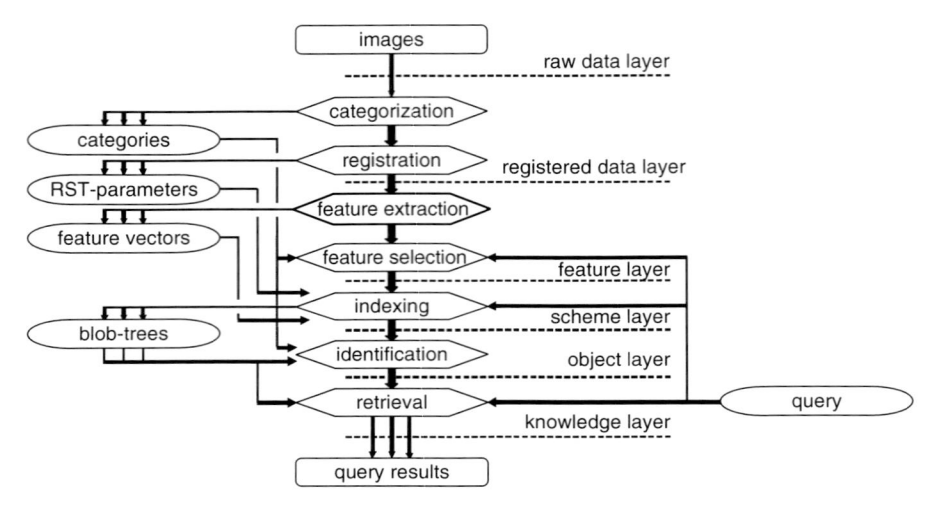

Fig. 1.36. *System architecture of the IRMA framework [16].* The processing steps in IRMA are shown in the middle column. Categorization is based on global features and classifies images in terms of imaging modality, view direction, anatomic region, and body system. According to its category, the image geometry and contrast are registered to a reference. The abstraction relies on local features, which are selected specifically to context and query. The retrieval itself is performed efficiently on abstracted and thus information-reduced levels. This architecture follows the paradigm of image analysis (cf. Fig. 1.1). The in-between-representations as presented on the left describe the image increasingly abstract. The levels of abstraction (cf. Fig. 1.20) are named on the right side

1.10 Conclusion and Outlook

The past, present, and future paradigms of medical image processing are composed in Fig. 1.37. Initially (until approx. 1985), the pragmatic issues of image generation, processing, presentation, and archiving stood in the focus of research in biomedical image processing, because available computers at that time had by far not the necessary capacity to hold and modify large image data in memory. The former computation speed of image processing allowed only offline calculations. Until today, the automatic interpretation of biomedical images still is a major goal. Segmentation, classification, and measurements of biomedical images is continuously improved and validated more accurately, since validation is based on larger studies with high volumes of data. Hence, we focused this chapter on image analysis and the processing steps associated with it.

The future development is seen in the increasing integration of algorithms and applications in the medical routine. Procedures in support of diagnosis, treatment planning, and therapy must be easily usable for physicians and,

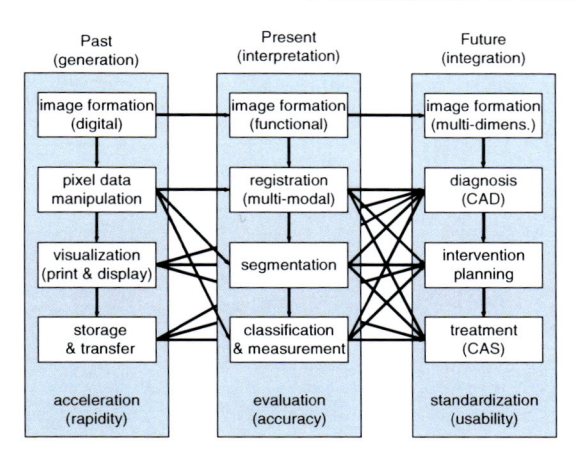

Fig. 1.37. *Changing paradigms in medical image processing [17].* Until now, formation, enhancement, visualization, and management of biomedical images have been in the focus of research. In future, integration, standardization, and validation are seen as major challenges for routine applications in diagnostics, intervention planning, and therapy

therefore, further standardized in order to ensure the necessary interoperability for a clinical use.

References

Related Textbooks

- Bankman IN (ed). Handbook of Medical Image Processing and Analysis. 2nd ed. New York: Academic Press; 2008. ISBN: 0-123-73904-7
- Beolchi L, Kuhn MH (ed). Medical Imaging: Analysis of Multimodality 2D/3D Images. IOS Press; 1995. ISBN: 9-051-99210-6
- Beutel J, Kundel HL, van Metter RL (eds). Handbook of Medical Imaging. Vol. 1: Physics and Psychophysics. Bellingham: SPIE Press; 2000. ISBN 0-819-43621-6
- Bryan RN (ed). Introduction to the Science of Medical Imaging. Cambridge: Cambridge University Press; 2010. ISBN: 978-0-521-74762-2
- Dougherty D. Digital Image Processing for Medical Applications. Cambridge: Cambridge University Press; 2009. ISBN: 0-521-86085-7
- Dougherty ER (ed). Digital Image Processing Methods. New York: CRC Press; 1994. ISBN: 978-0-824-78927-5
- Guy C, Ffytche D. Introduction to the Principles of Medical Imaging. London: Imperial College Press; 2005. ISBN: 1-860-94502-3
- Jan J. Medical Image Processing, Reconstruction and Restoration: Concepts and Methods. Boca Raton: CRC Press; 2005. ISBN: 0-824-75849-8
- Kim Y, Horii SC (eds). Handbook of Medical Imaging. Vol. 3: Display and PACS. Bellingham: SPIE Press; 2000. ISBN: 0-819-43623-2
- Meyer-Baese A. Pattern Recognition in Medical Imaging. San Diego: Academic Press; 2003. ISBN: 0-124-93290-8
- Preim B, Bartz D. Visualization in Medicine. Theory, Algorithms, and Applications. Amsterdam: Morgan Kaufmann; 2007. ISBN: 978-0-123-70596-9

- Rangayyan RM. Biomedical Image Analysis. New York: CRC Press; 2005. ISBN: 0-849-39695-6
- Sonka L, Fitzpatrik JM (eds). Handbook of Medical Imaging. Vol. 2: Medical Image Processing and Analysis. Bellingham: SPIE Press; 2000. ISBN: 0-819-43622-4
- Suetens P. Fundamentals of Medical Imaging. Cambridge: Cambridge University Press; 2002. ISBN: 0-521-80362-4
- Tavares JM, Jorge RMN (eds). Advances in Computational Vision and Medical Image Processing: Methods and Applications. Berlin: Springer; 2008. ISBN: 1-402-09085-4
- Umbaugh SE. Computer Imaging: Digital Image Analysis and Processing. Boca Raton: CRC Press; 2005. ISBN: 0-849-32919-1

Citations

1. Lehmann TM, Oberschelp W, Pelikan E, et al. Bildverarbeitung für die Medizin: grundlagen, Methoden, Anwendungen. Springer, Heidelberg; 1997.
2. Morneburg H, editor. Bildgebende Systeme für die medizinische Diagnostik. 3rd ed. German: Siemens AG, Berlin & Publicis MCD Verlag, Erlangen; 1995.
3. Shannon CE. Communication in the presence of noise. Proc Inst Radio Eng. 1949;37(1):10–21; reprinted in Proc IEEE 1998;86(2):447–57.
4. Lehmann TM, Kaser A, Repges R. A simple parametric equation for pseudo-coloring grey scale images keeping their original brightness progression. Image Vis Comput. 1997;15(3):251–7.
5. Lehmann TM, Hiltner J, Handels H. Medizinische Bildverarbeitung. Chapter 10 In: Lehmann TM, editor. Handbuch der Medizinischen Informatik. Hanser, Munich; 2005.
6. Wagenknecht G, Kaiser HJ, Büll U. Multimodale Integration, Korrelation und Fusion von Morphologie und Funktion: Methodik und erste klinische Anwendungen. Rofo. 1999;170(1):417–6.
7. Pommert A, Höhne KH, Pflesser B, et al. Ein realistisches dreidimensionales Modell der inneren Organe auf der Basis des Visible Human. Munich: Hanser; 2005. p. 72–76.
8. Metzler V, Bienert H, Lehmann TM, et al. A novel method for geometrical shape analysis applied to biocompatibility evaluation. ASAIO J. 1999;45(4):264–1.
9. König S, Hesser J. Live-wires using path-graphs. Methods Inf Med. 2004;43(4):371–5.
10. Kass M, Witkin A, Terzopoulos D. Snakes: active contour models. Int J Comput Vis. 1988;1(5):321–31.
11. Metzler V, Bredno J, Lehmann TM, et al. A deformable membrane for the segmentation of cytological samples. Proc SPIE. 1998;3338:1246–7.
12. Bredno J. Höherdimensionale Modelle zur Quantifizierung biologischer Strukturen. in German: PhD Thesis, RWTH Aachen University, Aachen, Germany; 2001.
13. Cootes TF, Taylor CJ, Cooper DH, et al. Active shape models: their training and application. Comput Vis Image Underst. 1995;61(1):38–59.
14. Lehmann TM, Schmitt W, Horn H, et al. IDEFIX: Identification of dental fixtures in intraoral X-rays. Proc SPIE. 1996;2710:584–5.

15. Lehmann TM, Bredno J, Metzler V, et al. Computer-assisted quantification of axosomatic boutons at the cell membrane of motoneurons. IEEE Trans Biomed Eng. 2001;48(6):706–7.
16. Lehmann TM, Güld MO, Thies C, et al. Content-based image retrieval in medical applications. Methods Inf Med. 2004;43(4):354–61.
17. Lehmann TM, Meinzer HP, Tolxdorff T. Advances in biomedical image analysis: past, present and future challenges. Methods Inf Med. 2004;43(4):308–14.

Part I

Image Formation

2

Fusion of PET and MRI for Hybrid Imaging

Zang-Hee Cho, Young-Don Son, Young-Bo Kim, and Seung-Schik Yoo

Summary. Recently, the development of the fusion PET-MRI system has been actively studied to meet the increasing demand for integrated molecular and anatomical imaging. MRI can provide detailed anatomical information on the brain, such as the locations of gray and white matter, blood vessels, axonal tracts with high resolution, while PET can measure molecular and genetic information, such as glucose metabolism, neurotransmitter-neuroreceptor binding and affinity, protein–protein interactions, and gene trafficking among biological tissues. State-of-the-art MRI systems, such as the 7.0 T whole-body MRI, now can visualize super-fine structures including neuronal bundles in the pons, fine blood vessels (such as lenticulostriate arteries) without invasive contrast agents, in vivo hippocampal substructures, and substantia nigra with excellent image contrast. High-resolution PET, known as High-Resolution Research Tomograph (HRRT), is a brain-dedicated system capable of imaging minute changes of chemicals, such as neurotransmitters and –receptors, with high spatial resolution and sensitivity. The synergistic power of the two, i.e., ultra high-resolution anatomical information offered by a 7.0 T MRI system combined with the high-sensitivity molecular information offered by HRRT-PET, will significantly elevate the level of our current understanding of the human brain, one of the most delicate, complex, and mysterious biological organs. This chapter introduces MRI, PET, and PET-MRI fusion system, and its algorithms are discussed in detail.

2.1 Introduction

Among the modern medical imaging technologies, Positron Emission Tomography (PET) and Magnetic Resonance Imaging (MRI) are considered to be the most powerful diagnostic inventions. In the 1940s, modern medical imaging technology began with advancements in nuclear medicine. In the early 1970s, by combining the diagnostic properties of X-rays with computer technology, scientists were able to construct 3D images of the human body in vivo for the first time, prompting the birth of the Computed Tomography (CT). The emergence of CT was an important event that motivated scientists to invent PET and MRI. These imaging tools were not based on simple modifications

T.M. Deserno (ed.), *Biomedical Image Processing*, Biological and Medical Physics,
Biomedical Engineering, DOI: 10.1007/978-3-642-15816-2_2,
© Springer-Verlag Berlin Heidelberg 2011

of existing techniques or devices. Instead, they are the new medical imaging modalities that were the result of the combined effort of numerous scientific disciplines such as physics, mathematics, chemistry, computer science, biology, medicine, and pharmacology.

Initially, PET was based on somewhat primitive form of positron imaging device created in 1953 by Gordon Brownell et al. Massachusetts Institute of Technology (MIT) and then newly born concept of the CT. The first modern PET device was developed by two groups of scientist. One was at University of California at Los Angeles (UCLA) in the mid 1970s by Cho et al [1] and the other was by Ter-Pogossian & Phelps at Washington university, St. Louis [2]. Subsequently, a new detector material, Bismuth-Germanate (BGO) was introduced for use in high-resolution imaging [3]. Today, most commercial PET scanners have adopted the ring-type detector system, based on the use of BGO or Cerium-doped Lutetium Oxyorthosilicate (LSO) scintillators [4]. These PET systems now have spatial resolutions of 5–6 mm at FWHM (Fig. 2.1).

The development of MRI was based on Nuclear Magnetic Resonance (NMR), explored in 1940s by Felix Bloch at Stanford University and Edward Mills Purcell at Harvard University. The first principles of MRI were proposed in 1973 by Paul Lauterbur, and necessary image reconstruction algorithm were developed in the mid 1970s by Richard Ernst. For their achievements, Ernst received 1990 the Nobel Prize in Chemistry, and in 2003, Lauterbur and Mansfield won Nobel Prize in Physiology or Medicine. Much of the MRIs

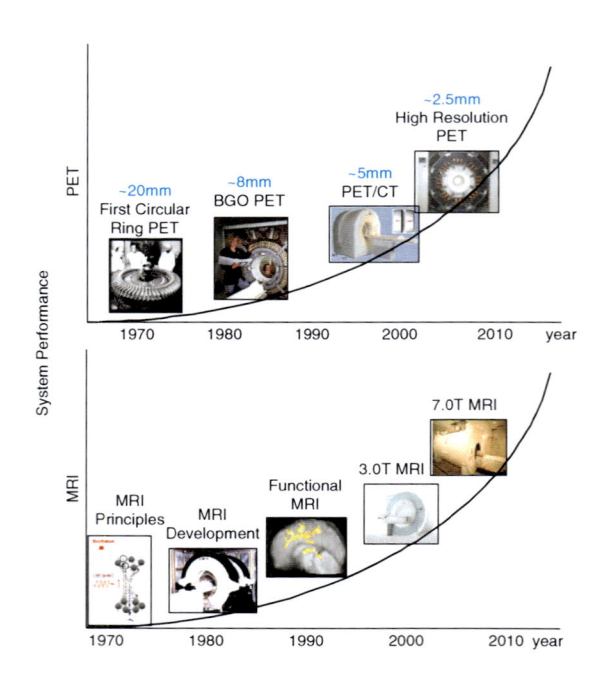

Fig. 2.1. *Historical development of PET and MRI.* The exponential increase of system performance is visualized. The ordinate shows resolution and field strength for PET (*top*) and MRI (*bottom*), respectively

rapid success can be attributed to its non-invasiveness and tissue discrimination capability in the brain. Continuous hardware and software advancements have followed, and new MRI scanners boast sub-millimeter resolution with excellent contrast. The strength of the magnetic field used in the device is described with a unit of Tesla (T) or Gauss G. 1 T is equal to 10,000 G. Although currently available MRI in the market for humans use is up to 8 T, only 4 T is permitted for clinical use by the Food and Drug Administration (FDA) of the United States.

The invention of PET and MRI changed the scene of modern medicine and was perhaps one of the greatest achievements in medicine and the biology. This chapter provides a brief introduction of the basic principles of MRI and PET, followed by an overview of state-of-the-art PET and MRI systems. Subsequently, we described the complementary use of these two devices and technical aspects related to the new PET/MRI fusion system, which has been recently developed, and potential applications are discussed.

2.2 Positron Emission Tomography

2.2.1 Basic Principles

PET is an imaging system that detects two annihilation photons or gamma rays originating from the tracer compounds labeled with positron-emitting radionuclides, which are injected or administered into the subject. Many proton-rich radioisotopes may decay via positron β^+-decay, in which a proton in the nucleus decays to a neutron by emission of a positron and a neutrino. The decay product has one atomic number less than the parent. Examples of radionuclides which undergo decay via positron emission are shown in Table 2.1 [5].

Positron-emitting radionuclides possess an important physical property that makes PET a unique high-resolution molecular imaging device. That is the directionality or collinearity of two simultaneously emitted photons by the annihilation process. When the emitted positron collides with a nearby electron, they annihilate and produce two annihilation photons of 511 keV. The two annihilation photons, which are identical to two gamma photons with 511 keV of energy, then travel in nearly exact opposite directions of

Isotope	Half life (min)	Positron energy maximum (MeV)	Positron range in water (mm)
^{11}C	20.30	0.96	1.1
^{13}N	9.97	1.19	1.4
^{15}O	2.03	1.70	1.5
^{18}F	109.80	0.64	1.0
^{68}Ga	67.80	1.89	1.7
^{82}Rb	1.26	3.15	1.7

Table 2.1. *Radionuclides and their physical properties.* The positron range is given at FWHM

each other. This near collinearity of the two annihilation photons allows to identify the location of the annihilation event or the existence of positron emitters through the detection of two photons by detectors poised exactly on opposite sides of the event, which are contacted by the photons at nearly the same time. This simultaneity also plays an important role for the coincident detection.

A set of detectors converts the high-energy photons into electrical signals that are subsequently processed by signal processing electronics. For detection of these two annihilation 511 keV photons, two scintillation detectors are coupled to an individual Photo-Multiplier Tube (PMT), pulse timer and amplitude analyzer. The detectors of current PET systems are made of inorganic materials called scintillation detectors. Scintillators convert the incident gamma quantum into a large number of light photons. The scintillator must be made of a highly dense material with high atomic number to maximize the gamma photon absorption. In the early development of PET and up to the late 1970s, NaI(Tl) was a commonly used for scintillation detectors. Currently, most modern PET use BGO [3]. While BGO has larger absorption power, LSO has a faster response time and more light output. The light, visible photons, from these scintillators are converted to electrical signals by the PMT or equivalent device. The PMT multiplies the weak signals from the scintillation detectors to electrically detectable signals with both pulse timing and amplitude. Although PMT is the most widely used light photon amplifier, more recently semiconductor type PMTs, such as an Avalanche Photodiode (APD) and the Silicon Photomultiplier (SiPM) have been developed and are in use. A semiconductor type PMT has the advantage over conventional PMTs due to its non-magnetic properties, which supports use in MRI environments, but it has several disadvantages in rigidity and stability (Fig. 2.2).

The amplified electrical signals from PMTs, as electrical pulses, are analyzed to determine when the signal occurred and whether the signal is above a certain threshold. When the valid PET signal is generated by annihilation photons of 511 keV that pass the energy threshold, time information is recorded and used for coincident time analysis. The detected pulses are then

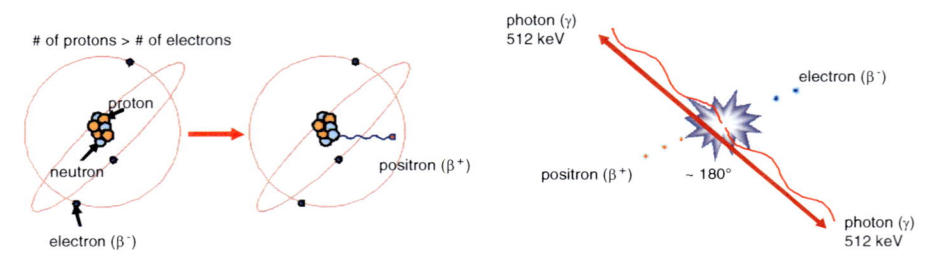

Fig. 2.2. *Positron generation and annihilation.* Left: positron emission from a radionuclide; Right: positron annihilation, which generates two annihilation photons or gamma photons

fed to a coincidence module, which examine whether two pulses are truly due to the annihilation process. The smaller the difference between the two pulses, the closer the detection stems from the true annihilation event. Modern electronics, however, can measure time with a resolution of 10^{-8} s or larger. As a result, the event is registered as coincidence only if a pair of detectors (opposite to each other) detects the signal simultaneously within a certain time window. The coincident time window used is in the range of 10 ns.

2.2.2 Image Reconstruction

In a typical PET scan, 10^6–10^9 events (decays) are detected depending on the radioactivity injected and the time of measurement. Coincidence events are saved as the special data set called sinogram. The sinogram is the line-integral projection data obtained from a large number of detectors at different views surrounding the entire object. The sinogram data is used to reconstruct an image through mathematical algorithms such as analytical or iterative reconstruction methods. Analytical methods calculate the radionuclide tracer distribution directly from the measured sinogram data. Backprojection and Filtering (BF) or Filtered Backprojection (FB) are typical algorithms used for analytical methods. They require less computational burden than statistical methods such as the Expectation Maximization (EM) algorithm. Analytic approaches, however, often suffer from an artifact known as the streak artifact, which arises from the physical gaps existing between the detectors. In contrast, iterative methods like EM reconstruct the image in an iterative fashion using the measured sinogram. Iterative methods, therefore, are often more robust to noise, such as streak artifacts, and can provide better Signal to Noise Ratio (SNR) at a given spatial image resolution. Although iterative methods require much more computational burden, due to the recent improvement of computing technologies and algorithms, EM algorithm is now widely used as the main stream method of PET image reconstruction.

2.2.3 Signal Optimization

In addition to the basic image reconstruction, it is important to note that there are several physical phenomena, such attenuation, scattering, and random coincidences, which are necessary to correct when more quantitatively accurate PET images are required. Interactions within the body with incident photons, which result in scattering and attenuation, are known as scatter and coincidence events, respectively, and require correction. The scattered coincidence events and attenuation corrections are two major problems together with accidentally-occurring coincidence events. In addition, the efficiency of each detector may vary between each detector and influence the measured data. Therefore, various correction techniques have been developed to correct the effect of attenuation, scatters, and random events. Only when these corrections schemes are completely incorporated into the main reconstruction algorithm,

Fig. 2.3. *PET system components.* The simplified diagram shows the ring detector system and the coincident board, which is fundamental for image reconstruction in the computer

Fig. 2.4. *PET image examples.* The PET images of dopamine transporter and receptor binding using ^{18}F-FP-CIT and ^{11}C-Raclopride radionuclides are displayed as examples of how PET can be used for the measurements of neural status of the brain

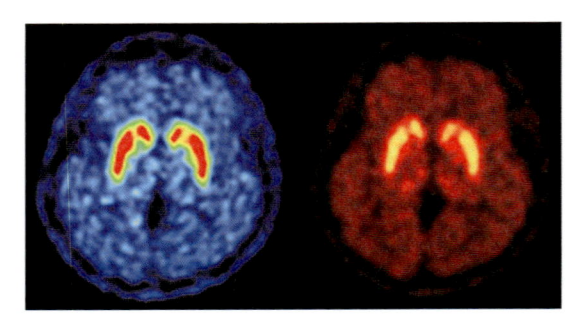

PET can quantitatively image various radio-labeled ligands binding to specific receptors, transporters, and enzymes. This quantitatively calibrated molecular imaging is one of the strengths of PET imaging. For example, the interaction and distribution of the dopamine transporters or receptors in the brain can be measured using a PET with ^{18}FluoroPropyl-CarbomethoxyIodophenyl-norTropane (F-FP-CIT) or ^{11}C-Raclopride (Figs. 2.3, 2.4).

2.2.4 High-Resolution Research Tomograph

One of the most advanced PET scanners is the High-Resolution Research Tomograph (HRRT), which has been introduced by Siemens. HRRT-PET is designed to obtain the highest spatial resolution and the highest sensitivity known in human brain PET imaging [6]. In contrast to the commercial PET, which usually has a system diameter of more than 80 cm to accommodate the whole body, the system diameter of HRRT-PET is only 46.7 cm, which is only suitable for a human brain scan. This small system diameter improved each detector's solid angle and, therefore, the sensitivity. In addition, the HRRT-PET has a longer axial Field-of-View (FOV) of 25.2 cm, as compared with the conventional PET, which has only 17 cm of axial FOV. The shorter system diameter and the longer axial FOV provide a dramatically improved detection efficiency, and thereby enhance the overall system sensitivity. With this increased sensitivity, HRRT-PET provides a considerably high

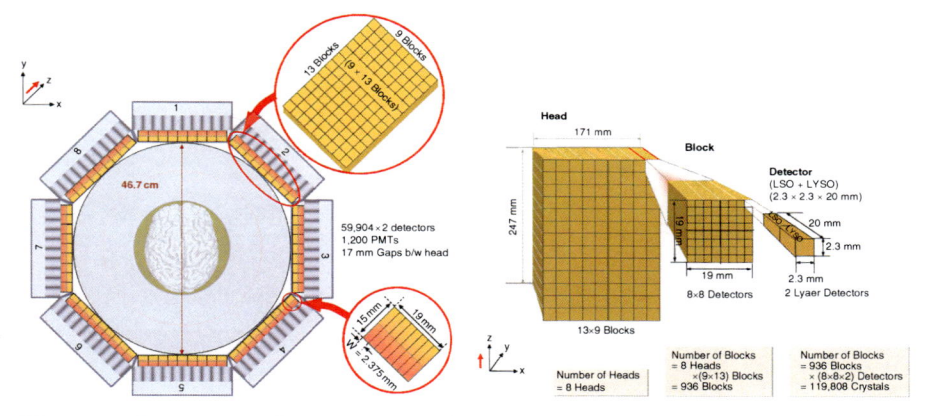

Fig. 2.5. *Detector configuration of HRRT-PET.* The HRRT-PET detector system consists of eight heads (*left*). Each head has 13 × 9 detector blocks build with two layers of 8 × 8 detectors. This results in a total of 119,808 scintillation detector crystals with dimension of 2.3 × 2.3 × 10 mm^3. These small detector crystals are packed in the form of a mosaic panel (*right*)

spatial resolution together with the small detector size, which has dimensions of 2.3 mm × 2.3 mm for width and height, respectively (Fig. 2.5).

This high spatial resolution combined with enhanced sensitivity [7, 8] makes the HRRT-PET the most advanced PET scanner for human brain studies [9]. Simultaneously, the improved spatial resolution reduces partial-volume effects, thereby improving quantification of metabolic rates in the brain such as the regional Cerebral Metabolic Rate of Glucose (rCMRGlc) [10]. In addition, transmission images are also obtained, supporting a more accurate attenuation and scatter correction.

These improvements in performance by the brain optimized configuration of the HRRT provided better imaging and allowed us to image smaller Region of Interest (ROI) than the previously available. In addition to PET's ability to measure and visualize metabolism, the distribution of neuroreceptors and neurotransporter in the brain can be measured. Also, HRRT-PET now allows the ability to measure the specific distribution of different ligands in various neurodegenerative disorders [10]. According to recent report, many small brain structures can be studied due to the availability of HRRT-PET [10,11]. These structures include:

1. The dopamine transporter-enriched nuclei in the midbrain where the dopaminergic cell bodies are located [12].
2. The substantia nigra, from which dopaminergic neurons projecting to the dorsal striatum.
3. The ventral tegmental area, from where neurons project to the limbic regions and the cerebral cortex [13].

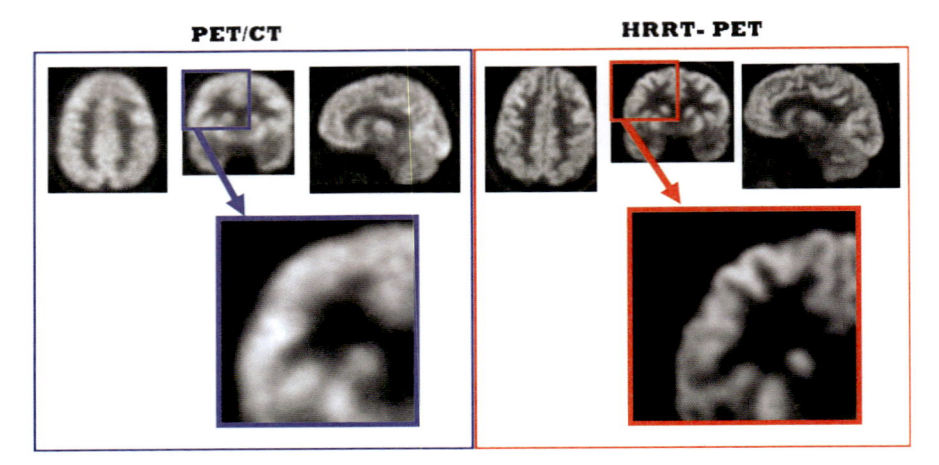

Fig. 2.6. *Resolution of HRRT-PET.* A section of the brain is captured with currently available PET scanner (*left*) and the HRRT-PET (*right*), which shows much more details and higher resolution

Other small regions such as the ventral striatum are actively pursued in search of potential ROIs for HRRT-PET research. Even arterial blood sampling in the human brain using the HRRT-PET is being studied by several groups [9, 13].

The previously mentioned features made HRRT-PET have the one of the highest sensitivities and spatial resolutions among any currently available PET scanners. In Fig. 2.6, two comparative ^{18}F-Fludeoxygloucose (FDG) PET images emphasize the clear advantage of HRRT-PET over a conventional PET. In these images, the cortical gyri are seen much clearer with HRRT-PET than with PET/CT, suggesting that the HRRT can more accurately localize the molecular interactions in the brain than any other PET system available today.

2.3 Magnetic Resonance Imaging

2.3.1 Basic Principles

The main components of the MRI system are the main magnet, the Radio Frequency (RF) system, the gradient coil, the shim system, and the computer. The main magnet generates a strong magnetic field, which determines the imaging power of MRI. Permanent magnets and resistive magnets can be used to produce the external magnetic field; however, they are unable to produce high magnetic fields and are only used for the low field MRI. Today, the main magnetic field is commonly produced by a superconducting magnet maintained at a very low temperature. The superconducting electromagnet consists of a coil that has been made super-conductive by a cooling system, which often consists of liquid helium surrounded by liquid nitrogen. By cooling, the superconductor becomes resistance free, which means large amounts of current can flow through the coil to produce high magnetic fields.

Once the coil becomes a superconducting magnet, it is capable of producing strong and stable magnetic fields suitable for MRI applications. The superconducting wire is usually made of Niobium-Titanium (NbTi), a rigid material that is simple to handle. Once a magnet is constructed with the strong magnetic field, one can insert an object which has spins of nuclei such as water protons. The spins in the object then will be aligned either parallel or anti-parallel to the main magnetic field. A slightly larger fraction of these protons will be oriented in the anti-parallel form and lead to a net magnetization. In a given magnetic field, all the spins precess with the specific frequency known as the Larmor frequency, which is specific to the strength of the magnetic field. If an external magnetic field or energy oscillating at the Larmor frequency is applied to the spins, the spins absorb the applied energy and are excited to the high energy status due to the magnetic resonance absorption phenomena.

The external energy is usually delivered by RF coils, which transmit RF energy to the object with a specific resonance frequency, often within a certain bandwidth that has a center frequency equivalent to the Larmor frequency. The precessing spins with the corresponding frequency at the non-excited or resting states are then flipped to a higher excited state where they last for a certain time depending on the relaxation properties of the object. The excited spins will return to the steady state and give off energy in the form of an electromagnetic signal or radiation, which is referred to as the Free Induction Decay (FID) signal. During the spin flips, the large numbers of small electrical dipoles, which are proportional to the resonant proton density, induce current on the RF coils that are surrounding the object. The simplest form of an RF system is composed of a transmitter coil and receiver coil. The RF system is an antenna which is sending excitatory RF pulses to the object or brain and also receiving the signals generated from the object. RF coils are one of the key components that determine the SNR of images. The development of specific purpose RF coils, therefore, is one of the central themes of MRI research.

2.3.2 Image Reconstruction

Each signal received by the antenna or RF coil contains information of the total sum of the object signals, but they are not encoded to produce an image yet. Formation of MRI requires the magnetic gradients to encode spatial information to the object. The gradient system has various functions such as slice selection, spatial encoding, spoiling, rewinding, echo production, and pre-saturation, among others. Among them, slice selection and spatial encoding are the most essential functions of the gradients system to spatially localize the magnetic resonance signal. There are three gradient coils located within the bore of the magnet, which are wrapped along three orthogonal axes. The gradient is called according to the axis along which they act when switched on. For example, G_x is assigned for the horizontal axis of the magnet bore and alters the field along the x-axis.

These magnetic gradients are referred to as slice-selection gradient (G_z), phase-encoding gradient (G_y), and frequency-encoding frequency-encoding gradient or readout gradient (G_x), respectively. Slice selection gradients are usually applied at the RF excitation period so that only spins within a slice corresponding to the specific RF bandwidth are excited. Frequency-encoding or phase-encoding gradients are typically applied during or before data acquisition. This encoding scheme encodes spatial information into the RF signal. The received signal is in the spatial frequency domain, what is called k-space, equivalent to the 2D or 3D Fourier transform of the object.

2.3.3 Signal Optimization

The raw field produced by a superconducting magnet is approximately 1,000 parts per million (ppm) or worse, thus the magnetic field has to be corrected or shimmed. The shim system is used to correct field inhomogeneity and optimize for each imaging session. Field homogeneity is measured by examining an FID signal in the absence of field gradients. Shimming is important for a number of imaging applications. Most modern MRI techniques such as Echo Planar Imaging (EPI) and Chemical Shift Imaging (CSI) require homogeneous magnetic fields to be less than 3.5 ppm over the imaging volume. Usually this is accomplished by a combination of current loops (active or dynamic shim) and ferromagnetic material (passive or fixed shim). Gradient coils are used to provide a first-order shim. Since the introduction of a patient also distorts the magnetic field, often an active shim correction is made before scanning.

The signals that are detected via the RF coils are recorded in the computer system, and an image is reconstructed using a mathematical algorithm, such as the Fourier transform. The complexity of modern MRI arises mainly due to the many physical parameters involved such as spin relaxations of different kinds, for example spin-lattice and spin-spin relaxation times (T1 and T2), respectively. Most of the conventional imaging utilizes these magnetic properties, such as T1, T2, and susceptibility. In simple terms, the T1 value is the recovery time of the flipped spins and determines the interactions between spins and its surrounding lattice (tissue). The T2 value is the dephasing time of the in-phased spins due to spin-to-spin interaction, and the susceptibility is a spin dephasing factor due to surrounding magnetic fields. These magnetic properties can be weighted in the image by adjusting the pulse sequence and related imaging parameters.

In summary, MRI is a multi-purpose medical imaging instrument utilizing those intrinsic parameters mentioned and offers exquisite spatial resolution often more than an order of magnitude better than PET (Fig. 2.7). MRI,

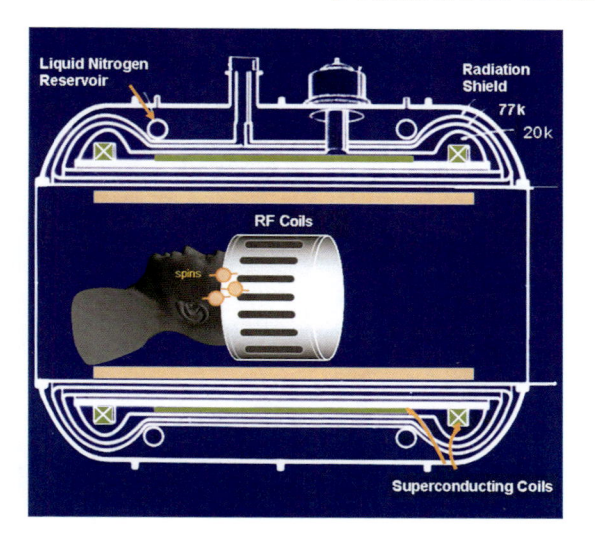

Fig. 2.7. *Diagram of MRI gantry.* The super-conduction coils are cooled with liquid nitrogen. The inner temperature is only 25 K

however, lacks molecular specificity, although a number of new techniques are being developed in combination with spectroscopic techniques or, more recently, with nano-particles.

2.3.4 High-Field MRI

The magnetic resonance signal is commonly proportional to the volume of data acquired and magnetic field strength. Therefore, high-field MRI systems provide an image with a higher SNR. In higher magnetic fields, it is possible to decrease the volume or voxel without sacrificing the SNR. It means that a high-field MRI system makes it possible to obtain higher spatial resolution and sensitivity than low-field MRI. Structural, metabolic, and functional assessments of an intact, living brain can be made using high-field MRI systems.

The 7.0 T MRI system, which uses an ultra-high field magnet, currently exists for human imaging with the high performance gradient coil set and RF coils. It provides us with many exquisite high-resolution images with an extremely high SNR. Recently, many ultra high-resolution images were obtained from 7.0 T MRI (Figs. 2.8, 2.9). Many fine structures, which were once thought impossible to image using MRI, were observed in the brain in vivo. The structures include the substantia nigra, red nucleus, and cerebral peduncle or crus cerebri in the midbrain. As demonstrated in Fig. 2.8a, the red nucleus and substantia nigra are clearly visible. In addition, the image of the cerebral peduncle surrounding the substantia nigra shows not only the fine vascular structures but also the fibers, suggesting that the details of ultra-fine high-resolution images can be of great help in identification of various neurological disorders and in the planning of surgical operations in a

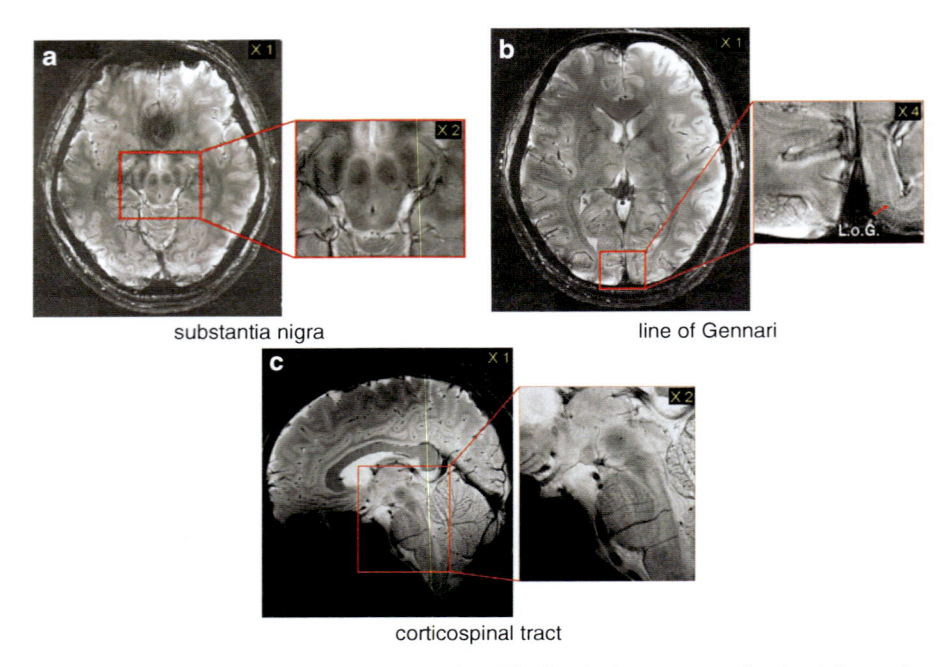

Fig. 2.8. *High-field MRI image examples.* The brain images are obtained from the 7.0 T MRI. In contrast to previous MRI systems, details of brain substructures can now be observed

totally non-invasive manner. The line of gennari in the visual cortex (Fig. 2.8b) and the perforating arteries and the corticospinal tracts in the pontine area (Fig. 2.8c) are also visualized in the 7.0 T MRI. Note the details of the thalamic area and the structures of the deep gray and white matter areas, like the anterior commissure, the mammillary body, and the red nucleus.

The substantia nigra is an important region in the area of Parkinson's disease research. Figure 2.9a and b are images of the central midbrain areas obtained from the same subject using 1.5 T and 7.0 T, respectively. As seen, 7.0 T MRI images are far superior and clearer than 1.5 T, particularly, in the boundary between the substantia nigra and surrounding tissues in 7.0 T MRI. Since the substantia nigra is believed to include iron, it is darker than other regions due to T2-related signal reduction. Likewise, the structure of the hippocampus and the parahippocampal regions, major ROI in the Alzheimer's studies, were clearly visualized by T2-weighted imaging in vivo by using 7.0 T MRI (not shown here). 7.0 T MRI began to show possible visualization of micro-vascular structures, such as the Lenticulostriate Arterie (LSA) in the human brain, which would be extremely useful for clinical purposes. Recently, we have reported regarding the advancements in micro-vascular imaging, such as the in vivo visualization of LSAs, which was once thought to be impossible [14].

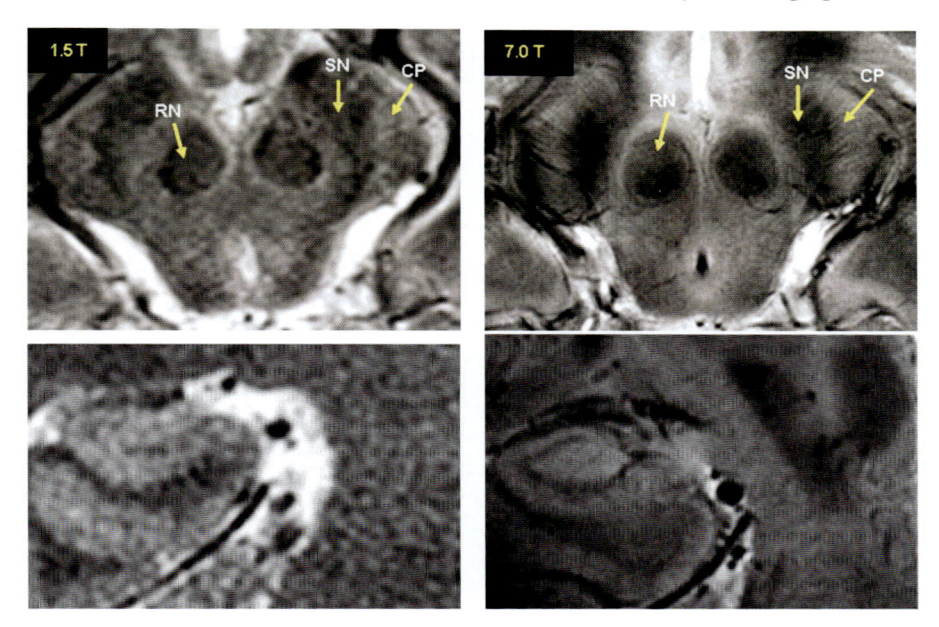

Fig. 2.9. *High- and low-field MRI.* Resolution in 1.5 T images (*left*) is lower as compared to 7.0 T (*right*), as it can be seen in the substantia nigra (*top*) and hippocampal region (*bottom*)

2.4 Hybrid PET Fusion System

PET is one of the most widely used imaging tools in both clinical areas as well as neuroscience research, especially for its ability to perform non-invasive, in vivo imaging of biochemical changes. PET can show how well tissues are working by the consumption of the amount of nutrients, neurotransmitter bindings, and blood flow within the tissue. In addition, there have been many new developments in radiopharmaceutical ligands and probes. PET uses various radioactive pharmaceuticals as tracers, which make it possible to detect molecular changes down to the pico-molar range. It has allowed us to look at in vivo physiology as well as the molecular chemistry of living humans non-invasively and has opened up modern quantitative molecular neuroscience. PET applications also expanded to the study of amino acid metabolism ([Methyl-^{11}C]-L-Methionine) and gene proliferation ^{18}F-L-Thymidine (FLT). PET has changed classical nuclear imaging concepts and has led to an entirely new domain of molecular imaging. Various radionuclide tracers of PET are listed in Table 2.2. Depending on the radionuclides and their labeled compounds, various PET imaging techniques are available [15–22].

Although PET provides direct information about tracer uptake into the cell for specific tissues, the spatial resolution of PET is poor in comparison to CT or MRI. As of today, 2.5 mm FWHM is the best spatial resolution that

Table 2.2. *Radionuclide tracers for PET [23].* The generation of the nuclides ^{11}C, ^{13}N, ^{15}O, ^{18}F, and ^{124}I requires a cyclotron

Hemodynamic parameters	H$_2$ ^{15}O, ^{15}O-butanol, ^{11}CO, ^{13}NH$_3$
Hypoxia or angiogenesis	^{18}FMSIO, ^{64}Cu-ATSM, ^{18}F-Galacto-RGD
Substrate metabolism	^{18}F-FDG, ^{15}O$_2$, ^{11}C-acetate
Protein synthesis	^{11}C-Methionine, ^{11}C-leucine, ^{11}C-tyrosine
DNA synthesis	^{18}F-FLT, ^{11}C-thymidine, ^{18}F-fluorouracil
Drugs	^{11}C-Cocaine, ^{13}N-cisplatin, ^{18}F-fluorouracil
Receptor affinity	^{18}F-FESP, ^{18}F-FP-Gluc-TOCA, ^{18}F-FES
Gene expression	^{18}F-FHBG, ^{18}F-Penciclovir, ^{18}I-FIAU
Antibodies	^{124}I-CEA mimibody, ^{64}Cu-DOTA Her2/*neu* minibody

PET can have [24]. This resolution is still relatively poor for the localization of many delicate organs in the brain, such as the sub-regions in hippocampus.

In order to overcome its limitations, combining PET images with other high-resolution morphological imaging modalities such as radiography, CT and MRI has been studied [25–27]. Combining two or more imaging modalities is probably the best solution, especially in the field of neurological imaging.

In the past, to combine two different modalities, software registration has been used, and it works well in some studies where resolution requirement is relatively low.

2.4.1 PET/CT Systems

PET/CT is the first successful product in this series of research. Although it was the simple overlay of two images based on a mechanically calibrated shuttle bed or table, high-resolution anatomical images from CT partially aided the PET image, which has poor spatial resolution. The hybrid PET/CT imaging system can provide the functional image of PET with the superior anatomical delineation of CT. For example, PET/CT provides better distinction between cancerous tissue and healthy tissue in the diagnosis of cancer and the planning of radiation therapy.

2.4.2 PET/MRI Systems

On the other hand, magnetic resonance has much greater soft tissue contrast than CT, making it especially useful in neurological, musculoskeletal, cardiovascular, and oncological imaging. Unlike CT, it uses no ionizing radiation. Instead, it uses a powerful magnetic field to align the magnetization of hydrogen atoms in the body and provides excellent tissue contrasts in both brain and body imaging.

MRI has many advantages, such as its nonhazardous nature, high-resolution capability, potential for chemically specified imaging, capability of obtaining cross-sectional images in any desired directions, ability to use a large variety of high tissue contrasts, diffusion imaging capabilities, flow-related imaging capabilities, and ability to perform functional MRI [28]. MRI is, therefore, preferred to CT in the new fusion imaging system.

In contrast to PET/CT, which provides the simple combination of functional and anatomical images, PET/MRI could provide the complex combination of various functional information, such as PET, functional MRI (fMRI) or Magnetic Resonance Spectroscopy (MRS), and detailed morphological information through using soft tissue contrasts, for example. Thus, the PET/MRI system allows us to complement quantitative biological functional information from PET, such as metabolisms and binding potentials, with the high-resolution morphological information or other functional information from MRI. When PET/MRI fusion images are available, perhaps unique biochemical and molecular information with high resolution will be obtained from our body, especially where high-resolution imaging is of utmost importance, such as the brain.

Fusion Concepts

A major obstacle in developing a fusion PET/MRI system is that conventional PET uses PMTs for detector components. Because PMTs are very vulnerable to magnetic fields, especially in ultra high-field MRI such as 7.0 T, the unacceptably large stray magnetic fields from MRI practically prohibit any close positioning of the PET to the MRI.

In order to alleviate this problem, two types of approaches have been suggested [29–40]:

1. Fiber optics is used to relay the scintillation light from detection crystals to the PET modules, which would be located outside the magnetic field of the MRI. Since the scintillation crystals and optic fibers are not sensitive to the magnetic field, this arrangement would be suited for the PET/MRI combination. Fiber optics, however, attenuate the optical signals and, therefore, degrade the overall sensitivity and spatial resolution.
2. APD and a semiconductor-type PMT in the PET/MRI fusion system is used. APDs can replace PMTs since they are insensitive to magnetic fields. Although APDs have been used successfully on small scale PET scanners for animal use, APD-based PET appears to suffer from long term stability.

Extensive research is still ongoing and some progress have been be achieved. In 2008, Judenhofer et al. developed an APD-based PET and MRI hybrid imaging system for animal use [41]. In three-dimensional (3D) animal PET, the APD-based and magnetically compatible scanner can be inserted into an animal 7.0 T MRI system to simultaneously acquire functional and morphological PET/MRI images from living mice. With this PET/MRI system, they have found a tumor hidden in tissue through using high-resolution magnetic resonance data and simultaneously determined whether it is malignant by functional PET data.

A PET/MRI hybrid system for humans using this APD-based PET insertion is being developed by Siemens, Inc. and the Max Plank Institute in Germany. This human PET/MRI consists of an APD based PET-insert and a low field MRI (3.0 T MRI) system. Other companies and institutions are also developing integrated PET/MRI systems using a 3 T or 1.5 T MRI [36–39]. The major obstacle of this integrated PET/MRI system appears to be the stability of APD circuitry due to interference with the magnetic fields and radiofrequency waves from the MRI unit. It has been reported that there is significant variation of the timing resolution of the APDs, thereby increasing the coincidence timing window up to 40 ns, compared to less than 10 ns in most of the current PET systems [36].

2.4.3 High-Resolution Fusion

In order to avoid these interferences and to fully utilize the molecular imaging capability of PET and the anatomical imaging capability of UHF-MRI such as 7.0 T in human imaging, each scanner can be operated separately and connected using a high precision mechanical shuttle bed. This configuration shares many characteristics with the current PET/CT configuration [42, 43]. In this manner, one can conveniently avoid any possible artifacts due to the magnetic field interference incurred with PET-inserts or magnetic field influencing the PET operation. The major drawback of this approach is that image acquisition is performed sequentially rather than simultaneously.

In 2007, Cho et al. [23] developed this kind of PET/MRI hybrid system using shuttle bed system. One of the major differences of the system with others is the coupling of two high-end systems, i.e., HRRT-PET and 7.0 T MRI, was achieved without any compromise. For molecular imaging, HRRT-PET is used and provides a spatial resolution of 2.5 mm FWHM, the highest resolution among the human PET systems, and 7.0 T MRI system for the highest resolution anatomical imaging. These two devices together will provide the highest sensitivity and resolution molecular information, further aided by sub-millimeter resolution 7.0 T MRI imaging.

The conceptual design of the new fusion PET/MRI is shown in Fig. 2.10. PET and MRI are installed as closely as possible. The two are connected by a shuttle system composed of a bed and its guided rails. Proper magnetic and RF shielding was designed both in the PET side as well as shuttle bed to avoid interference of strong magnetic fields of the 7.0 T MRI. The shuttle system is designed to fulfill the mechanical precision of less than 0.05 mm and is able to operate under a high magnetic field, such as 7.0 T. This precision is sufficient to meet the spatial resolution of the 7.0 T MRI so that HRRT-PET images are precisely guided into the desired neuroanatomical region(s). The major advantage of this type PET/MRI system is that it allows the exploitation of the best qualities of the two systems, i.e., the available resolution and sensitivity of HRRT-PET and UHF 7.0 T MRI without any compromise and interference.

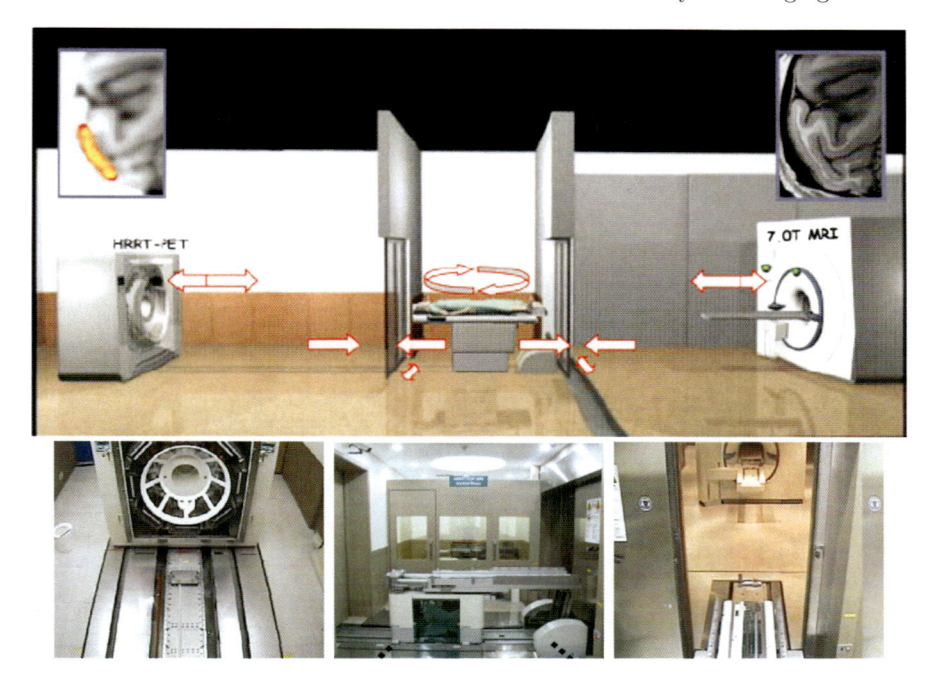

Fig. 2.10. *PET/MRI fusion system.* The system combines by two separate high-end imaging devices (*top*), the HRRT-PET (*left*) and the 7.0 T MRI (*right*) by means of a high-precision shuttle railway system (*bottom*)

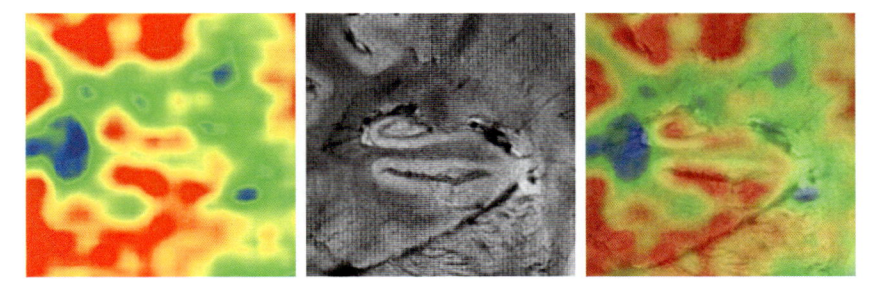

Fig. 2.11. *PET/MRI fusion image example.* The functional HRRT-PET image (*left*) and the high-resolution 7.0 T MRI image (*middle*) are overlaid (*right*) providing full diagnostic information

Critical issues for the success of this approach are the design and development of precision mechanics including the shuttle and railway, proper magnetic shield, and image fusion algorithm. The calibration method is also one of the important components of this system to correctly align the image coordinates of both imaging systems.

2.4.4 PET/MRI Fusion Algorithm

Two approaches are under development to integrate anatomical and molecular information. One is the image visualization approach and the other is the image restoration approach.

Image Visualization Approach

Most of the functional or molecular images, by themselves, cannot easily localize their signal origin. Overlaying the molecular information onto the structural image greatly improves the ability to characterize the desired anatomical location in the brain or organs [25–27, 44–46]. Advancement in image fusion methods allow for blending of higher spatial information of anatomical images with the higher spectral information of functional or molecular images (Fig. 2.11). These methods are classified into two categories by the fusion domain:

- Principal Component Analysis (PCA) [47], the Brovey method, and the Hue Intensity Saturation (HIS) method [48] are fused in the spatial domain.
- Discrete Wavelet Transform (DWT) and "A-trous" wavelet methods are fused in the transform domain [49, 50].

There are several ways to represent the color information depending on the color models: Red, Green, Blue (RGB), Cyan, Magenta, Yellow (CMY), and Intensity, Hue, Saturation (IHS). In some cases, the color transform between color models, such as RGB and IHS, is useful to combine the information from the multiple sensors.

$$C_{\text{IHS}} = M \cdot C_{\text{RGB}} \quad \text{or} \quad C_{\text{RGB}} = M - 1 \cdot C_{\text{IHS}} \tag{2.1}$$

where

$$C_{\text{RGB}} = \begin{bmatrix} R \\ G \\ B \end{bmatrix}, \quad C_{\text{IHS}} = \begin{bmatrix} I \\ v_1 \\ v_2 \end{bmatrix} \tag{2.2}$$

and

$$M = \begin{bmatrix} \frac{1}{3} & \frac{1}{3} & \frac{1}{3} \\ -\frac{\sqrt{2}}{6} & -\frac{\sqrt{2}}{6} & \frac{\sqrt{2}}{3} \\ \frac{1}{\sqrt{2}} & -\frac{1}{\sqrt{2}} & 0 \end{bmatrix}, \quad M^{-1} = \begin{bmatrix} 1 & -\frac{1}{\sqrt{2}} & \frac{1}{\sqrt{2}} \\ 1 & -\frac{1}{\sqrt{2}} & \frac{1}{\sqrt{2}} \\ 1 & \sqrt{2} & 0 \end{bmatrix} \tag{2.3}$$

Hue and saturation information is obtained from the intermediate variables v_1 and v_2 as following

$$H = \tan^{-1}\left(\frac{v_2}{v_1}\right) \quad \text{and} \quad S = \sqrt{v_1^2 + v_2^2} \tag{2.4}$$

The IHS model separates the image information I into the spatial information, such as intensity, and spectral information, such as hue H and saturation S.

The most common method of image fusion using HIS model is to substitute the whole or partial information of intensity channel from the lower resolution I_L to the higher resolution I_H.

Brovey's method simply modifies the brightness information by multiplying the intensity ratio of the higher resolution and the lower resolution.

$$C'_{\text{RGB}} = \gamma \cdot C_{\text{RGB}} \quad \text{where} \quad \gamma = \frac{I_H}{I_L} \tag{2.5}$$

The PCA technique transforms the inter-correlated variables to the uncorrelated variables.

$$C_{\text{PCA}} = \Phi \cdot C_{\text{RGB}} \quad \text{where} \quad C_{\text{PCA}} = \begin{bmatrix} PC1 \\ PC2 \\ PC3 \end{bmatrix}, \quad \Phi = \begin{bmatrix} \varphi_{11} & \varphi_{12} & \varphi_{13} \\ \varphi_{21} & \varphi_{22} & \varphi_{23} \\ \varphi_{31} & \varphi_{32} & \varphi_{33} \end{bmatrix} \tag{2.6}$$

The hue and saturation is obtained in a similar way as the HIS method.

$$H = \tan-1\left(\frac{PC3}{PC2}\right) \quad \text{and} \quad S = \sqrt{PC2^2 + PC3^2} \tag{2.7}$$

The primary component $PC1_H$ of the higher resolution image replaces the one $PC1_L$ of the lower resolution in the image fusion algorithm. The PCA fusion has the advantage of minimally distorting the spectral characteristics.

The wavelet transform method can be applied to the image fusion based on the multi-resolution analysis approach.

$$C_{\text{RGB}} = R_{\text{RGB}} + [W_{\text{RGB}}]_n \tag{2.8}$$

where

$$R_{\text{RGB}} = \begin{bmatrix} R_r \\ G_r \\ B_r \end{bmatrix}, \quad [W_{\text{RGB}}]_n = \begin{bmatrix} \sum_{k=1}^{n} W_{R,k} \\ \sum_{k=1}^{n} W_{G,k} \\ \sum_{k=1}^{n} W_{B,k} \end{bmatrix} \tag{2.9}$$

The C_{RGB} image is composed of a multi-resolution wavelet plane, $[W_{\text{RGB}}]_n$, and the residual multi-spectral images, R_{RGB}. R_{RGB} and $[W_{\text{RGB}}]_n$ contain the lower and higher spatial frequency of the image, respectively. For the image fusion, replacing wavelet coefficients $[W_{\text{RGB}}]_{n,L}$ of the lower resolution image to the ones $[W_{\text{RGB}}]_{n,H}$ of the higher resolution image.

Image Restoration Approach

Most of the functional or molecular images have lower resolution than anatomical images. For example, the intrinsic resolution of PET images is substantially poorer than MRI images. The intrinsic resolution of the PET system is determined mainly by the scintillation detector or crystal size and is approximately half of the detector width. The actual measured resolution is worse than the intrinsic resolution due to other image blurring factors, such as the source size, the positron range, the traveling distance, the penetration effect due to the adjacent detectors, and the angular jitters due to variations in the angles of the annihilation photons. These blurring factors can be mathematically modeled as the spatially invariant system. Among the blurring factors, the source size and the positron range are source-dependent and spatially invariant factors, while the others are spatially variant.

For simplicity, PET system can be assumed to be the spatially invariant system and the system blurring can be defined as a Point Spread Function (PSF). Once the exact PSF of the system is properly estimated, it can be used for an image deblurring operation or deconvolution. The PET image deblurring operation can be performed using a number of parameters that are extractable from neurochemical and molecular information, as well as image resolution information obtainable from MRI. For example, it is well-known that glucose utilization within cells takes place in the gray matter rather than in the white matter or in the Cerebrospinal Fluid (CSF). To execute the PET/MRI image fusion process, an appropriate segmentation of MRI images is essential to separate the corresponding tissues (gray matter) from the others such as white matter and CSF [51]. From an actual image processing point of view, it is equivalent to deconvolution of the measured PET sinogram with a Gaussian PSF derived from the MRI data to enhance the resolution of the PET image [51–53]. It is often accomplished in an iterative fashion (Fig. 2.12). The resolution of molecular image of PET, which is usually much poorer than that of MRI, can be enhanced by combining the high-resolution image of MRI with additional information such as neurochemical or molecular a priori information, such as the potential neurophysiological location of

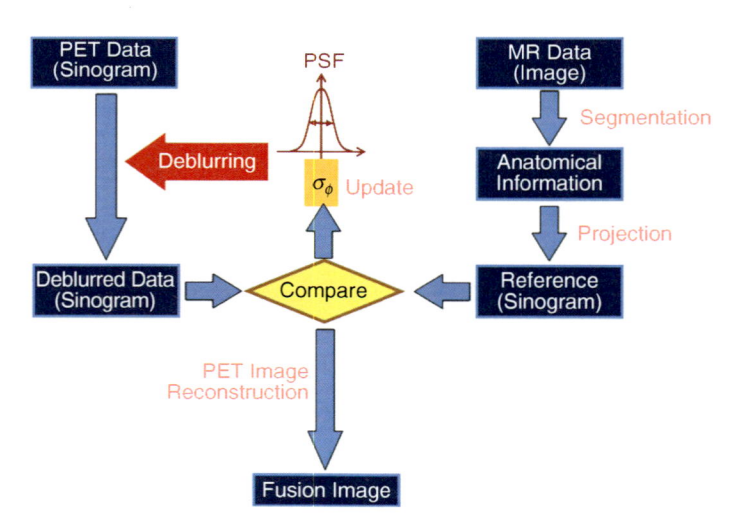

Fig. 2.12. *Iterative fusion algorithm.* The anatomical information of MRI data is utilized iteratively to confine the blurred PET image

receptor distribution for a particular ligand. In this fusion scheme, the segmented anatomical information from the MRI image is important and must be strongly correlated to the molecular information from the PET image.

Mismatched a priori information may cause over- or under-estimation of the PSF and the resultant fusion image may be biased. In order to utilize the anatomical image as a priori information, the contrast of the target tissue is also important. For instance, we know that the neuronal activities are generally confined to the gray matter rather than white matter, and segmentation of gray matter provides important morphological structures that can be registered with cortical activity (confined to the gray matter) as detected by FDG PET.

A preliminary human study was conducted to validate the usefulness of our fusion algorithm. A human brain image obtained by PET/MRI and the fusion algorithm is shown in Fig. 2.13, demonstrating that PET images can indeed be confined and localized with help of MRI, especially with an ultra high-resolution MRI such as the 7.0 T system.

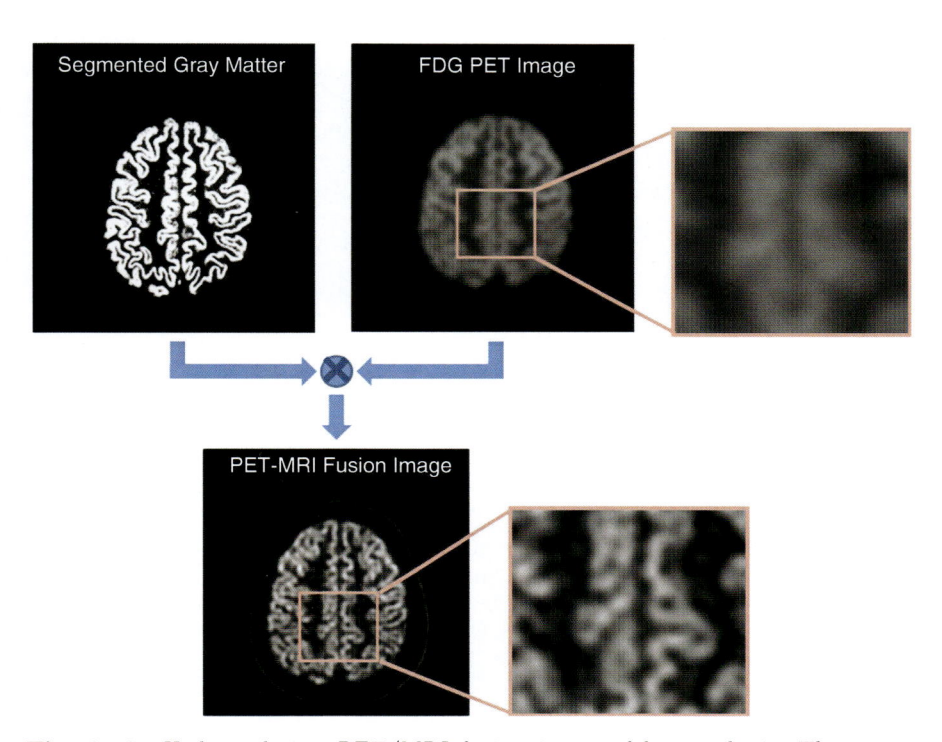

Fig. 2.13. *High-resolution PET/MRI fusion image of human brain.* The source images are obtained with the author's new PET/MRI fusion system. The spatial resolution of the PET is improved via the deblurring process based on the anatomical information of the MRI. The image data is processed in the sinogram domain

2.5 Conclusions

PET and MRI have been the most promising diagnostic tools among medical imaging tools, especially in the area of neuroscience. PET delivers information on molecular activities of human brain in vivo including enzymes and receptor distributions with resolutions down to 2.5 mm FWHM. On the other front, the MRI can obtain images with sub-millimeter resolution (down to 250 µm) and allows us to visualize the entire brain including the brain stem areas as well as other cortical and sub-cortical areas. For advanced and accurate diagnosis, these two systems are combined to overcome their limitations.

Although a few problems still remain, the current PET/MRI fusion system produces the highest quality images of molecular activities of the human brain in vivo and provides unprecedented molecular activity matched high-resolution images, which represent highly correlated molecular information to anatomically well established organs. This new PET/MRI fusion system, for the first time, began to provide anatomically well-defined molecular activities in the brain hitherto unavailable by any other imaging devices. This molecular fusion imaging system would be an important and essential tool for studying cognitive neurosciences and neurological diseases, such as the Parkinson's and Alzheimer's diseases. A mathematical image processing strategy that integrates anatomical and molecular information together is a still unfinished challenge in the field of medical image processing. We hope that the technology will provide novel and unique information to clinicians and research scientists in the field of neuroscience.

Acknowledgment

This work was supported by Basic Science Research Program through the National Research Foundation (NRF) funded by the Ministry of Education, Science and Technology (R11-2005-014 & 2008-04159).

References

1. Cho Z, Chan J. Circular ring transverse axial positron camera for 3-dimensional reconstruction of radionuclides distribution. IEEE Trans Nucl Sci. 1976;23(1):613–22.
2. Ter-Pogossian M, Phelps M. A positron-emission transaxial tomograph for nuclear imaging PET. Radiology. 1975;114(1):89–98.
3. Cho Z, Farukhi M. Bismuth germanate as a potential scintillation detector in positron cameras. Soc Nucl Med. 1977;18:840–44.
4. Melcher C, Schweitzer J. A promising new scintillator: cerium-doped lutetium oxyorthosilicate. Nucl Instrum Methods Phys Res. 1992;314:212–14.
5. Cho Z, Chan J. Positron ranges obtained from biomedically important positron-emitting radionuclides. Soc Nucl Med. 1975;16:1174–76.
6. Knoß C. Evaluation and Optimization of the High Resolution Research Tomograph HRRT. RWTH Aachen University, Aachen, Germany; 2004.

7. Eriksson L, Wienhard K, Eriksson M, et al. The ECAT HRRT: NEMA NEC evaluation of the HRRT system, the new high-resolution research tomograph. IEEE Trans Nucl Sci. 2002;49(5):2085–88.
8. De Jong H, Velden F. Performance evaluation of the ECAT HRRT: an LSO-LYSO double layer high resolution, high sensitivity scanner. Phys Med Biol. 2007;52(5):1505–26.
9. Mourik J, Velden F. Image derived input functions for dynamic high resolution research tomograph PET brain studies. Neuroimage. 2008;43(4):676–86.
10. Heiss W, Habedank B. Metabolic rates in small brain nuclei determined by high-resolution PET. J Nucl Med. 2004;45(11):1811–15.
11. Willeit M, Ginovart N. High-affinity states of human brain dopamine D2/3 receptors imaged by the agonist [11C]-(+)-PHNO. Biol Psychiatry. 2006;59(5):389–94.
12. Jucaite A, Fernell E. Reduced midbrain dopamine transporter binding in male adolescents with attention-deficit/hyperactivity disorder: association between striatal dopamine markers and motor hyperactivity. Biol Psychiatry. 2005;57(3):229–38.
13. Hirvonen J, Johansson J. Measurement of striatal and extrastriatal dopamine transporter binding with high-resolution PET and [(11)C]PE2I: quantitative modeling and test-retest reproducibility. J Cereb Blood Flow Metab. 2008;28(5):1059–69.
14. Cho Z, Kang C. Observation of the lenticulostriate arteries in the human brain in vivo using 7.0 T MR angiography. Stroke. 2008;39(5):1604–06.
15. Tjuvajev J, Stockhammer G. Imaging the expression of transfected genes in vivo. Cancer Res. 1995;55(24):6126–32.
16. Gambhir S, Barrio J. Imaging of adenoviral-directed herpes simplex virus type 1 thymidine kinase reporter gene expression in mice with radiolabeled ganciclovir. Soc Nuclear Med. 1998;39:2003–011.
17. Gambhir S, Bauer E. A mutant herpes simplex virus type 1 thymidine kinase reporter gene shows improved sensitivity for imaging reporter gene expression with positron emission tomography. Natl Acad Sci. 2000;97:2785–90.
18. Harrington K, Mohammadtaghi S. Effective targeting of solid tumors in patients with locally advanced cancers by radiolabeled pegylated liposomes. Proc Am Assoc Cancer Res. 2001;7:243–54.
19. Jacobs A, Braunlich I. Quantitative kinetics of [124I] FIAU in cat and man. J Nucl Med. 2001;42(3):467–75.
20. Jacobs A, Voges J. Positron-emission tomography of vector-mediated gene expression in gene therapy for gliomas. Lancet. 2001;358(9283):727–29.
21. Yaghoubi S, Barrio J. Human pharmacokinetic and dosimetry studies of [18F] FHBG: a reporter probe for imaging herpes simplex virus type-1 thymidine kinase reporter gene expression. Soc Nucl Med. 2001;42:1225–34.
22. Tjuvajev J, Doubrovin M. Comparison of radiolabeled nucleoside probes FIAU, FHBG, and FHPG for PET imaging of HSV1-tk gene expression. J Nucl Med. 2002;43(8):1072–83.
23. Cho Z, Son Y. A fusion PET-MRI system with a high-resolution research tomograph-PET and ultra-high field 7 T-MRI for the molecular-genetic imaging of the brain. Proteomics. 2008;8(6):1302–23.
24. Wienhard K, Schmand M. The ECAT HRRT: performance and first clinical application of the new high resolution research tomograph. IEEE Trans Nucl Sci. 2002;49(1):104–10.

25. Makela T, Pham Q. A 3-D model-based registration approach for the PET, MR and MCG cardiac data fusion. Med Image Anal. 2003;7(3):377–89.
26. Terence ZW, Timothy G. PET and brain tumor image fusion. Cancer J. 2004;10:234–42.
27. Borgwardt L, Hojgaard L. Increased fluorine-18 2-fluoro-2-deoxy-D-glucose FDG uptake in childhood CNS tumors is correlated with malignancy grade: a study with (FDG) positron emission tomography/magnetic resonance imaging coregistration and image fusion. J Clin Oncol. 2005;23(13):3030–37.
28. Cho Z, Jones J. Foundations of Medical Imaging. New York: Wiley-Interscience; 1993.
29. Garlick P, Marsden P. PET and NMR dual acquisition PANDA: applications to isolated, perfused rat hearts. NMR Biomed. 1997;10(3):138–42.
30. Shao Y, Cherry S. Simultaneous PET and MR imaging. Phys Med Biol. 1997;42(10):1965–70.
31. Shao Y, Cherry S. Development of a PET detector system compatible with MRI/NMR systems. IEEE Trans Nucl Sci. 1997;44(3):1167–71.
32. Farahani K, Slates R. Contemporaneous positron emission tomography and MR imaging at 1.5 T. J Magn Reson Imaging. 1999;9(3):497–500.
33. Slates R, Farahani K. A study of artefacts in simultaneous PET and MR imaging using a prototype MR compatible PET scanner. Phys Med Biol. 1999;44:2015–28.
34. Slates R, Cherry S. Design of a small animal MR compatible PET scanner. IEEE Trans Nucl Sci. 1999;46(3):565–70.
35. Marsden P, Strul D. Simultaneous PET and NMR. Br J Radiol. 2002;75:S53.
36. Catana C, Wu Y. Simultaneous acquisition of multislice PET and MR images: initial results with a MR-compatible PET scanner. J Nucl Med. 2006;47(12):1968–76.
37. Grazioso R, Zhang N. APD-based PET detector for simultaneous PET/MR imaging. Nucl Instrum Methods Phys Res. 2006;569(2):301–05.
38. Raylman R, Majewski S. Initial tests of a prototype MRI-compatible PET imager. Nucl Instrum Methods Phys Res. 2006;569(2):306–09.
39. Raylman R, Majewski S. Simultaneous MRI and PET imaging of a rat brain. Phys Med Biol. 2006;51(24):6371–80.
40. Raylman R, Majewski S. Simultaneous acquisition of magnetic resonance spectroscopy MRS data and positron emission tomography PET images with a prototype MR-compatible, small animal PET imager. J Magn Reson Imaging. 2007;186(2):305–10.
41. Judenhofer M, Wehrl H. Simultaneous PET-MRI: a new approach for functional and morphological imaging. Nat Med. 2008;14(4):459–65.
42. Kinahan P, Townsend D. Attenuation correction for a combined 3D PET/CT scanner. Med Phys. 1998;25(10):2046–53.
43. Beyer T, Townsend DW. A combined PET/CT scanner for clinical oncology. J Nucl Med. 2000;41(8):1369–79.
44. Barillot C, Lemoine D. Data fusion in medical imaging: merging multimodal and multipatient images, identification of structures and 3D display aspects. Eur J Radiol. 1993;17(1):22–27.
45. Rehm K, Strother SC. Display of merged multimodality brain images using interleaved pixels with independent color scales. J Nucl Med. 1994;35(11):1815–21.

46. Pietrzyk U, Herholz K. Clinical applications of registration and fusion of multimodality brain images from PET, SPECT, CT, and MRI. Eur J Radiol. 1996;21(3):174–82.
47. Gonzalez-Audicana M, Saleta J. Fusion of multispectral and panchromatic images using improved IHS and PCA mergers based on wavelet decomposition. IEEE Trans Geosci Remote Sens. 2004;42(6):1291–99.
48. Tu T, Su S. A new look at IHS-like image fusion methods. Info Fusion. 2001;2(3):177–86.
49. Wang A, Sun H. The application of wavelet transform to multi-modality medical image fusion. Proc IEEE ICNSC. 2006; p. 270–274.
50. Zheng Y, Essock E. A new metric based on extended spatial frequency and its application to DWT based fusion algorithms. Info Fusion. 2007;8(2):177–92.
51. Baete K, Nuyts J. Anatomical-based FDG-PET reconstruction for the detection of hypo-metabolic regions in epilepsy. IEEE Trans Med Imaging. 2004;23(4):510–19.
52. Gindi G, Lee M. Bayesian reconstruction of functional images using anatomical information as priors. Med Imaging. 1993;12(4):670–80.
53. Teo BK, Seo Y. Partial-volume correction in PET: validation of an iterative postreconstruction method with phantom and patient data. J Nucl Med. 2007;48(5):802–10.

3
Cardiac 4D Ultrasound Imaging

Jan D'hooge

Summary. Volumetric cardiac ultrasound imaging has steadily evolved over the last 20 years from an electrocardiography (ECC) gated imaging technique to a true real-time imaging modality. Although the clinical use of echocardiography is still to a large extent based on conventional 2D ultrasound imaging it can be anticipated that the further developments in image quality, data visualization and interaction and image quantification of three-dimensional cardiac ultrasound will gradually make volumetric ultrasound the modality of choice. In this chapter, an overview is given of the technological developments that allow for volumetric imaging of the beating heart by ultrasound.

3.1 The Role of Ultrasound in Clinical Cardiology

Ultrasound (US) imaging is the modality of choice when diagnosing heart disease. This is due to fact that it is non-invasive; does not show adverse biological effects; has an excellent temporal resolution; is portable (and can thus be applied bed-side) and is relatively cheap when compared to other imaging modalities. As such, US imaging has become an indispensable tool for daily management of cardiac patients.

Historically, cardiac ultrasound started with acquiring a single image line as a function of time, which is referred to as motion mode (M-mode). It allowed studying basic morphological properties of the heart such as estimating the dimension of the left ventricular cavity or the segmental wall thickness. In addition, the motion of the heart during the cardiac cycle could be monitored which can give information on cardiac performance. However, as the field-of-view of this imaging approach remained very limited, correct navigation through the heart and interpretation of the recordings was difficult.

Hereto, two-dimensional (2D) ultrasound imaging (brightness mode (B-mode)) was introduced by mechanically moving (i.e., tilting), the transducer between subsequent line acquisitions. This mechanical motion of the transducer was replaced by electronic beam steering in the late sixties when phased array transducer technology was introduced. As such, cross-sectional images of the heart could be produced in real-time at typical frame rates of about 30 Hz. Although continuous improvements in image quality and image resolution were obtained in the following decades, imaging a 2D cross-section of a complex 3D organ such as the heart continued to have intrinsic pitfalls. As such, three-dimensional (3D) US imaging of the heart has been a topic of research for several decades.

3.2 Principles of Ultrasound Image Formation

The fundamental principle of echocardiography is relatively simple: an US pulse is transmitted into the tissue and the reflections that occur while the wave propagates (due to local inhomogeneities in mass density or regional elasticity) are detected by the same transducer as a function of time. As the velocity of sound in tissue is known, the time at which a reflection is detected and the distance at which this reflection took place are linearly related. As such, the reflected signal can be used to reconstruct a single line in the ultrasound image giving information on the tissue reflectivity (i.e., its acoustic properties) as a function of depth. In order to generate a 2D or 3D image, the above measurement is repeated by transmitting ultrasound in different directions either by mechanically translating/tilting the transducer or by electronic beam steering.

3.2.1 The Pulse-Echo Measurement

The basic measurement of an ultrasound device can shortly be summarized as follows:

1. A short electric pulse is applied to a piezoelectric crystal. This electric field re-orients the (polar) molecules of the crystal and results in a change of its shape. The crystal will thus deform.
2. The sudden deformation of the piezoelectric crystal induces a local compression of the tissue with which the crystal is in contact (Fig. 3.1a).
3. This local compression will propagate away from the piezoelectric crystal (Fig. 3.1b). This compression wave (i.e., the acoustic wave) travels at a speed of approximately 1,530 m/s in soft tissue through the interaction of tissue elasticity and inertia. Indeed, a local compression is counteracted upon by the tissue elasticity which results in a return to equilibrium. However, due to inertia, this return to equilibrium is too large resulting in a local rarefaction (i.e., de-compression), which in turn is counteracted upon by tissue elasticity. After a few iterations, depending on the tissue characteristics and of the initial compression, equilibrium is reached since each iteration is accompanied by damping, i.e., attenuation. The rate of compression/decompression determines the frequency of the wave and is typically 2.5–8 MHz for diagnostic ultrasound imaging. As these frequencies cannot be perceived by the human ear, these waves are said to

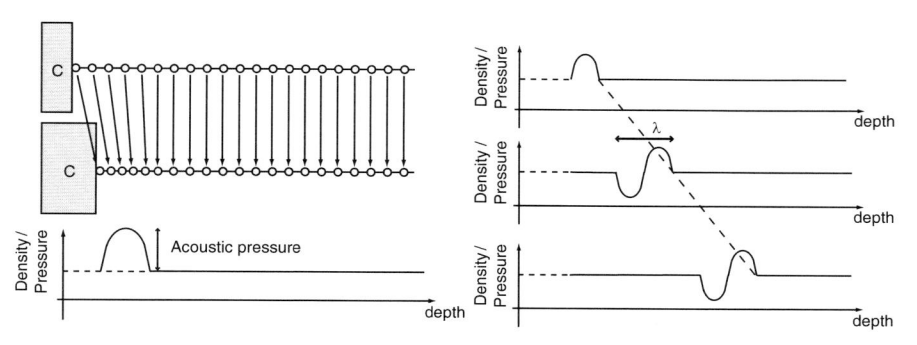

Fig. 3.1. *Generation and propagation of sound waves. Left*: Local tissue compression due to deformation of the piezoelectric crystal; *Right*: The generated compression propagates away from the transducer

   be ultra-sonic. Typically, the higher the ultrasound frequency, the more attenuation and therefore – for a given amplitude – less penetration.
4. Inhomogeneities in tissue density or tissue elasticity will result in a disturbance of the propagating wave. They will cause part of the energy in the wave to be scattered, i.e., re-transmitted in all possible directions. The part of the scattered energy re-transmitted back into the direction of origin of the wave is called backscatter. At interfaces between different types of tissue (i.e., blood and cardiac muscle), part of the acoustic wave is reflected (i.e., specular reflections). Both specular and backscatter reflected waves propagate back towards the piezoelectric crystal.
5. When the reflected (compression) waves impinge upon the piezoelectric crystal, the crystal deforms which results in the generation of an electric signal. The amplitude of this electric signal is proportional to the amount of compression of the crystal, i.e., the amplitude of the reflected or backscattered wave. This electric signal is called Radio Frequency (RF) signal (Fig. 3.2).

For diagnostic frequencies used in cardiac ultrasound, the above pulse-echo measurement typically takes about 250 µs.

3.2.2 Gray Scale Encoding

A single pulse-echo measurement results in a single line in the US image. The RF signal is further processed:

1. *Envelope detection*: The high frequency information of the RF signal is removed by detecting the envelope of the signal (Fig. 3.3). This is most commonly done by using the Hilbert transform.
2. *Grayscale encoding*: As a function of time, the signal is sub-divided in small intervals (i.e., pixels). Each pixel is attributed a number defined by the local amplitude of the signal. Usually, these gray scales range between

Fig. 3.2. *Radio frequency signal.* Reflected waves (i.e., the echos) are detected using the same transducer resulting in a signal in the radio-frequency range (a few MHz) as a function of time. Two strong specular reflections are detected at 50 and 83 µs while scatter reflections are received within this time interval

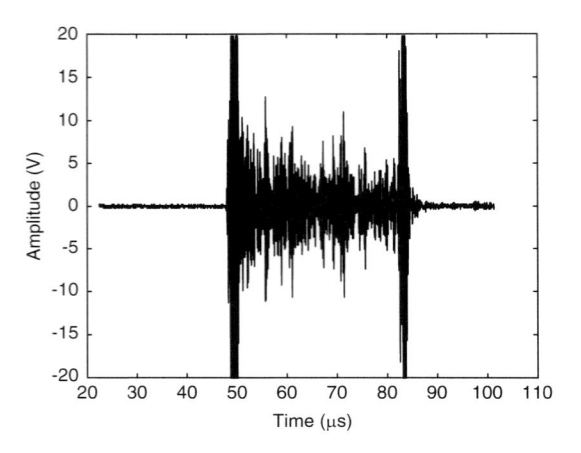

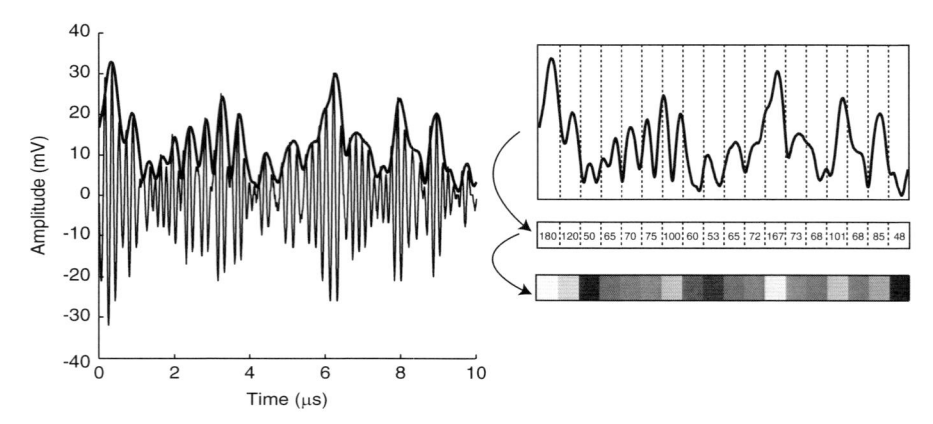

Fig. 3.3. *Ultrasound demodulation.* The RF signal is demodulated in order to detect its envelope (*left*). This envelope signal (*bold*) is color encoded based on the local signal amplitude (*right*)

0 (black) and 255 (white). By definition, bright pixels thus correspond to high amplitude reflections (Fig. 3.3).

3. *Attenuation correction*: As wave amplitude decreases with propagation distance due to attenuation, reflections from deeper structures are intrinsically smaller in amplitude and would thus show less bright. Identical structures should have the same gray value however and, consequently, the same reflection amplitudes. To compensate for this effect, the attenuation is estimated and compensated for. Since time and depth are linearly related in echocardiography, attenuation correction is often called time or depth gain compensation. Sliders on the ultrasound scanner allow for a manual correction of this automatic compensation in case it fails to correct appropriately.

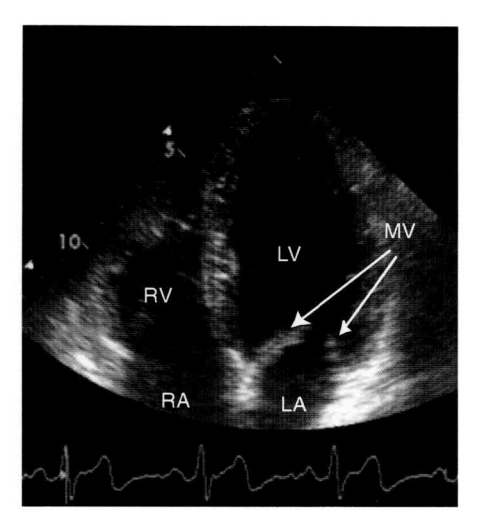

Fig. 3.4. *Cross-sectional ultrasound image of the heart.* The four cardiac chambers Left Ventricle (LV), Right Ventricle (RV), Left Atrium (LA), and Right Atrium (RA) are visualized dark – as blood is little reflective – together with the leaflets of the Mitral Valve (MV) and the cardiac muscle, which appear as gray region around the dark cavities

4. *Logarithmic compression*: In order to increase the contrast in the dark regions of the image (as the RF signal typically has a large dynamic range), gray values in the image are re-distributed according to a logarithmic curve.

3.2.3 Gray Scale Imaging

In order to obtain an ultrasound image, the gray scale encoding procedure is repeated. For B-mode imaging, the transducer is either translated or tilted within a plane (conventional 2D imaging) or in space (3D imaging) between two subsequent pulse-echo experiments. In this way, a cross-sectional image can be constructed (Fig. 3.4).

Typically, a 2D cardiac image consists of 120 lines spread over an angle of 90°. The construction of a single image thus takes about 120×250 μs equaling (approximately) 30 ms. Per second, about 33 images can therefore be produced which is sufficient to look at motion (e.g., standard television displays only 25 frames per second). However, a straight forward extension of this approach to 3D will result in a frame rate below 1 Hz which is unacceptable for cardiac applications. Therefore, cardiac volumetric imaging requires other approaches.

3.2.4 Phased Array Transducer Technology

Rather than mechanically moving or tilting the transducer, modern US devices make use of electronic beam steering. Hereto, an array of piezoelectric crystals is used. By introducing time delays between the excitation of different crystals in the array, the US wave can be send in a particular direction without mechanical motion of the transducer (Fig. 3.5a). Similarly, the ultrasound

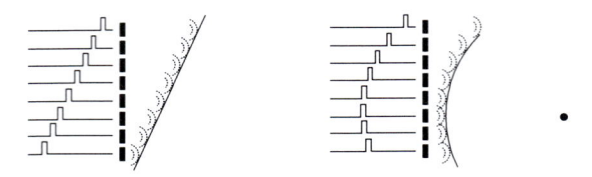

Fig. 3.5. *Phased array transducer.* An array of crystals can be used to steer (left) and/or focus (right) the ultrasound beam electronically by introducing time delays between the activation of individual elements in the array

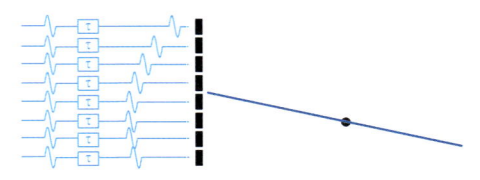

Fig. 3.6. *Receive focusing.* By applying the appropriate time delays on the received RF signals of the individual array elements, the receive "beam" can be steered and focused in a similar way as the transmit beam as reflections arriving from the chosen direction/position will constructively interfere

wave can be focused in a specific point by making sure that the contributions of the individual elements arrive simultaneously in this point (Fig. 3.5b).

The detected RF signal for a particular transmitted (directed, focused) pulse is then simply the sum of the RF signals received by the individual elements. These individual contributions can be filtered, amplified and time-delayed separately before summing. This process is referred to as beam forming and is a very crucial aspect for obtaining high-quality images. For example, by introducing the proper time delays between the contributions of the individual crystals prior to summation, beam focusing can also be achieved during receive. As the sound velocity in tissue is known, the depth from which reflections can be expected at a certain moment after transmit can be estimated. As such, the focus point can dynamically be moved during reception by dynamically changing the time delays upon reception. This approach is referred to as dynamic (receive) focusing which has a significant impact on image quality (Fig. 3.6).

3.3 Limitations of 2D Cardiac Ultrasound

Despite of the fact that significant advances in 2D ultrasound image resolution (both spatially and temporally) and quality, i.e., Signal to Noise Ratio (SNR), Contrast to Noise Ratio (CNR) have been made over the years, 2D imaging to visualize a moving 3D structure such as the heart has intrinsic pitfalls.

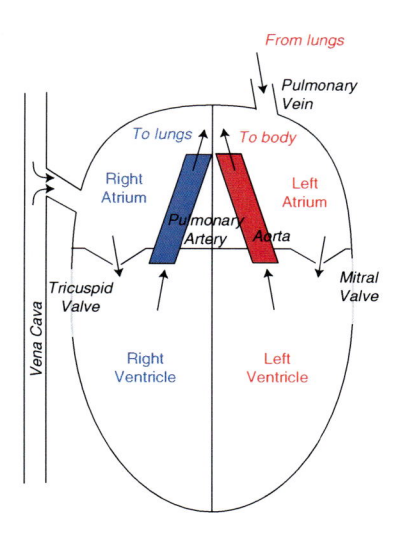

Fig. 3.7. *Scheme of the heart and big vessels.* The heart can be seen as a four compartment box connected with big vessels

3.3.1 Complex Anatomy (Congenital Heart Disease)

The heart can roughly be seen as a four compartment box in which two atria sit on top of two ventricles (Fig. 3.7).

The right atrium pumps de-oxygenated blood coming from the peripheral organs (through the inferior and superior vena cava) to the right ventricle that – in turn – propels the blood through the lungs into the left atrium. The – now oxygenated – blood then passes through the mitral valve into the left ventricle in order to be pumped into the aorta and to the rest of the body. The different chambers are separated by valves in order to avoid retrograde flow to occur and therefore make sure that pumping function is efficient.

Although most cardiac patients do indeed have a normal cardiac anatomy, a significant amount of patients are born with cardiac malformations, i.e., congenital heart diseases. In such patients, cardiac anatomy and morphology deviates from normality. Moreover, some of the big vessels (e.g., aorta, pulmonary artery) may be inappropriately connected to the cardiac chambers. For example, a connection may exists between the left and right ventricles due to a defect in the inter-ventricular or inter-atrial septum, i.e., Ventricular Septum Defect (VSD) or Atrial Septal Defect (ASD), respectively. In some patients, the ventricular septum is totally absent (i.e., uni-ventricular hearts). Others have the aorta implanted on the right ventricle while the pulmonary artery is implanted on the left ventricle (i.e., malposition of the big vessels). The abnormal anatomy of these hearts can have a significant impact on both the function and the morphology of each of the cardiac chambers. As such, the congenital heart is typically very complex in shape and structure. Reconstructing and understanding the exact 3D anatomy of such a heart needs expert training as this reconstruction needs to be done in the operator's mind by navigating through a number of 2D cross-sections. Even for highly skilled experts, this remains challenging.

3.3.2 Geometric Assumptions to Assess Volumes

An important clinical parameter for a patient's prognosis is the volume of the cardiac chambers. As such, the assessment of the left ventricular volume is part of all routine clinical echo examinations. However, as conventional ultrasound equipment only allows measuring distances (M-mode, B-mode) or areas (B-mode), assumptions have to be made on the geometry of the left ventricle in order to assess its volume.

Different models of different complexities have been proposed. Most often the assumption is made that the left ventricle can be modeled as an ellipsoid of revolution. Measuring its long and short axis dimensions (on a 2D US image) is then sufficient to estimate its volume (Fig. 3.8). Moreover, if the additional assumption is made that the long-to-short axis ratio of this ellipsoid is known, a single measurement of the diameter of the left ventricular cavity on an M-mode image is sufficient to estimate the entire volume. Although this method relies strongly on the geometrical assumptions made, this approach is the most commonly used clinical method for assessing left ventricular volume. By making this measurement both at end-diastole and at end-systole (end of the ejection period of the ventricle) the amount of blood ejected by the ventricle into the aorta during one cardiac cycle (i.e., the stroke volume), and the ejection fraction, i.e., the stroke volume normalized to the end-diastolic volume, can also be estimated.

Although a normal left ventricle roughly resembles half an ellipsoid of revolution the diseased ventricle does not necessarily. For example, after myocardial infarction, an aneurysm can form resulting in regional bulging of the myocardial wall. Similarly, during pulmonary hypertension the septum will bulge into the left ventricular cavity due to the high blood pressure in the right ventricle. Obviously, in such cases, the boundary conditions to be able to use the geometric model do not hold and the estimated volumes can be seriously biased.

Similarly, the shape of the other cardiac chambers (right ventricle, left and right atrium) is much more complex (even in the normal heart) and good geometrical models are not available. As such, a correct and reliable estimation of their volume using US imaging remains challenging.

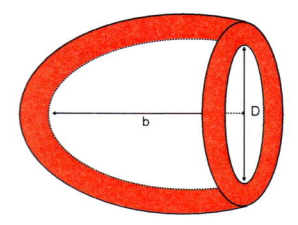

Fig. 3.8. *Ellipsoid model for volume estimation.* Assuming that the left ventricle can be modeled as an ellipsoid of revolution, its volume can be calculated based on the long and short axis dimensions of this ellipsoid. Those can be assessed by 2D ultrasound

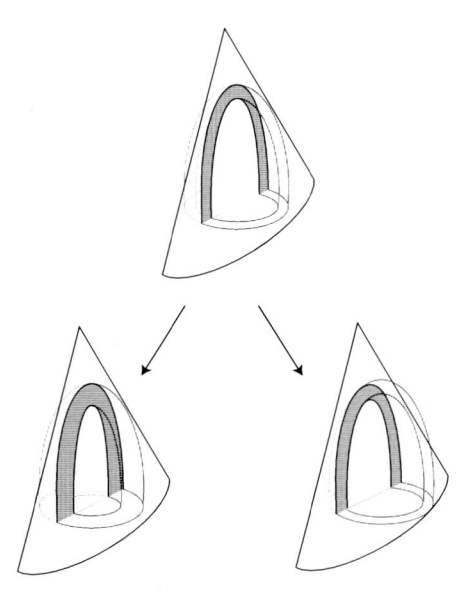

Fig. 3.9. *Principle of foreshortening.* Myocardial inwards motion and thickening during the cardiac cycle (*left*) are visually scored in daily clinical practice in order to assess regional myocardial function. However, out-of-plane motion of the heart (*right*) can result in oblique cross-sections that can easily be misread as inward motion or wall thickening (especially near the highly curved apex). This effect is referred to as foreshortening

3.3.3 Out-of-Plane Motion and Foreshortening

Coronary artery disease is a major cardiovascular problem in the Western world. Echocardiography is typically used to evaluate the impact of the coronary narrowing on segmental function of the cardiac muscle (as the narrowing will result in a hypo-perfusion of the distal myocardium which will as such not be able to contract normally). For that, the heart is visualized (sometimes while stressing the heart pharmaceutically) in order to detect regions of abnormal wall thickening and motion.

Although cardiac US does indeed allow visualization of segmental wall motion and deformation, a common pitfall is the problem of foreshortening due to out-of-plane motion of the heart during the cardiac cycle. As a result, the image plane might cut correctly through the heart at end-diastole (i.e., along the long axis of the left ventricle) while the end-systolic cross-section cuts in a parallel (anatomical) plane. As such, segmental wall motion and thickening may appear normal while they are in fact abnormal (Fig. 3.9). It remains a major difficulty for stress echocardiography and its interpretation.

3.4 Approaches Towards 3D Cardiac Ultrasound

Given the intrinsic difficulties with 2D ultrasound to study a complex 3D organ such as the heart, volumetric ultrasound imaging would offer obvious advantages. However, as mentioned above, a straight forward implementation of the typical 2D imaging concepts for a 3D ultrasound system would result in a frame rate below 1 Hz. This is clearly unacceptable given that a complete cardiac cycle typically takes less than a second. Several approaches have therefore been proposed to solve this problem.

3.4.1 Freehand 3D Ultrasound

The most straight forward way of acquiring volumetric US cardiac data at an acceptable temporal resolution is by merging different 2D US acquisitions taken over subsequent cardiac cycles into one 3D volumetric data set. By tracking the position and orientation of the transducer by an electromagnetic tracking device while moving it freely manually, the 3D + t volume can be filled with data points [1].

Obviously such an approach is cumbersome and time consuming. Moreover, it can have problems with motion artifacts as a result of patient motion and breathing and the 3D + t volume will typically be sampled in an irregular manner. Finally, the tracking devices have limited accuracy which will result in misalignment of the collected 2D data in 3D space.

3.4.2 Prospective Gating

In order to make the volumetric acquisition process more reproducible and to fill the 3D + t space more homogeneously, the motion of the transducer can be guided by an external device that rotates the transducer by a certain angle at every heart beat (prospective gating). In this way, a full volume data set can be constructed as a fan of US images through ECG gating as illustrated in Fig. 3.10. Here, the ECG signal is used to synchronize data acquisition over several heart beats.

Although fully applicable, this approach has several disadvantages. Firstly – as in any gating technique – acquisition time is significantly prolonged. It is directly proportional to the number of image planes and inversely proportional to the heart rate. Given that a rotation of 180° is sufficient to cover the entire volume (as the 2D ultrasound sector image is symmetric), a 2° rotation

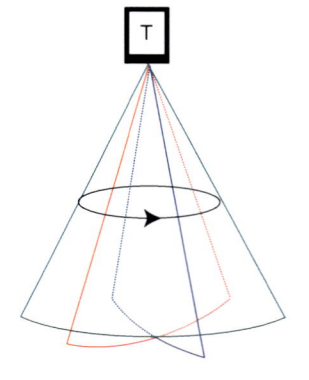

Fig. 3.10. *Rotating transducer.* Rotating the 2D ultrasound image allows filling 3D space in a regular way. In order to obtain a sufficiently high frame rate 2D image data need to be merged over several heart beats

step will require 90 2D acquisitions, i.e., heart beats. This corresponds to a typical acquisition time of 1–1.5 min. Secondly, as acquisition time becomes significant, this approach is sensitive to motion and respiratory artifacts that will result in a spatial misalignment of the image planes within the 3D volume. Finally, this approach can only be applied in patients in sinus rhythm. Any heart rate variability, which is physiologic and, therefore, always present (even in a healthy individual) will result in a temporal misalignment of some of the image data.

Obviously, this 3D imaging approach requires an off-line reconstruction of the volumetric image. This implies that no volumetric feedback can be given to the operator which makes the acquisition of high quality data sets more cumbersome. Indeed, during rotation of the 2D image plane, drop-outs (e.g., due to rib artefacts) can occur which may require to re-start the acquisition process after re-positioning the transducer on the patient's chest. Although such problems could be avoided by taking proper scouting images, this scouting again prolongs the acquisition process.

Despite these potential pitfalls, this approach has been introduced and validated in several clinical studies (e.g., [2]).

3.4.3 Retrospective Gating

In order to avoid some of the problems met with ECG-gating, a fast-rotating array transducer has been proposed [3]. In this approach, the transducer rotates very quickly around its axis (approx. 400–500 rpm) so that it makes several revolutions (approx. 8) per heart beat. Obviously, one cannot use a conventional ultrasound transducer as the electrical cable connecting the transducer to the ultrasound system would quickly wind up and break. Djao et al. have proposed to use a slip-ring system whereby the piezoelectric elements of the array transducer remain connected to their electrodes in a sliding manner [3].

The major advantage of this system over the approach described in the previous sections is that volumetric data becomes available in real-time (16 vols./s) and can thus be shown to the operator as visual feedback in order to optimize the volumetric data acquisition. Moreover, ECG gating is not strictly required as the volume can simply be updated continuously at a rate of 16 vols./s. However, if volumetric data at higher temporal or spatial resolution are required, (retrospective) ECG-gating – with its associated pitfalls – remains a must.

The major disadvantage of this approach is that a dedicated array transducer is needed. Although the slip-ring principle has been proven to work well, the mechanical interconnection comes at the expense of noise. Moreover, volumetric image reconstruction becomes quiet challenging. Indeed, given the high rotation speed of the transducer and given that a 2D image is constructed by acquiring subsequent image lines (Sect. 3.3) the image planes are curved (Fig. 3.11). In addition, spatial sampling becomes rather irregular and

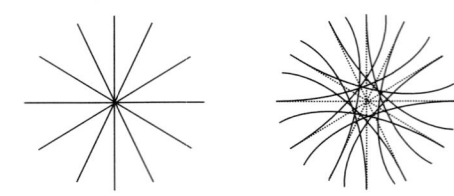

Fig. 3.11. *Prospective and retrospective gating.* Although prospective ECG gating results in equidistant image plane (*left*) retrospective gating combined with a fast rotating array results in curved image planes (*right*). The latter makes reconstruction of the volumetric data more challenging

is dependent on the heart rate to rotation rate ratio. Reconstruction of 3D volume data from this irregular sampled space is not trivial but several methods to do so have been presented [4].

3.4.4 Two-Dimensional Arrays

The concept of a phased array transducer can be extended towards volumetric imaging. Indeed, if a 2D matrix of piezoelectric elements can be constructed, the US beam can be steered electronically both in plane (as for the 1D array transducer, cf. Sect. 3.4) and out-of-plane (as required for 3D imaging). Although this is theoretically feasible, its practical implementation is not straight forward due to the large number of elements required in a 2D phased array transducer. Indeed, a regular 1D phased array transducer typically consists of 64 up to 128 elements. As such, a full 2D array transducer will contain $64 \times 64 = 4,096$ elements up to $128 \times 128 = 16,384$ elements. Each of these elements needs to be electrically isolated from its neighbors and needs wiring. As the footprint of the transducer needs to be limited in size for cardiac applications (approx. 1.5×1 cm; it has to fit in between ribs) this becomes very challenging in terms of dicing the piezoelectric materials and wiring/cabling.

Sparse 2D Array Transducers

Because of the wiring issues, the first 2D array transducers had only a limited amount of active elements (about 512). As such, they are referred to as "sparse 2D array transducers". Although these transducers can be used for volumetric imaging, the major challenge lies in finding the optimal configuration of the active elements, i.e., which elements should be used in transmitting and receiving in order to obtain the optimal volumetric image quality. Several element distributions have been investigated and compared in numerous publications [5,6].

Although the 2D (sparse) phased-array transducer can avoid mechanical rotation or motion of the transducer, it does not solve the problem of acquisition rates as a line-per-line acquisition of a 3D volume of US data at a spatial resolution comparable to what is used in conventional 2D imaging would require about 1 s. This is unacceptable for cardiac applications.

The simplest solution to this problem is reducing the field-of-view and/or the spatial resolution of the volumetric data set. For example, lowering the opening angle from $90° \times 90°$ to $40° \times 40°$ and reducing the line density to 1 per 2 degrees instead of 1 per degree would result in a volume rate of about 10 Hz. Obviously, this increased temporal resolution then comes at the expense of spatial resolution and/or the field-of-view.

Parallel Beam Forming

As an alternative solution, parallel beam forming has been proposed [7]. In this technique, multiple image lines are constructed for each transmit pulse. As electronic beam steering implies that time delays have to be introduced between different elements in the array, the same is done during receive (i.e., for the received beam, see Sect. 3.4). Indeed, for a certain US transmit, all crystals in the array receive an echo RF signal. Before summing these signals – to obtain a single RF signal (and then after reconstruction a single image line) – they are time-delayed in order to steer and focus the received beam in the same direction as the transmitted beam. In this way, the transmit-receive (i.e., two-way) beam profile is optimal.

In parallel beam forming, the RF signals from the individual elements will be summed by introducing time delays to direct the receive beam a bit to the side of the original transmit beam. Using the same RF signals, this summation will also be done by time delaying for a receive beam oriented a bit to the other side (Fig. 3.12). As such, two image lines can be constructed for a single ultrasound transmit. This implies that an US image can be constructed twice as fast.

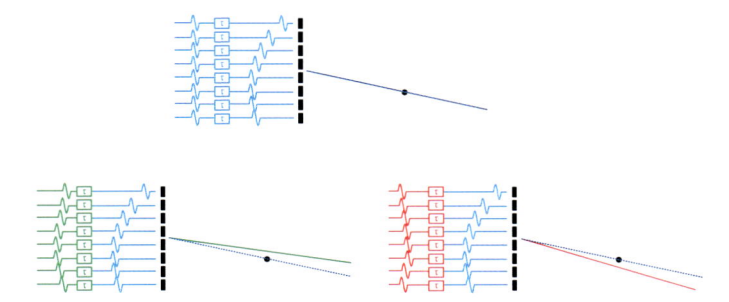

Fig. 3.12. *Conventional and parallel beam forming.* The reflections were originating from a reflector (*black dot*). In conventional beam forming (*top*) time delays are applied in order to focus the receive beam in the same direction as the transmit beam. In parallel beam forming (*bottom*) the same RF signals (*blue*) are time delayed in order to focus the receive beam slightly to the *left* and *right*. As such, they are not longer perfectly aligned after time delaying (*green and red*)

Parallel beam forming was initially introduced for conventional 2D imaging in order to increase the temporal resolution of these systems further. In this way, current high-end systems have default frame rates of approximately 70 Hz. The flip-side of the medal is that the beam former of such systems needs to be doubled. As this is an expensive part of an US device, this has a significant impact on its cost. The same principle can be used to have quadruple receive beams (by including 4 beam formers in the system) or more. As such, the first true volumetric ultrasound device, developed at Duke University, had 16 parallel receive lines which allowed real-time 3D scanning at acceptable volume rates [8, 9]. The first commercial real-time 3D ultrasound system using 2D phased array technology was available at the end of the 1990's and was produced by Volumetrics Medical Imaging (Durham, North Carolina). The system was based on technology developed in the ultrasound research group at Duke University and was later acquired by Philips Medical Systems.

Second Generation 2D Array Transducers

Although the design of the sparse 2D arrays has been optimized for image quality, intrinsically, image quality would be better when more elements of the 2D array could be used. Indeed, the more elements that are used for transmit/receive, the better the US beam characteristics for imaging. As such, a need remained to improve the 2D array transducer technology in order to allow more transmit/receive elements (and thus improve image quality).

An important step forward was made by Philips Medical Systems (Best, the Netherlands) by putting part of the beam former of the imaging system in the housing of the US transducer. In such a system, certain elements of the 2D transducer array form a sub-group of elements that are beam formed (i.e., time delayed, amplitude weighted and potentially filtered) both in transmit and in receive in the transducers handle. Subsequently, these pre-beam formed signals are transferred to the US system where they are combined (i.e., beam formed) into a single RF signal. The beam forming process thus becomes a two-step procedure in which signals from certain elements are combined inside the transducer handle while combining these pre-beam formed signals is done inside the US system. As such, the number of cables going to the transducer from the US scanner can be limited while many more elements in the transducer can effectively be used for imaging (Fig. 3.13). Clearly, optimizing the design of such a 2D transducer setup is not obvious as there are a lot of degrees of freedom (e.g., which elements to combine in groups, how to pre-beam form the groups of elements, how to beam form the remaining signals).

The pre-beam forming concept combined with parallel beam forming allows current state-of-the-art real-time volumetric scanners to produce narrow volumes (approx. $25° \times 90°$) at a frame rate of 25 Hz with acceptable image resolution. In order to increase the field-of-view to a full volume of $90° \times 90°$, data needs to be collected over multiple cardiac cycles (typically four) and merged into a single full-volume.

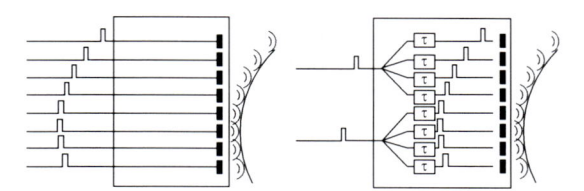

Fig. 3.13. *Pre-beam forming.* In the first generation of 2D array transducers each element was linked directly to the US scanner for beam forming (*left*). In the second generation of 2D arrays, part of the beam forming is done inside the transducer handle which reduces the number of cables going to the system significantly (*right*). This in turn allows to use more elements in the array for a given number of channels

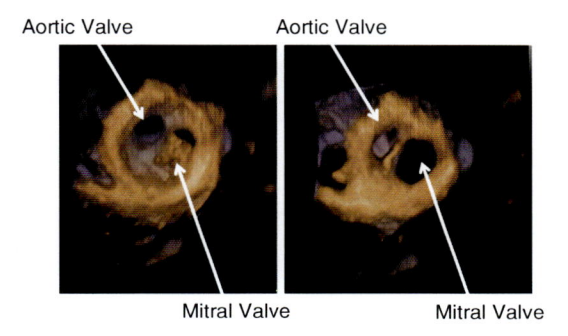

Fig. 3.14. *Volumetric ultrasound using state-of-the-art clinical equipment.* Looking from the apex of the heart towards the atria, one clearly sees both aortic and mitral valves throughout the cardiac cycle. *Left*: aortic valve opened; mitral valve closed; *Right*: aortic valve closed; mitral valve opened

Today, all major vendors of cardiac US equipment have a 3D system commercially available that allows full-volume imaging (i.e., 90° × 90°) at frame rates of about 25 Hz by gating over four to six cardiac cycles. True real-time volumetric scanning typically requires a reduction of the field-of-view however. An example of a volumetric data set acquired using this (gated) real-time 3D technology is given in Fig. 3.14.

3.5 Validation of 3D Cardiac Ultrasound Methodologies

The validation of 3D ultrasound systems for the assessment of cardiac chamber volumes can be done:

- *In-vitro*: Typically, latex balloons of known volume are scanned using 3D ultrasound (and sometimes using an alternative imaging technique as reference).
- *In-vivo*: Most often, another medical imaging modality is used as reference (e.g., Cardiac CT, Cardiac MRI).

- *Ex-vivo*: In the special case of heart transplantation, the pre-transplantation (myocardial) volume measured by 3D ultrasound in-vivo has been compared to post-transplantation volumes measurements [10].

Both in-vitro and in-vivo validations of the free-hand and real-time 3D approaches for all cardiac chambers (left and right ventricle and left and right atrium) have been presented in numerous publications, e.g., [11–16]. Although volumes measured by 3D ultrasound typically show a bias with respect to cardiac MRI as it systematically measures smaller (cavity) volumes, all studies show excellent corrections with this reference technique. The underestimation of LV cavity volume has been attributed to the fact that the spatial resolution of the 3D systems remains limited making it hard to distinguish trabeculae from myocardium. It can be expected, however, that this bias will gradually disappear with the further improvements of volumetric US systems.

3.6 Emerging Technologies

3.6.1 Transesophageal 3D Imaging

In Transesophageal Echocardiography (TEE), an US transducer is inserted into the oesophagus through the mouth of the patient in order to be able to approach the heart more closely. This in turn allows using US waves of higher frequency (as less penetration of the US wave is required) which improves image resolution. TEE has become a well-established clinical technique (Fig. 3.15).

Although ECG-gated TEE imaging does allow reconstructing 3D volumes [17], a potentially more favourable approach has more recently been introduced by Philips Healthcare (Best, the Netherlands). Indeed, Philips Healthcare was able to build a transesophageal 2D matrix array transducer

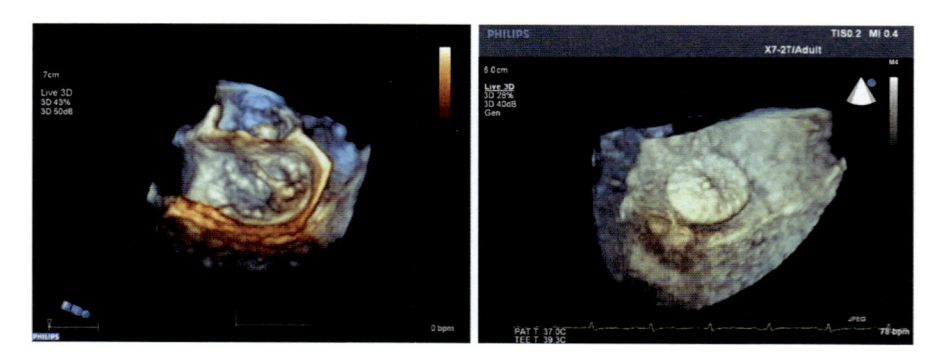

Fig. 3.15. *Transesophageal echocardiography.* Left: mitral valve; Right: device inserted into the heart in order to close an atrial septum defect. (Courtesy: Philips Healthcare)

that allows volumetric transesophageal imaging [18]. Given that the probe needs to be swallowed by a patient, its size has to be limited and additional safety precautions need to be taken which implies that building such a probe is a true technological achievement. Numerous articles published in the last two years show that this new imaging approach is very promising and seems to be particularly useful in the operation theatre for monitoring and guiding of interventions [19]. For some interventions, this new real-time volumetric imaging technique may replace other guiding techniques completely (most often X-ray fluoroscopy) as it does not bring radiation hazard to the patient and/or operator and is relatively cheap and easy to use.

3.6.2 True Real-Time Volumetric Imaging

Full volume (i.e., $90° \times 90°$) 3D imaging still requires merging data acquired over four to six heart beats. As such, this is not true real-time volumetric imaging yet and so-called "stitching artifact" (i.e., the registration errors between the sub-volumes) can occur and need to be accounted for in the volume reconstruction [20].

Very recently, Siemens Medical Solutions (Mountain View, CA) introduced their newest US system that combines 2D array technology (including transducer pre-beam forming) with a system embedding 64 parallel beam formers. In this way, their system is able to provide full volume data sets ($90° \times 90°$) in real-time at a frame rate of approximately 50 Hz. As such, this system allows true real-time volumetric imaging (Fig. 3.16). Another true real-time 3D system was also recently introduced by GE Health Care.

However, obtaining good image quality for such a highly parallelized system remains challenging. The above mentioned system thus remains to be validated both in-vitro and in-vivo.

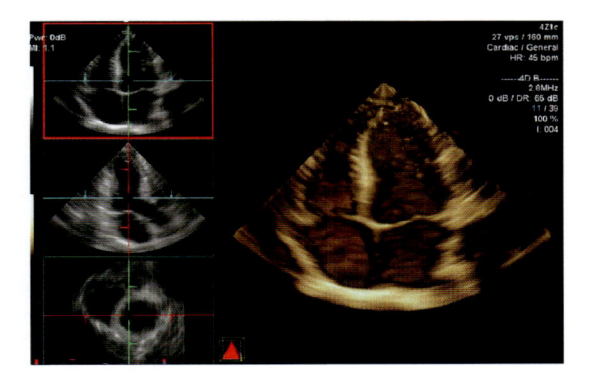

Fig. 3.16. *Real-time full volume imaging.* The new system architecture introduced by Siemens Medical Solutions allows true full volume imaging in real-time at acceptable frame rates (Courtesy: Shane Williams, Siemens Medical Solutions)

3.7 Remaining Challenges in 4D Cardiac Ultrasound

3.7.1 Resolution

Although spatial and temporal resolution of 3D ultrasound systems have steadily improved over the last few years by the introduction of parallel beam forming and pre-beam forming, both spatial and temporal resolution remain inferior to what is currently available in 2D ultrasound. The increased field-of-view of the volumetric systems thus comes at the expense of both spatial and temporal resolution of the US data set. As such, 3D systems have not yet replaced conventional 2D systems in clinical routine but are rather used in specialized centers as an add-on to 2D echocardiography.

Temporal Resolution

Temporal resolution of the US data set is mostly restricted by the speed of sound. Indeed, as an US wave propagates at finite speed (approximately 1,530 m/s) through soft tissue, a pulse-echo experiment for imaging up to 15 cm intrinsically takes about 200 µs. Per second, a maximum of about 5,000 pulse-echo measurements can thus be made; this limit is set by physics. These 5,000 measurements need to fill up space-time as much as possible. It thus remains a challenge to find the optimal balance between line density (i.e., spatial resolution), the temporal resolution (i.e., number of scans per second) and the field of view.

Spatial Resolution

Clearly, adding an additional spatial dimension in volumetric imaging will intrinsically result in a reduction in sampling in the other dimensions. Parallel beam forming has increased the amount of information available in four-dimensional (4D) space-time significantly. However, parallelization not only comes at an economical cost but also at the cost of image quality. As such, massive parallelization (i.e., reconstructing a full volume for a single transmit) may be theoretically possible but would come at the expense of image quality unless adapted reconstruction algorithms are developed. In this context, a complete different imaging paradigm for US image formation has been proposed, which would theoretically allow producing volume data at acceptable spatial resolution at a frame rate of 5 kHz [21].

Spatial resolution of the volumetric US data set is mostly determined by the number of 2D array elements that can effectively be used and by aperture size (i.e., the active area of the transducer). Indeed, both system characteristics are essential for proper steering and focusing of the US beam. For example, when using few transducer elements for an US transmit, not only the total amount of energy in the US wave will be reduced (and thus the signal-to-noise ratio of the resulting image will decrease) but also part of the US energy will "leak" into other directions (referred to as "side lobes" or "grating lobes"). These will reduce the SNR further and can result in significant image artifacts [22]. The technological improvements in wiring and cabling in combination with more advanced and optimized pre-beam forming techniques will play an important role in this context.

In order to allow for larger US transducers (to improve focusing and therefore image quality), the introduction of silicon-based US transducers is likely an important step forward as they allow building conformable arrays [23]. These transducer systems can – in theory – be positioned as a flexible path on top of the patient's chest in order to increase the active area of the transducer. Beam steering and focusing with these transducers becomes however non-trivial and requires specialized image reconstruction algorithms [24]. Nevertheless, one can anticipate that these conformal arrays will play an important role in future volumetric US systems.

3.7.2 Image Quality

The current approach towards increasing the field-of-view of the US system without compromising line density and/or temporal resolution dramatically is the implementation of parallel beam forming . Although parallel beam forming does indeed allow reconstructing multiple image lines for a single US transmit, it meets several drawbacks.

Amplitude Reduction and Inhomogeneity

The sensitivity of the US system is determined by its two-way beam profile, i.e., the product of the transmit and the receive beam characteristics (Fig. 3.17). In conventional 2D imaging, the transmit focus and receive focus are identical which results in a maximal narrowing of the US two-way beam profile and therefore an optimal image resolution (for a given transducer/system setup). However, for parallel beam forming the receive beams

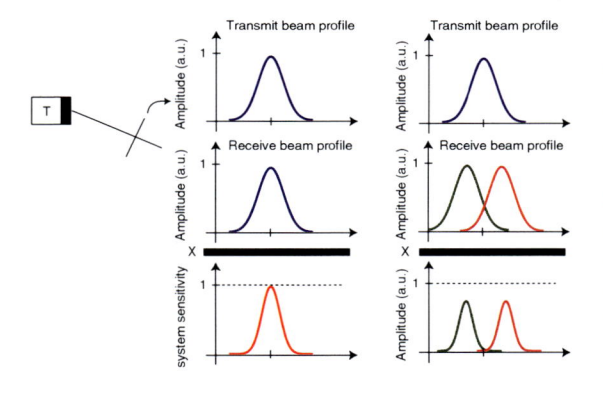

Fig. 3.17. *Two-way beam profile*. The US beam profile refers to its amplitude as a function of distance orthogonal to the image line (*left*). Due to parallel beam forming the sensitivity of the system is intrinsically reduced (*right*) resulting in lower SNR of the reconstructed images

will be positioned around the transmit beam focal point. As such, the amplitude of a reflected signal will be reduced (with respect to conventional imaging) which has a direct (negative) impact on the SNR of the reconstructed image. Moreover, when more than two lines are reconstructed in parallel (e.g., four), the sensitivity of the two inner beams and the two outer beams is different. After image reconstruction, this would result in two brighter and two darker lines respectively (i.e., strike artifacts) unless this amplitude difference is corrected for.

Although this correction can indeed be done, it requires an accurate knowledge of the transmit beam profile characteristics. Unfortunately, this beam profile not only depends on geometrical parameters of the transducer design but also on the properties of the tissue that is being imaged. As such, this is not a trivial task. In any case, the SNR of these image lines remains lower.

Beam Broadening

As a possible solution to the above mentioned amplitude inhomogeneity problem, the transmit beam can be broadened. Indeed, by broadening the transmit beam, the spatial variations in transmit beam amplitude are reduced. As a result, the two-way beam profiles become more homogeneous. This principle is used in current volumetric US scanners in order to reduce the inhomogeneity in sensitivity between image lines as much as possible. However, a broader transmit beam intrinsically results in a broader two-way beam profile and therefore in an imaging system with decreased spatial resolution.

Beam Distortion

A straight forward implementation of parallel beam forming would position the receive beam centrally around the transmit beam. However, this implies that the two-way beam profiles are "pulled" towards the transmit beam so that the effective angle between parallel beam formed lines and other ones is different. This effect is called beam warping and results in strike artifacts in the reconstructed images (Fig. 3.18). A simple theoretical solution to this problem is steering the receive beams further out so that the effective two-way beam profiles are regularly sampled. However, as the width of the US beam is spatially variant (and dependent on the medium in which the US wave is traveling), this may require dynamic beam steering (i.e., reconstructing curved receive beams rather than straight beams) [25]. Similarly, steering the receive beams away from the focus point will results in an asymmetric two-way beam profile with mirrored asymmetry for parallel beam formed lines. This effect is referred to as "beam skewing" and, again, results in typical image artifacts.

These problems show that a straight forward implementation of parallel beam forming intrinsically is associated with a reduction in image resolution and contrast. The higher the degree of parallelization, the more difficult the

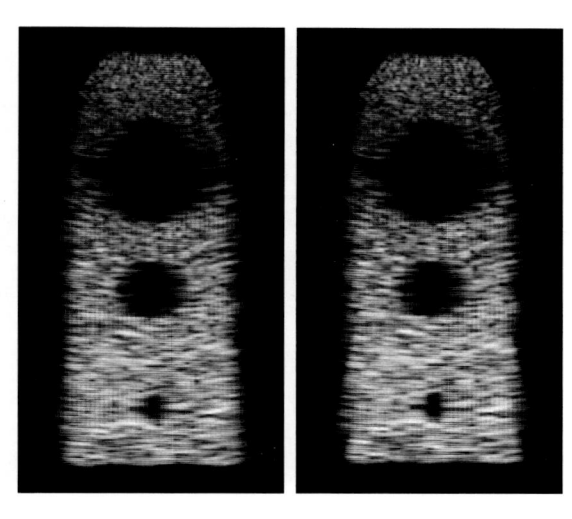

Fig. 3.18. *Beam warping.* Conventional 2D ultrasound image of an US phantom containing three cystic regions (*left*). In a straight forward implementation of parallel beam forming, beam warping results in strike artifacts (*right*)

associated image artifacts and the more ingenious solutions need to be developed. As such, further research is required in order to optimally use massive parallel beam forming for volumetric US imaging at preserved image quality and resolution.

3.7.3 Data Visualization and Interaction

A major difficulty with volumetric imaging (independent of the imaging modality used) is data visualization and interaction. Indeed, standard monitors remain two dimensional which implies that volume rendering techniques are essential for display. Unfortunately, some of these rendering techniques neutralize the volume information of the object making navigation and image interpretation difficult.

Although experiments have been carried out with stereoscopic vision of cardiac volumetric data through colored/polarized glasses or 3D screens [26], volume rendering using a depth-encoding color maps remains the current standard (Fig. 3.15). Similarly, holograms have been used to visualize the 3D cardiac data sets but only in a research setting [27].

As for data visualization, data interaction is typically not straight forward in a volume data set most importantly due to the lack of a good depth perception. Novel data visualization and interaction tools will thus become important for the further introduction of volumetric US in clinical cardiology (Fig. 3.19).

3.7.4 Segmentation/Automated Analysis

The visual presentation of the volumetric US data can give useful information about the heart anatomy. In particular, cardiac surgeons gain valuable

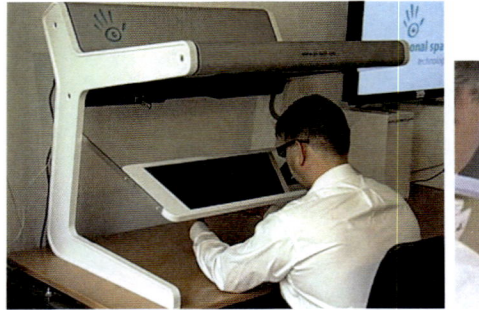

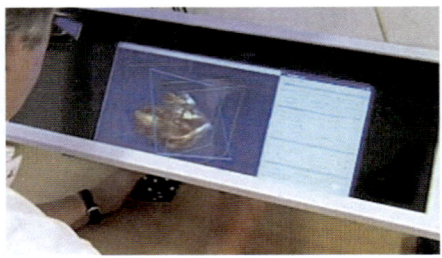

Fig. 3.19. *3D visualization.* New data interaction tools will be important for the further introduction of volumetric US imaging in clinical cardiology (Courtesy: Personal Space Technologies, Amsterdam)

information on valve leaflets and shape prior to surgery for valve reconstruction or replacement [19]. Nevertheless, quantification of chamber volumes and the myocardial volume (i.e., mass) is important for prognosis, diagnosis and treatment follow-up. Obviously, this requires segmentation of the relevant structures in the US volume data set.

To date, myocardial segmentation is most often done manually or involves at least a significant amount of manual user interaction. As such, analysis times can be significant making the clinical application of these methods less practical and decreasing reproducibility. Hereto, several approaches towards automated segmentation are developed [28, 29]. They will be important to move volumetric echocardiography to the clinical routine. Toshiba Medical Systems recently combined such an automated segmentation tool with 3D motion estimation algorithms which allows obtaining functional information of the heart with minimal user interaction [30].

References

1. Barratt DC, Davies AH, Hughes AD, et al. Optimisation and evaluation of an electromagnetic tracking device for high-accuracy three-dimensional ultrasound imaging of the carotid arteries. Ultrasound Med Biol. 2001;27(7):957–68.
2. Pandian NG, Roelandt J, Nanda NC, et al. Dynamic three-dimensional echocardiography: methods and clinical potential. Echocardiography. 1994;11(3):237–59.
3. Djoa KK, Jong N, Egmond FC, et al. A fast rotating scanning unit for real-time three-dimensional echo data acquisition. Ultrasound Med Biol. 2000;26(5): 863–9.
4. Bosch JG, Stralen M, Voormolen MM, et al. Novel spatiotemporal voxel interpolation with multibeat fusion for 3D echocardiography with irregular data distribution. Proc SPIE. 2006; 6147:OQ

5. Yen JT, Steinberg JP, Smith SW. Sparse 2-D array design for real time rectilinear volumetric imaging. IEEE Trans Ultrason Ferroelectr Freq Control. 2000;47(1):93–110.
6. Austeng A, Holm S. Sparse 2-D arrays for 3-D phased array imaging-design methods. IEEE Trans Ultrason Ferroelectr Freq Control. 2002;49(8):1073–86.
7. Shattuck DP, Weinshenker MD, Smith SW, et al. Explososcan: a parallel processing technique for high speed ultrasound imaging with linear phased arrays. J Acoust Soc Am. 1984;75(4):1273–82.
8. Smith SW, Pavy HR, Ramm OT. High-speed ultrasound volumetric imaging system. IEEE Trans Ultrason Ferroelectr Freq Control. 1991;38(2):100–8.
9. Ramm OT, Smith SW, Pavy HR. High-speed ultrasound volumetric imaging system. IEEE Trans Ultrason Ferroelectr Freq Control. 1991;38(2):109–15.
10. Gopal AS, Schnellbaecher MJ, Shen Z, et al. Freehand three-dimensional echocardiography for measurement of left ventricular mass: in vivo anatomic validation using explanted human hearts. J Am Coll Cardiol. 1997;30(3):802–10.
11. Gopal AS, Schnellbaecher MJ, Shen Z, et al. Freehand three-dimensional echocardiography for determination of left ventricular volume and mass in patients with abnormal ventricles: comparison with magnetic resonance imaging. J Am Soc Echocardiogr. 1997;10(8):853–61.
12. Rusk RA, Mori Y, Davies CH, et al. Comparison of ventricular volume and mass measurements from B- and C-scan images with the use of real-time 3-dimensional echocardiography: studies in an in vitro model. J Am Soc Echocardiogr. 2000;13(10):910–7.
13. Keller AM, Gopal AS, King DL. Left and right atrial volume by freehand three-dimensional echocardiography: in vivo validation using magnetic resonance imaging. Eur J Echocardiogr. 2000;1(1):55–65.
14. Lu X, Nadvoretskiy V, Bu L, et al. Accuracy and reproducibility of real-time three-dimensional echocardiography for assessment of right ventricular volumes and ejection fraction in children. J Am Soc Echocardiogr. 2008;21(1):84–9.
15. Sugeng L, Mor-Avi V, Weinert L, et al. Quantitative assessment of left ventricular size and function: side-by-side comparison of real-time three-dimensional echocardiography and computed tomography with magnetic resonance reference. Circulation. 2006;114(7):654–61.
16. Bosch AE, Robbers-Visser D, Krenning BJ, et al. Real-time transthoracic three-dimensional echocardiographic assessment of left ventricular volume and ejection fraction in congenital heart disease. J Am Soc Echocardiogr. 2006;19(1):1–6.
17. Handke M, Jahnke C, Heinrichs G, et al. New three-dimensional echocardiographic system using digital radiofrequency data–visualization and quantitative analysis of aortic valve dynamics with high resolution: methods, feasibility, and initial clinical experience. Circulation. 2003;107(23):2876–9.
18. Pothineni KR, Inamdar V, Miller AP, et al. Initial experience with live/real time three-dimensional transesophageal echocardiography. Echocardiography. 2007;24(10):1099–104.
19. Scohy TV, Cate FJT, Lecomte PV, et al. Usefulness of intraoperative real-time 3D transesophageal echocardiography in cardiac surgery. J Card Surg. 2008;23(6):784–6.
20. Brekke S, Rabben SI, Stoylen A, et al. Volume stitching in three-dimensional echocardiography: distortion analysis and extension to real time. Ultrasound Med Biol. 2007;33(5):782–96.

21. Lu JY. Designing limited diffraction beams. IEEE Trans Ultrason Ferroelectr Freq Control. 1997;44(1):181–93.
22. Jong N, Souquet J, Faber G, et al. Transducers in medical ultrasound: part two: vibration modes, matching layers and grating lobes. Ultrasonics. 1985;23(4):176–82.
23. Singh RS, Culjat MO, Vampola SP, et al. Simulation, fabrication, and characterization of a novel flexible, conformal ultrasound transducer array. Proc IEEE Ultrasonics Symp. 2007; p. 1824–7.
24. He Z, Ma Y. Optimization of transmitting beam patterns of a conformal transducer array. J Acoust Soc Am. 2008;123(5):2563–9.
25. Bjastad TG. High frame rate ultrasound imaging using parallel beam forming. PhD thesis, Norwegian Institute of Science and Technology. 2009.
26. Littlefield RJ, Heiland RW, Macedonia CR. Virtual reality volumetric display techniques for three-dimensional medical ultrasound. Stud Health Technol Inform. 1996;29:498–510.
27. Bol GR, Koning AH, Scohy TV, et al. Virtual reality 3D echocardiography in the assessment of tricuspid valve function after surgical closure of ventricular septal defect. Cardiovasc Ultrasound. 2007;5:8.
28. Angelini ED, Homma S, Pearson G, et al. Segmentation of real-time three-dimensional ultrasound for quantification of ventricular function: a clinical study on right and left ventricles. Ultrasound Med Biol. 2005;31(9):1143–58.
29. Hansegard J, Urheim S, Lunde K, et al. Semi-automated quantification of left ventricular volumes and ejection fraction by real-time three-dimensional echocardiography. Cardiovasc Ultrasound. 2009;7:18.
30. Nesser HJ, Mor-Avi V, Gorissen W, et al. Quantification of left ventricular volumes using three-dimensional echocardiographic speckle tracking: comparison with MRI. Eur Heart J. 2009;30(13):1565–73.

Part II

Image Enhancement

4

Morphological Image Processing Applied in Biomedicine

Roberto A. Lotufo, Leticia Rittner, Romaric Audigier, Rubens C. Machado, and André V. Saúde

Summary. This chapter presents the main concepts of morphological image processing. Mathematical morphology has application in diverse areas of image processing such as filtering, segmentation and pattern recognition, applied both to binary and gray-scale images. Section 4.2 addresses the basic binary morphological operations: erosion, dilation, opening and closing. We also present applications of the primary operators, paying particular attention to morphological reconstruction because of its importance and since it is still not widely known. In Sect. 4.3, the same concepts are extended to gray-scale images. Section 4.4 is devoted to watershed-based segmentation. There are many variants of the watershed transform. We introduce the watershed principles with real-world applications. The key to successful segmentation is the design of the marker to eliminate the over-segmentation problem. Finally, Sect. 4.5 presents the multi-scale watershed to segment brain structures from diffusion tensor imaging, a relatively recent imaging modality that is based on magnetic resonance.

4.1 Introduction

There are several applications of Morphological Processing (MP) in diverse areas of biomedical image processing. Noise reduction, smoothing, and other types of filtering, segmentation, classification, and pattern recognition are applied to both binary and gray-scale images. As one of the advantages of MP, it is well suited for discrete image processing and its operators can be implemented in digital computers with complete fidelity to their mathematical definitions. Another advantage of MP is its inherent building block concept, where any operator can be created by the composition of a few primitive operators.

This text introduces and highlights the most used concepts applied to real situations in biomedical imaging, with explanations based on a morphological intuition of the reader, whenever possible. However, although the lack of full details, the text is as coherent as possible to the mathematical theory and the motivated reader is invited to investigate the many texts, where

T.M. Deserno (ed.), *Biomedical Image Processing*, Biological and Medical Physics,
Biomedical Engineering, DOI: 10.1007/978-3-642-15816-2_4,
© Springer-Verlag Berlin Heidelberg 2011

these details and formalisms are treated in depth [1, 2]. In this chapter, only
the main mathematical equations are given. Source codes of implementations
of these equations are available on the toolbox IA870 in the Adessowiki[1]
project [3]. Other text with example applications from image microscopy using
this toolbox is also available [4].

Morphological image processing is based on probing an image with a
structuring element and either filtering or quantifying the image according
to the manner in which the structuring element fits (or does not fit) within
the image. By marking the locations at which the structuring element fits
within the image, we derive structural information concerning the image. This
information depends on both the size and shape of the structuring element.
Although this concept is rather simple, the majority of operations presented
in this chapter is based on it: erosion, dilation, opening, closing, morphological
reconstruction, etc., applied both for binary and gray-scale images.

In this chapter, only symmetric structuring elements will be used. When
the structuring element is asymmetric, care must be taken as some properties
are valid for a reflected structuring element. Four structuring elements types
will be used in the illustrations and demonstrations throughout this chapter:

- *cross*: the elementary cross is a 3×3 structuring element with the central pixel and its four direct neighbors (4-neighborhood).
- *box*: the elementary box is a 3×3 structuring element with all nine pixels, the central and its eight neighbors (8-neighborhood).
- *disk*: the disk of a given radius is a structuring element with all pixels that are inside the given radius.
- *line*: the linear structuring element can be composed for a given length and orientation.

4.2 Binary Morphology

4.2.1 Erosion and Dilation

The basic fitting operation of mathematical morphology is erosion of an image
by a structuring element. The erosion is computed by scanning the image with
the structuring element. When the structuring element fits completely inside
the image, the scanning position is marked. The erosion consists of all scanning
locations where the structuring element fits inside the image. The erosion of
set A by set B is denoted by $A \ominus B$ and defined by

$$A \ominus B = \{x : B_x \subset A\}, \tag{4.1}$$

where $\subset$ denotes the subset relation and $B_x = \{b + x : b \in B\}$ the translation
of set B by a point x.

---

[1] http://www.adessowiki.org

4 Morphological Image Processing Applied in Biomedicine

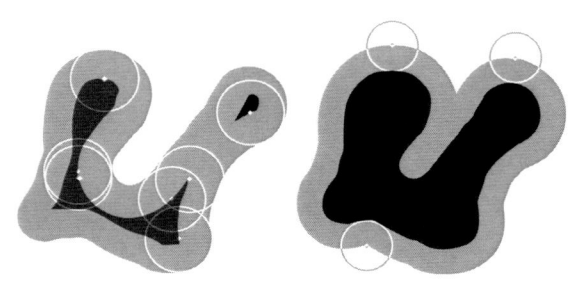

Fig. 4.1. *Erosion and dilation. Left*: input image in black and gray and erosion in black (region where the center of the robot can move); *Right*: input image in black and dilation in black and gray

A binary image is formed by foreground and background pixels. In morphology, for every operator that changes the foreground, there is a dual operator that changes the background. The dual operator for the erosion is the dilation. Since dilation involves a fitting into the complement of an image, it represents a filtering on the outside, whereas erosion represents a filtering on the inside (Fig. 4.1). For intuitive understanding, the structuring element can be seen as a moving robot.

Formally, the dilation of set A by B, denoted by $A \oplus B$, is defined by

$$A \oplus B = (A^c \ominus \widetilde{B})^c \qquad (4.2)$$

where A^c denotes the set-theoretic complement of A and $\widetilde{B} = \{-b : b \in B\}$ is the reflection of B, i.e., a 180°-rotation of B about the origin. Foreground is generally associated to white color while background is associated to black color. But note that in impression works, the inverse convention is sometimes used.

Another alternative equivalent way to compute the dilation is by "stamping" the structuring element on the location given by every foreground pixel in the image. Formally, the dilation can also be defined by

$$A \oplus B = \bigcup_{a \in A} B_a. \qquad (4.3)$$

Dilation has the expected expanding effect, filling in small intrusions into the image (Fig. 4.1, right) and erosion has a shrinking effect, eliminating small extrusions (Fig. 4.1, left).

As dilation by a disk expands an image and erosion by a disk shrinks an image, both can be used for finding boundaries for binary images. The three possibilities are:

1. *External boundary*: dilation minus the image.
2. *Internal boundary*: the image minus the erosion.
3. *Combined boundary*: dilation minus erosion.

The latter straddles the actual Euclidean boundary and is known as the morphological gradient, which is often used as a practical way of displaying the boundary of the segmented objects.

4.2.2 Opening and Closing

Besides the two primary operations of erosion and dilation, there are two important operations that play key roles in morphological image processing, these being opening and its dual, closing.

The opening of an image A by a structuring element B, denoted by $A \circ B$, is the union of all the structuring elements that fit inside the image (Fig. 4.2, left):

$$A \circ B = \bigcup \{B_x : B_x \subset A\} \quad \text{or} \tag{4.4}$$

$$A \circ B = (A \ominus B) \oplus B. \tag{4.5}$$

It can also be defined as an erosion followed by a dilation (4.5) and has its dual version called closing (Fig. 4.2, right), which is defined by

$$A \bullet B = (A^c \circ \widetilde{B})^c \quad \text{or} \tag{4.6}$$

$$A \bullet B = (A \oplus B) \ominus B. \tag{4.7}$$

Note that whereas the position of the origin relative to the structuring element has a role in both erosion and dilation, it plays no role in opening and closing. However, opening and closing have two important properties [5]:

1. Once an image has been opened (closed), successive openings (closings) using the same structuring element produce no further effects.
2. An opened image is contained in the original image which, in turn, is contained in the closed image (Fig. 4.2).

As a consequence of this property, we could consider the subtraction of the opening from the input image, called opening top-hat, and the subtraction of the image from its closing, called closing top-hat, respectively, defined by

$$A \hat{\circ} B = A - (A \circ B) \quad \text{and} \tag{4.8}$$

$$A \hat{\bullet} B = (A \bullet B) - A. \tag{4.9}$$

Fig. 4.2. *Opening and closing. Left*: input image in black and gray and opening in black (region where the robot can move); *Right*: input image in black and closing in black and gray

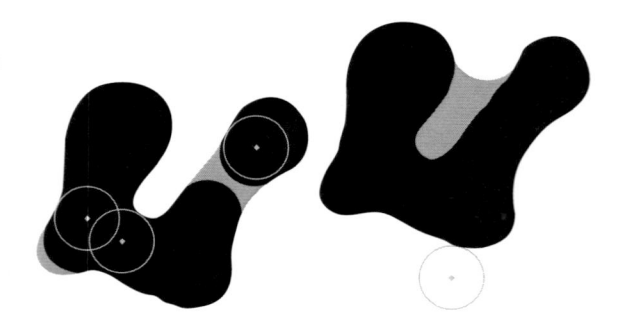

Opening top-hat and closing top-hat correspond to the gray parts of Fig. 4.2 left and right, respectively.

As a filter, opening can clean the boundary of an object by eliminating small extrusions; however, it does this in a much finer manner than erosion, the net effect being that the opened image is a much better replica of the original than the eroded image (compare left parts of Figs. 4.2 and 4.1). Analogous remarks apply to the closing, the difference being the filling of small intrusions (compare right parts of Figs. 4.2 and 4.1).

When there is both union and subtractive noise, one strategy is to open to eliminate union noise in the background and then close to fill subtractive noise in the foreground. The open-close strategy fails when large noise components need to be eliminated but a direct attempt to do so will destroy too much of the original image. In this case, one strategy is to employ an Alternating Sequential Filter (ASF). Open-close (or close-open) filters are performed iteratively, beginning with a very small structuring element and then proceeding with ever-increasing structuring elements.

The close-open filter is given by

$$ASF^n_{oc,B}(S) = (((((S \bullet B) \circ B) \bullet 2B) \circ 2B) \ldots \bullet nB) \circ nB \quad (4.10)$$

and the open-close filter by

$$ASF^n_{co,B}(S) = (((((S \circ B) \bullet B) \circ 2B) \bullet 2B) \ldots \circ nB) \bullet nB, \quad (4.11)$$

where $nB = B + B + \ldots + B$ (n times).

4.2.3 Morphological Reconstruction from Markers

One of the most important operations in morphological image processing is reconstruction from markers, the basic idea being to mark some image components and then reconstruct that portion of the image consisting of the marked components.

Given a neighborhood relationship, a region (collection of pixels) is said to be connected if any two pixels in the region can be linked by a sequence of neighbor pixels also in the region. 4-neighborhood and 8-neighborhood are usual neighborhoods that include vertically and horizontally adjacent pixels and, only for the latter one, diagonally adjacent pixels.

Every binary image A can be expressed as the union of connected regions. If each of these regions is maximally connected, which means that it is not a proper subset of a larger connected region within the image, then the regions are called connected components of the image. The union of all connected components C_k recovers the input image A and the intersection of any two connected components is empty.

To find all the connected components of an image, one can iteratively find any pixel of the image, use it to reconstruct its connected component, remove the component from the image, and iteratively repeat the same extraction

Fig. 4.3. *Reconstruction from markers. Left*: input image, *Middle*: marker image; *Right*: reconstructed image

until no more pixels are found in the image. This operation is called labeling (cf. panel (c) in Fig. 4.5). The labeling decomposes an image into its connected components. The result of the labeling is usually stored in a numeric image with each pixel value associated to its connected component number.

The morphological reconstruction of an image A from a marker M (a subset of A) is denoted by $A \triangle M$ and defined as the union of all connected components of image A that intersect marker M. This filter is also called component filter:

$$A \triangle M = \bigcup \{C_k : C_k \cap M \neq \emptyset\}. \tag{4.12}$$

In addition to the input image and the marker, the reconstruction operation also requires a connectivity. The marker informs which component of the input image will be extracted, and the connectivity can be specified, in some software packages, by a structuring element, usually the elementary cross for 4-neighborhood or the elementary box to specify 8-neighborhood.

An example of reconstruction from markers, based on 8-connectivity, is shown in Fig. 4.3. The input image is a collection of grains. The markers are made of a central vertical line intersecting the grains. The reconstruction from the markers extracts the three central components from the original image.

There are typically three ways to design the marker placement for the component filter:

1. A-priori selection,
2. Selection from the opening, or
3. Selection by means of some rather complex operation.

The edge-off operation, particularly useful to remove objects touching the image frame, combines reconstruction and top-hat concepts. The objects touching the frame are selected by reconstructing the image from its frame as an a priori marker. The objects not connected to the image frame are selected by subtracting the input image from the reconstructed image.

4.2.4 Reconstruction from Opening

With marker selection by opening, the marker is found by opening the input image by a structuring element. The result of the reconstruction detects all connected components where the structuring element fits.

Using the same mechanism of the reconstruction from opening to detect objects with particular geometric features, more complex techniques can be designed to find the markers from combined operators. At the last step, the reconstruction reveals the objects that exhibit those features.

The following biomedical application (Fig. 4.4) detects overlapping chromosomes. To identify overlapping chromosomes (Fig. 4.4, panel (b)) only the shapes (connected components) are chosen that all the four linear structuring elements can fit. This is achieved by intersecting the four reconstructions from opening using four linear structuring elements: vertical, horizontal, 45°, and −45°, as visualized in Fig. 4.4 on panels (c), (d), (e), and (f), respectively.

The top-hat concept can be applied to reconstruction by opening producing the reconstruction from opening top-hat (i.e., the image minus its reconstruction). In this case, the operator reveals the objects that do not exhibit a fitting criterion. For instance, to detect thin objects, one can use a disk of diameter larger than the thickest of the thin objects.

Another common criterion for selection of connected component is its area. This is achieved by the area opening which removes all connected component C_i with area less than a specified value α:

$$A \circ (\alpha)_E = \bigcup \{C_i, \ \mathrm{area}(C_i) \geq \alpha\}. \tag{4.13}$$

The next demonstration targets cytogenetic images of meta-phase human cells. This is a classification application of area opening. The task is to preprocess the image by segmenting the chromosomes from the nuclei, stain debris and the background. Figure 4.5 shows the input image (a), the thresholded (b), the labeling (c) identifying the connected components, and the result (d) with the components classified by area. The components with area less than

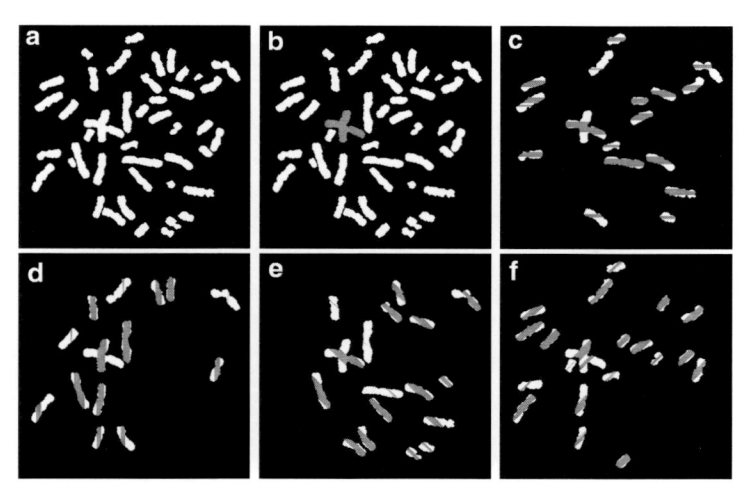

Fig. 4.4. *Detecting overlapping chromosomes.* (**a**) Input image; (**b**) intersection (*in gray*) of four reconstruction from openings; (**c**) opening (*in gray*) by horizontal line and its reconstruction; (**d**) opening (*in gray*) by vertical line and its reconstruction; (**e**) opening (*in gray*) by 45°-line and its reconstruction; and (**f**) opening (*in gray*) by −45°-line and its reconstruction

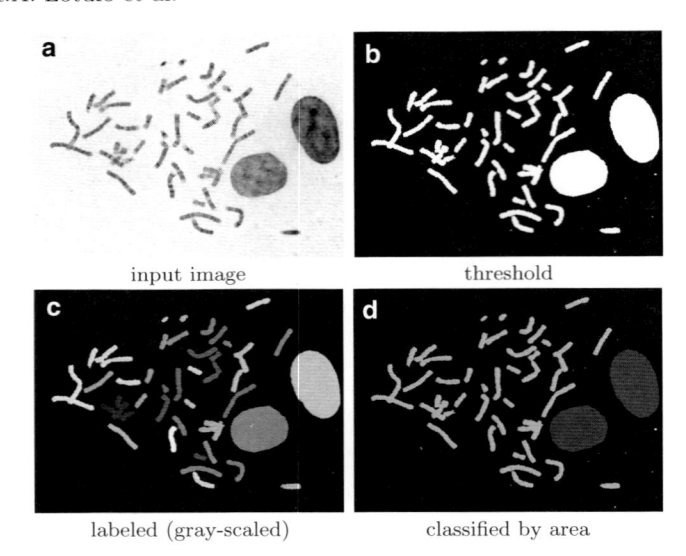

Fig. 4.5. *Chromosome spreads and area opening.* Residues are coded in white (area less than 100), chromosomes in light gray (area between 100 and 10,000), and nuclei in dark gray (area larger than 10,000)

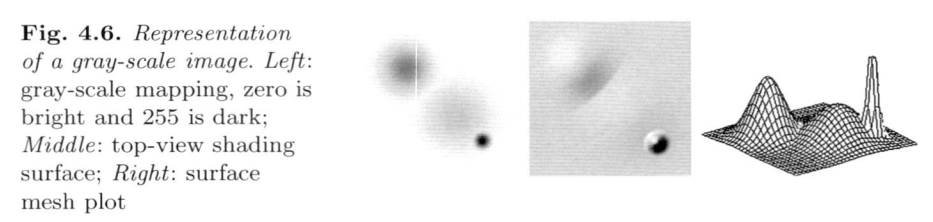

Fig. 4.6. *Representation of a gray-scale image. Left*: gray-scale mapping, zero is bright and 255 is dark; *Middle*: top-view shading surface; *Right*: surface mesh plot

100 pixels are background noise, the ones with area larger than 10,000 pixels are nuclei (dark gray) and the rest are the chromosomes (light gray).

So as not to be restricted to openings, analogous dual concepts can be developed to form sup-reconstruction from closing, sup-reconstruction from closing top-hat and area closing.

4.3 Gray-Scale Operations

It is useful to look at a gray-scale image as a surface. Figure 4.6 shows a gray-scale image made of three Gaussian-shape peaks of different heights and variances. The image is depicted in three different graphical representations: (a) the inverted print, where the pixel values are mapped in a gray scale: low values are bright and high values are dark gray tones; (b) a top-view shading surface; and (c) a mesh plot of the surface.

A gray-scale image can also be seen as the cardboard landscape model, i.e., a stack of flat pieces of cardboard. Thus, the threshold decomposition of a gray-scale image f is the collection of all the threshold sets $X_t(f)$ obtained at each gray level t:

$$X_t(f) = \{z : f(z) \geq t\}. \tag{4.14}$$

The image can be characterized uniquely by its threshold decomposition collection and can be recovered from its threshold sets by stack reconstruction:

$$f(x) = \max\{t : x \in X_t(f)\}. \tag{4.15}$$

In all gray-scale operations presented hereafter, we will use flat structuring elements, i.e., structuring elements that have no gray-scale variation, the same used in the binary case. Although they are the same structuring elements, we will use the term flat structuring elements not to confuse with their gray-scale versions. This restriction has many simplifications in the definition, characterization and use of the gray-scale operators as an extension from the binary operators. Care must be taken however, when the reader uses a gray-scale structuring element, as the erosion (dilation) is not a moving minimum (moving maximum) filter, the threshold decomposition property does not hold for the primitive operators nor for gray-scale morphological reconstruction. Moreover, as we said before, only symmetric structuring element will be used.

4.3.1 Erosion and Dilation

Gray-scale erosion (dilation) of an image f by a flat structuring element D is equivalent to a moving minimum (moving maximum) filter over the window defined by the structuring element. Thus, erosion $f \ominus D$ and dilation $f \oplus D$ in this case are simply special cases of order-statistic filters:

$$(f \ominus D)(x) = \min\{f(z) : z \in D_x\} \text{ and} \tag{4.16}$$
$$(f \oplus D)(x) = \max\{f(z) : z \in D_x\}. \tag{4.17}$$

An example of gray-scale erosion by a disk on a gray-scale image is shown in Fig. 4.7. The two images on the left, input and eroded, are represented in gray-scale shades and the two on the right are the same images represented by their top-view surfaces. Note how well the term "erosion" applies to this illustration. The eroded surface appears as being created by a pantograph engraving machine equipped with a flat disk milling cutter. The pantograph is guided to follow the original surface while shaping the eroded surface using the flat disk milling cutter.

The geometric intuition for erosion is the following: slide the structuring element along beneath the surface and at each point record the highest altitude the location of the structuring element can be translated while fitting beneath the surface.

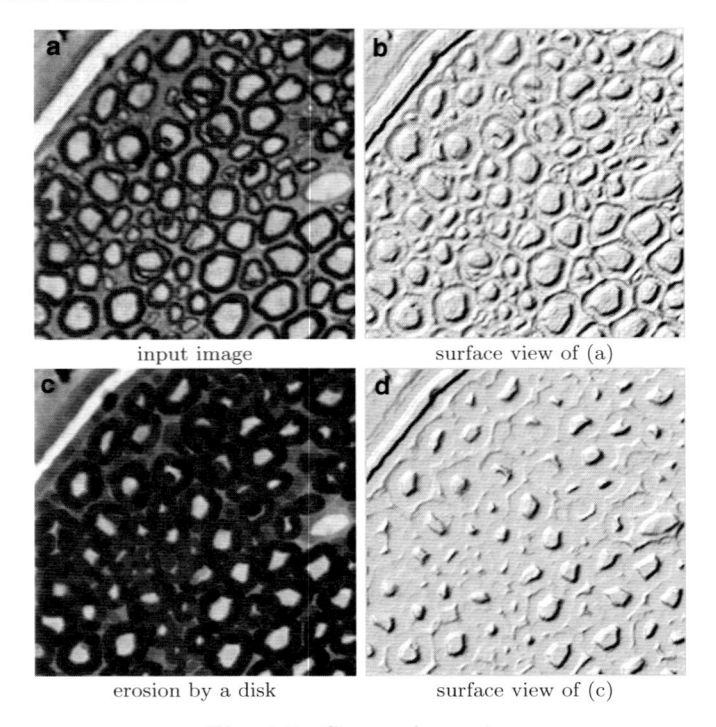

a (input image)	b (surface view of (a))
c (erosion by a disk)	d (surface view of (c))

Fig. 4.7. *Gray-scale erosion*

Alternatively, one can simply compute the erosion (dilation) of a gray-scale image by computing the stack decomposition of the image, applying binary erosion (dilation) on the threshold sets, and finally stack reconstruction.

Figure 4.8 illustrates the discrete gray-scale erosion by means of threshold decomposition. At the right of the gray-scale images (original and eroded), there are three threshold sets at gray levels 80, 120, and 180. Note that the binary images shown in (f), (g), and (h) are eroded versions of the binary images shown in (b), (c), and (d).

Observe that the morphological gradient, described for binary pictures, is directly extensible to gray-scale with gray-scale erosions and dilations (dilation minus erosion). At each point, the morphological gradient yields the difference between the maximum and minimum values over the neighborhood at the point determined by the flat structuring element.

4.3.2 Opening and Closing

As an extension of the binary case, gray-scale opening (closing) can be simply achieved by stack decomposition, binary opening (closing), and stack reconstruction.

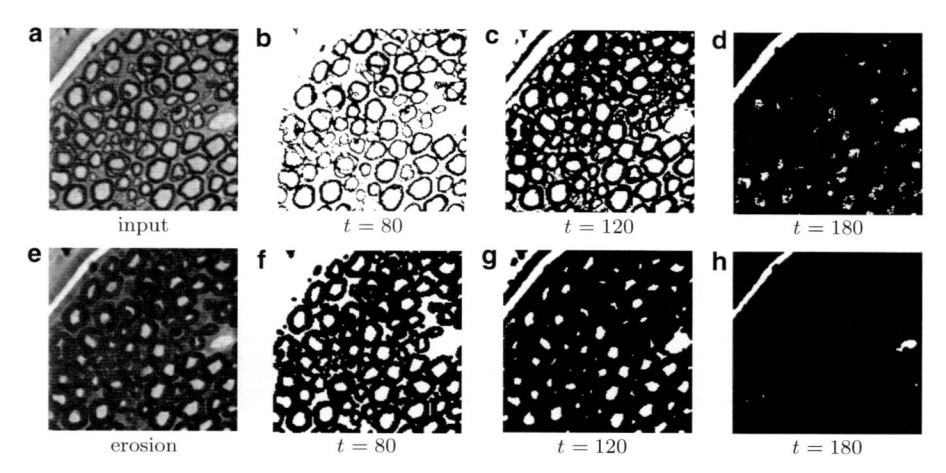

Fig. 4.8. *Gray-scale erosion by threshold decomposition.* The threshold sets are indicated by the threshold t

The geometric intuition for opening is the following: slide the structuring element along beneath the surface and at each point record the highest altitude the structuring element can be translated while fitting beneath the surface. The position of the origin relative to the structuring element is irrelevant. Note the slight difference between the opening and the erosion: while in the opening the highest altitude is recorded for all points of the structuring element, in the erosion, only the location of the structuring element is recorded.

Geometric intuition regarding closing can be obtained from the duality relation. Whereas opening filters from below the surface, closing filters from above; indeed, by duality, closing is an opening of the negated image.

The gray-scale opening and closing have the same properties of their binary equivalents [6]. The top-hat concept is also valid: gray-scale opening top-hat is given by the subtraction of the opened image from the input image, and the gray-scale closing top-hat is the subtraction of the image from its closing.

Open top-hat is very useful as a preprocessing step to correct uneven illumination before applying a threshold, thereby acting as an adaptive thresholding technique. The following application illustrates this.

Figure 4.9 shows the open top-hat transform applied to a Fluorescent In-Situ Hybridization (FISH) image. FISH is a key technique for molecular diagnosis in which labeled hybridizing agents (such as Deoxyribonucleic Acid (DNA) or Ribonucleic Acid (RNA) probes) are exposed to intact tissue sections. The probe hybridizes to a defined target nucleotide sequence of DNA in the cell. Each chromosome containing the target DNA sequence will produce a fluorescent signal (spot) in every cell when the specimen is illuminated with suitable excitation. Hence, FISH is an excellent method for detection of

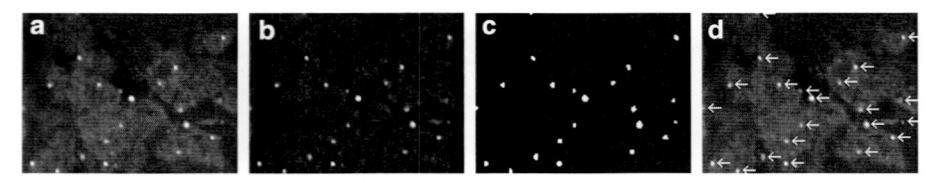

Fig. 4.9. *Gray-scale open top-hat.* (**a**) Input image; (**b**) opening top-hat by a disk of radius 4; (**c**) thresholded area open (by 2 pixels); and (**d**) dilation of centroids by arrow, for illustration, overlaid on original

input image   single close-open   3-stage ASF

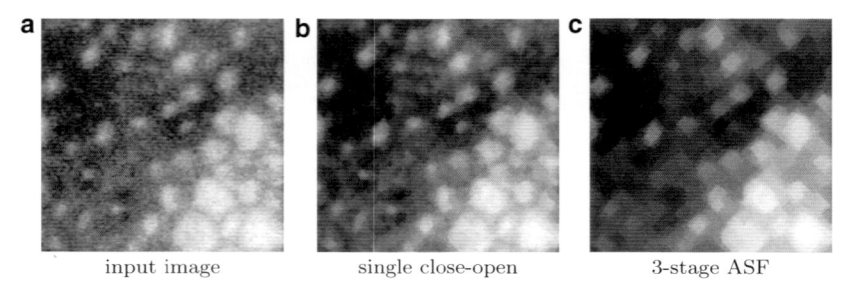

Fig. 4.10. *Gray-scale alternating sequential filtering (ASF)*

gene copy number alterations in cancer and other diseases, and the task of image processing is automatic detection of these spots. Owing to noise, the top-hat methodology typically yields very small spots in the thresholded top-hat image. These can be eliminated by area open. So, the image is filtered by a gray-scale area opening of two. In Fig. 4.9d, the arrows were overlaid automatically by a dilation of the centroids of the detected spots by an arrow-shape structuring element, with the origin slightly translated from the arrow tip so as not to disturb the visualization of the original spots.

Gray-scale opening can be employed to filter noise spikes lying above the signal and the closing can be used to filter noise spikes beneath the image. Typically noise is mixed, there being noise spikes both above and below the signal. So long as these noise spikes are sufficiently separated, they can be suppressed by application of either a composition of opening and closing or of closing and opening. Selection of an appropriately sized structuring element is crucial. In addition, if there is a mixture of unevenly spatially sized noise spikes and they are not sufficiently dispersed, one can employ an ASF, which is a composition of alternating openings and closings with increasingly wide structuring elements (Fig. 4.10). A single stage close-open filter results from closing followed by opening using a 3×3 diamond structuring element. For the second stage, another close-open is concatenated using a 5×5 diamond structuring element. In Fig. 4.10c, a three stage ASF was applied with the last stage being processed by a 7×7 diamond structuring element.

4.3.3 Component Filters and Morphological Reconstruction

The concept of connected component filter built from morphological reconstruction from markers, reconstruction from openings and area open introduced in the binary morphology section can be extended for gray-scale. These binary operators can be used to construct a correspondent gray-scale operator by using the notion of threshold decomposition. The level component in a gray-scale image is defined as a connected set of the pixels in a threshold set of the image at a given level. When modeling the gray-scale image by a cardboard landscape, the cardboard is first cut into the shape of each iso line (i.e., lines of same height), then all pieces are stacked up to form the topography. Each cardboard piece is a connected component of a level set of the gray-scale image. A gray-scale component filter is an operator that selects only a few level components (cardboard pieces) in such a way that the stack reconstruction is not disturbed, i.e., a level component is removed only if all the level components above it are also removed.

Since the notion of connected components is intrinsically related to this section, it is wise to recall that the definition of all component filters require the specification of a connectivity, which is normally the 4- or 8-neighborhood. One important property of a component filter is that it never introduces a false edge, so it is part of the so-called family of edge-preserving smoothing filters.

Morphological reconstruction is one of the most used tools to build component filters [7–9]. As with binary reconstruction, gray-scale reconstruction proceeds from markers. The morphological reconstruction of an image from a marker can be obtained by stack decomposition of the image and the marker, binary reconstructions at each gray-level, and stack reconstruction of the results. This can be intuitively explained by using the cardboard landscape model of the image. Imagine that the cardboard pieces are stacked but not glued. The markers are seen as needles that pierce the model from bottom to top. If one shakes the model while holding the needles, every cardboard pieces not pierced by the needles will be removed. The remaining cardboards constitute a new model that corresponds to the gray-scale morphological reconstruction, also called Inf-reconstruction. By duality, the sup-reconstruction works in the same manner but on the negated image. Observe that the marker can be a gray-scale image, the pixel intensity corresponding to the height the needles reach when they pierce the model.

Gray-scale area opening is another component filter [10]. It is defined analogously to the binary case. The size α area opening of a gray-scale image can be modeled as a stack filter in which at each level, only binary connected components containing at least α pixels are passed: it removes all cardboard pieces whose area is less than α.

An important class of component filters is composed of those generated from alternating reconstructions from openings and closings. They are called Alternating Sequential Component Filter (ASCF).

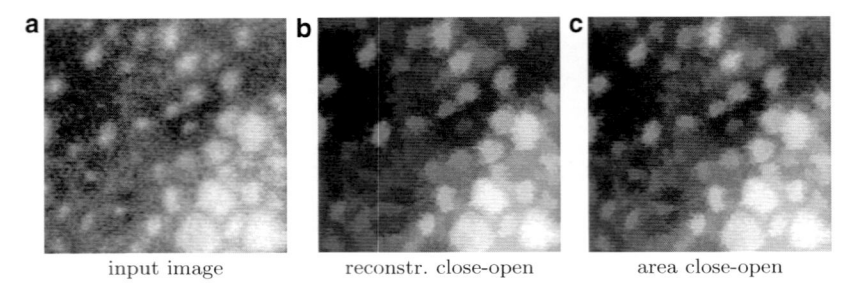

a	b	c
input image	reconstr. close-open	area close-open

Fig. 4.11. *Gray-scale alternating sequential component filtering (ASCF).* The reconstructive and area close-open was performed with a 3 × 3 diamond of stage 3 and an area parameter of 30 pixels, respectively

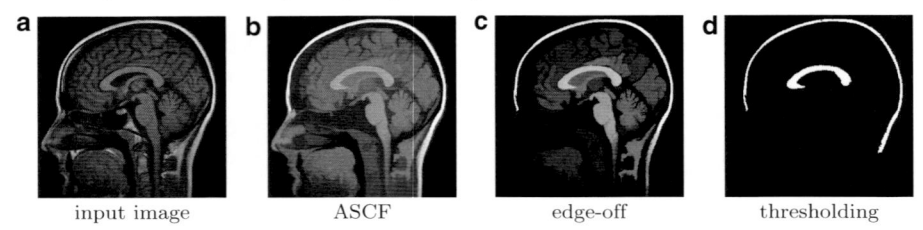

a	b	c	d
input image	ASCF	edge-off	thresholding

Fig. 4.12. *Segmenting the corpus callosum with gray-scale component filters.* In panel (**b**), an area close-open ASCF is applied

Figure 4.11 shows two examples of gray-scale ASCF using the same input image as in Fig. 4.10. A three stage close-open filter is performed by sup-reconstruction from closing followed by a reconstruction from opening using a 3 × 3, 5 × 5, and 7 × 7 diamond structuring element in the first, second and last stage, respectively. An ASCF area close-open with area parameter of 30 pixels yields similar effects (Fig. 4.10c). Note the high fidelity of the edges maintained by component filters.

The gray-scale edge-off operator can be easily derived from the binary case and it is very useful in many situations. As in the binary case, the edge-off is the top-hat of the reconstruction from a marker placed at the image frame (case where the marker is placed a priori). In the cardboard landscape model (threshold decomposition), all the cardboard pieces that touch the image frame are removed, leaving only the cardboard pieces that form domes inside the image.

The following application illustrates the use of area close-open ASCF as a preprocessing filter followed by the edge-off operator to segment the corpus callosum from an Magnetic Resonance Imaging (MRI) of the brain (Fig. 4.12). The area close-open ASCF is used with area parameter 1,500 pixels, first filling cardboard holes of less than this size, and removing cardboards with area less than 1,500 pixels afterwards. After applying the edge-off operator, a hard segmentation can be obtained by thresholding at level 40.

The reconstruction of an image f from itself subtracted by h is called h-maxima: $\text{HMAX}_{h,D}(f)$. It is a component filter that removes any dome with height less than or equal h and decreases the height of the other domes by h. For a geometric interpretation of the h-maxima based on the threshold decomposition concept, the height attribute associated to a level component (cardboard piece) is one plus the maximum number of levels that exist above it. The h-maxima filter removes all the cardboard pieces with height attribute below or equal to h.

The dual operator of h-maxima is called h-minima: $\text{HMIN}_{h,D}(f)$ fills in any basin of image f with depth less than h and decreases the depth of the other basins by h.

4.3.4 Regional Maxima

Considering the threshold decomposition of a gray-scale image, regional maxima are level components with height attribute equal to one, i.e., there is no other level component above them. These are the cardboard pieces that are on top of a dome. For instance in Fig. 4.18, panel (c) shows the regional maxima of the image in panel (b). All regional maxima of an image f can be found by subtracting the h-maxima with $h = 1$ from image f:

$$\text{RMAX}_D(f) = f - \text{HMAX}_{1,D}(f). \tag{4.18}$$

By duality, a regional minimum is a flat connected region that is at the bottom of a basin. Regional maxima and minima are generically called extrema of the image. Due to the inherent noise associated with the acquisition process, a biomedical image typically presents a large number of regional maxima. If the regional maximum operator is applied to a gradient image, then the situation is even worse. Dynamics provide a tool for selecting significant extrema with respect to a contrast criterion [11]. The dynamics of a regional maximum is the height we have to climb down from the maximum to reach another maximum of higher altitude. The dynamics of a minimum is the minimum height a point in the regional minimum has to climb to reach a lower regional minimum. Filtering the domes of the image also removes regional maxima. Dome filtering can be accomplished using opening, reconstruction from opening, area open, or h-maxima. The choice of appropriate filter is part of the design strategy.

Figure 4.13 shows the regional maxima of the input image following different filters. Note how the over-segmentation given by the direct regional maxima was reduced. These markers can be used to segment the image using the watershed transform. One of the crucial steps in the watershed transform is marker extraction. A marker must be placed in a representative sample of the region of the object to be extracted. The marker finding using the regional maxima (minima) of filtered images is a powerful method since it is independent of gray-scale thresholding values.

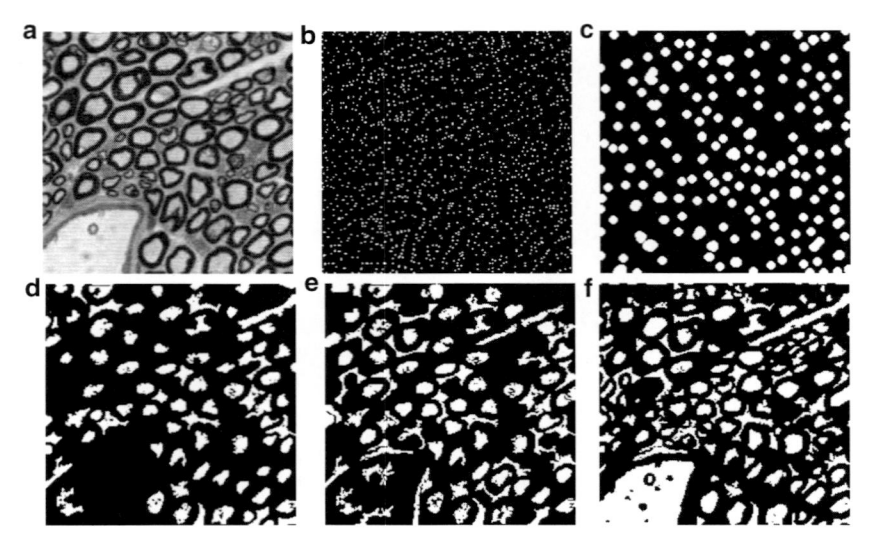

Fig. 4.13. *Regional maxima of filtered image.* (**a**) input image; (**b**)–(**f**) regional maxima; (**c**) after opening by a disk of radius 3; (**d**) after reconstruction from opening by the same disk; (**e**) after area open of 100 pixels; and (**f**) after h-maxima filtering with $h = 40$

4.4 Watershed Segmentation

The watershed transform is a key building block for morphological segmentation [12]. In particular, a gray-scale segmentation methodology results from applying the watershed to the morphological gradient of an image to be segmented. The watershed methodology has become highly developed to deal with numerous real-world contingencies, and a number of implementation algorithms have been developed [13,14]. There are many watershed algorithms in the literature. Here, we only cover the basic methodology.

4.4.1 Classical Watershed Transform

The most intuitive formulation of the watershed transform is based on a flooding simulation. Consider the input gray-scale image as a topographic surface. The problem is to produce the watershed lines on this surface. To do so, holes are punched in each regional minimum of the image. The topography is slowly flooded from below by letting water rise from each regional minimum at a uniform rate across the entire image. When the rising water coming from distinct minima is about to merge, a dam is built to prevent the merging. The flooding will eventually reach a stage when only the tops of dams are visible above the water surface, and these correspond to the watershed lines. The final regions arising from the various regional minima are called the catchment basins.

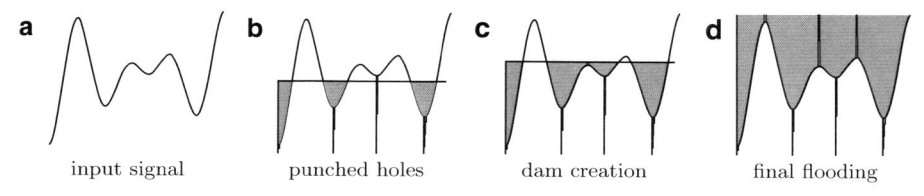

Fig. 4.14. *Flooding simulation of the watershed transform*

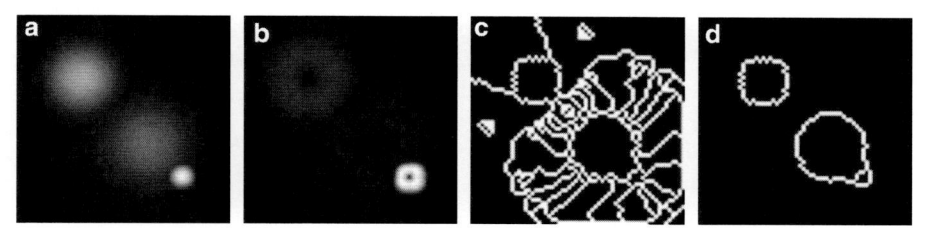

Fig. 4.15. *Classical watershed, regional minima filtering.* (**a**) small synthetic input image (64 × 64); (**b**) morphological gradient; (**c**) watershed on the morphological gradient; (**d**) watershed on the h-minima ($h = 9$) filtered morphological gradient

Figure 4.14 illustrates this flooding process on a signal with four regional minima generating four catchment basins. At first, holes are punched at the minima (cf. panel (b)). Then, the flooding process is initialized. A dam is created whenever water from different minima are about to merge (cf. panel (c)). The final flooding yielding three watershed lines and four catchment basins (cf. panel (d)).

For image segmentation, the watershed is usually (but not always) applied on a gradient image. As real digitized images present many regional minima in their gradients, this typically results in a large number of catchment basins, the result being called watershed over-segmentation.

4.4.2 Filtering the Minima

A solution to cope with the over-segmentation problem is to filter the image, in order to reduce the number of regional minima, creating less catchment basins. Figure 4.15 shows the typical application of the classical watershed transform. Although the image is synthetic, due to the discretization process there are several regional minima in the image, each one generating one catchment basins. By filtering the gradient image using the h-minima with parameter $h = 9$, the watershed gives the desired result. This kind of filtering is very subtle for our eyes as we cannot distinguish the difference from the original morphological gradient and the filtered one, despite the difference between their number of regional minima.

In Fig. 4.16, watershed over-segmentation is reduced by filtering the minima of the input image with a close by a disk of radius 3 (cf. panel (c));

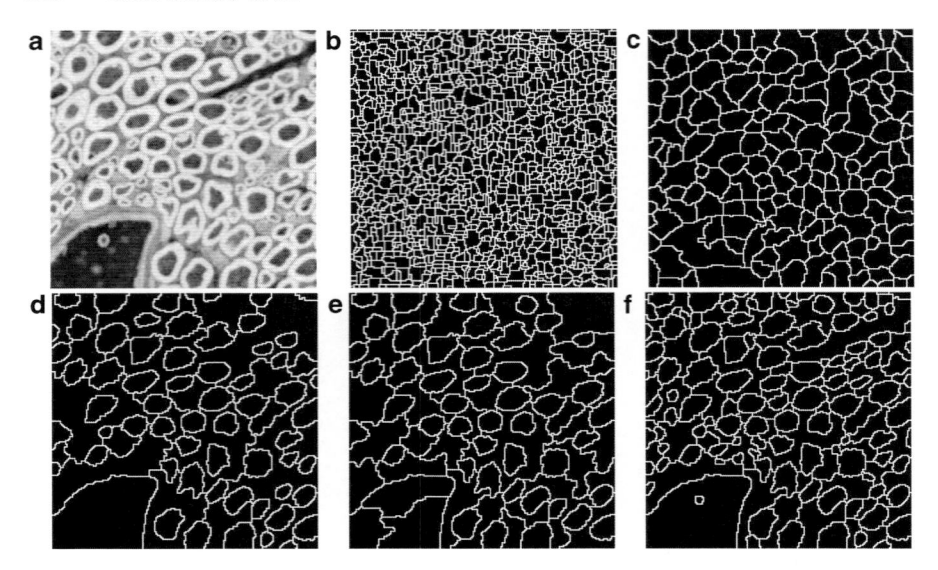

Fig. 4.16. *Reducing over-segmentation by minima filtering.* (**a**) Input image; (**b**) watershed of the input image; (**c**) watershed of the input image after closing by a disk of radius 3; (**d**) watershed of the input image after sup-reconstruction from closing by the same disk; (**e**) watershed of the input image after area open of 100 pixels; and (**f**) watershed of the input image after h-minima filtering with $h = 40$

with sup-reconstruction from closing by the same disk (cf. panel (c)); with area closing (cf. panel (d)); and with h-minimum (cf. panel (f)). This example is equivalent to the regional maxima simplification shown in Fig. 4.13. If we compute the regional minima of the filtered images, we get the same results of that figure. Note that to filter regional minima, the filters used are those that operate with the valleys such as closings and h-minima. Applying filters that operate on domes do not result in any simplification on the number of minima or on the number of catchment basins of the watershed.

4.4.3 Watershed from Markers

Markers are very effective to reduce over-segmentation if one knows how to place the markers within the objects to be segmented. The watershed from markers can also be described as a flooding simulation process (Fig. 4.17). At first, holes are punched at the marker regions and each marker is associated with a color. The topography is then flooded from below by letting colored water rise from the hole associated with its color, this being done for all holes at a uniform rate across the entire image. If the water reaches a catchment basin with no marker in it, then the water floods the catchment basin without restriction. However, if the rising waters of distinct colors are about to

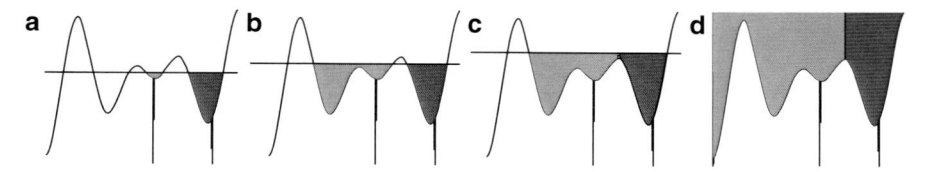

Fig. 4.17. *Flooding simulation of the watershed from markers.* (**a**) punched holes at markers and initial flooding; (**b**) flooding a primitive catchment basin without marker; (**c**) a dam is created when waters coming from different markers are about to merge; (**d**) final flooding, only one watershed line

merge, then a dam is built to prevent the merging. The colored regions are the catchment basins associated with the various markers. To differentiate these catchment basins from the ones obtained with the classical watershed transform, we call the latter primitive catchment basins.

The classical watershed transform can be constructed using the watershed from markers and vice-versa. If we place the markers for the watershed from markers at the regional minima of the input image, then we get the classical watershed transform. To get the watershed from markers from the standard watershed transform is a bit more complicated: we need to apply the classical watershed transform on the sup-reconstruction of the image from the markers.

4.4.4 Inner and Outer Markers

A typical watershed-based segmentation problem is to segment cell-like objects from a gray-scale image. The general approach is threefold:

1. Preprocessing using a smoothing connected filter.
2. Extracting object (inner marker) and background markers (outer marker).
3. Obtaining watershed lines of the morphological gradient from the markers.

Usually, the most crucial part is the extraction of object markers (Step 2): if an object is not marked properly, then it will be missed in the final segmentation.

To illustrate watershed segmentation using inner and outer markers, we consider a poor quality microscopic image of a cornea tissue (Fig. 4.18). The cell markers are extracted by the regional maxima of the opening by a disk of the input image. The criterion used with regional maxima is mainly topological. We can model each cell as a small hill and we want to mark the top of each hill that has a base larger than the disk used in the opening. The regional maxima constitute the inner markers, and the outer markers are obtained by a watershed on the negated input image. After labeling the markers, the morphological gradient is computed. Although it is a very noisy gradient, the final watershed lines provide a satisfactory segmentation.

While it is often the case that the watershed transform is applied to a gradient image, a top-hat image or a distance function image, in some other cases, the input image itself is suitable for application of the watershed.

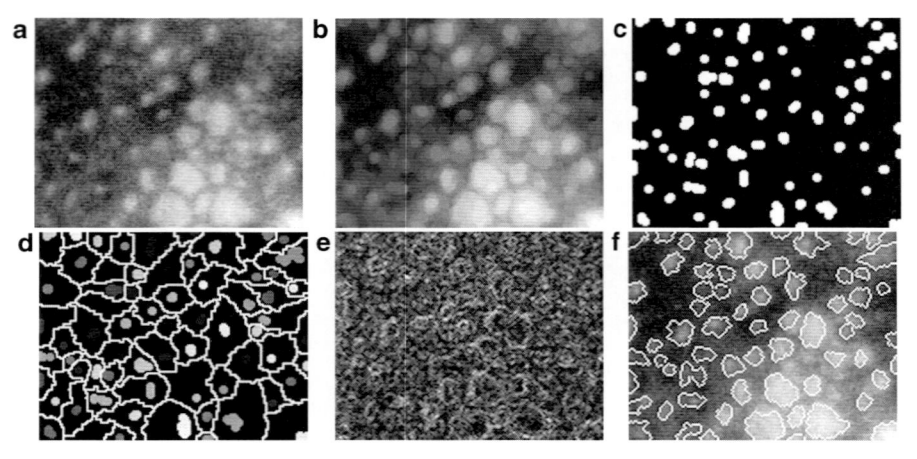

Fig. 4.18. *Segmentation of cornea cells.* (**a**) input image; (**b**) filtered by open; (**c**) regional maxima of the open (inner markers); (**d**) inner and outer markers (watershed lines of the negated input image from the inner markers); (**e**) morphological gradient of the original image; (**f**) final watershed lines overlaid on input image

4.5 Segmentation of Diffusion MRI

Diffusion Tensor Imaging (DTI) is a MRI-based modality able to characterize water molecules diffusion inside tissues and useful to demonstrate subtle brain abnormalities caused by diseases, such as stroke, multiple sclerosis, dyslexia, and schizophrenia [15,16]. However, while most traditional imaging methods in medicine produce gray scale images, DTI produces a tensor valued image (i.e., each pixel or voxel contains a tensor), demanding new, and usually complex, visualization and/or processing methods, cf. Chap. 16, page 403.

One example of DTI can be seen in Fig. 4.19. Each voxel contains a tensor, represented by an ellipsoid, where the main hemiaxis lengths are proportional to the square roots of the tensor eigenvalues λ_1, λ_2, and λ_3 ($\lambda_1 \geq \lambda_2 \geq \lambda_3$) and their directions correspond to the respective eigenvectors.

Several approaches for DTI-based segmentation, where regions of similar diffusion characteristics must be delineated, have been proposed in the last decade [17–22]. One possible solution is to segment DTI using concepts from mathematical morphology, such as the morphological gradient and the watershed transform. Instead of adapting the watershed to work with tensorial images, a tensorial morphological gradient is defined, which translates relevant information from tensors to a scalar map. The desired segmentation is achieved by applying the watershed over this scalar map.

Let $E = \mathbb{Z} \times \mathbb{Z}$ be the set of all points in the tensorial image f. The Tensorial Morphological Gradient (TMG) is defined by

$$\nabla_B^T(f)(x) = \bigvee\nolimits_{y,z \in B_x} d_n(\boldsymbol{T}_y, \boldsymbol{T}_z), \tag{4.19}$$

4 Morphological Image Processing Applied in Biomedicine   127

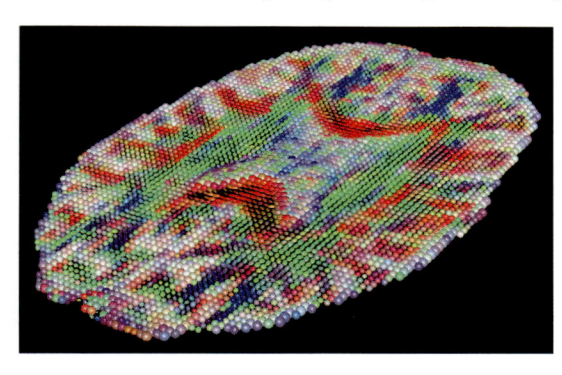

Fig. 4.19. *DTI.* One slice of a diffusion tensor image of the brain. The color of each ellipsoid is related to the principal eigenvector direction of the tensor associated to that voxel

a   b

original image   TMG

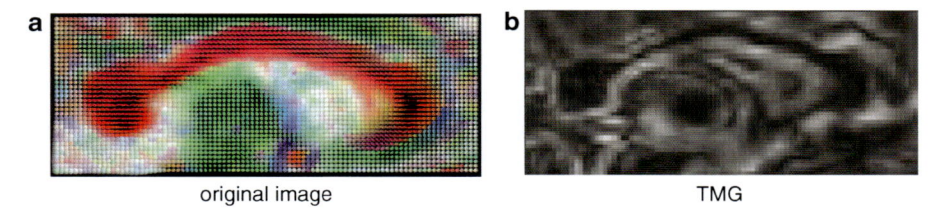

Fig. 4.20. *Tensorial morphological gradient (TMG)*

$\forall x \in E$, where $\bigvee$ is the supremum of a subset, $B \subset E$ is a structuring element centered at the origin of E, d_n represents any of the similarity measures presented in DTI literature [23,24], $\boldsymbol{T}_y$ is the tensor that represents the diffusion in y, and $\boldsymbol{T}_z$ is the tensor that represents the diffusion in z (y and z are in the neighborhood of x, defined by E). ∇_B^T denotes the TMG.

The TMG, as the morphological gradient presented in Sect. 4.2.1, can be used to display the boundary of objects (Fig. 4.20).

Once the tensorial information from DTI is mapped to a gray-scale image, all morphological operators described in Sect. 4.3 can be used to process it. Since the classical watershed transform would lead, in this case, to an over-segmentation, the best choice is to use the hierarchical watershed or Multi-Scale Watershed (MSW) transform, which creates a set of nested partitions. The MSW presented here can be obtained by applying the watershed from markers to a decreasing set of markers. The watershed at scale 1 (finest partitioning) is the classical watershed, made of the primitive catchment basins (perhaps an over-segmentation). As the scale increases, less markers are involved and the coarsest partition is the entire image obtained from a single marker at the regional minimum of largest dynamics.

Figure 4.21 illustrates the MSW transform applied to the TMG from Fig. 4.20b. The decreasing sets of markers are obtained by applying h-minima filter, i.e., dynamics of the minima are used. We show two levels in the hierarchy with markers as the regional minima with the n highest dynamics. With $n = 60$, we are able to correctly segment the corpus callosum. With $n = 150$, the corpus callosum is already subdivided into four regions.

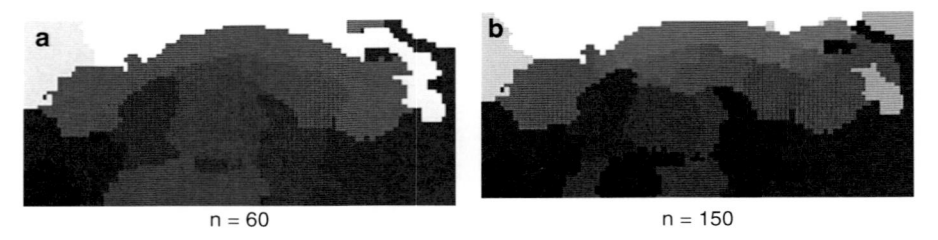

Fig. 4.21. *Multi-scale watershed (MSW).* The MSW transform is computed using the n regional minima with highest dynamics as markers

4.6 Conclusions

Morphological tools are very powerful and useful for biomedical image processing, both for binary and gray-scale images. Unfortunately, they are still not widely known for the majority of the researchers in the field. This chapter illustrated that there are difficult real-world problems that can be solved just using morphological tools. Particularly, there are two important techniques well established such as the watershed-based segmentation and the morphological reconstruction as the primitive operator of a family of component filters.

References

1. Dougherty ER, Lotufo RA. Hands-on Morphological Image Processing. vol. TT59. Bellingham, USA: SPIE Press; 2003.
2. Soille P. Morphological Image Analysis: Principles and Applications. Secaucus, NJ, USA: Springer New York, Inc.; 2003.
3. Lotufo RA, Machado RC, Körbes A, et al. Adessowiki On-line Collaborative Scientific Programming Platform. In: WikiSym '09: Proceedings of the 5th International Symposium on Wikis and Open Collaboration. New York: ACM; 2009. p. 1–6.
4. Lotufo RA, Audigier R, Saúde A, et al. Morphological image processing. In: Microscope Image Processing. Burlington: Academic; 2008. p. 113–158.
5. Serra J. Morphological filtering: an overview. Signal Process. 1994;38(1):3–11.
6. Haralick RM, Sternberg SR, Zhuang X. Image analysis using mathematical morphology. IEEE Trans Pattern Anal Mach Intell. 1987;9(4):532–50.
7. Vincent L. Morphological grayscale reconstruction in image analysis: applications and efficient algorithms. IEEE Trans Image Process. 1993;2(2):176–201.
8. Salembier P, Serra J. Flat zones filtering, connected operators, and filters by reconstruction. IEEE Trans Image Process. 1995;4(8):1153–60.
9. Salembier P, Oliveras A, Garrido L. Anti-extensive connected operators for image and sequence processing. IEEE Trans Image Process. 1998;7(4):555–70.
10. Vincent L. Gray scale area opening and closing, their efficient implementation and applications. In: Serra J, Salembier P, editors. Mathematical Morphology and Its Applications to Signal Processing. Barcelona: UPC Publications; 1993. p. 22–27.

11. Grimaud M. A new measure of contrast: the dynamics. Proc SPIE. 1992; 1769:292–305.
12. Beucher S, Meyer F. The morphological approach to segmentation: the watershed transformation. In: Mathematical Morphology in Image Processing. New York: Marcel Dekker; 1992. p. 433–481.
13. Vincent L, Soille P. Watersheds in digital spaces: an efficient algorithm based on immersion simulations. IEEE Trans Pattern Anal Mach Intell. 1991;13(6):583–98.
14. Lotufo RA, Falcão AX. The ordered queue and the optimality of the watershed approaches. In: Goutsias J, Vincent L, Bloomberg DS, editors. Mathematical Morphology and Its Applications to Image and Signal Processing. vol. 18. Dordrecht: Kluwer; 2000. p. 341–350.
15. Symms M, Jager HR, Schmierer K, et al. A review of structural magnetic resonance neuroimaging. J Neurol Neurosurg Psychiatry. 2004;75(9):1235–44.
16. Dong Q, Welsh RC, Chenevert TL, et al. Clinical applications of diffusion tensor imaging. J Magn Reson Imag. 2004;19(1):6–18.
17. Zhukov L, Museth K, Breen D, et al. Level set modeling and segmentation of DT-MRI brain data. J Electron Imaging. 2003;12:125–133.
18. Wang Z, Vemuri B. DTI segmentation using an information theoretic tensor dissimilarity measure. IEEE Trans Med Imaging. 2005;24(10):1267–77.
19. Jonasson L, Bresson X, Hagmann P, et al. White matter fiber tract segmentation in DT-MRI using geometric flows. Med Image Anal. 2005;9(3):223–36.
20. Weldeselassie Y, Hamarneh G. DT-MRI segmentation using graph cuts. Proc SPIE. 2007;6512:1K-1–9.
21. Lenglet C, Rousson M, Deriche R. A statistical framework for DTI segmentation. Proc IEEE ISBI. 2006;794–97
22. Niogi SN, Mukherjee P, McCandliss BD. Diffusion tensor imaging segmentation of white matter structures using a reproducible objective quantification scheme (ROQS). NeuroImage. 2007;35:166–74.
23. Pierpaoli C, Basser PJ. Toward a quantitative assessment of diffusion anisotropy. Magn Reson Med. 1996;36(6):893–906.
24. Alexander D, Gee J, Bajcsy R. Similarity measures for matching diffusion tensor images. Proc Br Mach Vis Conf. 1999;1:93–102.

5
Medical Image Registration

Daniel Rueckert and Julia A. Schnabel

Summary. Over the last decade, image registration has emerged as one of key technologies in medical image computing with applications ranging from computer assisted diagnosis to computer aided therapy and surgery. This chapter introduces the theory of medical image registration, its implementation and application. In particular, the three key components of any image registration algorithm are described: transformation models, similarity measures, and optimization techniques. The evaluation and validation of image registration is crucial in clinical applications and the chapter discusses techniques for the validation of image registration algorithms. Finally, the chapter illustrates the use of image registration techniques for image segmentation, shape modeling and clinical applications.

5.1 Introduction

Medical image registration [1] plays an increasingly important role in many clinical applications including Computer-assisted Diagnosis (CAD), Computer-aided Therapy (CAT) and Computer-assisted Surgery (CAS). A particular reason for the importance of image registration is the growing and diverse range of medical imaging modalities: Some modalities provide anatomical information about the underlying tissues such as the X-ray attenuation coefficient from X-ray Computed Tomography (CT) and proton density or proton relaxation times from Magnetic Resonance Imaging (MRI). Such images allow clinicians to quantify and visualize the size, shape and spatial relationship between anatomical structures and any pathology, if present. Other imaging modalities provide functional information such as the blood flow or glucose metabolism from Positron Emission Tomography (PET) or Single Photon Emission Computed Tomography (SPECT), and permit clinicians to study the relationship between anatomy and physiology. Finally, optical images acquired either in-vivo or ex-vivo (in form of histology) provide another important source of information which depicts structures at microscopic levels of resolution.

T.M. Deserno (ed.), *Biomedical Image Processing*, Biological and Medical Physics,
Biomedical Engineering, DOI: 10.1007/978-3-642-15816-2_5,
© Springer-Verlag Berlin Heidelberg 2011

The goal of image registration is to find corresponding anatomical or functional locations in two or more images. Image registration can be applied to images from the same subject acquired by different imaging modalities (multi-modal image registration) or at different time points (serial image registration). Both cases are examples of intra-subject registration since the images are acquired from the same subject. Another application area for image registration is inter-subject registration where the aim is to align images acquired from different subjects, e.g., to study the anatomical variability within or across populations.

In general, the process of image registration involves finding the optimal geometric transformation which maximizes the correspondences across the images. This involves several components:

- A *transformation model* which defines a geometric transformation between the images. There are several classes of transforms including rigid, affine and deformable transforms. The question, which model is appropriate, is strongly linked to the application.
- A *similarity metric* or registration basis, which measures the degree of alignment between the images. In cases where features such as landmarks, edges or surfaces are available, the distances between corresponding features can be used to measure the alignment. In other cases, the image intensities can be used directly to determine the alignment.
- An *optimization method* which maximizes the similarity measure. Like many other problems in medical imaging, image registration can be formulated as an optimisation problem whose goal it is to maximize an associated objective function.
- A *validation protocol* which measures the performance of the registration techniques in general terms such as accuracy and robustness as well as in application-specific terms such as clinical utility.

In this chapter, we will discuss these components in more detail. We will also illustrate how image registration can be used in clinical applications. An extensive survey of registration techniques can be found in Zitova and Flusser [2]. Throughout this chapter, we will use $\mathbf{X}$, $\mathbf{x}$, and x to indicate a matrix, vector, or scalar quantity, respectively.

5.2 Transformation Model

The goal of image registration is to find a transformation $\boldsymbol{T} : (x, y, z) \mapsto (x', y', z')$, which maps any point in the source (or floating) image into the corresponding point in the target (or reference) image. There are several transformation models, ranging from simple (e.g., rigid or affine transformations) to more complex transforms (e.g., deformable transforms). The spectrum of simple to complex transforms is usually characterized by an increasing number of parameters that describe the Degrees of Freedom (DoF) of the transformation.

There are several key considerations which must be taken into account when choosing a transformation model:

1. It must be able to characterize the geometric transformation between the images. For example, if there is a significant amount of non-rigid deformation between the images, e.g., respiratory motion in images of the liver, then a transformation model only allowing for rigid or affine transforms is inappropriate as it cannot represent the transforms required to align the images.
2. It should be a simple as possible. For example, if two images are related by a simple rotation, then a non-rigid transformation is inappropriate as it has many parameters, which are not necessary to describe the solution.

5.2.1 Rigid Transformation

A common assumption in medical image registration is that both images are related by a rigid transformation. For example, for images of the head the rigid-body assumption is normally justified as the skull is rigid and constrains the motion of the brain sufficiently. In three dimensions, a rigid transformation involves six DoFs: three rotations and three translations. Using homogeneous coordinates [3], the rigid transformation can be expressed in matrix form

$$\boldsymbol{T}_{\text{rigid}}(x,y,z) = \begin{pmatrix} x' \\ y' \\ z' \\ 1 \end{pmatrix} = \begin{pmatrix} r_{11} & r_{12} & r_{13} & t_x \\ r_{21} & r_{22} & r_{23} & t_y \\ r_{31} & r_{32} & r_{33} & t_z \\ 0 & 0 & 0 & 1 \end{pmatrix} \begin{pmatrix} x \\ y \\ z \\ 1 \end{pmatrix} \quad (5.1)$$

where t_x, t_y, t_z define the translations along the axes of the coordinate system, while the coefficients r_{ij} are the result of the multiplication of three separate rotation matrices, which determine the rotations about each coordinate axis.

5.2.2 Affine Transformation

In some cases, it is necessary to correct not only for rigid transformation but also for scaling. This additional scaling can be expressed in matrix form as

$$\boldsymbol{T}_{\text{scale}} = \begin{pmatrix} s_x & 0 & 0 & 0 \\ 0 & s_y & 0 & 0 \\ 0 & 0 & s_z & 0 \\ 0 & 0 & 0 & 1 \end{pmatrix} \quad (5.2)$$

where s_x, s_y, and s_z define the scaling along the different coordinate axes. In some cases, it may also be necessary to correct for shears, for example caused by the gantry tilt of CT scanners. A shear in the (x,y) plane can be expressed as

$$\boldsymbol{T}_{\text{shear}}^{xy} = \begin{pmatrix} 1 & 0 & h_x & 0 \\ 0 & 1 & h_y & 0 \\ 0 & 0 & 1 & 0 \\ 0 & 0 & 0 & 1 \end{pmatrix} \qquad (5.3)$$

Combining the rigid transformation matrix with the scaling and shearing matrices yields an affine transformation

$$\boldsymbol{T}_{\text{affine}}(x,y,z) = \boldsymbol{T}_{\text{shear}} \cdot \boldsymbol{T}_{\text{scale}} \cdot \boldsymbol{T}_{\text{rigid}} \cdot (x,y,z,1)^T \qquad (5.4)$$

whose twelve DoF represent rotations, translations, scaling and shears. Like rigid or affine transforms, global transforms affect the entire image domain. Higher-order global transforms such as tri-linear (24 DoF) or quadratic (30 DoF) transforms can be modeled in a similar fashion [4].

5.2.3 Projective Transformation

Projective transforms play an important role in applications involving the alignment of 3D volumes like CT or MRI to 2D images such as radiography and photography. Different types of projections including parallel or perspective projections can be used depending on the application [3]. However in most cases, the transformation that relates the 3D and 2D images is a combination of a projective with a rigid transformation, which determines the pose of the 3D volume relative to the camera. Often it is possible to determine the perspective transformation parameters using either knowledge about the internal geometry of the camera, or by camera calibration techniques [5], in which case the problem is reduced to rigid registration.

5.2.4 Non-Rigid Transformation: Parametric Models

The transformation models discussed so far can be characterized by a small number of parameters (six for rigid and twelve for affine transforms). While such transformation models are frequently used for the registration of anatomical structures like the brain or bones, they are not applicable in the case where significant deformation is expected, e.g., in soft tissues like the liver or breast. In these cases, deformable or non-rigid transforms are required. These non-rigid transforms can be modeled either using parametric or non-parametric models. In the case of non-parametric transformation models, a dense displacement field is stored which describes the deformation at every voxel. In contrast to this, parametric transformation models are controlled by a set of parameters (typically, the number of parameters is much smaller than the number of voxels).

Basis Functions

A common approach for modeling non-rigid transforms is to describe the transformation as a linear combination of basis functions θ_i:

$$\boldsymbol{T}(x,y,z) = \begin{pmatrix} x' \\ y' \\ z' \\ 1 \end{pmatrix} = \begin{pmatrix} a_{00} & \cdots & a_{0n} \\ a_{10} & \cdots & a_{1n} \\ a_{20} & \cdots & a_{2n} \\ 0 & \cdots & 1 \end{pmatrix} \begin{pmatrix} \theta_1(x,y,z) \\ \vdots \\ \theta_n(x,y,z) \\ 1 \end{pmatrix} \quad (5.5)$$

A common choice is to represent the deformation field using a set of (orthonormal) basis functions such as Fourier (trigonometric) [6, 7] or wavelet basis functions [8]. In the case of trigonometric basis functions this corresponds to a spectral representation of the deformation field where each basis function describes a particular frequency of the deformation. Restricting the summation in (5.5) to the first N terms (where $1 < N < n$) has the effect of limiting the frequency spectrum of the transformation to the N lowest frequencies.

Splines

Many registration techniques using splines are based on the assumption that a set of corresponding points or landmarks can be identified in the source and target images (see also the topic point-based registration in Sect. 5.3.1). These corresponding points are often referred to as control points. At these control points, spline-based transforms either interpolate or approximate the displacements, which are necessary to map the location of the control point in the target image into its corresponding counterpart in the source image. Between control points, they provide a smoothly varying displacement field. For each control point, the interpolation condition can be written as

$$\boldsymbol{T}(\phi_i) = \phi_i' \quad i = 1, \ldots, n \quad (5.6)$$

where ϕ_i and ϕ_i' denote the location of the control point in the target and the source image, respectively. There are a number of different ways of how the control points can be determined. For example, anatomical or geometrical landmarks which can be identified in both images can be used [9]. In addition, Meyer et al. [10] suggested to update the location of control points by optimization of a voxel similarity measure such as mutual information. Alternatively, control points can be arranged with equidistant spacing across the image forming a regular mesh [11]. In this case, the control points are only used as a parameterization of the transformation and do not correspond to anatomical or geometrical landmarks. Therefore, they are often referred to as pseudo- or quasi-landmarks.

Thin-Plate Splines

Thin-plate splines are part of a family of splines which are based on radial basis functions. They have been formulated by Duchon [12] and Meinguet [13] for the surface interpolation of scattered data. In recent years, they have been widely used for image registration [9,14,15]. Radial basis function splines can be defined as a linear combination of n radial basis functions $\theta(s)$:

$$t(x,y,z) = a_1 + a_2 x + a_3 y + a_4 z + \sum_{j=1}^{n} b_j \theta(|\phi_j - (x,y,z)|) \quad (5.7)$$

Defining the transformation as three separate thin-plate spline functions $\boldsymbol{T} = (t_1, t_2, t_3)^T$ yields a mapping between images in which the coefficients $\boldsymbol{a}$ and $\boldsymbol{b}$ characterize the affine and the non-affine part of the transformation, respectively. The interpolation conditions in (5.6) form a set of $3n$ linear equations. To determine the $3(n+4)$ coefficients uniquely, twelve additional equations are required. These twelve equations guarantee that the non-affine coefficients b sum to zero and that their crossproducts with the x, y, and z coordinates of the control points are likewise zero. In matrix form, this can be expressed as

$$\begin{pmatrix} \Theta & \Phi \\ \Phi^T & 0 \end{pmatrix} \begin{pmatrix} b \\ a \end{pmatrix} = \begin{pmatrix} \Phi' \\ 0 \end{pmatrix} \quad (5.8)$$

Here $\boldsymbol{a}$ is a 4×3 vector of the affine coefficients a, $\boldsymbol{b}$ is a $n \times 3$ vector of the non-affine coefficients b, and Θ is the kernel matrix with $\Theta_{ij} = \theta(|\phi_i - \phi_j|)$. Solving for $\boldsymbol{a}$ and $\boldsymbol{b}$ using standard algebra yields a thin-plate spline transformation, which will interpolate the displacements at the control points.

The radial basis function of thin-plate splines is defined as

$$\theta(s) = \begin{cases} |s|^2 \log(|s|) & \text{in 2D} \\ |s| & \text{in 3D} \end{cases} \quad (5.9)$$

There are a wide number of alternative choice for radial basis functions including multi-quadrics and Gaussians [11,16]. Modeling deformations using thin-plate splines has a number of advantages. For example, they can be used to incorporate additional constraints such as rigid bodies [17] or directional constraints [18] into the transformation model. They can be extended to approximating splines where the degree of approximation at the landmark depends on the confidence of the landmark localization [19].

Free-Form Deformations and B-Splines

In the late 1980s a number of techniques for modeling deformations emerged in the computer graphics community. In particular, Sederberg and Parry developed Free-Form Deformations (FFD) [20] as a powerful modeling tool for 3D deformable objects. The basic idea of FFDs is to deform an object by manipulating an underlying mesh of control points. The resulting deformation controls the shape of the 3D object and produces a smooth and continuous transformation. In the original paper by Sederberg and Parry [20], tri-variate Bernstein polynomials were used to interpolate the deformation between control points. A more popular choice is to use tri-variate B-spline tensor products as the deformation function [21,22]. The use of FFDs based on B-splines for image registration was first proposed by Rueckert et al. [23,24]. Over the last decade, the use of FFDs for image registration has attracted significant interest [25–28].

To define a spline-based FFD, we denote the domain of the image volume as $\Omega = \{\mathbf{p} = (x,y,z) \mid 0 \leq x < X, 0 \leq y < Y, 0 \leq z < Z\}$. Let Φ denote a $n_x \times n_y \times n_z$ mesh of control points $\phi_{i,j,k}$ with uniform control point spacing δ (Fig. 5.1). Then, the FFD can be written as the 3D tensor product of the familiar 1D cubic B-splines:

$$T_{\text{local}}(\mathbf{p}) = \sum_{l=0}^{3}\sum_{m=0}^{3}\sum_{n=0}^{3} B_l(u)B_m(v)B_n(w)\phi_{i+l,j+m,k+n} \tag{5.10}$$

where $i = \lfloor \frac{x}{\delta} \rfloor - 1, j = \lfloor \frac{y}{\delta} \rfloor - 1, k = \lfloor \frac{z}{\delta} \rfloor - 1, u = \frac{x}{\delta} - \lfloor \frac{x}{\delta} \rfloor, v = \frac{y}{\delta} - \lfloor \frac{y}{\delta} \rfloor, w = \frac{z}{\delta} - \lfloor \frac{z}{\delta} \rfloor$ and where B_l represents the l-th basis function of the B-spline [21,22]:

$$B_0(u) = (1-u)^3/6$$
$$B_1(u) = (3u^3 - 6u^2 + 4)/6$$
$$B_2(u) = (-3u^3 + 3u^2 + 3u + 1)/6$$
$$B_3(u) = u^3/6$$

In contrast to thin-plate splines [15] or elastic-body splines [11], B-splines are locally controlled, which makes them computationally efficient even for a large number of control points. In particular, the basis functions of cubic B-splines have a limited support, i.e., changing control point $\phi_{i,j,k}$ affects the transformation only in the local neighborhood of that point.

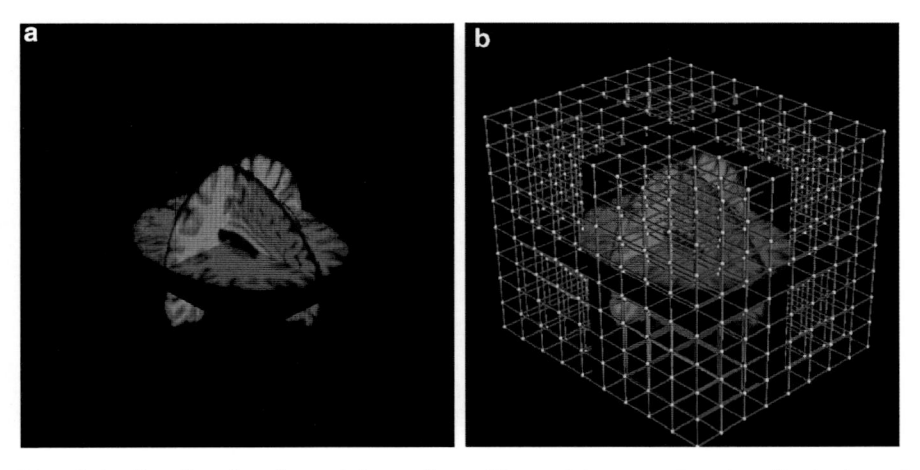

Fig. 5.1. *B-spline free-form deformations.* The axial, coronal and sagittal view of a 3D MRI of the brain is shown without control point mesh (**a**) and with control point mesh (**b**)

5.2.5 Non-Rigid Transformation: Non-Parametric Models

In contrast to parametric transformation models, non-parametric models characterize the deformation at every voxel. These models offer the greatest amount of flexibility in describing the transformation but are also expensive in terms of memory storage.

Small Deformation Models

Small deformation models encode the transformation at every voxel directly as a displacement vector. In computer vision, these models are closely related to the notion of optical flow [29], while in medical image registration their first use has been proposed in the context of elastic models [30]: The idea is to model the deformation of the source image into the target image as a physical process, which resembles the stretching of an elastic material such as rubber. This physical process is governed by two forces: The first term is the internal force, which is caused by the deformation of elastic material (i.e., stress) and counteracts any force, which deforms the elastic body from its equilibrium shape. The second term corresponds to the external force which acts on the elastic body. As a consequence, the deformation of the elastic body stops if both forces acting on the elastic body form an equilibrium solution. The behavior of the elastic body is described by the Navier linear elastic Partial Differential Equation (PDE):

$$\mu \nabla^2 \mathbf{u}(x,y,z) + (\lambda + \mu)\nabla(\nabla \cdot \mathbf{u}(x,y,z)) + \boldsymbol{f}(x,y,z) = 0 \qquad (5.11)$$

Here $\mathbf{u}$ describes the displacement field, $\boldsymbol{f}$ is the external force acting on the elastic body; ∇ and ∇^2 denote the gradient operator and the Laplace operator, respectively. The parameters μ and λ are Lamé's elasticity constants, which describe the behavior of the elastic body. These constants are often interpreted in terms of Young's modulus E_1, which relates the strain and stress of an object, and Poisson's ratio E_2, which is the ratio between lateral shrinking and longitudinal stretching:

$$E_1 = \frac{\mu(3\lambda + 2\mu)}{(\lambda + \mu)}, \qquad E_2 = \frac{\lambda}{2(\mu + \lambda)} \qquad (5.12)$$

The external force $\boldsymbol{f}$ is the force which acts on the elastic body and drives the registration process. A common choice for the external force is the gradient of a similarity measure such as a local correlation measure based on intensities [30], intensity differences [31] or intensity features such as edge and curvature [32], however, in principle any of the similarity metrics described in Sect. 5.3 can be used as a force.

The PDE in (5.11) may be solved by finite differences and Successive Over-Relaxation (SOR) [33]. This yields a discrete displacement field for each voxel. Alternatively, the PDE can be solved for only a subset of voxels which correspond to the nodes of a finite element model [32, 34]. These nodes form a set of points for which the external forces are known. The displacements at other voxels are obtained by finite element interpolation. An extension of the elastic registration framework has been proposed by Davatzikos [35] to allow for spatially varying elasticity parameters. This enables certain anatomical structures to deform more freely than others.

Large Deformation Model

In the small deformation model, the displacements are stored at each voxel with respect to their initial position. Since the regularization in (5.11) acts on the displacement field, highly localized deformations cannot be modeled since the deformation energy caused by stress increases proportionally with the strength of the deformation. In the large deformation model, however, the displacement is generated via a time dependent velocity field $\mathbf{v}$ [36]. The relationship between $\mathbf{v}$ and the displacement $\mathbf{u}(x, y, z, 1)$ is given by

$$\mathbf{u}(x, y, z, 1) = \int_0^1 \mathbf{v}(\mathbf{u}(x, y, z, t)) \, dt \qquad (5.13)$$

with $\mathbf{u}(x, y, z, 0) = (x, y, z)$. The resulting transformations are *diffeomophic* as they provide a smooth one-to-one (invertible) mapping. One of the first approaches to use such a representation of the deformations coined the term fluid registration [37]. Here the deformations are characterized by the Navier–Stokes partial differential equation

$$\mu \nabla^2 \mathbf{v}(x, y, z) + (\lambda + \mu) \nabla (\nabla \cdot \mathbf{v}(x, y, z)) + \boldsymbol{f}(x, y, z) = 0 \qquad (5.14)$$

similar to (5.11) except that differentiation is carried out on the velocity field $\mathbf{v}$ rather than on the displacement field $\mathbf{u}$ and is solved for each time step. Christensen [37] suggested to solve (5.14) using SOR [33]. However, the resulting algorithm is rather slow and requires significant computing time. A faster implementation has been proposed by Bro-Nielsen et al. [38]. Here, (5.14) is solved by deriving a convolution filter from the eigenfunctions of the linear elasticity operator. Bro-Nielsen et al. [38] also pointed out that this is similar to a regularization by convolution with a Gaussian as proposed in a non-rigid matching technique by Thirion [39], in which the deformation process is modeled as a diffusion process. However, the solution of (5.14) by convolution is only possible if the viscosity is assumed constant, which is not always the case.

5.3 Registration Basis

The second component of a registration algorithm is the registration basis, which measures the degree of alignment of the images. The two main approaches are feature-based and voxel-based similarity measures.

5.3.1 Feature-Based Registration

Feature-based registration approaches are usually based on points, lines or surfaces and aim at minimizing the distance between corresponding features in the two images. This requires the extraction of the features as well as the estimation of correspondences.

Points

One of the most intuitive registration criteria is the proximity of corresponding point features, which can be identified in the two images. Given two point sets **p** and **q**, we can define a similarity measure based on the squared distance of the points:

$$\mathcal{S} = -\sum_i ||\mathbf{q}_i - \boldsymbol{f}(\mathbf{p}_i)||^2 \tag{5.15}$$

In (5.15), we assume that the correspondences across the point sets are known a priori (we will see later how to formulate this problem if correspondences are unknown, e.g., aligning surfaces described by point clouds). For a rigid transformation, three or more non collinear landmarks are sufficient to establish the transformation between two 3D images. The optimization problem defined in (5.15) can be solved in closed-form [40] and does not require any iterative optimization techniques: It involves the alignment of centroids of the two sets of points followed by a rotation to minimize the sum of the squared displacements between source and destination points. This is achieved by simple matrix manipulation using the method of Singular Value Decomposition (SVD) [33].

Points that can be used for registration can be intrinsic, anatomical landmarks which correspond to point-like structures (e.g., apical turn of the cochlea) or to points can be unambiguously defined (e.g., junction of the vertebral arteries or center of the orbit of the eyes). Other possibilities include the identification of geometrical features of anatomical structures such as a maximum of 2D curvature (e.g., cartoid syphon) or a maximum of 3D curvature (e.g., occipital pole of the brain). A disadvantage in using intrinsic landmarks is the sparse nature and the difficulty of automatic identification and extraction of these landmarks, even though several approaches exist [19]. Alternatively, one can use extrinsic landmarks such as pins or markers fixed to the patient and visible on each image. These markers may be attached to the skin or screwed into the bone. The latter can provide very accurate registration but are more invasive and uncomfortable for the patient. Skin markers on the other hand can easily move by several millimeters due to the mobility of the skin and are difficult to attach firmly. However, well-designed, extrinsic markers can be accurately localized with sub-voxel precision, for instance, by computing the center of gravity of spherical or cylindrical markers [41]. The task of identifying the markers can be automated using computer vision techniques [42].

Surfaces

Point features are relatively sparse and for many tasks, in particular for non-rigid registration, more dense features are required. Another feature that can be used for the registration of images are the boundaries of corresponding anatomical structures. This requires the segmentation of a contour or surface representing the boundary of the anatomical structure in both images and can be achieved using an interactive or automated procedure. The resulting contours or surfaces are often represented as point sets, which can be registered by minimizing the distance between corresponding points of both sets [43,44]. However, since the correspondences between both point sets are not known a priori, these need to be estimated at the same time. Besl and McKay proposed a generic registration algorithm for point sets, called Iterative Closest Point (ICP) algorithm [45], which assumes that there is a correspondence between each point in the first set and its closest point in the second set. In this approach, the similarity measure can be defined as

$$\mathcal{S} = -\sum_i ||\mathbf{y}_i - \boldsymbol{T}(\mathbf{p}_i)||^2 \qquad (5.16)$$

where

$$\mathbf{y}_i = \min_{\mathbf{q}_j \in \mathbf{q}} \left\{ ||\mathbf{q}_j - \boldsymbol{T}(\mathbf{p}_i))||^2 \right\} \qquad (5.17)$$

is the closest point in $\mathbf{q}$. To register two surfaces in form of point sets we first estimate correspondences between the point sets using (5.17). Then, the resulting point sets can be registered by minimizing (5.16), e.g., using the point-based technique discussed in the previous section. The process is repeated until convergence is achieved. In practice, the computational cost can be significantly reduced by using efficient spatial data structures such as octrees [46] or distance transforms [47,48] to pre-calculate the distance to the closest point.

5.3.2 Voxel-Based Registration

The advantage of feature-based registration is that it can be used for both mono- and multi-modality registration. However, the disadvantage of feature-based registration is the need for feature extraction in form of landmark detection or segmentation. Moreover, while features can be extracted manually or automatically, any error during the feature extraction stage will propagate into the registration and cannot be recovered at a later stage. To avoid these errors, it is possible to use the image intensities directly without the need for feature extraction. In this case, voxel-based similarity measures aim at measuring the degree of shared information of the image intensities. This is relatively simple in the case of mono-modality registration, but more complex for multi-modality registration. Over the last decade, voxel-based similarity measures have been come the method of choice for measuring image alignment, largely due to their robustness and accuracy.

Statistical Similarity Measures

The simplest statistical measure of image similarity is based on the Sum of Squared Differences (SSD) between the intensities in the images $\mathcal{I}_A$ and $\mathcal{I}_B$,

$$\mathcal{S}_{\text{SSD}} = -\frac{1}{n} \sum (\mathcal{I}_A(\mathbf{q}) - \mathcal{I}_B(\boldsymbol{T}(\mathbf{p})))^2 \quad (5.18)$$

where n is the number of voxels in the region of overlap. This measure is based on the assumption that both imaging modalities have the same characteristics. If the images are correctly aligned, the difference between them should be zero except for the noise produced by the two modalities. If this noise is Gaussian distributed, it can be shown that the SSD is the optimal similarity measure [49]. Since this similarity measure assumes that the imaging modalities are identical, it is restricted to mono-modal applications such as serial registration [50, 51].

In a number of cases, the assumption of identical imaging modalities is too restrictive. A more general assumption is that of a linear relationship between the two images. In these cases, the similarity between both images can be expressed by the Normalized Cross Correlation (NCC)

$$\mathcal{S}_{\text{NCC}} = \frac{\sum (\mathcal{I}_A(\mathbf{q}) - \mu_A)(\mathcal{I}_B(\boldsymbol{T}(\mathbf{p})) - \mu_B)}{\sqrt{(\sum \mathcal{I}_A(\mathbf{q}) - \mu_A)^2 (\sum \mathcal{I}_B(\boldsymbol{T}(\mathbf{p})) - \mu_B)^2}} \quad (5.19)$$

where μ_A and μ_B correspond to average voxel intensities in the images $\mathcal{I}_A$ and $\mathcal{I}_B$, respectively. Nevertheless, the application of this similarity measure is largely restricted to mono-modal registration tasks.

Information-Theoretical Measures

There has been significant interest in measures of alignment based on the information content or entropy of the registered image. An important step to understanding these methods is the feature space of the image intensities which can also be interpreted as the joint probability distribution: A simple way of visualizing this feature space is by accumulating a 2D histogram of the co-occurrences of intensities in the two images for each trial alignment. If this feature space is visualized for difference degrees of image alignment, it can be seen that the feature space disperses as misalignment increases, and that each image pair has a distinctive feature space signature at alignment.

In an information theoretic framework the information content of images can be defined as the Shannon–Wiener entropy, $H(\mathcal{I}_A)$ and $H(\mathcal{I}_B)$ of images

$\mathcal{I}_A$ and $\mathcal{I}_B$:

$$H(\mathcal{I}_A) = - \sum_{a \in \mathcal{I}_A} p(a) \log p(a) \qquad (5.20)$$

and

$$H(\mathcal{I}_B) = - \sum_{b \in \mathcal{I}_B} p(b) \log p(b) \qquad (5.21)$$

where $p(a)$ is the probability that a voxel in image $\mathcal{I}_A$ has intensity a and $p(b)$ is the probability that a voxel in image $\mathcal{I}_B$ has intensity b. The joint entropy $H(\mathcal{I}_A, \mathcal{I}_B)$ of the overlapping region of image $\mathcal{I}_A$ and $\mathcal{I}_B$ may be defined by

$$H(\mathcal{I}_A, \mathcal{I}_B) = - \sum_{a \in \mathcal{I}_A} \sum_{b \in \mathcal{I}_B} p(a,b) \log p(a,b) \qquad (5.22)$$

where $p(a, b)$ is the joint probability that a voxel in the overlapping region of image $\mathcal{I}_A$ and $\mathcal{I}_B$ has values a and b, respectively.

To derive a measure of image alignment one can use concepts from information theory such as Mutual Information (MI) [52,53]. In terms of entropies, MI is defined as

$$\mathcal{S}_{MI}(\mathcal{I}_A; \mathcal{I}_B) = H(\mathcal{I}_A) + H(\mathcal{I}_B) - H(\mathcal{I}_A, \mathcal{I}_B) \qquad (5.23)$$

and should be maximal at alignment. Mutual information is a measure of how one image "explains" the other but makes no assumption of the functional form or relationship between image intensities in the two images. It is also not independent of the overlap between two images [54]. To avoid any dependency on the amount of image overlap, Normalised Mutual Information (NMI) has been suggested as a measure of image alignment [54]:

$$\mathcal{S}_{\mathrm{NMI}}(\mathcal{I}_A; \mathcal{I}_B) = \frac{H(\mathcal{I}_A) + H(\mathcal{I}_B)}{H(\mathcal{I}_A, \mathcal{I}_B)} \qquad (5.24)$$

Similar forms of normalised mutual information have been proposed by Maes et al. [55].

Information-theoretic voxel similarity measures are based on the notion of the marginal and joint probability distributions of the two images. These probability distributions can be estimated in two different ways: The first method uses histograms whose bins count the frequency of occurrence (or co-occurrence) of intensities. Dividing these frequencies by the total number of voxels yields an estimate of the probability of that intensity. The second method is based on generating estimates of the probability distribution using Parzen windows [56], which is a non-parametric technique to estimate probability densities. The Parzen-based approach has the advantage of providing a differentiable estimate of mutual information, which is not the case for the histogram-based estimate of mutual information.

5.4 Optimization

Like many other problems in medical image analysis, the problem of image registration can be formulated as an optimization problem whose goal is to maximize an objective function. In general, the objective function can be written as combination of two terms: The first term aims at maximizing the similarity of the images and is the driving force behind the registration process. The second term aims at minimizing the cost associated with particular transformations. In the case of rigid or affine registration, the second term usually plays no role and is often omitted. In this case, maximizing the objective function is equal to maximizing the similarity metric. However, in the case of non-rigid registration, the second term can act as a regularization or penalty function which can be used to constrain the transformation relating both images: For example, in elastic or fluid registration the regularization term (linear-elasticity model) forms an integral part of the registration. Other regularization models are the Laplacian or membrane model [6, 57] and the biharmonic or thin-plate model [15, 58]. Both models have an intuitive physical interpretation: While the Laplacian model approximates the energy of a membrane (such as a rubber sheet), which is subjected to elastic deformations, the biharmonic term approximates the energy of a thin plate of metal which is subjected to bending deformations [59]. From a probabilistic point of view, the regularization term can be interpreted as a prior which represents a priori knowledge about the expected transformation, whereas the similarity metric can be viewed as a likelihood term that expresses the probability of a match between the images.

5.5 Validation of Registration

Prior to clinical use, medical image registration algorithms need to be validated. However, the validation of registration performance usually suffers from the lack of knowledge as to whether, how much, and where patient movement has occurred between and even during scanning procedures, and whether such movement affects the clinical usefulness of the data. To maintain clinical usefulness, and inherently improve patient treatment and health care, it is, therefore, mandatory to ensure that registration is successful [60].

A registration method can be assessed in an independent evaluation in the absence of a ground truth. An initial visual inspection allows for a qualitative assessment of registration performance, which can be complemented by quantitative checks for robustness and consistency. The former establishes the measurement precision by testing the bias sensitivity when adding noise or choosing different starting estimates [61]. The latter assesses the capability of a registration technique to find circular transforms based on a registration circuit, but can be sensitive to bias and may not be applicable to non-invertible transforms generated by many non-rigid registration methods. Nonetheless, consistency checks have been successfully used for intra-modality rigid body registration applications (e.g., serial MRI of the brain [62]). In addition, an expert observer can perform a visual assessment of the registration performance. This can involve the inspection of subtraction images, contour or segmentation overlays, alternate pixel displays, or viewing anatomical landmarks. These approaches have been applied to rigid registration [63], and since they involve inspection of the entire volume domain of the image pair, they can be extended to non-rigid registration [64]. For non-rigid registration, visual assessment is an important step toward clinical acceptance and routine use of a registration method, but may be compromised by locally implausible deformations which may not be readily picked up by observers [65]. Nonetheless, visual assessment often forms the first and last line of defense of any image registration validation.

In the absence of ground truth, registration accuracy can be studied by setting up a gold standard transformation. For example, the Retrospective Registration Evaluation Project (RREP) used skull-implanted markers in patients undergoing brain surgery to derive a gold standard for multi-modal rigid-body image registration of the head to compare different established rigid registration methods [66]. For non-rigid registration validation, extrinsic markers cannot easily be used as they would need to be implanted into soft tissue. In a recent study by Koshani et al. [67], markers were inserted into a CT lung phantom, which was deformed and imaged under controlled conditions. In an alternative approach [68], a bio-mechanical motion simulator was introduced that allowed the simulation of physically plausible deformations for clinically realistic motion scenarios, with application to contrast-enhanced Magnetic Resonance Mammography (MRM). This motion simulator was designed to be independent of the image registration and transformation model used. Similar approaches have been proposed for the simulation of X-ray type compression for MRM [69] and Alzheimer's Disease [70].

Finally, the segmentation of anatomical structures provide the means to measure the structure overall or surface distances before and after registration, but does not provide any insight as to whether the registration is accurate within the structure, or along its outline. If, however, the objective of the registration is to propagate (and hence automate) segmentation, the segmentation quality can be used as a surrogate measurement. For example in [71], a number of non-rigid registration methods were compared for inter-subject brain registration including segmentation quality. A number of carefully annotated image databases[1] are emerging, which can be used to establish the accuracy of non-rigid registration methods on the basis of carefully delineated image structures.

---

[1] http://www.nirep.org/

5.6 Application

There are numerous applications of image registration ranging from clinical applications and trials to image registration as a tool for other image analysis tasks such as segmentation or shape modeling.

5.6.1 Intra-Subject Registration

Intra-subject registration applications can be divided into registering images of the same subject acquired using different imaging modalities (multi-modal registration or fusion) or at different time points (serial registration).

Multi-Modal Image Fusion

Medical imaging modalities can be broadly distinguished into two categories: anatomical and functional. In many clinical applications, imaging modalities from both categories must be combined for an effective diagnosis. One such example is the registration of PET with either CT or MRI for applications in oncology. Here, the PET is used to determine the tumor activity and the CT/MRI allows accurate localization of this activity. Similarly, in many neurological diseases the metabolism in the brain is altered by the disease. This can be detected using PET and localized using CT/MRI. Multi-modal registration is usually based on features such as points and surfaces [43] or uses information theoretic similarity measures such as mutual information [72, 73].

Quantification of Changes

A key application of image registration is the alignment of images from the same subject acquired at different time points. The difference between successive time points can range from fractions of a second (e.g., cardiac motion) to several years (e.g., longitudinal growth, atrophy). The comparison of images across time enables the quantification of the anatomical differences between the time points of measurement. These differences can have a number of reasons, such as bulk patient motion, organ motion (e.g., respiratory or cardiac motion), growth, atrophy, or disease progression.

In neuroimaging, image registration has been widely used for the detection of changes in the human brain: For example, the rigid registration of serial MRI is well suited for the visualization of subtle changes in the brain [50, 51]. Moreover, it is also possible to quantify the amount of changes by using non-rigid registration of serial MRI [74, 75]: When two images are related to each other by a non-rigid transformation, the target image is subjected to local deformation, which locally changes the volume of regions in the coordinate space. The local volume change in an infinitesimally small neighborhood around any given point is represented by the local Jacobian determinant of the coordinate transformation at that point

$$|J(x,y,z)| = \det \begin{pmatrix} \frac{\partial T_x}{\partial x} & \frac{\partial T_x}{\partial y} & \frac{\partial T_x}{\partial z} \\ \frac{\partial T_y}{\partial x} & \frac{\partial T_y}{\partial y} & \frac{\partial T_y}{\partial z} \\ \frac{\partial T_z}{\partial x} & \frac{\partial T_z}{\partial y} & \frac{\partial T_z}{\partial z} \end{pmatrix} \qquad (5.25)$$

If there is no local volume change, $J = 1$. If there is a local volume decrease or increase, $J < 1$ or $J > 1$, respectively. Thus, local volume change can be estimated directly from the deformation field that aligns baseline and follow-up images. In such a way, growth patterns have been mapped in 4D (3D plus time) in older children [76,77], and clinically useful information has been extracted in conditions such as Alzheimer's Disease [78,79], and imaginative studies of neurogenetics and development have been undertaken [80].

5.6.2 Inter-Subject Registration

In cross-sectional studies, non-rigid registrations between images of different subjects can be used to characterize the differences between populations or the differences between an individual and a reference image. Approaches that focus on the transformations in this way have been referred to as Voxel-Based Morphometry (VBM), Deformation-Based Morphometry (DBM) or Tensor-Based Morphometry (TBM).

In a simple cross-sectional study, for example, where two clinical groups are to be separated, it is possible to non-rigidly register all the images to a reference image. If the non-rigid registration only accounts for global shape, differences methods such as voxel-based morphometry [81] can be used to investigate the differences in the probabilistic segmentations. If, however, the non-rigid registration also accounts for local shape differences, the differences between the subjects are encoded in the resulting transforms and their properties can be used to identify group differences. Such approaches usually use properties such as the Jacobian of the deformation fields as defined in eq. (5.25) to analyze these differences and is often referred to as DBM or TBM [82].

Segmentation

The amount of data produced by imaging increasingly exceeds the capacity for expert visual analysis, resulting in a growing need for automated image analysis. In particular, accurate and reliable methods for segmentation (classifying image regions) are a key requirement for the extraction of information from images, cf. Chap. 11, page 279. A common approach to automatic segmentation is atlas-based segmentation in which an anatomical atlas is registered to an individual's anatomical image [83,84]. By transforming the segmentation of the atlas into the coordinate system of the subject, one can obtain a segmentation of the subject's image. Errors in the registration process will affect the accuracy of the propagated segmentation.

More recently, it has been shown that using multiple atlases and combining the multiple segmentations obtained by registration, random errors in the registration can be compensated for, resulting in an improved segmentation [85]. Using this method, each atlas is registered to the subject in question. The resulting transformation is then used to transformation the atlas segmentation into the subject's coordinate system. By applying decision fusion techniques at every voxel in subject space, e.g., using the vote rule [86] or Simultaneous Truth And Performance Level Estimation (STAPLE) [87], an average segmentation can be generated (Fig. 5.2).

Shape Modeling

Statistical models of shape variability have been successfully applied to perform various analysis tasks in 2D and 3D images. In particular, their application for image segmentation in the context of Active Shape Models (ASMs) has been very successful [88]. In building those statistical models, a set of segmentations of the shape of interest is required as well as a set of landmarks that can be unambiguously defined in each sample shape. However, the manual identification of such correspondences is a time consuming and tedious task. This is particularly true in 3D where the amount of landmarks required to describe the shape accurately increases dramatically as compared

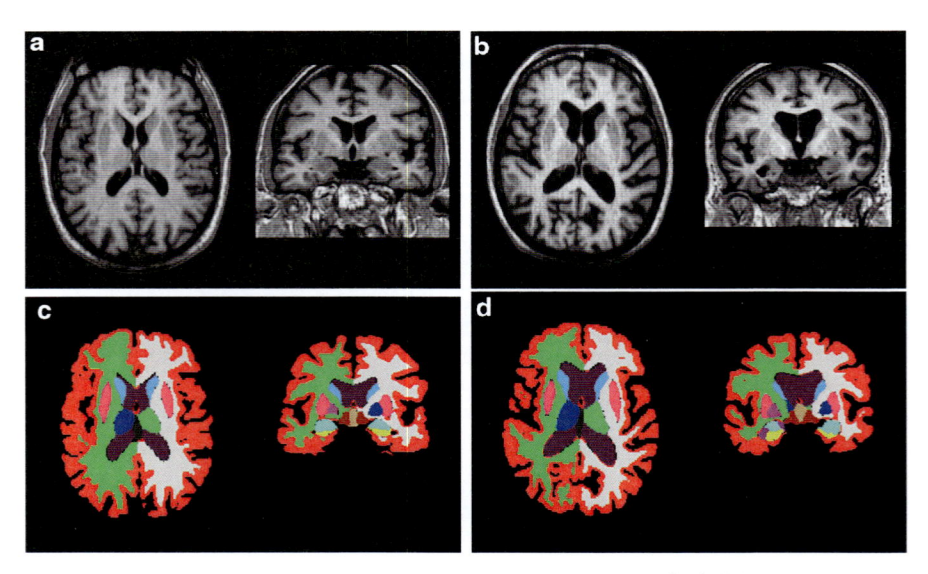

Fig. 5.2. *Example of multi-atlas segmentation in brain MRI [85].* (**a**) Normal elderly subject and (**c**) corresponding segmentation. (**b**) subject with Alzheimer's Disease and (**d**) corresponding segmentation. Despite the significant differences between normal and pathologic anatomy, the segmentation produced by multi-atlas fusion is stable

to 2D applications. While several approaches have been proposed that aim at addressing this problem by automating the process of correspondence estimation [89–91], the problem of finding correspondences for the construction of shape models can be solved via registration. For example, Frangi et al. [92] have proposed an procedure in which an atlas or reference shape is registered to all other shapes and the landmarks of the atlas are propagated to all other shapes. The registration used for this purpose can be either surface-based [93] or image-based registration [92].

A complementary approach to building statistical shape models of anatomical variability is the concept of statistical deformation models, which is closely related to the rapidly developing discipline on computational anatomy pioneered by the work of Grenander and Miller [94]. One of the key ideas here is to carry out a statistical analysis directly on the deformation fields that describe a dense correspondence between the anatomies. There are two distinct advantages of this approach: First, the resulting statistical models are not limited to single anatomical structures but can instead describe the intra- and inter-structure variability of the anatomy across a population. Second, the deformation fields can be obtained by non-rigid registration, thereby eliminating the need for any segmentation. Such an approach has been developed by several groups [95–97].

5.7 Summary and Conclusions

In this chapter, we have presented a number of theoretical and practical aspects of medical image registration. Medical image registration is widely used, both in clinical applications (e.g., image fusion, image-guided surgery) as well as a tool for biomedical research (e.g., to study populations in clinical trials). The components of a registration method very much depend on the clinical application at hand: The majority of applications in clinical practice uses rigid or affine registration where automated solutions have been shown to be accurate, robust and fast. The transformation model in rigid/affine registration is obviously fixed, but current developments focus on finding better registration bases (similarity measures, in particular for multi-modal or contrast-enhanced imaging), as well as faster optimization techniques for automated real-time image guidance. There is an increasing interest in non-rigid registration for clinical applications. However, automated solutions have not yet reached the same degree of maturity as for rigid or affine registrations. non-rigid registration is still an area of on-going research and most algorithms are in the stage of development and evaluation.

A particular development focus is on diffeomorphic transforms, which allow a one-to-one (invertible) mapping between different subjects, and more accurate intra-subject registration in dynamic imaging, such as contrast-enhanced MRI. For physically more plausible transforms, local constraints are investigated in an attempt to preserve volume locally, and ensure smooth deformation fields. Biomechanical motion models are limited by lack of a priori knowledge about patient-specific tissue properties, and are generally computationally very complex. The lack of a generic gold standard for assessing and evaluating the success of non-rigid registration algorithms is mostly performed using volume overlaps, sparse landmark distances or surface distances of delineated structures, but non-rigid validation methodologies have been emerging using biomechanical motion or disease simulations [68,70] or physical phantoms [98] for specific clinical applications.

References

1. Hajnal JV, Hill DLG, Hawkes DJ, et al. Medical Image Registration. Boca Raton: CRC Press; 2001.
2. Zitová B, Flusser J. Image registration methods: a survey. Image Vis Comput. 2003;21(11):977–1000.
3. Foley J, Dam A, Feiner S, et al. Computer Graphics. 2nd ed. Reading: Addison Wesley; 1990.
4. Szeliski R, Lavallée S. Matching 3-D anatomical surfaces with non-rigid deformations using octree-splines. In: IEEE Workshop Biomed Image Anal. 1994; p. 144–53.
5. Tsai RY. A versatile camera calibration technique for high-accuracy 3D machine vision metrology using off-the-shelf TV cameras and lenses. IEEE J Rob Autom. 1987;3(4):323–44.
6. Amit Y, Grenander U, Piccioni M. Structural image restoration through deformable templates. J Am Stat Assoc. 1991;86(414):376–87.
7. Ashburner J, Friston KJ. Nonlinear spatial normalization using basis functions. Hum Brain Mapp. 1999;7:254–66.
8. Amit Y. A nonlinear variational problem for image matching. SIAM J Sci Comput. 1994;15(1):207–24.
9. Bookstein FL. Thin-plate splines and the atlas problem for biomedical images. Inf Process Med Imaging. 1991;326–42.
10. Meyer CR, Boes JL, Kim B, et al. Demonstration of accuracy and clinical versatility of mutual information for automatic multimodality image fusion using affine and thin-plate spline warped geometric deformations. Med Image Anal. 1997;1(3):195–207.
11. Davis MH, Khotanzad A, Flamig DP, et al. A physics-based coordinate transformation for 3-D image matching. IEEE Trans Med Imaging. 1997;16(3):317–28.
12. Duchon J. Interpolation des fonctions de deux variables suivant les principes de la flexion des plaques minces. RAIRO Anal Numérique. 1976;10:5–12.
13. Meinguet J. Multivariate interpolation at arbitrary points made simple. Z Angew Math Phys. 1979;30:292–304.
14. Goshtasby A. Registration of images with geometric distortions. IEEE Trans Geosci Remote Sens. 1988;26(1):60–4.
15. Bookstein FL. Principal warps: thin-plate splines and the decomposition of deformations. IEEE Trans Pattern Anal Mach Intell. 1989;11(6):567–85.
16. Arad N, Dyn N, Reisfeld D, et al. Image warping by radial basis functions: application to facial expressions. Comput Vis Graph Image Process. 1994;56(2):161–72.

17. Little JA, Hill DLG, Hawkes DJ. Deformations incorporating rigid structures. Comput Vis Image Underst. 1997;66(2):223–32.
18. Bookstein FL, Green WDK. A feature space for edgels in images with landmarks. J Math Imaging Vis. 1993;3:231–61.
19. Hartkens T, Rohr K, Stiehl HS. Evaluation of 3D operators for the detection of anatomical point landmarks in mr and ct images. Comput Vis Image Underst. 2002;86(2):118–36.
20. Sederberg TW, Parry SR. Free-form deformation of solid geometric models. Proc SIGGRAPH. 1986;20(4):151–160.
21. Lee S, Wolberg G, Chwa KY, et al. Image metamorphosis with scattered feature constraints. IEEE Trans Vis Comput Graph. 1996;2(4):337–54.
22. Lee S, Wolberg G, Shin SY. Scattered data interpolation with multilevel b-splines. IEEE Trans Vis Comput Graph. 1997;3(3):228–44.
23. Rueckert D, Hayes C, Studholme C, et al. Non-rigid registration of breast MR images using mutual information. Lect Notes Comput Sci. 1998; 1496:1144–52.
24. Rueckert D, Sonoda LI, Hayes C, et al. Non-rigid registration using free-form deformations: application to breast MR images. IEEE Trans Medical Imaging. 1999;18(8):712–21.
25. Rohlfing T, Maurer CR. Nonrigid image registration in shared-memory multiprocessor environments with application to brains and breasts and and bees. IEEE Trans Inf Technol Biomed. 2003;7(1):16–25.
26. Rohlfing T, Maurer CR, Bluemke DA, et al. Volume-preserving nonrigid registration of MR breast images using free-form deformation with an incompressibility constraint. IEEE Trans Medical Imaging. 2003;22(6):730–41.
27. Kybic J, Unser M. Fast parametric elastic image registration. IEEE Trans Image Process. 2003;12(11):1427–42.
28. Mattes D, Haynor DR, Vesselle H, et al. PET–CT image registration in the chest using free-form deformations. IEEE Trans Med Imaging. 2003;22(1):120–28.
29. Horn BKP, Schnuck BG. Determining optical flow. Artif Intell. 1981;17:185–203.
30. Bajcsy R, Kovačič S. Multiresolution elastic matching. Computer Vis Graph Image Process. 1989;46:1–21.
31. Christensen GE. Deformable Shape Models for Anatomy. Washington University; 1994.
32. Gee JC, Barillot C, Briquer LL, et al. Matching structural images of the human brain using statistical and geometrical features. In: Proc VBC. 1994; p. 191–204.
33. Press WH, Flannery BP, Teukolsky SA, et al. Numerical Recipes in C. 2nd ed. Cambridge: Cambridge University Press; 1989.
34. Gee JC, Haynor DR, Reivich M, et al. Finite element approach to warping of brain images. Proc SPIE. 1994;2167:18–27.
35. Davatzikos C. Spatial transformation and registration of brain images using elastically deformable models. Comput Vis Image Unders. 1997;66(2):207–22.
36. Beg MF, Miller MI, L Younes AT. Computing metrics via geodesics on flows of diffeomorphisms. Int J Comput Vis. 2005;61(2):139–57.
37. Christensen GE, Rabbitt RD, Miller MI. Deformable templates using large deformation kinematics. IEEE Trans Image Process. 1996;5(10):1435–47.
38. Bro-Nielsen M, Gramkow C. Fast fluid registration of medical images. In: Proc VBC. 1996; p. 267–76.
39. Thirion JP. Image matching as a diffusion process: an analogy with Maxwell's demons. Med Image Anal. 1998;2(3):243–60.

40. Arun KS, Huang TS, Blostein SD. Least squares fitting of two 3D point sets. IEEE Trans Pattern Anal Mach Intell. 1987;9:698–700.
41. Maurer CR, Fitzpatrick JM, Wang MY, et al. Registration of head volume images using implantable fiducial markers. IEEE Trans Med Imaging. 1997;16:447–62.
42. Wang MY, Maurer CR, Fitzpatrick JM, et al. An automatic technique for finding and localizing externally attached markers in CT and MR volume images of the head. IEEE Trans Biomed Eng. 1996;43:627–37.
43. Pelizzari CA, TY CG, Spelbring DR, et al. Accurate three dimensional registration of CT and PET and/or MR images of the brain. J Comput Assisted Tomography. 1989;13:20–6.
44. Jiang H, Robb RA, Holton KS. A new approach to 3-D registration of multimodality medical images by surface matching. In: Proc VBC. 1992; p. 196–213.
45. Besl PJ, McKay ND. A method for registration of 3-D shapes. IEEE Trans Pattern Anal Mach Intell. 1992;14(2):239–56.
46. Wilhelms J, Gelder AV. Octrees for faster isosurface generation. ACM Trans Graph. 1992;11(3):201–27.
47. Borgefors G. Distance transformations in digital images. Computer Vis Graph Image Process. 1986;34:344–71.
48. Maurer CR, Qi R, Raghavan V. A linear time algorithm for computing exact euclidean distance transforms of binary images in arbitrary dimensions. IEEE Trans Pattern Anal Mach Intell. 2003;25(2):265–70.
49. Viola P. Alignment by maximization of mutual information. Int J Comput Vis. 1997;24(2):134–54.
50. Hajnal JV, Saeed N, Oatridge A, et al. Detection of subtle brain changes using subvoxel registration and subtraction of serial MR images. J Comput Assist Tomogr. 1995;19(5):677–91.
51. Hajnal JV, Saeed N, et al. A registration and interpolation procedure for subvoxel matching of serially acquired MR images. J Comput Assist Tomogr. 1995;19(2):289–96.
52. Collignon A, Maes F, Delaere D, et al. Automated multimodality image registration using information theory. Proc Int Conf Inform Process Med Imaging (IPMI). 1995; p. 263–74.
53. Viola P, Wells WM. Alignment by maximization of mutual information. In: Proc 5th International Conference on Computer Vis (ICCV'95). 1995; p. 16–23.
54. Studholme C, Hill DLG, Hawkes DJ. An overlap invariant entropy measure of 3D medical image alignment. Pattern Recognit. 1998;32(1):71–86.
55. Maes F, Collignon A, Vandermeulen D, et al. Multimodality image registration by maximization of mutual information. IEEE Trans Med Imaging. 1997;16(2):187–98.
56. Duda RO, Hart PE. Pattern Classification and Scene Analysis. New York: Wiley; 1973.
57. Terzopoulos D. Regularization of inverse visual problems involving discontinuities. IEEE Trans Pattern Anal Mach Intell. 1986;8(4):413–24.
58. Wahba G. Spline models for observational data. Soc Ind Appl Math; 1990.
59. Courant R, Hilbert D. Methods of Mathematical Physics vol 1. London: Interscience; 1953.
60. Fitzpatrick JM. Detecting failure and assessing success. Proc Med Image Regist. 2001; p. 117–39.

61. Studholme C, Hill DLG, Hawkes DJ. Automated 3D registration of MR and PET brain images by multi-resolution optimisation of voxel similarity measures. Med Phys. 1997;24(1):25–35.
62. Holden M, Hill DLG, Denton ERE, et al. Voxel similarity measures for 3-D serial MR brain image registration. IEEE Trans Med Imaging. 2000;19(2):94–102.
63. Fitzpatrick JM, Hill DLG, Shyr Y, et al. Visual assessment of the accuracy of retrospective registration of MR and CT images of the brain. IEEE Trans Med Imaging. 1998.
64. Denton ERE, Sonoda LI, Rueckert D, et al. Comparison and evaluation of rigid and non-rigid registration of breast MR images. J Comput Assist Tomogr. 1999;23:800–5.
65. Tanner C, Schnabel JA, Chung D, et al. Volume and shape preservation of enhancing lesions when applying non-rigid registration to a time series of contrast enhancing MR breast images. Lect Notes Comput Sci. 2000;1935:100–10.
66. West JB, Fitzpatrick JM, Wang MY, et al. Comparison and evaluation of retrospective intermodality image registration techniques. J Comput Assist Tomogr. 1997;21:554–66.
67. Koshani R, Hub M, Balter J, et al. Objective assessment of deformable image registration in radiotherapy: a multi-institution study. Med Phys. 2009;35(12):5944–53.
68. Schnabel JA, Tanner C, Castellano-Smith AD, et al. Validation of non-rigid image registration using finite element methods: application to breast MR images. IEEE Trans Med Imaging. 2003;22(2):238–47.
69. Hipwell JH, Tanner C, Crum WR, et al. A new validation method for x-ray mammogram registration algorithms using a projection model of breast x-ray compression. IEEE Trans Med Imaging. 2007;26(9):1190–200.
70. Camara O, Schnabel JA, Ridgway GR, et al. Accuracy assessment of global and local atrophy measurement techniques with realistic simulated longitudinal Alzheimer's disease images. NeuroImage. 2008;42(2):696–709.
71. Klein A, Andersson J, Ardekani BA, et al. Evaluation of 14 nonlinear deformation algorithms applied to human brain MRI registration. NeuroImage. 2009;46(3):786–802.
72. Studholme C, Hill DLG, Hawkes DJ. Multiresolution voxel similarity measures for MR-PET registration. Proc Int Conf Inform Process Med Imaging (IPMI). 1995; p. 287–98.
73. Studholme C, Hill DLG, Hawkes DJ. Automated 3-D registration of MR and CT images of the head. Med Image Anal. 1996;1(2):163–75.
74. Freeborough PA, Fox NC. Modeling brain deformations in Alzheimer disease by fluid registration of serial 3D MR images. J Comput Assist Tomogr. 1998;22(5):838–843.
75. Rey D, Subsol G, Delingette H, et al. Automatic detection and segmentation of evolving processes in 3D medical images: application to multiple sclerosis. Proc Int Conf Inform Process Med Imaging (IPMI). 1999; p. 154–67.
76. Paus T, Zijdenbos A, Worsley K, et al. Structural maturation of neural pathways in children and adolescents: in vivo study. Science. 1999;283:1908–11.
77. Thompson PM, Giedd JN, Woods RP, et al. Growth patterns in the developing brain detected by continuum mechanical tensor maps. Nature. 2000;404:190–3.
78. Thompson PM, Vidal C, Giedd JN, et al. Mapping adolescent brain change reveals dynamic wave of accelerated grey matter loss in very early-onset schizophrenia. Nature. 2001;98(20):11650–5.

79. Fox NC, Crum WR, Scahill RI, et al. Imaging of onset and progression of Alzheimer's disease with voxel-compression mapping of serial magnetic resonance images. Lancet. 2001;358:201–5.
80. Thomson PM, Cannon TD, Narr KL, et al. Genetic influences on brain structure. Nat Neurosci. 2001;4(12):1–6.
81. Ashburner J, Friston KJ. Voxel-based morphometry: the methods. NeuroImage. 2000;11(6):805–21.
82. Ashburner J, Hutton C, Frackowiak R, et al. Identifying global anatomical differences: deformation-based morphometry. Hum Brain Mapp. 1998;6:638–57.
83. Miller M, Christensen GE, Amit Y, et al. Mathematical textbook of deformable neuroanatomies. Proc Natl Acad Sci USA 1993;90:11944–8.
84. Collins DL, Evans AC. Animal: validation and applications of nonlinear registration-based segmentation. Int J Pattern Recognit Artif Intell. 1997;11:1271–94.
85. Heckemann RA, Hajnal JV, Aljabar P, et al. Automatic anatomical brain MRI segmentation combining label propagation and decision fusion. Neuroimage. 2006;33(1):115–26.
86. Rohlfing T, Maurer CR. Multi-classifier framework for atlas-based image segmentation. Pattern Recognit Lett. 2005;26(13):2070–79.
87. Warfield SK, Zhou KH, Wells WM. Simultaneous truth and performance level estimation (STAPLE): an algorithm for the validation of image segmentation. IEEE Trans Med Imaging. 2004;23(7):903–21.
88. Cootes TF, Taylor CJ, Cooper DH, et al. Active shape models: their training and application. Comput Vis Image Underst. 1995;61(1):38–59.
89. Davies RH, Twining CJ, Cootes TF, et al. 3D Statistical shape models using direct optimization of description length. Proc Eur Conf Comput Vis (ECCV). 2002; p. 3–20.
90. Fleute M, Lavallee S. Nonrigid 3-D/2-D registration of images using statistical models. Lect Notes Comput Sci. 1999;1679:879–87.
91. Kelemen A, Székely G, Gerig G. Elastic model-based segmentation of 3-D neurological data sets. IEEE Trans Med Imaging. 1999;18(10):828–39.
92. Frangi AF, Rueckert D, Schnabel JA, et al. Automatic construction of multiple-object three-dimensional statistical shape models: application to cardiac modeling. IEEE Trans Med Imaging. 2002;21(9):1151–66.
93. Yuhui Y, Bull A, Rueckert D, et al. 3D statistical shape modelling of long bones. Lect Notes Comput Sci. 2003;2717:306–14.
94. Grenander U, Miller MI. Computational anatomy: an emerging discipline. Q Appl Math. 1998;56(4):617–94.
95. Joshi SC. Carge Deformation Diffeomorphisms and Gaussian Random Fields for Statistical Characterization of Brain Sub-manifolds. Wash University; 1998.
96. Gee JC, Bajcsy RK. Elastic matching: continuum mechnanical and probabilistic analysis. In: Toga AW, editor. Brain Warping. San Diego: Academic; 1999. p. 183–97.
97. Rueckert D, Frangi AF, Schnabel J. Automatic construction of 3d statistical deformation models of the brain using non-rigid registration. IEEE Trans Med Imaging. 2003;22(8):1014–25.
98. Koshani R, Hub M, Balter J, et al. Objective assessment of deformable image registration in radiotherapy: a multi-institution study. Med Phys. 2007;35(12):5944–53.

Part III

Feature Extraction and Selection

6
Texture in Biomedical Images

Maria Petrou

Summary. An overview of texture analysis methods is given and the merits of each method for biomedical applications are discussed. Methods discussed include Markov random fields, Gibbs distributions, co-occurrence matrices, Gabor functions and wavelets, Karhunen–Loève basis images, and local symmetry and orientation from the monogenic signal. Some example applications of texture to medical image processing are reviewed.

6.1 Introduction

Texture is variation of the data at scales smaller than the scales of interest. For example, if we are interested in identifying the human brain in a Magnetic Resonance Imaging (MRI) image, any variation in the gray values of the imaged brain may be thought of as texture. According to this definition, even variation due to noise may be thought of as texture. In the computer graphics community this is acceptable, but among image processing and pattern recognition researchers, texture is a property intrinsic to the imaged object and not something caused by the instrument with which the image has been captured, like noise is. Once we accept texture to be an intrinsic property of the imaged object, then texture becomes a valuable cue in relation to the recognition of this object. For example, the texture created by the sulci helps us to identify the brain as such.

The most important characteristic of texture is that it is scale dependent. Different types of texture are visible at different scales. For example, if we look at a section of the human brain through a microscope, we are going to see different structure of the tissue than the sulci mentioned above. In order to be able to use texture to identify different types of tissue or different human organs, we must be able to measure it in a repeatable and reliable way. In other words, it is necessary to be able to characterize texture in an objective way, independent of human perception and visual abilities. In this chapter, we first present a brief overview of texture quantification methods and discuss

T.M. Deserno (ed.), *Biomedical Image Processing*, Biological and Medical Physics,
Biomedical Engineering, DOI: 10.1007/978-3-642-15816-2_6,
© Springer-Verlag Berlin Heidelberg 2011

the peculiarities of texture analysis in relation to biomedical images. Then some examples of successful use of texture cues in medical applications are presented.

The simplest problem one tries to solve using texture is to identify the class of a texture swatch. In this case, one assumes the availability of a library of textures of different classes, all captured under the same imaging conditions as the texture swatch in question. The issue then is to quantify the query texture in the same way as the library textures have been quantified, and compare the extracted values with those of the library textures, to assess similarity or divergence. The query texture is identified as of the same class as the texture in the library with the most similar numerical values.

The second, and most difficult problem one has to deal with, is that of simultaneous isolation and characterization of the texture. This problem arises very often, as in the same image, different depicted objects may be characterized by different textures. The problem becomes complicated as soon as one realizes that the borders between the different textures are often diffuse, with the textures blending gradually into each other, and that the shapes of the different texture patches are not necessarily regular and sometimes the patches are not even extensive enough for the texture to be easily characterizable by automatic methods.

Next, we shall discuss these two problems in turn.

6.2 Characterizing the Texture of Swatches

6.2.1 From Grammars to Markov Random Fields

Texture may be characterized by structural or statistical methods. A structural method identifies a texture primitive pattern and the rules by which this is repeated to create the viewed appearance. This approach leads to the so called grammars for texture description. Such methods are of little interest to biomedical applications, but they tend to be appropriate for man-made textures that present some regularity. Statistical approaches are of more interest. Figure 6.1 shows schematically how natural the transition from deterministic methods to statistical methods is. In this example, we may easily infer in the first case that a possible texture primitive is a 1 surrounded by four 0s in its four immediate neighboring positions. In the second case, we may work out the pattern by inspecting the instantiations of this type of neighborhood we

1	0	1	0	1	0	1
0	1	0	1	0	1	0
1	0	1	0		0	1
0	1	0	1	0	1	0
1	0	1	0	1	0	1
0	1	0	1	0	1	0
1	0	1	0	1	0	1

1	0	1	0	1	0	1
0	1	0	1	0	1	0
2	0	1	0		0	1
0	2	0	2	0	1	0
2	0	1	0	2	0	1
0	1	0	2	0	2	0
2	0	1	0	1	0	1

Fig. 6.1. *Deterministic versus probabilistic context dependence. Left*: It is obvious that the missing value should be 1. *Right*: The missing value has probability 58% to be 1 and probability 42% to be 2

have. To avoid border effects, we may ignore the pixels in the periphery of the image. We can see then that, in 7 out of the 12 cases, the cross arrangement of four 0s indicates the presence of a 1 in the middle, while in the remaining 5 cases, it indicates the presence of a 2. We may then say that the missing value is a 2 with probability 42% and a 1 with probability 58%. Such texture modeling is known as Markov Random Field (MRF) modeling.

In MRF, we model the probability with which the values of the neighbors determine the value of the pixel in question. The parameters of this modeling are used to characterize the texture. As the statistics have to be performed on neighbor value combinations, and as pixels, particularly in medical images, take up a large number of gray values, one needs a very large sample of the texture in order to work out these parameters. It is often the case that the gray values of the pixels are re-quantized in fewer bins, in order to be able to work out the local patterns.

6.2.2 From Markov Random Fields to Fractals

One of the issues in MRF texture approaches is the definition of the size of the neighborhood. The neighborhood we considered in the above example was the smallest possible: only the four immediate neighbors in the grid. However, neighborhoods may be much larger, to include pixels further away from the focal pixel, and they may also consist of disconnected pixels, i.e., pixels that are not physical neighbors in the grid. At the extreme case, all other pixels in the image may be thought of as neighbors. In that case, the model explicitly models the dependence of the value of a pixel on the values of all other pixels. It is as if we are assuming that the value of a pixel is *directly* influenced by the values of all other pixels in the image. This is like saying that no matter how far a pixel is in the image, it has a saying in dictating the value of the pixel in question. In other words, all scales of influence are important.

Such a model leads to the fractal modeling of texture. Fractal models assume that at whatever scale the pattern is viewed, it will look the same. Such models may be relevant in modeling the texture created by vascularity, since the branches of the vascular tree create smaller branches and those create smaller branches and so on, so that a snapshot of this pattern looks the same at whatever scale it is viewed. Indeed, fractal models have been used in medical applications to model such structures. However, they are not very appropriate as generic texture descriptors, as patterns rarely show such statistical regularity for a wide range of scales.

6.2.3 From Markov Random Fields to Gibbs Distributions

Going back to MRF texture descriptors, and given that the value of a pixel is modeled as depending on the values of the neighbors, which have values that depend on their own neighbors, which also have values that depend on their own neighbors, and so on, the idea that the value of a pixel is influenced by the values of all other pixels does not seem that absurd. The difference is that fractal models model this global dependence *explicitly*, and with a model that has a particular form across the scales, while MRF models model this global dependence *implicitly*. It is obvious then that an MRF formulation may lead to a formulation that models the *joint* probability density function of a particular combination of values of the whole image to arise. Indeed, under certain conditions, an MRF modeling is equivalent to a Gibbs modeling that expresses exactly that: the joint probability density function of the whole grid of values.

This model is expressed in terms of the so called clique potentials. A clique is a set of grid positions that are neighbors of each other according to the Markov model, and, therefore, directly influence each other's value according to this model. In the example of Fig. 6.1, the Markov neighborhood we used consisted of the four immediate neighbors of the pixel. Then the only cliques one can have are pairs of pixels next to each other either vertically or horizontally. In the corresponding Gibbs formalism, there are parameters which multiply either the product or the difference of the pairs of values of neighboring pixels, to express their dependence. These parameters characterize the texture and they correspond to the Markov parameters of the MRF formalism.

6.2.4 Co-occurrence Matrices

Gibbs distributions as such have not been very helpful in texture description. They are, however, very helpful in other problems of medical image processing, like image restoration or image matching. What is important for us to consider here is how we moved from the local description (neighborhood dependent modeling) to a global (joint probability) and that from this global description, we may try to go back to local descriptors, but by a different route. Indeed, we may think of an MRF formalism as the calculation of a *marginal distribution*: from the global joint probability density function of the whole configuration of gray values, we calculate the local conditional probability of the values of a pixel, given the values of its neighbors. It is very logical then to define other marginals that may do even a better job than the MRFs. This naturally leads to the definition of the so called co-occurrence matrix. From the global joint probability density function, we may move to the joint probability of the values of a pair of grid cells, in a particular relative position from each other. This joint probability is computed directly from the data, in the form of a double entry table: each cell of this table corresponds to a pair of gray values; the value of the cell tells us how many pairs of these values in the particular relative position were identified in the image.

For the example of Fig. 6.1, this table will be 3×3 in size, as the possible values of a pixel are only three. Figure 6.2 shows some co-occurrence matrices

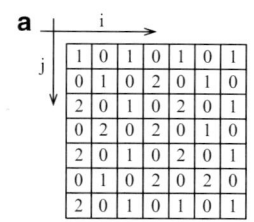

Fig. 6.2. *The concept of co-occurrence matrix.* Four co-occurrence matrices of the image at the top, for pixels in relative positions (**b**) $[(i, j), (i + 1, j)]$, (**c**) $[(i, j), (i, j + 1)]$, (**d**) $[(i, j), (i + 2, j - 2)]$, and (**e**) $[(i, j), (i + 2, j + 1)]$. In the bottom row the co-occurrence matrices normalized so that their elements add up to 1 and, so, they can be treated like discretized probability density functions

of this image, constructed for the following four relative positions of grid cells: (a) next door neighbors along the horizontal direction, (b) next door neighbors along the vertical direction, (c) two positions in the direction of the main diagonal once removed, and (d) pairs of positions expressed as (i, j) and $(i + 2, j + 1)$. These co-occurrence matrices are directional. It is possible to consider rotationally symmetric matrices, where the relative *distance* between two pixels is considered instead of their relative position. In either case, we can construct as many co-occurrence matrices as we like. These are bulky texture representations, that may not be used directly. It is a common practice to use them in order to compute image characteristic features from them. For example, if $C(i, j)$ is a co-occurrence matrix and we compute the sum $\sum_i \sum_j |i - j| C(i, j)$, we shall have a measure of image contrast for the relative positions used to construct the matrix, since indices i and j refer to the gray values in that pair of positions, and $C(i, j)$ tells us how frequently this pair of values is observed. For the co-occurrence matrices of Fig. 6.2, the values of this feature are: 1.41, 1.45, 0.38, and 1.47, respectively.

6.2.5 Generalized Co-occurrence Matrices

All the methods discussed so far utilize the gray values of the image. It is possible, however, to transform the image before any of the above methods is used. An image transform does not create any new information. It does, however, make explicit information that is only implicit there. A simple image transform is to compute the magnitude and orientation of the local gradient vector at each pixel position. So, each pixel now carries three numbers: (a) its gray value, (b) its gradient magnitude, and (c) its gradient orientation, measured with respect to some reference direction. One may envisage then marginals that measure the joint probability density function of any combination of these pixel values. This leads to the definition of the generalized co-occurrence matrices [1]. Such a matrix may present the number of pairs of pixels that are at a particular relative position from each other, one has this gray value, the other has that gray value, one has such gradient magnitude, the other has such gradient magnitude, and the relative orientation of their gradient vectors is such. The obvious consequence of such a generalization is that the co-occurrence matrix now is multi-dimensional. In this particular example, we need two indices for the gray values of the two pixels, we need two indices for their gradient magnitudes and we need one index for the relative gradient orientation of the two pixels. This means that the co-occurrence matrix is now five-dimensional. The issue is that gradient magnitude and orientation are continuous variables, so we have to decide what we mean by "index" corresponding to these variables. The way to deal with this problem is to quantize these values to a small number of levels, and use them as indices. The small number of levels is necessary also in order to keep the size of the co-occurrence matrix down.

6.2.6 Orientation Histograms

Alternatively, one may work out the marginal distribution of gradient vector orientations. This can be done counting the number of gradient vectors per orientation. Again, language must be used carefully: orientation is a continuous variable and one obviously has to use bins to quantize the orientations into a finite number. This way, the so called orientation histogram is created. For a 2D image the orientation histogram is 1D since one angle is enough to measure orientation. For 3D data, like the data obtained from various types of tomography, an orientation is defined by two angles, the polar and the azimuth angle [2]. The orientation histogram then is 2D. In this case care should be taken so that the solid angle represented in the 3D space by each bin is the same for all bins.

Often, for visualization purposes, the orientation histogram is presented in polar coordinates (Fig. 6.3). It is very easy then at a glance to identify data anisotropies and their direction. For example, if the texture is isotropic, the orientation histogram of the 2D image will resemble a circle and for 3D data a sphere. If the data are anisotropic, the orientation histogram will be elongated along the direction of maximum image variation.

In either case, features may be computed from the orientation histogram to characterize its shape and thus characterize the texture to which it corresponds. These features may be computed from the bin values, if we treat them as the samples of a function. So, their standard deviation or the ratio of the maximum over the minimum value of the function are useful. Other features may be the ratio of a particular pair of values of this histogram/function. An alternative way to see the histogram, when plotted in polar coordinates, is to view it as a shape that has to be characterized by a few numbers.

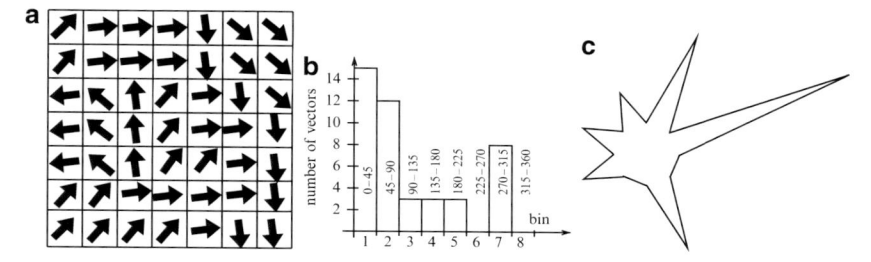

Fig. 6.3. *Orientation histogram.* (**a**) An image with the gradient vectors associated with its pixels. (**b**) The orientation histogram of this image with 8 bins. In each bin we count the number of gradient vectors that have orientation in the range of values of the bin. (**c**) The orientation histogram may be visualized as a function plotted in polar coordinates. The radius of the function along the orientation that corresponds to the center of each bin is equal to the value of the bin. Looking at this representation we immediately infer that this image is anisotropic, with strong variation along direction 22.5° with respect to the reference direction

These numbers, could be, for example, the coefficients of the expansion of this shape in terms of spherical harmonics, that could be used as texture features. Yet another way to view the orientation histogram is like a probability density function and compute its various moments as features of the texture it represents.

6.2.7 Textons

Gradient magnitude is a measure of the first order image differences. One may also use filters that respond to second image differences, i.e., they detect lines or ribbons. A combination of such filter outputs may be used and a statistical approach may followed to identify representative outputs, which are called textons [3]. The similarity then of the outputs of the same filters when applied to a test image with those of the prototypes identified may be used to characterize the image.

6.2.8 Features from the Discrete Fourier Transform

Another obvious transform before proceeding to quantify texture is to transform the image to the spatial frequency domain. Indeed, the Discrete Fourier Transform (DFT) identifies the frequency content of an image. In general, the DFT is complex, but we usually adopt its magnitude, which is known as the power spectrum. The power spectrum may help us identify the basic periodicities that are present in the texture. For example, if it contains spikes at some frequencies, the implication is that some basic structure, not necessarily discernible by the human eye, is repeated in the data with that periodicity.

We may wish to analyze the frequency content of the image in terms of different frequency ranges (bands). Then, it is necessary to isolate different bands in the frequency domain and work out image characteristics that correspond to these bands. This can be done with the help of some window, centered at a particular frequency, which isolates the required band. The most commonly used window is the Gaussian, because it does not create artifacts in the image (Fig. 6.4).

The DFT of the image is multiplied with the appropriate masking window to isolate the right frequency band, and the result is Fourier transformed back. The resulting "image" contains only characteristics that have frequencies in the selected bands. We may then compute features from them to characterize the texture. The most commonly used feature is the sum of the squares of the values of the output image in order to find its energy in the particular frequency band. Doing this systematically, for all bands that are needed to cover the full frequency domain of the input image, will yield a feature vector, the components of which are the components of the energy of the input image in the different bands.

We must remember that multiplication in the frequency domain corresponds to convolution with the inverse DFT of the used filter in the spatial domain. We must also recall that the inverse Fourier transform of a Gaussian is also a Gaussian, with standard deviation the inverse of the standard deviation of the Gaussian in the frequency domain. This immediately leads to the idea that the features we compute, by multiplying the DFT of the image with a Gaussian and inverting the result back to the spatial domain, are not only localized in frequency, but also in space, since convolution of the image with a Gaussian is effectively a way of isolating part of the image around one pixel at a time. This way of interpreting the approach naturally leads us to the second problem of texture, namely that of the simultaneous segmentation and recognition of textures.

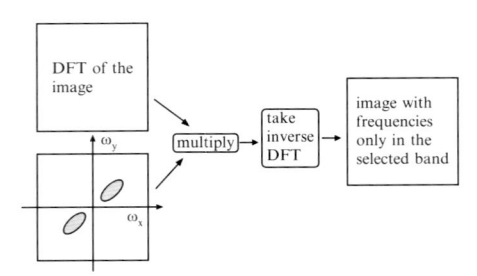

Fig. 6.4. *Frequency band splitting.* The DFT of an image may be multiplied with a masking filter to isolate the frequency content of the image in a certain band. The mask we use here is a Gaussian, with elliptical cross section. We have to make sure we use a pair of such masks symmetrically placed about the $(0, 0)$ frequency, so that, when we take the inverse DFT, we obtain a real, as opposed to a complex, output

6.3 Simultaneous Texture Segmentation and Recognition

The simultaneous segmentation and recognition of a texture requires the calculation of local features. The natural way to proceed is to consider small sub-images around each pixel and compute the features we wish to compute inside these subimages, while assigning the result to the central pixel. This brute force approach allows one to generalize easily all methods discussed in the previous section, for use in the calculation of local features.

One has to be careful, however, particularly if one is computing statistical features (as is usually the case): one needs a sufficiently large sub-image in order to have enough pixels to compute reliable statistics; one also needs a small enough sub-image in order to avoid interference between different textures. This is the well known aperture problem in image processing, for which there exists no solution. Every time, only a compromise in the choice of the size of the local window is reached.

Features that are considered inappropriate for such a treatment are features that require particularly intensive statistics, like MRFs and co-occurrence matrices. As the resolution required in medical applications is measured in millimeters, the decision whether we have enough samples for the statistical calculation depends on the resolution of the sensor. For example, a 2D Magnetic Resonance Imaging (MRI) image with 1 mm resolution per pixel allows us to have one hundred pixels inside $1\,\mathrm{cm}^2$. However, if the data is 3D, the same resolution allows us to have one thousand samples inside $1\,\mathrm{cm}^3$. Depending on the method we use, one hundred points may not be enough, but one thousand points may be plenty.

To improve the chances of reliable statistics, we re-quantize the number of gray levels. Typical medical images are 12-bit and often 16-bit. Typical images for other applications are 8-bit. Reducing the medical images to 8-bit is not adequate for computing reliable statistics for co-occurrence matrices. The number of gray levels has to reduce by at least another order of magnitude in order to have enough points to populate all cells we use. For example, if we have only ten gray levels, the co-occurrence matrix is 10×10, i.e., it consists of one hundred cells. In 2D, the hundred samples of $1\,\mathrm{cm}^2$ are far too few to populate one hundred cells, given that for a co-occurrence matrix of fixed relative position, at most one hundred pairs of pixels may be created. In 3D, the one thousand samples may be just about enough. For a co-occurrence matrix that considers pairs of pixels in fixed relative *distance* from each other, things are better, as each pixel may be used to create several pairs, depending on the distance considered. In the simplest of cases, where the distance is 1, each pixel contributes to two pairs (not four, because we do not want to count each pair twice) and this may produce just about acceptable statistics from the samples of $1\,\mathrm{cm}^2$.

Even with only ten gray levels, the MRF parameters cannot be estimated reliably: in the simplest of cases, where only the four immediate neighbors of a pixel are considered, there are 10^4 possible combinations of neighborhood values. For each one such combination, the statistics of the possible values of the central pixel have to be worked out. This is not possible, so MRF and directional co-occurrence matrix approaches are rather excluded from localizing the calculations. The approach that uses the orientation histogram may be localized, if we are careful in selecting the number of histogram bins. We have to make sure that the bins are not too many for the number of samples we have [4].

These problems do not arise if one uses frequency-based features to characterize the image locally. The use of a Gaussian filter in the frequency domain is known as expansion of the image spatio-frequency content in terms of Gabor functions. Once the process depicted in Fig. 6.4 has been repeated for several frequency bands, each image has been analyzed in a stack of new images, one for each band used. So, each pixel now carries a vector of values, one for each frequency band. These values may be used to cluster the pixels in a feature space and identify clumps of pixels with similar values, indicating that they belong to the same type of region.

This approach does not necessarily lead to spatially coherent clusters of pixels. Such clusters of pixels are recorded by the so called mean shift algorithm, which clusters the pixels in a spatio-feature space. In other words, if there are features constructed from L bands, the pixels are characterized by $L+2$ numbers, their values in the L bands plus their two spatial coordinates (or $L+3$ numbers in the case of volume data).

An alternative to Gabor functions are the wavelets. Wavelets also form bases for the spatio-frequency domain, allowing the analysis of an image in a stack of images that contain information localized in space and in frequency. The difference with Gabor functions is that wavelets allow the trade off between accuracy in the localization in space with accuracy in the localization in frequency [5–7].

6.3.1 From Spatio-Frequency to Spatio-Structural Space

All methods we discussed so far treated texture either as a statistical variation in the data, or as a variation with some repetition that may be characterized in the frequency domain. However, texture manifests itself also as a structure in the data. We touched upon this when we discussed the deterministic textures, where a texture primitive may be considered as repeated regularly, according to some rules (Fig. 6.1). The local structure of the data, however, may be captured by expanding the image locally in terms of some appropriate basis. The *frequency* content of the image is captured by expanding the image locally in terms of an appropriate *spatio-frequency* basis, either this is called Gabor functions or wavelets. In a similar way, the *spatio-structural* content of the image may be captured by using an appropriate basis too. In [5, 8], such a basis is constructed with the help of Walsh functions [9]. In fact, in [5] this space is referred to as the "what-looks-like-where" space. An incomplete such basis may be formed by Laws masks [10]. These bases measure the degree to which the image locally looks like a vertical edge from darker to brighter, a horizontal edge from darker to brighter, a horizontal edge from brighter to darker, a corner, etc.

Figure 6.5 shows a complete basis, in terms of which any 3×3 image neighborhood might be expanded, and another one, in terms of which any 5×5 image neighborhood might be expanded. The expansion of the image patch in terms of one such basis yields the coefficients of the expansion which tell us how much the local patch looks like the corresponding patch of the basis. Note that in both cases the first coefficient refers to the flat patch, i.e., it yields the local average of the image. This coefficient is of no relevance to texture, so it may be ignored, or used to normalize the other coefficients, so they become invariant to local image brightness. The normalized coefficients may be used as a feature vector to characterize the central pixel of the patch. They may even be combined in a unique way into a single scalar, so one does not have to deal with a multi-dimensional feature space. This was the case presented in [8], where the normalized coefficients were quantized and treated

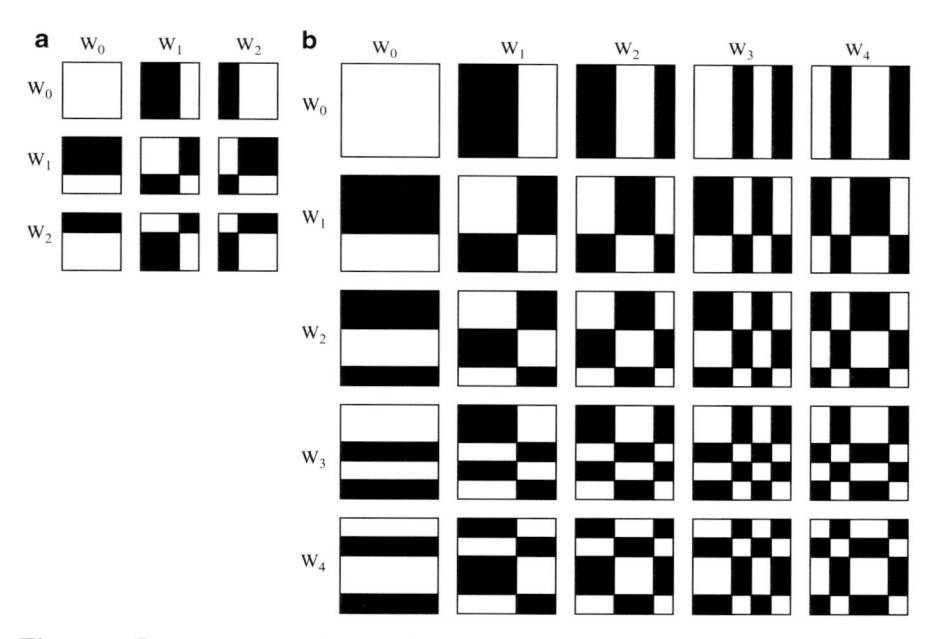

Fig. 6.5. *Bases constructed using the Walsh functions.* A complete basis in terms of which any (**a**) 3×3 or (**b**) 5×5 image patch may be expanded. The coefficients of the expansion may be used as texture features. Symbol W_x, with $x \in \{0, \ldots, 4\}$, indicates the Walsh function along the left that has to be multiplied with the Walsh function along the top, in vector outer product manner, to create the corresponding basis image

as the digits in a binary or a hexadecimal system, to form scalars. The result was called "structural image".

It is obvious that such an approach can be used either to classify swatches of texture, or to characterize an image locally so it can be segmented into textures. It is also obvious that the larger the local patch considered, the less appropriate this method becomes for medical images, as the basis images used contain several straight lines, that extent to the full width (or height) of the basis image, as well as corners, characteristics that are not often encountered in medical images (Fig. 6.5b). The question then arises: is it possible to construct a basis that is more appropriate for the type of data we have? The answer of course is yes, and this brings us back into the realm of statistics.

6.3.2 Statistical Spatio-Structural Space

If we have a collection of samples, we can always construct a basis in terms of which any sample of the collection can be represented as a linear combination. There are various methods for that. The simplest is to use clustering, as in [3]. The most well known method, however, is called Principal Component Analysis (PCA). This method is also known as Karhunen–Loève Transform (KLT). Other more complicated methods have also been developed, like, for example, Independent Component Analysis (ICA) and Non-negative Matrix Factorization (NMF). PCA constructs an orthogonal basis of *uncorrelated* components, while ICA constructs a basis of *independent* components. Once the basis has been constructed, any image patch of the same size may be expressed as a linear combination of the basis images. The coefficients of the expansion may be treated as the feature vector of the central pixel of the patch (Fig. 6.6).

This approach allows the creation of bases that are appropriate for the analysis of images of a certain type, e.g., mammograms. One may consider, however, the extraction of some more generic image characteristics, like, for example, quantities such as local contrast, local dominant orientation and local symmetry, which are not associated with a particular image type. Such

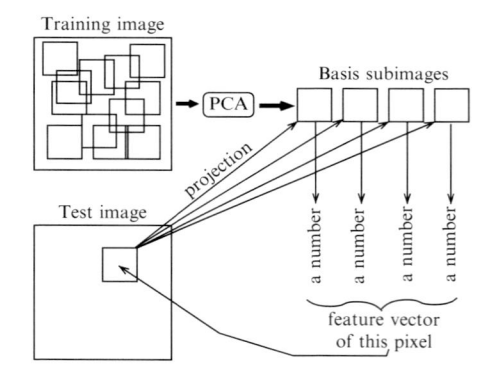

Fig. 6.6. *Constructing a basis using statistics.* Once a basis of local image patches has been identified, any patch of the test image may be expanded in terms of this basis. The coefficients of the expansion form the feature vector that is associated with the central pixel of the test patch

features may be extracted with the help of the frequency domain, although they are not frequency-type features. This leads to the use of the so called monogenic signal [11].

6.3.3 Monogenic Signal

The monogenic signal allows us to characterize an image locally, by its local contrast, local dominant orientation and local symmetry, without the need of statistics. Its definition appears to be complicated from the mathematical point of view, but it is really totally intuitive [9].

Let us start from the local contrast. This can be measured either by the magnitude of the local gradient vector, or by the second image derivative along various directions. The orientation may clearly be estimated from the relative strength of the components of the gradient vector. The local symmetry requires a little more thinking. A purely symmetric signal has a purely real Fourier transform, while a purely antisymmetric signal has a purely imaginary Fourier transform. One may then consider the creation of an appropriate basis in the Fourier domain, in terms of which the local signal is expanded. The ratio of the components of this expansion may be treated as a measure of local symmetry (or antisymmetry).

The first and second image derivative may be computed with the help of a pair of filters, designed for that job. The filters have to be appropriately normalized so that the sum of the squares of the elements of each is 1. An example of such a pair is $(1/\sqrt{2}, 0, -1/\sqrt{2})$ and $(-1/\sqrt{6}, 2/\sqrt{6}, -1/\sqrt{6})$. Note that the first one is an antisymmetric filter, so it has a purely imaginary Fourier transform, while the second is a symmetric filter, so it has a purely real Fourier transform. The sum of the squares of the elements of each filter is 1, so these two filters constitute an orthonormal basis in the complex Fourier domain. Such a pair of filters is known as a Hilbert transform pair. If we convolve a 1D signal with these filters, the output of the first one will be an estimate of the local first derivative of the signal, while the output of the second one will be an estimate of the local second derivative of the signal. The sum of the squares of the two outputs will be the total local signal contrast. The angle of the inverse tangent of the ratio of the two outputs will be a measure of local symmetry (Fig. 6.7).

Generalization to 2D is straightforward for the symmetric filter. However, in 2D we can have two antisymmetric filters, one in each direction. The outputs of these three filters constitute the monogenic signal. The outputs of the three filters squared and added will yield the local image contrast. The angle of the inverse tangent of the ratio of the outputs of the two antisymmetric filters will be the local orientation (naturally, since the outputs of the antisymmetric filter along the two orientations correspond to the components of the local gradient vector). The angle of the inverse tangent of the ratio of the square root of the sum of the squares of the outputs of the two antisymmetric filters (=gradient

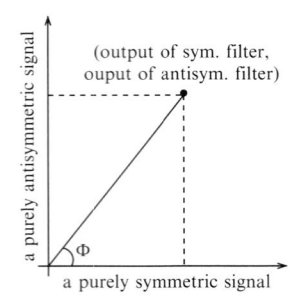

Fig. 6.7. *Measuring local image symmetry.* A signal may be neither purely symmetric, nor purely antisymmetric. Angle Φ may be used to characterise the local signal symmetry. This angle may be computed as $\Phi = \tan^{-1} \frac{A}{S}$, where A and S denote the output of convolution with the antisymmetric and symmetric filter, respectively. In 2D, as we have two antisymmetric filter outputs, one along each direction, the numerator in this expression is replaced by $\sqrt{A_1^2 + A_2^2}$, where A_1 and A_2 denote the output of the first and second antisymmetric filter, respectively

vector magnitude), over the output of the symmetric filter, will be a measure of local symmetry.

The monogenic signal is computed via the frequency domain, with the help of the Riez transform and it consists of three components that lead to the estimation of these three numbers for each pixel, namely local contrast, local orientation, and local symmetry.

6.3.4 From Monogenic Signal Back to Gabor Functions

We shall now close the loop by going back from the monogenetic signal to the Gabor functions: the two simple example Hilbert pair filters we saw in the previous section, for the calculation of the first and second order signal derivatives, are very specific. Local image contrast and local image symmetry may be computed in different ways too. For example, Fig. 6.8 shows a pair of symmetric-antisymmetric filters which may be used to characterize the signal locally in terms of symmetry (Fig. 6.7) and in terms of the presence of this type of ripple in the signal. Such filters may be created by considering specific bands in the frequency domain, isolated, for example, with the help of a Gaussian.

This leads us back to the use of Gabor functions to characterize the signal locally. The only difference now is that we consider pairs of such functions, one symmetric and one antisymmetric, that allow us not only to measure the presence or not of the particular type of ripple in the signal, but also to work out the local orientation of such a ripple and the local symmetry. This leads to an enhanced way of characterizing the image locally, in terms of frequency content (different pairs of filters will give us the content of the image in different bands), and in terms of its structural content, measured by

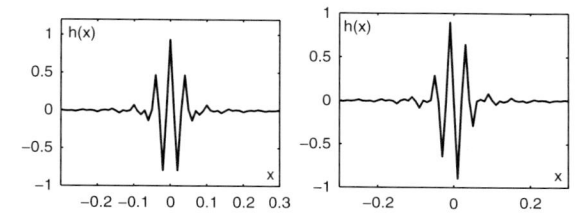

Fig. 6.8. *A Hilbert pair of filters.* A pair of symmetric-antisymmetric filters that can be used to estimate how much the local image structure resembles a ripple of such frequency, how much this ripple is symmetric or antisymmetric and what its local orientation is

the local contrast (also referred to as local image energy), local orientation, and local symmetry.

6.3.5 Beyond Spatial Patterns into Gray Value Distributions

All methods discussed above took into consideration the relative values of the pixels in brightness and in space or in spatial-frequency. If, however, we accept that the term texture means "variation at scales smaller than the scale of interest" [5], then we may consider also the case where the pixels do not show any spatial correlation to form recognizable spatial patterns, either deterministic or statistical. It is possible then, for different human organs or different types of tissue to be characterized by different distributions. Then, we may try to characterize texture from the gray value distributions only. In this case, neither frequency nor structural features are relevant: the only thing we have to go about is the gray level distribution of the pixels. This may be expressed by the local histogram of the pixels that make up a region, which may be thought of as the discretized version of the probability density function of the gray image values.

It is customary in ordinary image processing applications to assume that the gray values of a region are Gaussianly distributed. This, however, is not the case in medical images. For a start, given that negative values are not recorded by the sensors, even if there were a Gaussian distribution, its negative tail would have been curtailed, with the result to have the introduction of asymmetries in the distribution. Second, if we assume that there is natural variation of the recorded brightness values due to genuine tissue variation, there is no reason to expect this variation to be Gaussianly distributed.

One may speculate that subtle tissue changes may manifest themselves in the tail of the distribution, i.e., in the asymmetry of the gray value distribution and its departure from Gaussianity. Asymmetry of a distribution is measured by its third moment, otherwise known as skewness, while the bulginess of a distribution is measured by its fourth order moment, otherwise known as kurtosis. Note that for a Gaussian distribution, the value of the skewness

Fig. 6.9. *Statistical edge detection.* If we have some prior knowledge of the existence of a boundary, we may slide a bipolar window orthogonal to the hypothesized direction of the boundary and compute the exact location of the boundary as the place where the difference in the value of the computed statistic is maximal

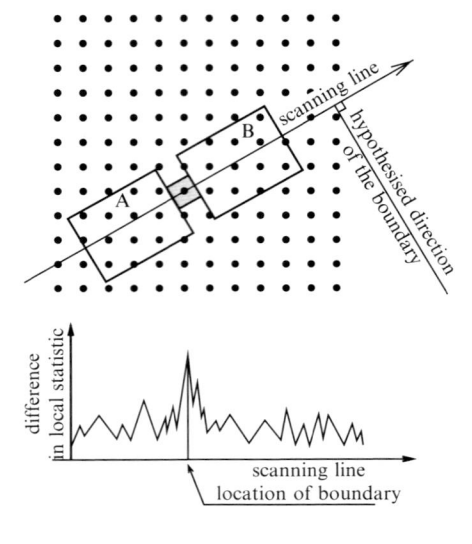

is zero and the value of the kurtosis is expressible in terms of the standard deviation of the distribution.

This leads to the idea of being able to characterize a tissue by the skewness or the kurtosis of the distribution of the gray values of the pixels that make it up, and use this to segment tissues of different types (Fig. 6.9).

As this method relies again on the calculation of statistics, and actually high order statistics that require several samples to be estimated reliably, one has to be careful on the number of pixels inside the local window used. So, to have enough points inside the local window, in order to compute reliably the high order statistic, we have to use 3D data, as 2D data will require far larger windows, leading to unacceptably low boundary resolutions. One way to reduce this requirement is to either interpolate spatially the data, so we create extra samples which carry values relevant to neighboring pixels, or to use a kernel-based approach. In a kernel-based method, we place a kernel at each sample we have, somehow to spread its effect. Summing up the continuous kernels associated with all samples leads to a much smoother representation of the data, from which reliable statistics may be computed (Fig. 6.10).

6.4 Examples of the Use of Texture Features in Biomedical Applications

6.4.1 Mammography

Texture has been used extensively in biomedical applications. For example, in mammography it has been used not only to help the recognition of tissue type, but also to help image registration [12–14], cf. Chap. 13, page 329. In [15] and

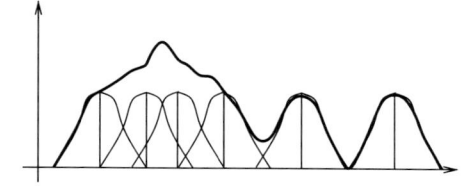

Fig. 6.10. *Kernel-based estimation.* Every straight line represents a datum. If we use directly the raw data, we have a spiky distribution. If we associate with each datum a smooth bell-shaped kernel and sum up at each point of the horizontal axis the values of the kernels that pass through that point, we get the thick black line, which is much smoother than the distribution of the original discrete points

[16] the method of textons was used to classify mammographic parenchyma patterns, while in [17] wavelet features were used to characterize breast density. Breast density was also characterized with the help of directional co-occurrence matrices [18], while in [19] breast density was characterized by the construction of gray level histograms from multiple representations of the same image at various scales.

6.4.2 Brain Image Data

In general, one may distinguish two broad approaches, which correspond to the two problems we discussed above, namely characterization of already segmented tissue, assumed to have uniform texture, and segmentation of tissue. In most papers the problem considered is the first one. This is particularly relevant to brain image data analysis, and more specifically to the characterization of degenerative conditions that affect the whole brain. The second approach has been used mainly to segment local tissue degeneration, like for example the growth of a tumor.

An early work in the characterization of degenerative brain states is that presented in [20]. In [21] the authors used texture features that were rotationally sensitive, computed directly from the PET sinograms of Alzheimer's patients, to distinguish between data that referred to patients and normal controls. The idea had come from an earlier study which had shown that the brains of patients with degenerative disorders tend to be more isotropic than the brains of normal people [2].

In [22] measures of texture anisotropy were used in order to characterize the brains of schizophrenics from MRI data. These authors showed that at the resolution scales of 1–2 mm, the inferior quarters of the brains of schizophrenics show statistically significant structural differences from the brains of normal controls, in the volume structure of the tissue of the gray matter. To show that, the authors used the 3D orientation histogram and they characterized its shape with the help of simple features, like the ratio of the value of the most populated bin over the value of the least populated bin. In [23] it was shown that there exists a strong correlation between the texture anisotropy features calculated from MRI data and the score of the patients (and normal controls) in the Mini Mental State examination. This is a remarkable result, as it shows a correlation between a purely image processing measurement and the score in a psychological test.

Such global methods, however, treat the whole brain as a unit. The authors in [22] were able to localize the structural differences between normal controls and schizophrenics to the inferior quarter of the brain by simply considering the corresponding slices of the 3D MRI data and calculating the 3D orientation histogram from them (after the brain parenchyma was manually segmented and the gradients associated with surface shape, i.e., with the sulci, were excluded). In the same paper, in order to localize the differences in the gray matter, a different method was used: the generalized co-occurrence matrices method, for pairs of voxels and the relative orientation of their gradient vectors. Once such a texture representation is created, and once these representations appear to be different between patients and controls, one can trace back the pairs of voxels that contributed to the cells of the co-occurrence matrix that made it different from the reference matrix. This way the texture features that contributed to the detected difference could be mapped back to locations in the data volume.

Leaving the characterization of manually segmented tissue and moving to the problem of tissue segmentation, the idea of using high order statistics, to characterize and segment tissue became possible with the availability of high resolution 3D data. In [6], the authors were able to show that boundaries present in the data, but not discernible by the human eye, could become visible by using the methodology of Fig. 6.9 and computing the skewness inside each half of the sliding window. These authors, using this approach, identified boundaries around glioblastomas. In absence of any ground truth, their only way of checking the validity of their results was to check for consistency: indeed, the data were consistent across modalities (MRI-T1 and MRI-T2) and the data showed consistency in the sign of the change as the scanning window was moving from the interior of the tumor outwards [7]. More recently, in [24] the authors were able to separate asymptomatic patients of Huntington's disease from normal controls, by using high order statistics computed from segmented layers of the patient's hippocampus. To enhance the accuracy of the calculation of the high order statistics, a kernel-based estimation method was used.

6.5 Discussion and Conclusions

The entire potential of texture-based methods has not yet been fully exploited in medical applications, although texture has been used in hundreds of papers on medical image processing.

A very promising direction of research is that of high order statistics. This is an important advance, as the human eye is known to have difficulties in discerning transitions in high order statistics, and the use of such analysis may help reveal data characteristics which otherwise are not visible. There is, however, a problem here: if we cannot see these changes in the data, how are we sure that what we compute is actually there? The quick answer to this is to perform tests of consistency as it was done in [6, 7]. The real answer will be given only if major studies are undertaken which follow patients for a few years and post-mortem pathology results are compared with image analysis data from the patient's medical history. Such studies are very challenging and they are the way forward.

One has to be careful, however, in using texture-based methods: the variation in the data observed may be due to inhomogeneities and imperfections in the process of data capture, rather than due to genuine tissue variation. Indeed, the inhomogeneity of the magnetic field in MRI data collection is well known. The only way that conclusions based on texture analysis can be drawn is if the data that refer to patients and controls have been captured by the *same* instrument and under the *same* operational protocol. Assuming that the generic inhomogeneities of the instrument are the same for all sets of data, comparison between them then is safe.

References

1. Kovalev VA, Petrou M. Multidimensional co-occurrence matrices for object recognition and matching. Graph Model Image Process. 1996;58:187–97.
2. Kovalev VA, Petrou M, Bondar YS. Texture anisotropy in 3D images. IEEE Trans Image Process. 1999;8:346–60.
3. Varma M, Zisserman A. Classifying images of materials achieving viewpoint and illumination independence. Proc ECCV. 2002;255–71.
4. Deighton M, Petrou M. Data mining for large scale 3D seismic data analysis. Mach Vis Appl. 2009;20:11–22.
5. Petrou M, Garcia-Sevilla P. Image Processing Dealing with Texture. London: Wiley; 2006.
6. Petrou M, Kovalev VA, Reichenbach JR. Three dimensional nonlinear invisible boundary detection. IEEE Trans Image Process. 2006;15:3020–32.
7. Petrou M, Kovalev VA. Statistical differences in the grey level statistics of T1 and T2 MRI data of glioma patients. Int J Sci Res. 2006;16:119–23.
8. Lazaridis G, Petrou M. Image registration using the Walsh transform. IEEE Trans Image Process. 2006;15:2343–57.
9. Petrou M, Petrou C. Image Processing, the Fundamentals. London: Wiley; 2010.
10. Laws KI. Texture image segmentation. Technical report, department of electrical engineering systems, University of Southern California. 1979;USCIPI:940.
11. Felsberg M, Sommer G. The monogenic signal. IEEE Trans Signal Process. 2001;49:3136–44.

12. Petroudi S, Brady M. Classification of mammographic texture patterns. Proc 7th Int Workshop of Digital Mammography, Chapel Hill, NC, USA. 2004.
13. Petroudi S, Brady M. Textons, contours and regions for improved mammogram registration. Proc 7th Int Workshop of Digital Mammography, Chapel Hill, NC, USA. 2004.
14. Rose CJ. Statistical Models of Mammographic Texture and Appearance. PhD thesis, The University of Manchester, United Kingdom. 2005.
15. Petroudi S, Kadir T, Brady M. Automatic classification of mammographic parenchymal patterns a statistical approach. Proc 25th Int Conf IEEE EMBS. 2003;1:798–801.
16. Tao Y, Freedman MT, Xuan J. Automatic categorization of mammographic masses using BI-RADS as a guidance. Proc SPIE. 2008;6915:26–1–26–7.
17. Bovis K, Singh S. Classification of mammographic breast density using a combined classifier paradigm. Proc 4th Int Workshop on Digital Mammography. 2002; p. 177–188.
18. Oliver A, Freixenet J, Zwiggelaar R. Automatic classification of breast density. Proc ICIP. 2005.
19. Muhimmah I, Zwiggelaar R. Mammographic density classification using multiresolution histogram information. Proc ITAB, Ioannina, Greece. 2006.
20. Freeborough PA, Fox NC. MR image texture analysis applied to the diagnosis and tracking of Alzheimer's disease. IEEE Trans Med Imaging. 1998;17:475–479 ISSN 0278–0062.
21. Sayeed A, Petrou M, Spyrou N, et al. Diagnostic features of Alzheimer's disease extracted from PET sinograms. Phys Med Biol. 2002;47:137–48.
22. Kovalev VA, Petrou M, Suckling J. Detection of structural differences between the brains of schizophrenic patients and controls. Psychiatry Res Neuroimaging. 2003;124:177–89.
23. Segovia-Martinez M, Petrou M, Crum W. Texture features that correlate with the MMSE score. In: Proc IEEE EMBS. 2001.
24. Kotoulas L, Bose S, Turkheimer F, et al. Differentiation of asymptomatic Huntington's disease from healthy subjects using high order statistics on magnetic resonance images. IEEE Trans Med Imaging. submitted (2011).

7

Multi-Scale and Multi-Orientation Medical Image Analysis

Bart M. ter Haar Romeny

Summary. Inspired by multi-scale and multi-orientation mechanisms recognized in the first stages of our visual system, this chapter gives a tutorial overview of the basic principles. Images are discrete, measured data. The optimal aperture for an observation with as little artefacts as possible, is derived from first principles and leads to a Gaussian profile. The size of the aperture is a free parameter, the *scale*. Convolution with the derivative of the Gaussian to any order gives *regularized* derivatives, enabling a robust differential geometry approach to image analysis. Features, invariant to orthogonal coordinate transformations, are derived by the introduction of gauge coordinates. The multi-scale image stack (the *deep structure*) contains a hierarchy of the data and is exploited by edge focusing, retrieval by manipulations of the singularities (*top-points*) in this space, and multi-scale watershed segmentation. Expanding the notion of convolution to group-convolutions, rotations can be added, leading to *orientation scores*. These scores are exploited for line enhancement, denoising of crossing structures, and contextual operators. The richness of the extra dimensions leads to a data explosion, but all operations can be done in parallel, as our visual system does.

7.1 Introduction

The notion of a multi-scale approach has given rise to many discussions. It may seem a loss of information to blur the image, the computational complexity increases, and the desired goal should obviously be as sharp as possible. In this chapter, we explain the notion carefully, based on mathematical and physical grounds, and give proper answers to such concerns.

There are strong indications that the first stage of our visual perception system has a large set of filter banks, where not only filters at multiple scales, but also multiple orientations are present. The receptive fields in the cortex can be modeled by many functions, e.g., the Gabor functions specifically model local spatial frequencies, and the Gaussian derivatives model local differential structure. In the following, we will focus on the Gaussian derivative model.

T.M. Deserno (ed.), *Biomedical Image Processing*, Biological and Medical Physics, Biomedical Engineering, DOI: 10.1007/978-3-642-15816-2_7,
© Springer-Verlag Berlin Heidelberg 2011

Fig. 7.1. *Zoom of an image.* Typically, pixels are measured with the wrong aperture function, such as squares, giving rise to spurious resolution, i.e., sharp edges and corners that are not in the original scene. Blurring (squeeze eye lashes) reduces the artefacts

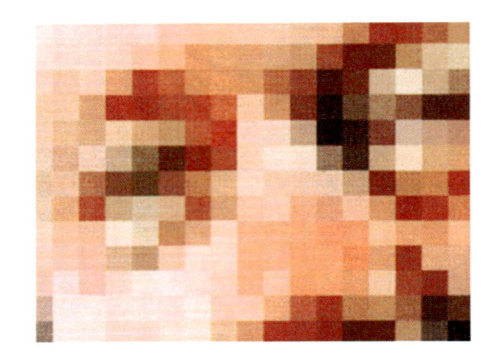

The expert in medical image analysis has to translate a clinical question about an image into a practical and robust algorithm. This is typically done by *geometrical reasoning*. The extra dimensions of scale and orientation give a neat opportunity to add extra "language" to this algorithm design process.

7.2 The Necessity of Scale

Multi-scale image analysis is a physical necessity. It is the necessary consequence of the fact that images are measurements (actually often millions of them). The content of each pixel or voxel is the result of a physical observation, done through an aperture, the physical opening that takes e.g. the incoming photons when we measure light. This aperture has to be finite and small, to have sufficient resolution. But it cannot be made infinitesimally small, as no photons could be measured anymore. The typical aperture shape in today's image acquisition equipment is square, as it is easy to fabricate on a detector chip, but Koenderink [1] already noted in the eighties that such a representation gives rise to spurious resolution, non-existing edges and corners. The effect appears most clearly when we zoom in (Fig. 7.1): the face of Einstein certainly had no square corners all over and sharp edge discontinuities. So what is the shape of the optimal aperture?

7.2.1 The Optimal Aperture Function

The optimal shape can be derived from first principles. The derivation below is based on Nielsen [2]:

- A measurement with a finite aperture is applied.
- All locations are treated similarly; this leads to translation invariance.
- The measurement should be linear, so the superposition principle holds.

These first principles imply that the observation must be a convolution (the example is for simplicity in 1D):

7 Multi-Scale and Multi-Orientation Medical Image Analysis

$$h(x) = \int_{-\infty}^{\infty} L(y) g(x-y) dy \tag{7.1}$$

$L(x)$ is the luminance in the outside world, at infinite resolution, $g(x)$ is the unknown aperture, $h(x)$ the result of the measurement. The following constraints apply:

A. The aperture function $g(x)$ should be a normalized filter:

$$\int_{-\infty}^{\infty} g(x)\, dx = 1 \tag{7.2}$$

B. The mean (first moment) of the filter $g(x)$ is arbitrary (and is taken 0 for convenience):

$$\int_{-\infty}^{\infty} x g(x) dx = x_0 = 0 \tag{7.3}$$

C. The width is the variance (second moment), and set to σ^2:

$$\int_{-\infty}^{\infty} x^2 g(x) dx = \sigma^2 \tag{7.4}$$

The entropy H of our filter is a measure for the amount of the maximal potential loss of information when the filter is applied, and is given by:

$$H = \int_{-\infty}^{\infty} -g(x) \ln g(x) dx$$

We look for the $g(x)$ for which the entropy is minimal given the constraints (7.2), (7.3) and (7.4). The entropy under these constraints with the Lagrange multipliers λ_1, λ_2 and λ_3 is:

$$\tilde{H} = \int_{-\infty}^{\infty} -g(x) \ln g(x) dx + \lambda_1 \int_{-\infty}^{\infty} g(x)\,dx + \lambda_2 \int_{-\infty}^{\infty} x g(x) dx + \lambda_3 \int_{-\infty}^{\infty} x^2 g(x) dx$$

and is minimum when $\frac{\partial \tilde{H}}{\partial g} = 0$. This gives

$$-1 - \log[g(x)] + \lambda_1 + x \lambda_2 + x^2 \lambda_3 = 0$$

from which follows

$$g(x) = e^{-1 + \lambda_1 + x \lambda_2 + x^2 \lambda_3} \tag{7.5}$$

λ_3 must be negative, otherwise the function explodes, which is physically unrealistic. The three constraint equations are now

$$\int_{-\infty}^{\infty} g(x)\,dx = 1,\ \lambda_3 < 0 \ \to\ e\sqrt{-\lambda_3} = e^{\lambda_1 - \frac{\lambda_2^2}{4\lambda_3}} \sqrt{\pi} \tag{7.6}$$

$$\int_{-\infty}^{\infty} x g(x) dx = 0,\ \lambda_3 < 0 \ \to\ e^{\lambda_1 - \frac{\lambda_2^2}{4\lambda_3}} \lambda_2 = 0 \tag{7.7}$$

$$\int_{-\infty}^{\infty} x^2 g(x) dx = \sigma^2,\ \lambda_3 < 0 \ \to\ \frac{e^{-1+\lambda_1 - \frac{\lambda_2^2}{4\lambda_3}} \sqrt{\pi} \left(\lambda_2^2 - 2\lambda_3\right)}{4(-\lambda_3)^{5/2}} = \sigma^2 \tag{7.8}$$

The three λ's can be solved from (7.6) to (7.8):

$$\left\{\lambda_1 = \frac{1}{4}\log\left[\frac{e^4}{4\pi^2\sigma^4}\right], \lambda_2 = 0, \lambda_3 = -\frac{1}{2\sigma^2}\right\}$$

These λ's now specify the aperture function $g(x)$ in (7.5), which is the Gaussian kernel:

$$g(x) = \frac{1}{\sqrt{2\pi}\sigma}e^{-\frac{x^2}{2\sigma^2}} \tag{7.9}$$

The Gaussian kernel has all the required properties. It is smooth, does not generate spurious resolution, is circular, and is the unique solution of this simple set of constraints. Many other derivations have been developed, based on other prerequisites, like causality or the non-generation of new maxima, with the same result. See for overviews [3] and [4]. When other constraints are added (or removed), other families of kernels are generated; e.g., preference for a certain spatial frequency leads to the Gabor family of kernels.

7.2.2 Derivatives of Sampled, Discrete Data, Such as Images

Now we have the Gaussian kernel as optimal sampling aperture, we obtain a discrete sampled dataset in two or more dimensions. It is a classical problem to take derivatives of discrete data [5], as we cannot apply the famous definition.

The Gaussian kernel is the Green's function of the diffusion equation:

$$\frac{\partial L}{\partial s} = \vec{\nabla}.\vec{\nabla}L \quad \text{or} \quad \frac{\partial L}{\partial s} = \frac{\partial^2 L}{\partial x^2} + \frac{\partial^2 L}{\partial y^2}$$

where $\vec{\nabla} = \{\frac{\partial}{\partial x}, \frac{\partial}{\partial y}\}$ is the so-called nabla or gradient operator.

In the next section, the notion of differential structure is discussed. The derivative of an observed entity is given by convolution with a Gaussian derivative

$$\frac{\partial}{\partial x}\{L_0(x,y) \otimes G(x,y;\sigma)\} = L_0(x,y) \otimes \frac{\partial}{\partial x}G(x,y;\sigma). \tag{7.10}$$

The Gaussian derivative functions are widely used in medical image analysis and are excellent models for the sensitivity profiles of the so-called simple cell receptive fields in the human visual cortex. Note that differentiation and observation are done in a single step: convolution with a Gaussian derivative kernel. A Gaussian derivative is a regularized derivative. Differentiation is now done by integration, i.e., by the convolution integral. It may be counterintuitive to perform a blurring operation when differentiating, but there is no way out, differentiation always involves some blurring. The parameter σ cannot be taken arbitrarily small; there is a fundamental limit given the order of differentiation, accuracy and scale [6]. A good rule of thumb is to not go smaller than $\sigma = 0.7\sqrt{n}$ pixels for n-th order derivatives.

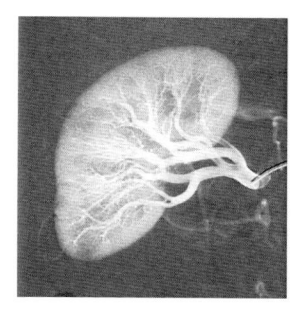

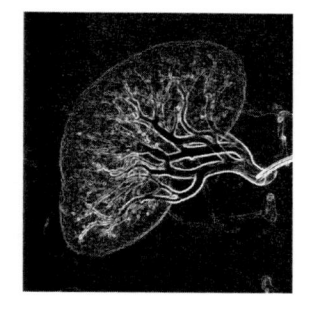

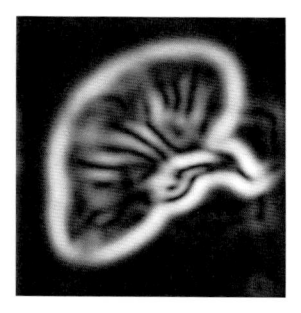

Fig. 7.2. *Edges at different scales give different sized details. Left*: Digital Subtraction Angiography (DSA) of a kidney in a scale showing the small vessels. *Middle*: gradient magnitude at $\sigma = 1$ pixel; this scale emphasizes the vessel tree. *Right*: $\sigma = 15$ pixels, which is the scale of the entire kidney (*outline*). The image resolution is 704×754 pixels

The parameter σ is a free parameter, we are free to choose it. The selection of the proper scale depends on the task. This is illustrated in the example of Fig. 7.2. In this example, we may aim at emphasizing the entire kidney, the vessel tree, or the small vessels themselves.

7.3 Differential Invariants

Derivatives with respect to x or y do not make much sense, as the position and direction of the coordinate system is completely arbitrary. We need to be invariant with respect to translation and rotation of the coordinate system. There are several ways to accomplish this. In this section, we discuss two methods (for 2D images): intrinsic geometry with gauge coordinates and tensor contraction [7, 8].

7.3.1 Gauge Coordinates

An elegant and often used way is to take derivatives with respect to a coordinate system, which is intrinsic, i.e. attached to the local image structure, in our case to the isophotes. Such coordinates are called "gauge coordinates". Isophotes (lines of constant intensity) fully describe the image. A first order gauge frame in 2D is defined as the local pair of unit vectors $\{v, w\}$, where v points in the tangential direction of the isophote, and w in the orthogonal direction, i.e., in the direction of the image gradient. So in every pixel we have a differently oriented $\{v, w\}$ frame attached to the image. The important notion is that any derivative with respect to v and w is invariant under translation and rotation, and so any combination of such gauge derivatives. So, $\frac{\partial L}{\partial w}$ is the gradient magnitude. And $\frac{\partial L}{\partial v} \equiv 0$, as there is no change in the luminance as we move tangentially along the isophote, and we have chosen this

Table 7.1. *Lowest order of differential invariants*

Order	Invariant
L_v	0
L_w	$\sqrt{L_x^2 + L_y^2}$
L_{vv}	$\dfrac{-2L_x L_{xy} L_y + L_{xx} L_y^2 + L_x^2 L_{yy}}{L_x^2 + L_y^2}$
L_{vw}	$\dfrac{-L_x^2 L_{xy} + L_{xy} L_y^2 + L_x L_y (L_{xx} - L_{yy})}{L_x^2 + L_y^2}$
L_{ww}	$\dfrac{L_x^2 L_{xx} + 2 L_x L_{xy} L_y + L_y^2 L_{yy}}{L_x^2 + L_y^2}$

direction by definition. However, we can only measure derivatives in our pixel grid along the x-axis and the y-axis (by convolution with the proper Gaussian derivatives), so we need a mechanism to go from gauge coordinates to Cartesis coordinates. This is derived as follows: writing derivatives as subscripts ($L_x = \frac{\partial L}{\partial x}$), the unit vectors in the gradient and tangential direction are

$$w = \frac{1}{\sqrt{L_x^2 + L_y^2}} \begin{pmatrix} L_x \\ L_y \end{pmatrix} \qquad v = \begin{pmatrix} 0 & 1 \\ -1 & 0 \end{pmatrix} . w$$

as v is perpendicular to w. The directional differential operators in the directions v and w are defined as $v.\nabla = v.(\frac{\partial}{\partial x}, \frac{\partial}{\partial y})$ and $w.\nabla = w.(\frac{\partial}{\partial x}, \frac{\partial}{\partial y})$. Higher order derivatives are constructed[1] through applying multiple first order derivatives, as many as needed. So L_{vv}, the second order derivative with respect to V is now

$$\left(\begin{pmatrix} 0 & 1 \\ -1 & 0 \end{pmatrix} \frac{1}{\sqrt{L_x^2 + L_y^2}} \begin{pmatrix} L_x \\ L_y \end{pmatrix} . \begin{pmatrix} \frac{\partial}{\partial x}, \frac{\partial}{\partial y} \end{pmatrix} \right)^2 f(x,y)$$

Table 7.1 shows the lowest order differential invariants. The second order gauge derivative L_{vv} is a well-known ridge detector. In Fig. 7.3, the ridges (centerlines) are extracted of a vascular pattern on the fundus of the eye.

7.3.2 Invariants from Tensor Contraction

In differential geometry, general derivatives are often denoted as (lower-script) indices, where the index runs over the dimensions, e.g., in 2D:

$$L_i = \begin{pmatrix} L_x \\ L_y \end{pmatrix}$$

When two similar indices occur in the same formula, they are summed over. The so-called Einstein convention means that in such a case the summation

---

[1] See for *Mathematica* code of the formulas in this chapter [6]

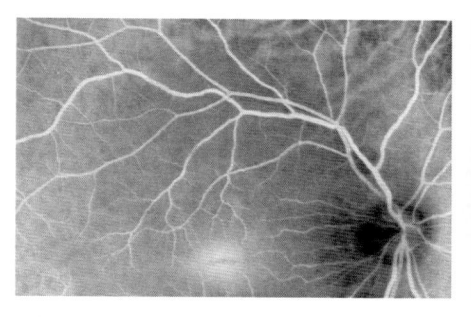

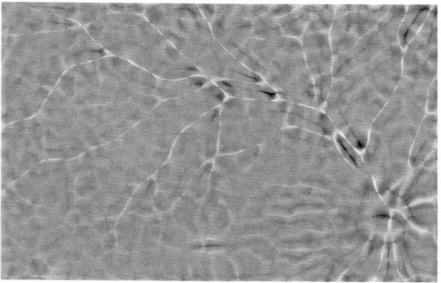

Fig. 7.3. *Ridge detector based on gauge coordinates*. *Left*: The fundus image of the eye has a resolution of 936 × 616 pixels. *Right*: The scale of the ridge operator L_{vv} is 14 pixels

Table 7.2. *Lower order examples of Einstein convention invariants*

Order	Invariant	Description
L	L	Intensity
$L_i L_i$	$L_x^2 + L_y^2$	Gradient magnitude square
L_{ii}	$L_{xx} + L_{yy}$	Laplacian
$L_i L_{ij} L_j$	$L_x^2 L_{xx} + 2 L_x L_y L_{xy} + L_y^2 L_{yy}$	Ridge strength
$L_{ij} L_{ij}$	$L_{xx}^2 + 2 L_{xy}^2 + L_{yy}^2$	Deviation from flatness

sign is left out (Table 7.2):

$$L_i L_i = \sum_{i=x}^{y} L_i L_i = L_x^2 + L_y^2$$

7.4 Second Order Image Structure and Features

7.4.1 Isophote Curvature

In gauge coordinates the Cartesian formula for isophote curvature is easily calculated by applying implicit differentiation twice. The definition of an isophote is

$$L(v, v(w)) = c$$

where c determines a constant. One time implicit differentiation with respect to v gives

$$L_v + L_{w(v)} w'(v) = 0$$

from which follows that $w'(v) = 0$ because $L_v = 0$ by definition. Using that, and second implicit differentiation gives:

$$L_{vv} + 2 L_{vw} w'(v) + L_{ww} w'(v)^2 + L_w w''(v) = 0$$

The isophote curvature is defined as $w''[v]$, the change of the tangent vector $w'[v]$ in the v-direction, so

$$\kappa = w''(v) = -\frac{L_{vv}}{L_w} = \frac{L_x^2 L_{yy} - 2L_x L_y L_{xy} + L_y^2 L_{xx}}{\left(L_x^2 + L_y^2\right)^{3/2}} \quad (7.11)$$

7.4.2 Flowline Curvature

The formula for isophote flowline curvature (flowlines are always perpendicular to the isophotes) is:

$$\lambda = -\frac{L_{vw}}{L_w} = \frac{L_x^2 L_{xy} - L_{xy} L_y^2 + L_x L_y \left(-L_{xx} + L_{yy}\right)}{\left(L_x^2 + L_y^2\right)^{3/2}} \quad (7.12)$$

7.4.3 Corners

A corner is defined as a location on a contour, i.e., on an edge, with high curvature (Fig. 7.4). Blom [9] derived an affine invariant formula for 'cornerness' by assuming a product of the edge strength L_w and the isophote curvature $\kappa = -\frac{L_{vv}}{L_w}$ with weight n for the edge strength: $K = L_w^n \kappa$. Affine invariance leads to the single choice $n = 3$, so the formula for cornerness Θ becomes

$$\Theta = -\frac{L_{vv}}{L_w} L_w^3 = -L_{vv} L_w^2 = L_x^2 L_{yy} - 2L_x L_{xy} L_y + L_{xx} L_y^2 \quad (7.13)$$

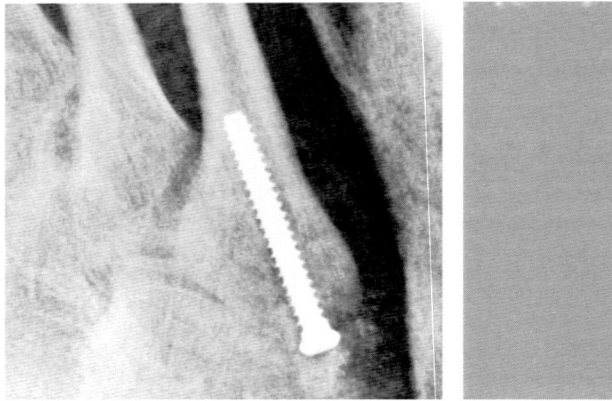

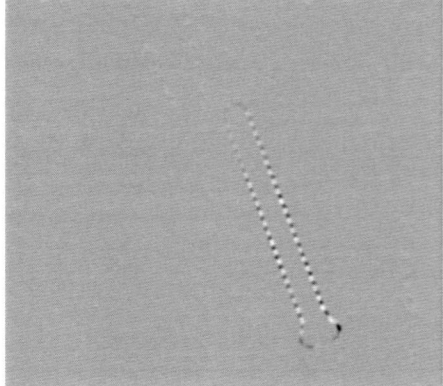

Fig. 7.4. *Corner detection for screw localization. Left*: The radiograph has a resolution of 469 × 439 pixels. *Right*: The scale operator is $\sigma = 2.5$ pixels

7.4.4 Principal Curvatures

The second order derivatives on the 2D isophote landscape form the second order structure matrix, or the Hessian matrix:

$$H = \begin{pmatrix} \dfrac{\partial^2 L}{\partial x^2} & \dfrac{\partial^2 L}{\partial x \partial y} \\ \dfrac{\partial^2 L}{\partial y \partial x} & \dfrac{\partial^2 L}{\partial y^2} \end{pmatrix}$$

The eigenvalues of the Hessian are found by solving the so called characteristic equation $|H - \kappa I| = 0$ for κ:

$$\kappa_1 = \frac{1}{2}\left(-\sqrt{-2L_{xx}L_{yy} + L_{xx}^2 + 4L_{xy}^2 + L_{yy}^2} + L_{xx} + L_{yy}\right) \quad (7.14)$$

$$\kappa_2 = \frac{1}{2}\left(\sqrt{-2L_{xx}L_{yy} + L_{xx}^2 + 4L_{xy}^2 + L_{yy}^2} + L_{xx} + L_{yy}\right) \quad (7.15)$$

They are the so-called principal curvatures. On every location on the isophote landscape, one can walk in many directions. For each of these directions the path has a local curvature, different in each direction. Gauss has proven that the smallest and largest curvature directions are always perpendicular to each other: the principal curvature directions. These directions are given by the eigenvectors of the Hessian matrix.

$$\boldsymbol{\kappa}_1 = \left(-\frac{-L_{xx} + L_{yy} + \sqrt{L_{xx}^2 + 4L_{xy}^2 - 2L_{xx}L_{yy} + L_{yy}^2}}{2L_{xy}}, 1\right) \quad (7.16)$$

$$\boldsymbol{\kappa}_2 = \left(\frac{L_{xx} - L_{yy} + \sqrt{L_{xx}^2 + 4L_{xy}^2 - 2L_{xx}L_{yy} + L_{yy}^2}}{2L_{xy}}, 1\right) \quad (7.17)$$

with $\boldsymbol{\kappa}_1 . \boldsymbol{\kappa}_2 = 0$ (the vectors are perpendicular). In Fig. 7.5 the local principal curvatures are plotted as ellipses with the long and short axes being defined by the Hessian eigenvectors.

The product of the principal curvatures is equal to the determinant of the Hessian matrix, and is called the Gaussian curvature (Fig. 7.6):

$$\mathcal{G} = \kappa_1 \kappa_2 = \det H = L_{xx}L_{yy} - L_{xy}^2 \quad (7.18)$$

The mean curvature is $\mathcal{H} = \frac{\kappa_1 + \kappa_2}{2}$. From (7.15) it can be seen that the two principal curvatures are equal when $4L_{xy}^2 + (L_{yy} - L_{xx})^2 = 0$. This happens in so-called umbilical points. In umbilical points the principal directions are undefined. The surface is locally spherical. The term $4L_{xy}^2 + (L_{yy} - L_{xx})^2$ can be interpreted as "deviation from sphericalness". The principal curvatures are often used to extract and enhance vascular structures. The notion of "vesselness", introduced by Frangi et al. [10] is a multi-scale shape analysis, based on a local zero and a high principal curvature, which indicates a cylindric shape.

Fig. 7.5. *Principal curvature ellipses.* The resolution of the MRI is 256 × 256 pixels

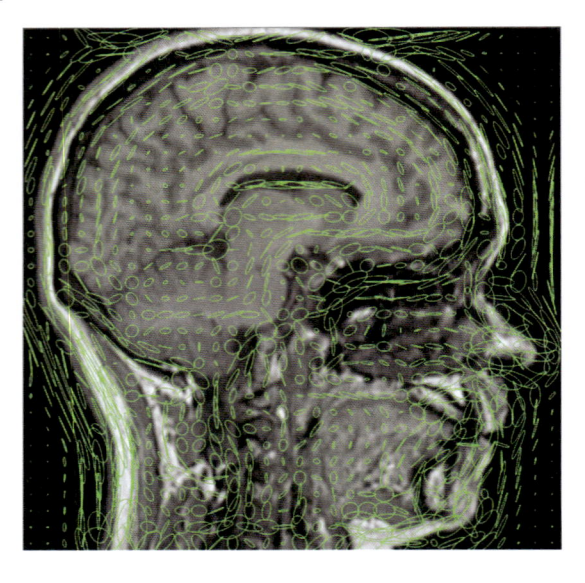

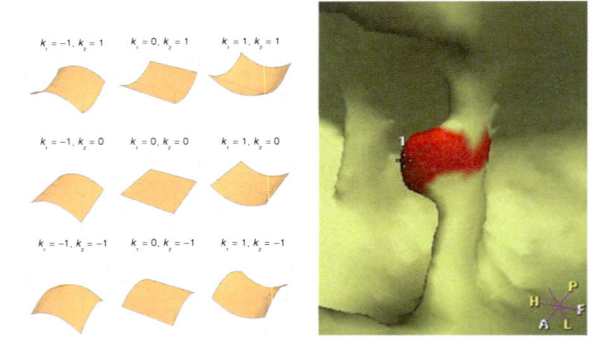

Fig. 7.6. *Gaussian curvature. Left*: Second order shapes. *Right*: an application is automatic polyp detection in virtual endoscopy. The highly curved surface of the polyp on the fold in the intestine is highlighted with a red color: high Gaussian curvature $\mathcal{G}$. (Courtesy of Philips Healthcare, Best, The Netherlands)

7.4.5 The Shape Index

When the principal curvatures κ_1 and κ_2 are considered coordinates in a 2D "shape graph", we see that all different second order shapes are represented. Each shape is a point on this graph. Table 7.3 summarizes possibilities for the local shape.

The shape index ζ_ϕ is defined as the angle of the shape vector in this graph [1]:

Table 7.3. *Local shapes.* The local shape is determined by the principle curvatures (κ_1, κ_2)

Principal curvature	Local shape
$\kappa_1 = 0, \kappa_2 = 0$	Flat
$\kappa_1 > 0, \kappa_2 > 0$	Convex
$\kappa_1 < 0, \kappa_2 < 0$	Concave
$\kappa_1 < 0, \kappa_2 > 0$	Saddle
$\kappa_1 = \kappa_2$	Spherical
$\kappa_1 = 0, \kappa_1 \neq 0$	Cylindrical

$$\zeta_\phi \equiv \frac{2}{\pi} \arctan \frac{\kappa_1 + \kappa_2}{\kappa_1 - \kappa_2} = \frac{2}{\pi} \arctan \frac{2}{\pi} \left(\frac{-L_{xx} - L_{yy}}{\sqrt{-2L_{xx}L_{yy} + L_{xx}^2 + 4L_{xy}^2 + L_{yy}^2}} \right) \quad (7.19)$$

for $\kappa_1 \geq \kappa_2$. The curvedness is defined as the length of the shape vector:

$$\zeta_s \equiv \frac{1}{2}\sqrt{L_{xx}^2 + 2L_{xy}^2 + L_{yy}^2} \quad (7.20)$$

7.5 Third Order Image Structure: T-Junctions

A nice example of geometric reasoning is the derivation of the formula for an invariant T-junction detector. Typically, at a T-junction the isophote curvature changes notably for small steps perpendicular to the isophotes, i.e., in the $\boldsymbol{w}$ gauge direction. So the T-junction detector $\mathcal{T} = \frac{\partial \kappa}{\partial w}$ is a good candidate. The formula below shows the third order spatial derivatives, and becomes quite complex:

$$\begin{aligned}
\mathcal{T} &= \frac{\partial \kappa}{\partial w} = \frac{\partial}{\partial w}\left(\frac{-L_{vv}}{L_w}\right) \\
&= \frac{1}{(L_x^2 + L_y^2)^3} \Big(L_{xxy}L_y^5 + L_x^4\left(-2L_{xy}^2 + L_x L_{xyy} - L_{xx}L_{yy}\right) \\
&\quad -L_y^4\left(2L_{xy}^2 - L_x(L_{xxx} - 2L_{xyy}) + L_{xx}L_{yy}\right) + L_x^2 L_y^2\left(-3L_{xx}^2 + 8L_{xy}^2\right. \\
&\quad \left.+L_x(L_{xxx} - L_{xyy}) + 4L_{xx}L_{yy} - 3L_{yy}^2\right) + L_x^3 L_y \left(6L_{xy}(L_{xx} - L_{yy})\right. \\
&\quad \left.+L_x(-2L_{xxy} + L_{yyy})\right) + L_x L_y^3 \left(6L_{xy}(-L_{xx} + L_{yy}) + L_x(-L_{xxy} + L_{yyy})\right)\Big)
\end{aligned} \quad (7.21)$$

The division by $\left(L_x^2 + L_y^2\right)^3$ gives problems at extrema and saddlepoints, where the gradient is zero, but this term may as well be taken out. The performance is quite good (Fig. 7.7).

7.6 Adaptive Blurring and Geometry-Driven Diffusion

In order to reduce noise in images, blurring is effective, but this also reduces the edges. Edge-preserving smoothing can be accomplished by an adaptive diffusion process. The strategy is simple: blur at homogeneous areas, and

Fig. 7.7. *T-junction detection. Left*: Image with T-junctions (at the red circles). *Right*: detection of points with a rapid change of isophote curvature in the gradient direction. This is an example of geometric reasoning for the design of specific features, given their geometric behavior

reduce the blurring at the locations of edges. Perona and Malik [11] pioneered this and proposed a modification of the diffusion equation $\frac{\partial L}{\partial s} = \vec{\nabla}.c\vec{\nabla}L$, where $c\vec{\nabla}L$ is the "flow" term, and c is a function of the gradient magnitude: $c = c(|\vec{\nabla}L|)$. Perona and Malik proposed $c = e^{-\frac{|\vec{\nabla}L|^2}{k^2}}$, so the non-linear diffusion equation in 2D becomes

$$\frac{\partial L}{\partial s} = \frac{e^{-\frac{L_x^2+L_y^2}{k^2}}\left(\left(k^2 - 2L_x^2\right)L_{xx} - 4L_xL_{xy}L_y + \left(k^2 - 2L_y^2\right)L_{yy}\right)}{k^2} \quad (7.22)$$

and in 3D:

$$\frac{\partial L}{\partial s} = \frac{1}{k^2}e^{-\frac{L_x^2+L_y^2+L_z^2}{k^2}}\left(\left(k^2 - 2L_x^2\right)L_{xx} + k^2L_{yy} - 2L_y^2L_{yy}\right. \\ \left. -4L_yL_{yz}L_z - 4L_x\left(L_{xy}L_y + L_{xz}L_z\right) + k^2L_{zz} - 2L_z^2L_{zz}\right) \quad (7.23)$$

Alvarez [12] introduced a flow, using only the *direction* of the gradient, i.e., he used the unit gradient vector as the flow. His non-linear diffusion equation (known as Euclidean shortening flow) becomes:

$$\frac{\partial L}{\partial s} = \left|\vec{\nabla}L\right|\vec{\nabla}.\left(\frac{\vec{\nabla}L}{\left|\vec{\nabla}L\right|}\right) = \frac{-2L_xL_{xy}L_y + L_{xx}L_y^2 + L_x^2L_{yy}}{L_x^2 + L_y^2} = L_{vv} \quad (7.24)$$

This is a striking result. Compared to normal diffusion, described by the normal isotropic diffusion equation $\frac{\partial L}{\partial s} = L_{vv} + L_{ww}$, the term L_{ww} is now missing, which indicates that the flow in the direction of the gradient w is penalized. The blurring is along the edges, not across the edges. Alvarez was so pleased with this result, that he coined his formula the "fundamental equation".

This non-linear evolution of the images leads to adaptive blurring, in this case to adaptive "edge-preserving" smoothing. The conductivity term c can contain *any* geometrical information, e.g., curvature terms, local structure tensor information, vesselness, etc. Hence, the general term "geometry-driven diffusion". See for a an early overview [6], and the excellent review by Weickert [13].

7.7 Edge Focusing

The deep multi-scale structure of images is rich in information. It contains the information of the scale of features, which can be exploited to establish their importance. Stated differently, it contains the *hierarchy* of the structures in the image. A good example is the extraction of larger edges from a noisy background by edge focusing. The disk visualized in Fig. 7.8 has a very low Signal-to-Noise Ratio (SNR). Blurring an image to reduce the noise destroys the localization.

The steepest point of an edge is given by the maximum of the gradient, which can easily be found by the zero crossing of the second order derivative. In Fig. 7.9 the zero crossings (black for down-going, white for up-going edges) are plotted along the image line profile as a function of scale. This is a so

Fig. 7.8. *Image of a disk at low SNR. Right*: gradient magnitude extraction with scales of $\sigma = 1, 2, 3, 4$ pixels. The SNR increases, the localization accuracy decreases. *Left*: intensity profile of the middle row of the noisy disk image (as indicated by the red line)

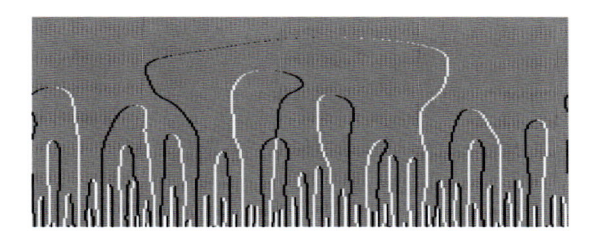

Fig. 7.9. *Multi-scale signature function of Fig. 7.8 (left).* The zero crossings of the second order derivatives are indicated as white (up-going edges) or black (down-going edges) dots, as a function of exponential scale (vertical). Horizontal is the x-direction. The two most important edges (of the disk) survive the blurring the longest time. The signature function generates the intrinsic hierarchy of structure

Fig. 7.10. *Top-points in scale-space.* The geodesic paths over scale (vertical direction) can be constructed from the intersections of $\partial_x L = 0$ and $\partial_y L = 0$ planes, the top-points (red dots) form from the intersections of the geodesics, and the plane $det H = 0$ where H is the Hessian matrix [15] (Courtesy: B. Platel)

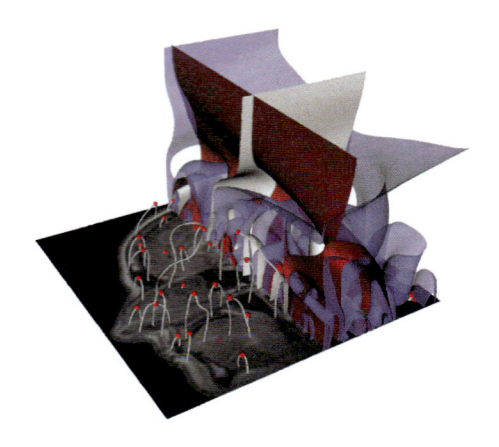

called "signature function". The edges follow geodesic tracks. Some edges survive the blurring for a long time, and they form the "important" edges. Note that a black geodesic annihilates with a white geodesic in a singularity, a so called top-point. Note also that the up- and down-going edges of the disk come together, indicating their intrinsic relation. From this we see important cues emerging from the deep structure analysis for the notion of symmetry and long-range contextual connections (Gestalt).

The same can be done in 2D. In this case, we follow specific singular points, i.e., maxima, minima and saddle points over scale. The geodesic tracks of these singular points can easily be calculated from the intersections of the zero-crossing surfaces of $\frac{\partial L}{\partial x}$, $\frac{\partial L}{\partial x}$. The top of the path is found where the determinant of the hessian $\frac{\partial^2 L}{\partial x^2}\frac{\partial^2 L}{\partial y^2} - (\frac{\partial^2 L}{\partial x \partial y})^2$ is zero. In Fig. 7.10, the surfaces are shown, as well as the top-points (red dots). The top-points generated from low contrast areas are unstable and can be detected and removed [14].

The top-points are highly descriptive for the structure in the image [16]. Duits [17] showed that the generating image can be reconstructed again from the top-points (not identically, but to a high degree of similarity). They have also been successfully exploited in Content-Based Image Retrieval (CBIR) of sub-scenes from larger complex scenes, despite scaling, rotation, and occlusion [15, 18]. As huge databases of medical images are now digitally accessible, this may have important applications. However, the variability of medical images for retrieval has not yet been mastered. Other, similar methods have been proposed for the generation of multi-scale singularity points, such as the Shift-Invariant Feature Transform (SIFT) [19].

7.8 Orientation Analysis

Recent discoveries in the mechanisms of the processing in the visual cortex are a rich source of inspiration for mathematical models. Optical measurements of the visual cortex that exploit voltage-sensitive dyes have revealed intricate structures, the so-called cortical columns. In each column, receptive fields (filters) are present at all orientations, highly organized in a pin-wheel fashion [20]. This inspires to the generalization of the notion of convolution [21,22], the so-called group-convolutions, where any group can be applied to the convolution kernel. The convential convolution is based on translation (shift integral):

$$\mathcal{W}_\psi[f](\boldsymbol{b}) = \int_{\mathbb{R}^n} \psi^*(\boldsymbol{x}-\boldsymbol{b}) f(\boldsymbol{x}) \, \mathrm{d}\boldsymbol{x}$$

When the kernel ψ is scaled by a dilation operation, we have the well-known family of wavelets:

$$\mathcal{W}_\psi[f](a,\boldsymbol{b}) = \int_{\mathbb{R}^n} \psi^*\left(\frac{\boldsymbol{x}-\boldsymbol{b}}{a}\right) f(\boldsymbol{x}) \, \mathrm{d}\boldsymbol{x}$$

The (Euclidean) group of rotations and translations gives the definition of "orientation bundles" or "orientation scores" [23]:

$$\mathcal{W}_\psi[f](a,\boldsymbol{b},\alpha) = \int_{\mathbb{R}^n} \psi^* \boldsymbol{R}_\alpha^{-1}\left(\frac{\boldsymbol{x}-\boldsymbol{b}}{a}\right) f(\boldsymbol{x}) \, \mathrm{d}\boldsymbol{x}$$

Just as the multi-scale deep structure, the orientation score is rich in information. Here, not the hierarchy, but the *contextual* relations between elongated structures emerge. Medical images (as any images) and their differential counterparts are rich in elongated structures, e.g., vessels, nerves, contours (in 2D edge maps), etc. The orientation score adds a new dimension to the realm of geometric reasoning, now with the notion of orientation (Fig. 7.11).

It is essential to have an *invertible* orientation score, i.e., to be able to reconstruct the image again from the orientation score, much like the Fourier transform has an inverse Fourier transform. The inverse transform is written as:

$$L(x) = \mathcal{G}_{\tilde{\psi}}^{-1}\left[\hat{L}_{\tilde{\psi}}(\boldsymbol{b},\alpha)\right](x) := \int_0^{2\pi} \int_{\mathbb{R}^2} \tilde{\psi}\left(R_\alpha^{-1}(x-b)\right) \tilde{L}_{\tilde{\psi}}(\boldsymbol{b},\alpha) db d\alpha$$

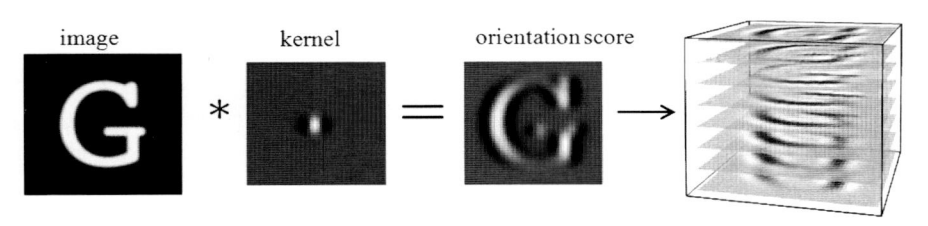

Fig. 7.11. *Generating an orientation score.* An image (*left*) is convolved with a rotating anisotropic kernel to generate a multi-orientation stack, the orientation score

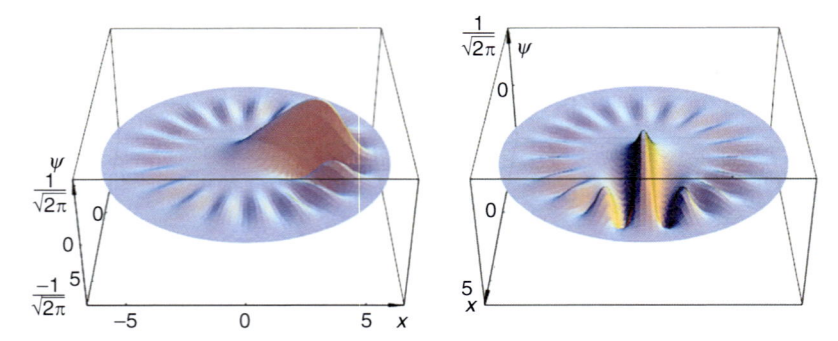

Fig. 7.12. *Spatial orientation filters.* The filters generate an invertible multi-orientation score by (7.25) for $m = 0$. Higher orders of m generate orientation derivative kernels $\partial_\phi^{(m)}$ [25]

The condition for invertibility on the kernels $\tilde{\psi}(\rho, \phi)$ is:

$$\int_0^{2\pi} \tilde{\psi}^*(\rho, \phi)\tilde{\psi}(\rho, \phi)\mathrm{d}\phi = \frac{1}{4\pi^2}$$

It can be shown that a special class of filters can be constructed giving *exact* invertibility [17], see also [22, 24, 25]:

$$\psi(\rho, \phi) = \sum_{n=0}^{\infty} \sum_{m=-\infty}^{\infty} a_{mn} \frac{e^{im\phi} \rho^{|m|} L_n^{|m|}(\rho^2)}{\sqrt{\pi}\sqrt{(n+1)_{|m|}}} e^{-\frac{\rho^2}{2}} \qquad (7.25)$$

A member of this class ($a_{nm} = 0$, $n > 0$) is depicted in Fig. 7.12.

The theory for orientation scores has recently been extended to 3D. It is interesting that the data structure of High Angular Resolution Diffusion Imaging (HARDI) in MRI is closely related to a 3D orientation score [26] (it is a factorial subgroup of the 3D Euclidean group).

7.9 Applications

7.9.1 Catheter Detection

Many interventional catherization procedures take a substantial time (some even up to hours), and the accumulation of the X-ray dose is a serious concern. However, if the dose is reduced, the SNR in the resulting navigation image seriously decreases. With the help of enhancement by the orientation based enhancement the visibility of dim catheters can be markedly improved (Fig. 7.13), and the dose reduced.

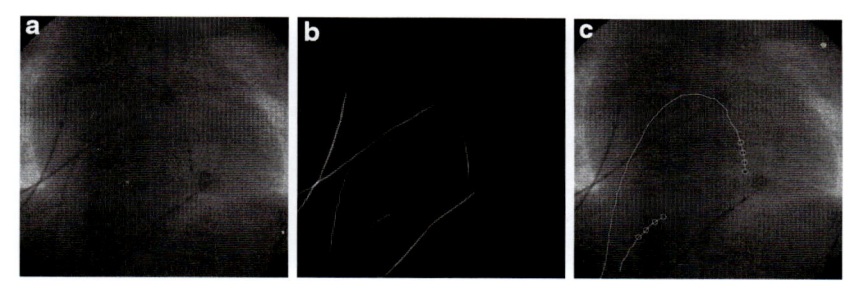

Fig. 7.13. *Catheter detection in fluoroscopy [27].* (**a**) Imaging with reduced dose leads to noisy images. (**b**) Detection of the elongated structures and suppressing the noise. (**c**) The right trace is detected and selected from all traces exploiting the markers at the tip

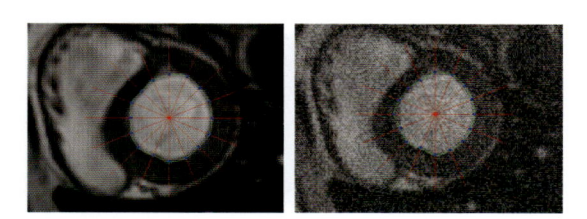

Fig. 7.14. *Segmentation of endocard contour in MRI.* The segmentation results are displayed without (*left*) and with adding of 30 % noise (*right*)

7.9.2 Endocard Contour Detection

To detect the endocard contour in MRI, profiles from a center of gravity point are sampled, and the main edges are found on each contour by coarse-to-fine edge focusing. The detection is robust, despite the addition of 30 % noise (Fig. 7.14).

7.9.3 Denoising of Crossing Lines

The orientation score *untangles* the local orientations (Fig. 7.15). Denoising of crossing line structures has typically been problematic, as at the crossing point it is not clear what direction an elongated filter should take. This is now handled elegantly in the orientation scores [22].

The evolution of an image in a non-linear scale-space, where geometry-adaptive constraints are built in with respect to the diffusion (leading to non-linear diffusion equations), can be integrated into the orientation scores. Figure 7.16 shows the result of denoising for a noisy microscopy image with many crossing fiber structures. This is another example of geometric reasoning for the design of specific features, given their orientation behavior.

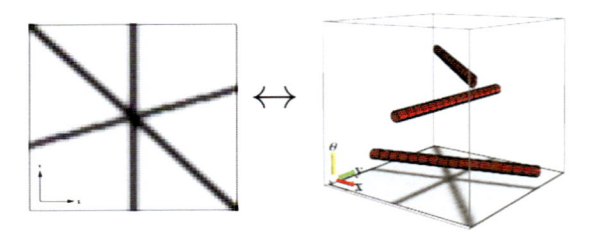

Fig. 7.15. *Orientation score.* Crossing lines (*left*) are untangled in the orientation score. Orientation adaptive non-linear denoising filters can now be applied easier at each orientation individually without being hampered by the other directions

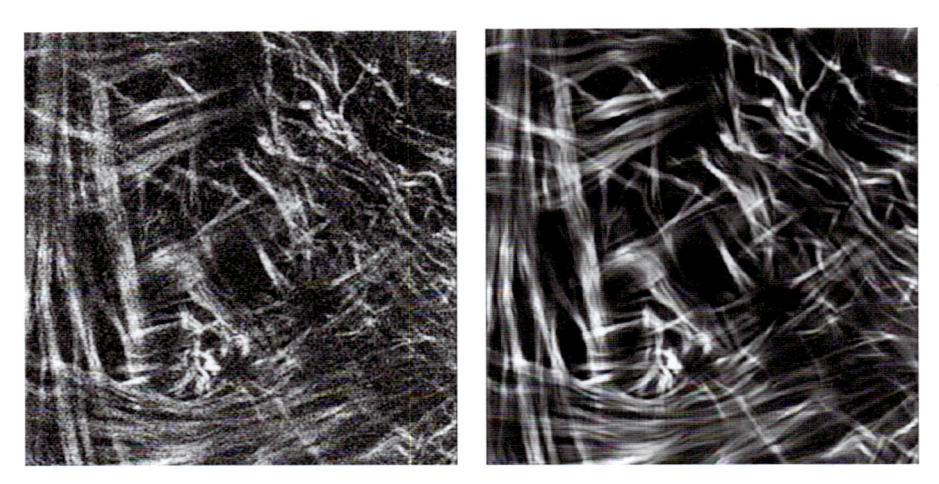

Fig. 7.16. *Denoising of crossing elongated structures in an orientation score [28].* *Left*: collagen fibers in tissue engineered cartilidge in a noisy 2-photon microscopy image. *Right*: Result after orientation-adaptive geometry-driven diffusion

7.10 Conclusion

This chapter discussed the mathematical theory of multi-scale and multi-orientation analysis of (medical) images. It is interesting to note that both approaches can be recognized in the filter banks formed by the receptive fields in the human front-end visual system. The Gaussian kernel emerges as a general smooth aperture kernel and its derivatives are the natural operators for taking derivatives of discrete data. They have the intrinsic property of regularization. The multi-scale nature ("deep structure") leads to the important notion of structural hierarchy, also in the differential structure such as edges ("edge focusing") and curvature. Also the long-range interaction between symmetric edges can be extracted from the geodesic paths in the deep structure. The multi-orientation nature of the analysis gives us short-range interactions between elongated structures. Contextual line completion (not described in this chapter) is now possible [25], which is useful in the analysis of dim and broken lines and contours in medical images. The angular directions at crossings are untangled in the orientation score. Non-linear diffusion in this orientation score gives a good denoising. The general approach of group-convolutions gives the theory a firm basis. The added dimensions of scale and orientation may seem computationally costly and counter-intuitive at start, but they are intrinsically suited for parallel implementation, and they give a rich new space for geometric reasoning.

References

1. Koenderink JJ. The structure of images. Biol Cybern. 1984;50:363–70.
2. M. Nielsen, L.M.J. Florack, R. Deriche: Regularization, scale-space and edge detection filters. J Math Imaging Vis. 1997;7(4):291–308.
3. Weickert JA, Ishikawa S, Imiya A. On the history of gaussian scale-space axiomatics. In: Sporring J, Nielsen M, Florack LMJ, et al., editors. Gaussian Scale-Space Theory. Vol. 8 of Computational Imaging and Vision Series. Dordrecht, The Netherlands: Kluwer; 1997. p. 45–59.
4. Duits R, Florack L, de Graaf J, et al. On the axioms of scale space theory. J Math Imaging Vis. 2004;20(3):267–98.
5. Hadamard J. Sur les problèmes aux dérivées partielles et leur signification physique. Bul Univ Princet. 1902;13:49–62.
6. Haar Romeny B.M.ter. Geometry-Driven Diffusion in Computer Vision, Kluwer Academic Publishers. Vol. 1. Dordrecht, The Netherlands; 1994.
7. Florack LMJ, ter Haar Romeny BM, Koenderink JJ, et al. Cartesian differential invariants in scale-space. J Math Imaging Vis. 1993;3(4):327–48.
8. Florack LMJ, Haar Romeny B.M.ter, Koenderink JJ, et al. Linear scale-space. J Math Imaging Vis. 1994;4(4):325–51.
9. Blom J. Topological and Geometrical Aspects of Image Structure. University of Utrecht, Department of Medical and Physiological Physics. Utrecht, The Netherlands; 1992.
10. Frangi AF, Niessen WJ, Vincken KL, et al. Muliscale vessel enhancement filtering. Lect Notes Comput Sci. 1998;1496:130–137.
11. Perona P, Malik J. Scale-space and edge detection using anisotropic diffusion. IEEE Trans Pattern Anal Mach. 1990;12(7):629–39.
12. Alvarez L, Guichard F, Lions PL, et al. Axioms and fundamental equations of image processing. Arch Rat Mech Anal. 1993;123:200–57.
13. Weickert JA. Anisotropic Diffusion in Image Processing. Stuttgart: Teubner; 1998.
14. Balmachnova E, Florack LMJ, Platel B, et al. Stability of top-points in scale space. Lect Notes Comput Sci. 2005;3459:62–72.
15. Platel B, Florack LMJ, Kanters FMW, et al. Using multiscale top points in image matching. Proc ICIP. 2004; p. 389–392.
16. Florack L, Kuijper A. The topological structure of scale-space images. J Math Imaging Vis. 2000;12(1):65–79.
17. Duits R, Felsberg M, Granlund G, et al. Image analysis and reconstruction using a wavelet transform constructed from a reducible representation of the euclidean motion group. Int J Comput Vis. 2007;72(1):79–102.

18. Platel B, Balmachnova E, Florack LMJ, et al. Using top-points as interest points for image matching. Lect Notes Comput Sci. 2005;3753:211–22.
19. Lowe DG. Distinctive image features from scale-invariant keypoints. Int J Comput Vis. 2004;60(2):91–110.
20. Bosking WH, Zhang Y, Schofield B, Fitzpatrick D. Orientation selectivity and the arrangement of horizontal connections in tree shrew striate cortex. J Neurosci. 1997;17(6):2112–2127.
21. Duits R, Burgeth B. Scale spaces on lie groups. Lect Notes Comput Sci. 2007;4485:300–12.
22. Duits R, Duits M, van Almsick M, et al. Invertible orientation scores as an application of generalized wavelet theory. Pattern Recognit Image Anal. 2007;17(1):42–75.
23. Duits R, van Almsick M, Duits M, et al. Image processing via shift-twist invariant operations on orientation bundle functions. Pattern Recognit Image Anal. 2005;15(1):151–56.
24. Kalitzin SN, Haar Romeny B.M.ter, Viergever MA. Invertible apertured orientation filters in image analysis. Int J Comput Vis. 1999;31(2/3):145–58.
25. Almsick MA, Duits R, Franken E, et al. From stochastic completion fields to tensor voting. Lect Notes Comput Sci. 2005;3753:124–34.
26. Franken EM, Duits R. Scale spaces on the 3D euclidean motion group for enhancement of HARDI data. Lect Notes Comput Sci. 1999;5567:820–31.
27. Franken EM, Rongen P, van Almsick M, et al. Detection of electrophysiology catheters in noisy fluoroscopy images. Lect Notes Comput Sci. 2006;4191:25–32.
28. Franken E. Enhancement of Crossing Elongated Structures in Images. Eindhoven University of Technology. Eindhoven, The Netherlands; 2008.

8

Feature Extraction and Selection for Decision Making

Agma J.M. Traina, Caetano Traina Jr., André G.R. Balan,
Marcela X. Ribeiro, Pedro H. Bugatti, Carolina Y.V. Watanabe,
and Paulo M. Azevedo-Marques

Summary. This chapter presents and discusses useful algorithms and techniques of feature extraction and selection as well as the relationship between the image features, their discretization and distance functions to maximize the image representativeness when executing similarity queries to improve medical image processing, mining, indexing and retrieval. In particular, we discuss the Omega algorithm combining both, feature selection and discretization, as well as the technique of association rule mining. In addition, we present the Image Diagnosis Enhancement through Associations (IDEA) framework as an example of a system developed to be part of a computer-aided diagnosis environment, which validates the approaches discussed here.

8.1 Introduction

The development of algorithms for gathering a set of representative and succinct features extracted from medical images that can truly represent an image remains a challenge for the researchers in the image processing and information retrieval fields. Mathematical and statistical methods aimed at extracting specific information from the images have been tailored in this regard. However, it is necessary to keep in mind that the process to get a summary of the image given by its features is driven by the need to store and retrieve images following specific intents. One of them concerns the Content-Based Image Retrieval (CBIR) systems, which have a strong potential to support the decision making process demanded by radiologists and health specialists, because they can retrieve similar cases analyzed in the past, bringing together every information associated to them. Another use for CBIR systems in health systems is to help teaching new physicians and the technical staff, presenting similar cases and providing a platform for learning and discussion [1]. This subject will be well explained and discussed in Chap. 21 of this book.

Images hold complex data, getting the most relevant information from them is a complex endeavor. Thus, one of the main challenges is how to automatically extract features from the images that are able to represent their "essence". The quest for features that can be automatically obtained comes from the huge load of images that are generated in the day to day activities in hospitals and medical centers. Usually, only low-level features can be automatically obtained, which are basically extracted as brightness, color, texture and shape distributions. These features are usually placed in a *feature vector*, also referred to as *signature*.

However, these features lack on conveying the image description expected by the user. This leads to the *semantic gap*, which is, basically, the difference between the user's image perception and what the extracted features actually represent. One could imagine that the way out of it should be getting as many features from the images as possible, trying to cover any aspect of the image. Such approach turns out to be a drawback, as it significantly increases the number of features. In fact, when the dimensionality is high, the indexing techniques collapse and do not help to execute retrieval operations. This effect is called *curse of dimensionality*. Therefore, to have an efficient system supporting clinical decision making, it is important to address the main aspects of the image with a small number of features. Comparing image features also demands a careful choice for a distance function (dissimilarity function) that allows a correct image separation. A proper distance function can largely improve the answer precision [2].

Similarity-based image mining techniques can help on turning a Computer-Aided Diagnosis (CAD) system into a more useful tool for the decision making process, improving the day-to-day activities in a radiology center. In this chapter, we also present and discuss the IDEA framework, which exemplifies how to deal with the aforementioned issues, and how an automatic system can help on providing a second-opinion suggestion to the radiologist analyzing mammograms.

8.2 Image Representation

The main features extracted from the images (feature vector or signature) are of foremost importance. The most common approach is computing image summarizing values and place them into a feature vector $F = \{f_1, \ldots f_E\}$, which represents each image as a point in an E-dimensional feature space, where E is the number of features extracted. Using this model, the image content can be compared using any existing distance metric, such as the well-known Euclidean one. Therefore, indexing techniques should be applied to speed up query answering.

There are two approaches to extract *global* features:

1. Compute features directly from the raw image data,
2. Compute features from transformed data.

A *local* approach is obtained from segmentation of relevant image regions, which are then indexed individually using the same techniques.

Color histograms are one of the most popular image features that are computed directly from image data. Transforming image data, on the other hand, allows computing important image features, such as texture and shape. Fourier and Wavelet transforms are well-known techniques employed to extract information from data in the frequency and spatial domains respectively.

Image segmentation is another important way to obtain image features. Segmentation enables finding elements that constitute a particular image and, consequently, add semantics to the feature vector, for example, allowing to find information about the number of elements in the image, their particular arrangement, their sizes, shapes, color, and so on. An example in the medical domain is the process to determine the presence of a specific lesion in a tissue. After having found a lesion, it is quite relevant to determine its characteristics. Thus, we can, for example, help a physician by automatically grouping hundreds of clinical images with similar diagnosis.

In the following section, we describe the most common low-level, generic features used to represent images from medical examinations.

8.2.1 Medical Image Segmentation and Feature Extraction

Many image segmentation methods were proposed in the literature, but there is a consensus that no such a method exists that is able to successfully segment every image type. In the medical domain, there is a strong tendency to consider that image data are generated from Random Variables (RVs) ruled by normal probability distributions. This is coherent for most of the imaging devices, and specially for magnetic resonance where a relatively larger number of tissues can be imaged. Hence, a well-grounded approach for image segmentation in medical domain is to model the probability distribution of image data, a non-parametric curve, as a parametric mixture of Gaussians, or, in other words, to employ a Gaussian Mixture Model (GMM).

It is important to have a prior information about the image we aim at segmenting in order to develop a segmentation algorithm based on a principled data modeling. Another model employed in medical image segmentation is the Markov Random Field (MRF). In this model, the random variables that "generate" image data (there is one RV per image pixel) are considered to have a local neighborhood interaction, meaning that the normal distribution of a given variable is, in some degree, influenced by the normal distribution of its neighbor variables. This property allows a segmentation algorithm to cope with the high frequency random noise inherent to many imaging technologies, including MRI. Also, this neighborhood interaction can be tuned in such a way that some texture patterns present in the images, mainly grainy patterns, can be successfully detected. Hence, texture-based image segmentation is achieved, in some degree, by the use of MRF in the data modeling.

The Expectation Maximization (EM) algorithm [3] is a well-known method for finding the parameters of a given GMM of an image. To combine the GMM and MRF statistical models, the EM algorithm is usually employed in combination with an optimization algorithm. The idea is to combine, via the Bayes rule, the parameters of the GMM and the parameters of the MRF to formulate a compound probability distribution called *posterior*. Then, the goal is to find a segmentation result that maximizes the posterior distribution, for which two strategies are usually employed: the Maximization A Posterior (MAP), and the Maximization of the Posterior Marginals (MPM).

The EM/MPM [4] is an image segmentation algorithm well suited to segment a wide range of medical image categories, since it combines the GMM and the MRF models. In this method, the EM algorithm is interleaved with the Gibbs sampler algorithm to achieve a maximization of the posterior marginals. However, depending on the noise level in an image, the EM/MPM can take a relatively long time to converge. In this case, extracting features from a large amount of images may be impractical. An optimized EM/MPM method was proposed by Balan et al. [5] to achieve faster convergence. The optimization consists of iteratively varying the spatial neighborhood interaction imposed by the MRF model. With little interaction at the beginning of the process, the EM algorithm has more freedom to change the parameters of the model and quickly get closer to an optimal segmentation solution. Then, as the process evolves, the spatial interaction increases, leading to a stable convergence and a noise-free segmentation result.

Figure 8.1 illustrates an image segmentation carried out by the EM/MPM method. The number of classes in the segmented image is a parameter to the algorithm. To obtain this figure, we set the number of classes to five, since we aim at finding four types of tissues and the background. In each class of Fig. 8.1, the segmented image has a predominant tissue (e.g., white matter, gray matter). Balan et al. [5] proposed an effective feature vector F based on image segmentation using the aforementioned technique. Each segmented image class is described by the following set of features: size, centroid (x and y), average gray level and fractal dimension. Thus, the number of features E is six times the number of classes found. The fractal dimension is a compact and effective shape feature, since it is capable of distinguishing objects formed by thin segments (Fig. 8.1f) from objects composed of larger compact blocks (Fig. 8.1e).

Fractal Dimension

A simple way of determining the fractal dimension of a region formed by N points $[p_1, \ldots, p_N]$ is to compute its correlation integral $C(r)$, such as follows

$$C(r) = \frac{1}{N^2} \sum_{i=1}^{N} \sum_{j=i+1}^{N} H(r - ||p_i - p_j||) \qquad (8.1)$$

where $r \in \mathbb{R}$ and $H(\alpha)$ is the heaviside function

$$H(\alpha) = \begin{cases} 0 & \text{if } \alpha \leq 0 \\ 1 & \text{if } \alpha > 0 \end{cases}$$

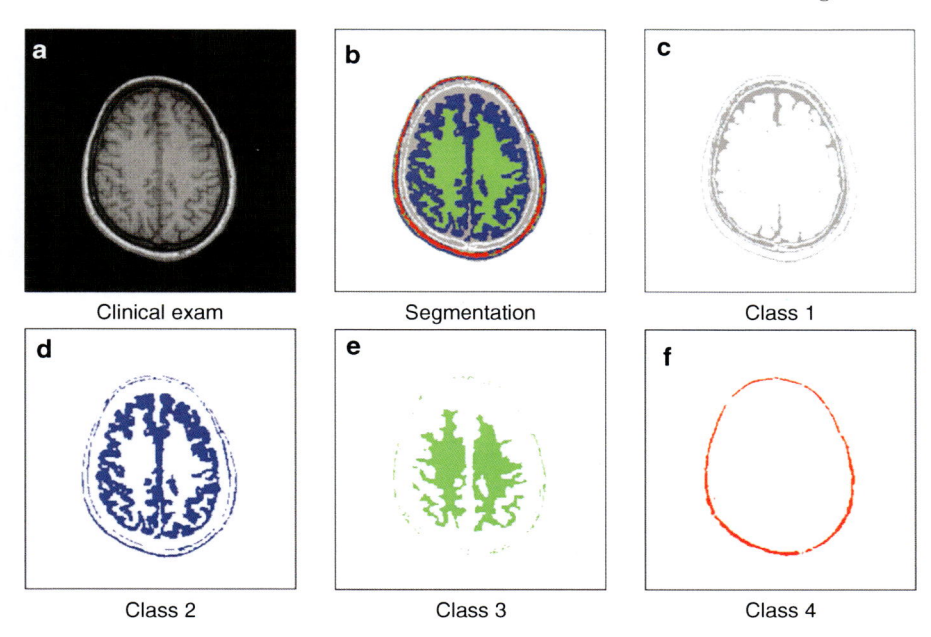

Fig. 8.1. *MRI segmentation using the EM/MPM method.* (**a**) Clinical exam, MRI head axial, slice thickness 1 mm; (**b**) Image segmentation with five classes; (**c**) Class 1, predominant tissue: cerebrospinal fluid; (**d**) Class 2, predominant tissue: gray matter; (**e**) Class 3, predominant tissue: white matter; and (**f**) Class 4, predominant tissues: dura, fat, and bone marrow

Then, the fractal dimension D_2 of the region is given by

$$D_2 = \lim_{r \to 0} \frac{\log C(r)}{\log r} \qquad (8.2)$$

In other words, D_2 is computed as the angular coefficient of a linear regression over of the plot $\log(C(r))$ *versus* $\log(r)$, and represents quite closely the correlation fractal dimension.

Figure 8.2 presents the plots used to compute the fractal dimensions of the four regions of Fig. 8.1. Observe that the region corresponding to class 3 (Fig. 8.1e) obtained the highest fractal dimension, meaning that it is more related to a plane than to a line. On the other hand, the region corresponding to class 4 (Fig. 8.1f) obtained the lowest fractal dimension, meaning that it is more related to a linear object.

8.2.2 Color Features

Histogram

One of the most common techniques used to represent an image regarding to the gray-level (color) content is the traditional histogram. It gives the

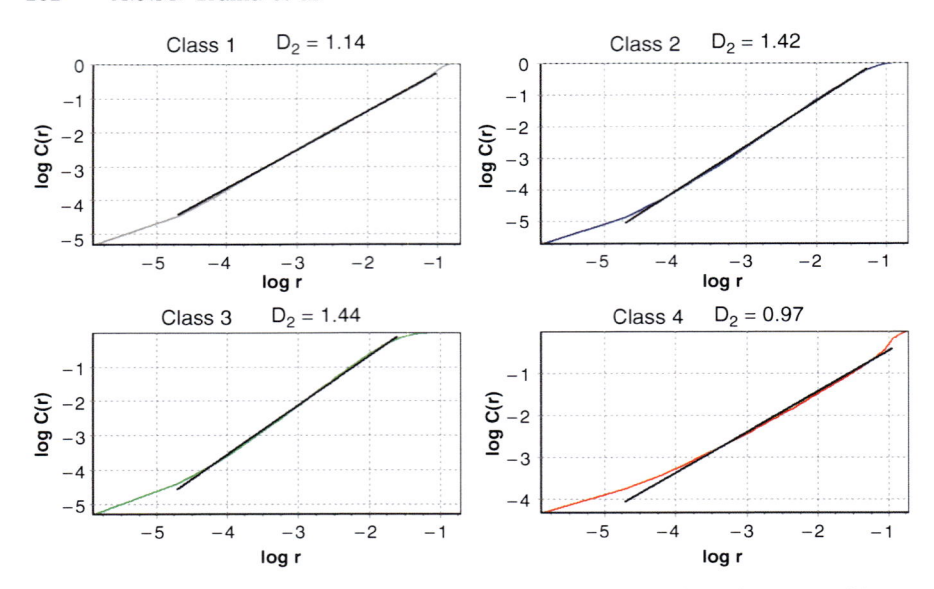

Fig. 8.2. *Plot of fractal dimensions.* The graphs correspond to the tissue Classes 1–4 in Fig. 8.1 (**c**)–(**f**), respectively

frequency of occurrences of a specific color histogram obtained from the pixels of the image, cf. Sect. 1.3.1, page 16. Its almost omni-presence in imaging systems is mostly due to its nice properties of linear cost to be obtained, as well as its invariance to rotation, translation and scale, for normalized histograms. It can also be used as a cheap first step when comparing images to select the most relevant one to answer a query, thus reducing the candidate set before applying a costlier feature extractor [1].

Metric Histogram

One of the drawbacks of traditional histograms is the high dimensionality, usually ranging from around 100 to more that 4,000 gray-levels for x-ray images (the Hounsfield units in CT). A histogram with 100 different colors represents an image as a point in a space with dimensionality $E = 100$. Most structures indexing datasets in this order of dimensionality are heavily affected by the dimensionality curse [6].

Traina et al. [7] proposed the metric histogram aiming at handling histograms and dealing with the dimensionality curse. This feature vector has a variable dimensionality to describe the images, maintaining only the gray-levels that really represents the image. In a metric histogram, the equivalent to a traditional histogram *bin* is called a *bucket*, and each one corresponds to a concatenation of one or more *bins* from the original histogram. Metric histograms bounds the histogram contour and thus reduce its dimensionality.

It creates histograms with varying dimensionality for distinct images, so it is well-suited for metric indexing structures.

8.2.3 Texture Features

Among the low-level features used to represent medical images in retrieval systems, the texture-based extractors stand out due to their effectiveness in discriminating tissues. The texture measures mainly capture the granularity and repetitive patterns in the pixels distribution.

The most widely used approach to describe region textures are the statistical ones, which describe textures as smooth, homogeneous, uniform, among other features. In this Chapter, we present the statistical moments of the intensity histogram of an image or image region [8]. A detailed introduction to texture analysis in medical images has already been given, cf. Chap. 6, page 157.

Co-Occurrence Matrix

The co-occurrence matrix, also called as Spatial Gray Level Dependence (SGLD) matrix [9], is the most popular techniques for texture feature representation. Its rows and columns represent the relative frequency of gray level of pixel pairs occurring at each direction and distance.

Let us consider an image f with L possible intensity levels. Let $P_{d,\phi}(i,j)$ be the co-occurrence matrix (Fig. 8.3), where each element (i,j) indicates the number of times that each pair of pixels with intensities i and j, $0 \leq i,j \leq L-1$ occurs in f in the position specified by a distance d and an angle ϕ. Formally, each element (i,j) represents the number of times that $p_1 = p_2(d\cos\theta, d\sin\theta)$, where p_1 and p_2 have intensity i and j, respectively. The number of possible intensity levels in the image determines the size of the co-occurrence matrix P for a given distance d and angle ϕ

Several features f_j can be extracted from a co-occurrence matrix for a feature vector F. Haralick [9] proposed 14, calling them descriptors. Table 8.1 presents the Haralick features most used in the literature, where

$$g_{ij} = \frac{P_{d,\phi}(i,j)}{n} \tag{8.3}$$

and n is the sum of the elements of $P_{d,\phi}$.

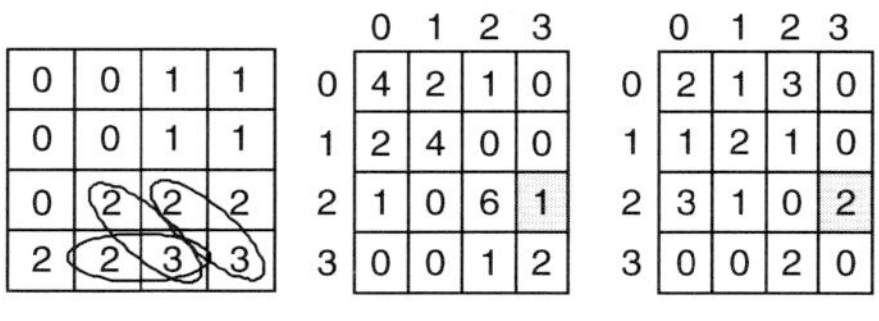

Fig. 8.3. *Co-occurrence matrix. Left*: image with $L = 4$, orientations $\phi = 0°$ and $\phi = 135°$ and $d = 1$ are indicated; *Middle*: co-occurrence matrix with $\phi = 0$ and $d = 1$; *Right*: co-occurrence matrix with $\phi = 135°$ and $d = 1$

Table 8.1. *Haralick descriptors.* From the 14 descriptors initially suggested by Haralick, we list the most used only

Feature	Formula
Contrast	$\sum_{i,j}(i-j)^2 g_{ij}$
Uniformity (or energy)	$\sum_i \sum_j g_{ij}^2$
Homogeneity	$\sum_i \sum_j \frac{g_{ij}}{1+|i-j|}$
Entropy (or suavity)	$-\sum_i \sum_j g_{ij} \log_2 g_{ij}$
Step (or intensity)	$\sum_i \sum_j g_{ij}$
Third moment (or distortion)	$\sum_i \sum_j (i-j)^3 g_{ij}$
Inverse of variance	$\sum_i \sum_j \frac{g_{ij}}{(i-j)^2}$

Fig. 8.4. *Tumors in mammographies.* The examples show a benign (*left*) and a malignant (*right*) tumor

8.2.4 Shape Features

Representing images based on shape is one of the most difficult problem to be dealt with due to its inherent subjectivity. The difficulty derives from the segmentation task, previously discussed in this chapter. Shape descriptors are important in several contexts. For instance, the shape and the size of tumors in mammograms are essential to classify them as benign or malignant. According to Alto et al. [10], tumors with irregular shape are usually malignant and tumors with regular shape are usually benign. Figure 8.4 shows an example of benign and malignant tumors.

A desirable property for shape features is that they must be invariant to scale, rotation and translation (geometric transformations), and, moreover, it can be able to describe the object's shape even when the image is noisy.

According to Zahn [11], we can describe shape and boundary by polynomial approximations, moment invariants and Fourier descriptors. In this section, we discuss the moment invariants and the Zernike moments, two shape-based extractors frequently used to provide shape-based features f_j for images.

Zernike Moments

Zernike moments are often used to build rotation invariant descriptors. Zernike polynomials form an orthogonal basis on a unit circle $x^2 + y^2 \leq 1$. Thus, they do not contain any redundant information and are convenient for image reconstruction [12]. The complex Zernike moments are expressed as $Z_{n,m}$, as follows

Table 8.2. *Orthogonal radial polynom.* The table shows the first six orthogonal radial polynomials

$R_{0,0}(\rho) = 1$	$R_{2,0}(\rho) = 2\rho^2 - 1$	$R_{3,1}(\rho) = 3\rho^3 - 3\rho$
$R_{1,1}(\rho) = \rho$	$R_{2,2}(\rho) = \rho^2$	$R_{3,3}(\rho) = \rho^3$

$$Z_{n,m} = \frac{n+1}{\pi} \sum_x \sum_y f(x,y)\, V^*_{n,m}(x,y), \qquad x^2 + y^2 \le 1 \qquad (8.4)$$

where $f(x,y)$ is an image, $n = 0, 1, 2, \ldots$ defines the Zernike polynomial order, $*$ denotes the complex conjugate, and m is an integer (either positive or negative) depicting the angular dependence, or rotation. Considering n and m, they must satisfy

$$n - |m| \quad \text{must be even and} \quad -n \le m \le n \qquad (8.5)$$

and $Z^*_{n,m} = Z_{n,-m}$ is true.

The Zernike polynomials $V_{n,m}(x,y)$ expressed in polar coordinates are

$$V_{n,m}(x,y) = V_{n,m}(\rho\cos(\theta), \rho\sin(\theta)) = R_{n,m}(\rho)e^{im\theta} \qquad (8.6)$$

where (ρ, θ) are defined over the unit disc, $\rho = \sqrt{x^2 + y^2}$, $\theta = \arctan(\frac{y}{x})$, $x = \rho\cos\theta$, $y = \rho\sin\theta$, $i = \sqrt{-1}$ and $R_{n,m}$ is the orthogonal radial polynomial defined as

$$R_{n,m}(\rho) = \sum_{s=0}^{\frac{n-|m|}{2}} (-1)^s \frac{(n-s)!}{s!\left(\frac{n+|m|}{2} - s\right)!\left(\frac{n-|m|}{2} - s\right)!} \rho^{n-2s} \qquad (8.7)$$

Notice that $R_{n,m}(\rho) = R_{n,-m}(\rho)$. If the conditions of (8.5) are not satisfied, then $R_{n,m}(\rho) = 0$. The first six orthogonal radial polynomials are given in Table 8.2.

8.3 Image Features and Distance Functions

Describing images by visual content relies on comparing image features using a distance function to quantify the similarity between them. A challenge to answer similarity queries is how to properly integrate these two key aspects. Plenty of research has been conducted on algorithms for image features extraction. However, little attention has been paid to the importance of selecting a well-suited distance function to compare the image features given by each image extractor. This section targets this problem. We show that a careful choice of a distance function improves in a great extent the precision of similarity queries even when the same features are used, and shows the strong relationship between these two key aspects.

8.3.1 Similarity Search and Metric Spaces

When working with image datasets, performing exact searches are not useful, since searching for the same data already under analysis usually has no meaning. Hence, the representation of complex data types, like images, is mainly performed considering similarity [13].

New query operators are required to compare data by similarity. As we saw in the previous section, feature vectors are often of very high dimensionality, rendering the multi-dimensional access methods useless. Moreover, there are also kinds of features that are "adimensional", when the number of features per image is not fixed, such as the metric histograms. Therefore, an ideal way to index images represented by feature vectors should be "dimensionality-independent". Considering that, whenever a function to compute the distances between the signatures is defined, both adimensional and dimensional data can be represented in a metric space. Thus, metric spaces turn out to be the best choice to represent these plethora of data, as they only require the elements and their pairwise distance [14].

A metric space is formally defined as a pair $\mathbb{M} = \langle \mathbb{S}, d \rangle$, where $\mathbb{S}$ denotes the universe of valid elements and d is the function $d : \mathbb{S} \times \mathbb{S} \to \mathbb{R}^+$, called a metric, that expresses the distance between elements in $\mathbb{S}$. To be a metric, a function d must satisfy the following properties for every $s_1, s_2, s_3 \in \mathbb{S}$:

1. *Symmetry*: $d(s_1, s_2) = d(s_2, s_1)$.
2. *Non-negativity*: $0 < d(s_1, s_2) < \infty$ if $s_1 \neq s_2$ and $d(s_1, s_1) = 0$.
3. *Triangular inequality*: $d(s_1, s_3) \leq d(s_1, s_2) + d(s_2, s_3)$.

Using a metric to compare elements, a similarity query returns the stored elements that satisfy a given similarity criterion, usually expressed in terms of one or more reference elements, which are called the *query center(s)*. The main comparison operators to perform similarity queries are:

- *Range query*: Given a dataset $S \in \mathbb{S}$, a query center $s_q \in \mathbb{S}$ and a radius $r_q \in \mathbb{R}^+$, a range query selects every element $s_i \in S$, such that $d(s_i, s_q) \leq r_q$.
- *k-NN query*: Given a dataset $S \in \mathbb{S}$, a query center $s_q \in \mathbb{S}$ and an integer value $k \geq 1$, the Nearest Neighbor (NN) query selects the k elements $s_i \in S$ that are at the shortest distances from s_q.

A k-nearest neighbor query example is: "Given the head x-ray of Jane Doe, find in the image database the 10 images most similar to it". A range query example is: "Given the head x-ray of Jane Doe, find in the image database the images that differ up to 5 units from it".

8.3.2 Distance Functions

The distance functions most widely employed to perform similarity queries over vector spaces are those of the Minkowski family (or L_p norm) [15], where the objects are identified with n real-valued coordinates. Considering two feature vectors $F = \{f_1, \ldots, f_E\}$ and $G = \{g_1, \ldots, g_E\}$, the L_p distances are defined as

$$L_p\left((f_1, \ldots, f_E), (g_1, \ldots, g_E)\right) = \sqrt[p]{\sum_{j=1}^{E} |f_j - g_j|^p} \quad (8.8)$$

Varying the value assigned to p we obtain the L_p family of distance functions. They are additive, in the sense that each feature contributes positive and independently to the distance calculation. The well-known Euclidean distance corresponds to L_2. The L_1 distance, also called city block or Manhattan Distance (MHD), corresponds to the sum of coordinate differences. The L_∞ distance, also known as infinity or Chebychev distance, corresponds to taking the limit of (8.8) when p tends to infinity. The result obtained computing the L_∞ distance is the maximum difference of any of its coordinates. A variation of the Minkowski family distance is the weighted Minkowski, where different weights are assigned to each feature from a given image. The idea of the weighted Minkowski distance is to emphasize the most important features. Different datasets require different weighting vectors, considering that some features present a higher or lower relevance regarding to others at a given application.

Besides the Minkowski family, several other distances are useful for comparing images. Considering the comparison of two feature vectors $F = \{f_1, \ldots, f_E\}$ and $G = \{g_1, \ldots, g_E\}$, we can define the:

- *Jeffrey divergence*, which is symmetric and presents a better numerical behavior than the Minkowski family. It is stable and robust regarding noise [16]

$$d_J(F, G) = \sum_{j=1}^{E} \left(f_j \log \frac{f_j}{m_j} + g_j \log \frac{g_j}{m_j} \right) \quad \text{where} \quad m_j = \frac{f_j + g_j}{2} \quad (8.9)$$

- *Statistic Value* χ^2, which emphasizes the elevated discrepancies between two feature vectors and measures how improbable the distribution is

$$d_{\chi^2}(F, G) = \sum_{j=1}^{E} \frac{(f_j - m_j)^2}{m_j} \quad \text{where} \quad m_j = \frac{f_j + g_j}{2} \quad (8.10)$$

- *Canberra distance*, which is a comparative Manhattan distance and the most restrictive distance function

$$d_C(F, G) = \sum_{j=1}^{E} \frac{|f_j - g_j|}{|f_j| + |g_j|} \quad (8.11)$$

- *Quadratic distance*, which takes into account the fact that certain pairs of attributes, correspond to features that are more important for the perceptual notion of similarity than others

$$d_Q(F,G) = \sqrt{(f-g)^T A(f-g)} \qquad (8.12)$$

where $A = [a_{ij}]$ is an $E \times E$ similarity matrix, and a_{ij} denotes the dissimilarity between the features i and j.

There is a close relationship between the features and the distance function used to compare the data, in order to return what the human beings would expect from such comparison. However, the majority of the works concerning indexing and retrieval of images overlooks this relationship and go for the most known and used distance functions, such as the Euclidean or other members of the L_p family, relegating the distance function to a secondary importance. However, it is important to highlight that the efficiency and the efficacy of a signature-based decision support system is significantly affected by the distance function ability on separating the data.

8.3.3 Case Study: Evaluating Distance Functions for Separating Data

In this section, we present experiments aimed at supporting our claim that a careful choice of a distance function considerably improves the retrieval of complex data [2].

Study Design

We performed k-NN queries on image datasets, using different distance functions and compared their retrieval ability. Each set of feature vectors was indexed using the Slim-tree metric access method [17] to accelerate executing the similarity query evaluation. To assess the distance function ability on properly separating the images, we have generated plots based on the Precision and Recall (P&R) approach [18], obtained executing the same sets of similarity queries using distinct metrics. A rule of thumb to read these plots is that the closer the curve to the top, the better the retrieval technique is. Therefore, the best combination of features and the distance function pushes the P&R curve nearest to the top.

To allow exploiting the ability of the several distance functions to identify the class of each image, we created a dataset mixing eight classes of images remarkably different, obtained from distinct body regions. In total 704 images were collected. The images have of 256×256 pixels and are represented by 8 bits, resulting in 256 gray-levels. Figure 8.5 illustrates an image example of each class.

All images of each dataset were employed as query centers to compose each query set. The feature vectors are composed of gray-level histograms, Zernike moments, Haralick descriptors, metric histograms, and the improved EM/MPM algorithm, all of them presented in Sect. 8.2.

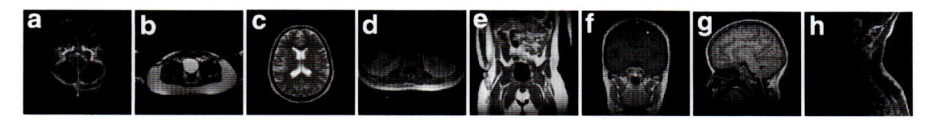

Fig. 8.5. *Examples of MRI slices from the dataset.* (**a**) MR angiography with 36 images; (**b**) axial pelvis with 86 images; (**c**) axial head with 155 images; (**d**) axial abdomen with 51 images; (**e**) coronal abdomen with 23 images; (**f**) coronal head with 36 images; (**g**) sagittal head with 258 images; and (**h**) sagittal spine with 59 images

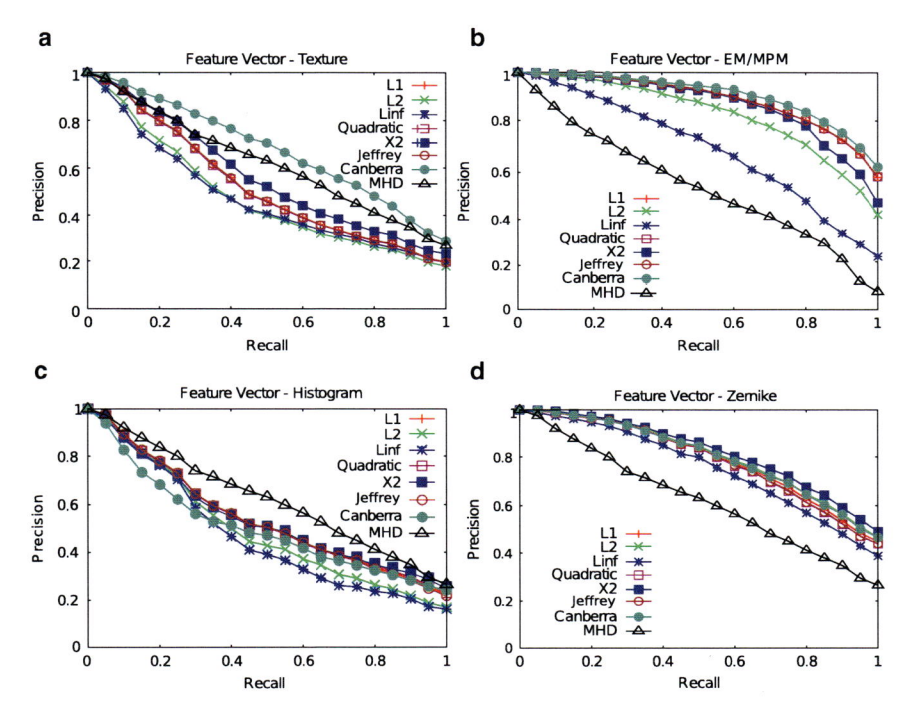

Fig. 8.6. *Precision and recall graphs.* The graphs illustrate the retrieval ability of several distance functions: (**a**) texture using Haralick descriptors; (**b**) shape/texture using the EM/MPM algorithm; (**c**) gray-level histogram; and (**d**) shape using Zernike features

Achieved Results

Since the dataset was composed of 704 MR images, for each distance function we posed 704 queries, using each image as the center of a query. The average values obtained from the P&R calculation was used to generate the plots of Fig. 8.6.

The plots in Fig. 8.6a show the results obtained using the Haralick feature vector. We can see that the Canberra distance allows a considerable gain in precision compared to the others, achieving approximately 80 % of precision at a 40 % of recall. The next one is the χ^2 distance, followed by the Jeffrey divergence and by L_1, obtaining a precision from 55 % up to 60 % at a 40 % recall level. The most commonly used Euclidean (L_2) and Chebychev (L_∞) distances presented the poorest results. The difference in precision reaches values of up to 92 % when the Canberra and the Chebychev functions are compared. This value would make a huge difference in the response set returned to the users.

The plots of the EM/MPM descriptor in Fig. 8.6b show that the Canberra distance presents the highest precision, achieving up to 95 % at a recall level of 60 %. The L_2 and L_∞ distances gave the worse results when compared to the other distances. It is important to note that the difference in precision when Canberra and Chebychev reaches values of approximately 36 % of precision at a 55 % of recall.

Analyzing the plots of Fig. 8.6c, we observe that the clear winner is the MHD distance, which is also the fastest distance to compute, being in average four times faster than L_1. The χ^2, the quadratic form, the Jeffrey divergence and the L_1 distance functions present almost the same behavior until 25 % of recall. The difference in precision when MHD and L_∞ are compared, reaches values of approximately 75 %. For these features, the Canberra distance presents initially the lower levels of precision, improving its behavior for recall levels above 35 %. We present the MHD distance also in the other graphs as a baseline to provide comparison among the features.

Finally, the plots of Fig. 8.6d illustrate the precision and recall values obtained using the Zernike features. We can notice that all distances but the L_∞ and the MHD distances almost tie. That is, the best distance when considering histograms became the worst for Zernike feature vectors. Notice that the gain for a careful choice of distance can reach 43 % when comparing the MHD distance and χ^2 distance at 45 % of recall.

Resumee

These results show that choosing of a proper distance function can improve to a great extent the precision of similarity queries (i.e. the quality of retrieving medical images). In general, we can see that Canberra would be a nice choice for both texture and Zernike features, remembering that it also presents low computational cost, whereas Euclidean and L_∞, the most frequently used metrics, never obtained good results, at least in these experiments.

8.4 Feature Selection

Images are traditionally described by a combination of raw features (recall Sect. 8.2), so different aspects of the image can be represented. This approach usually renders a feature vector $F = \{f_1, \ldots f_E\}$ with hundreds or even thousands of features.

8.4.1 Curse of Dimensionality

Differently from what common sense would tell, instead of helping on better separating the images, this approach actually introduces a problem. As the number E of features grows, the processes of indexing, retrieving, comparing and analyzing the images become more ineffective, since in high dimensionality the data tend to be not separable, and the comparisons are more time consuming. Moreover, in most cases a large number of features are correlated to each other, providing redundant information that actually disturbs the image's differentiation, and leads to the dimensionality curse [19]. In fact, a larger number of features only makes each feature less meaningful. Therefore, it is mandatory to keep the number of features small, establishing a trade-off between the discrimination power and the feature vector size.

8.4.2 Traditional Algorithm for Feature Selection

Feature selection can be seen as a mining process to access the most meaningful features to classify images. Recently, the mining of association rules has been successfully employed to perform feature selection. Traditional association rule mining is adequate when dealing with categorical (nominal) data items. However, image features consist of continuous attributes, so a type of association rule that considers continuous values is necessary. A recent type of continuous association rules is the statistical association rules, where the rules are generated using statistical measurements from images [20]. In this case, the mined association rules are used to weight the features according to their relevance, making a new and enhanced representation of the images. However, we first discuss some of the traditional algorithms for feature selection.

Relief Algorithm

One of the most well-known feature selection algorithm is Relief [21]. The Relief algorithm aims at measuring the quality of features according to how their values distinguish instances of different classes. One limitation of the Relief algorithm is that it works only for datasets with binary classes. This limitation is overcome by Relief-F [22] that also tackles datasets with multi-valued classes.

Relief-F finds 1-Nearest Neighbor (NN) of every instance E_1 from every class. Let C_{E1} be the class of the instance E. For each neighbor, Relief-F evaluates the relevance of every feature $f \in F$ updating the weight $W[f]$ in (8.13). The NN from the same class E_1 and from a different class C is referred to as hit H and miss $M(C)$, respectively

$$W[f] = W[f] - \text{dist}(f, E_1, H) + \sum_{C \neq C_{E1}} P(C) \times \text{dist}(f, E_1, M(C)) \quad (8.13)$$

The dist(f, E_1, E_2) is the distance (normalized difference module) of attribute f values between the instances E_1 and E_2. The prior probability of class $P(C)$ is employed in the summation.

One drawback of Relief-F is that it is unable of detect redundant attributes relations. For example, if there is a duplicated feature, the weight returned by Relief-F to both features is the same.

Decision Tree Method

Another well-known feature selection technique is the Decision Tree Method (DTM) [23]. DTM adopts a forward search to generate feature subsets, using the entropy criterion to evaluate them. DTM runs the C4.5 algorithm [24], an algorithm that builds a decision tree. Since a decision tree is a sequence of attributes that defines the state of an instance, DTM selects the features that appear in the pruned decision tree as the best subset, i.e., the features appearing in the path to any leaf node in the pruned tree are selected as the best subset.

Decision trees are hierarchical tree structures developed to perform data classification, where each intermediate node corresponds to a decision taking on the value of a feature (attribute) to divide data samples. Training samples are hierarchically split into smaller and usually more homogeneous groups, taking into account the value of one feature at each level of the decision tree, that is, the value of one feature is evaluated per node. The leaf nodes represent the pure classes of the samples.

Classification is performed by testing the samples through the hierarchical tree, while performing simple test on a feature at each step [25]. The decision tree construction leads to perform feature selection among the features available. The most discriminating features are the ones usually employed in the first levels of the decision tree. The least discriminating features tends to occur in the lower levels of the trees, which are generally responsible for classifier over-fitting. Such over-fitting may be partially overcome by employing pruning methods. One pruning technique is the backward pruning. This essentially involves growing the tree from a dataset until all possible leaf nodes have been reached and then removing specific subtrees. One way of pruning is measuring the number of instances that are misclassified from the test sample by propagating errors upwards from leaf nodes, at each node in a tree. Such error is compared to the error rate that would exists if the node was replaced by the most common class resulting from that node. If the difference is a reduction in the error rate, then the subtree at the node should be pruned [24].

The DTM consists in taking the (features) attributes that occurred in the nodes of the pruning decision tree as the relevant features, discarding the remaining ones from the dataset [23]. The usage of feature selection provided by DTM has been shown to increase the precision of content-based queries and the classifier accuracy, also speeding up the learning process [26].

8.4.3 Combined Feature Selection and Discretization

Feature selection is important to discard irrelevant features and to diminish the dimensionality curse problem. Feature discretization works as a filter of the feature values in the mining process: it reduces the number of possible features to be evaluated in the mining process (making it faster), adds some semantic to the features values and makes it possible to some mining algorithms designed to categorical data to work with continuous features. In fact, it is possible to bring together the tasks of feature selection and feature discretization in the same algorithm.

Omega Algorithm

As an example, we present the Omega algorithm that performs both tasks together. Omega [26] is a supervised algorithm that performs discretization of continuous values. Omega processes each feature separately and discretizes a range of N values following four steps, having linear cost on N. Let f_j be a feature from a feature vector F and $f_j(i)$ be the value of the feature type f_j in the image i. Omega uses a data structure that links each instance value $f_j(i)$ with the instance class label $c_j(i)$. Let an image instance $I(i)$ be a pair $(f_j(i), c_j(i))$. Let U_k and U_{k+1} be the limits of an interval T_k.

Lemma 1. *An instance $I(i) = (f_j(i), c_j(i))$ belongs to an interval $T_k = [U_k, U_{k+1}]$ if and only if $U_k < f_j(i) < U_{k+1}$.*

The Omega algorithm processes a range of N sorted values in four steps, having linear cost in N. In particular, the steps are:

1. Omega sorts the continuous values and defines the initial cut points. A cut point is placed before the smallest value and another cut point is placed after the highest value of the feature. Whenever a value is modified or the class label changes, a new cut point is created. This step produces pure bins, where the entropy is equal to zero, minimizing the inconsistencies created by the discretization process. However, the number of bins produced tends to be very large and susceptible to noise.
2. Omega restricts the minimum frequency that a bin must present, avoiding increasing the number of cut points too much. To do so, it removes the right cut points of the intervals that do not satisfy the minimum frequency restriction given by an input parameter $H_{\min}$. Only the last interval is allowed to not satisfy the minimum frequency restriction. The higher the value of $H_{\min}$, the fewer bins result from this step. However, the higher $H_{\min}$, the higher the inconsistencies generated by the discretization process. Figure 8.7 shows an example of the cut points found in the first and eliminated in the second step of Omega, using $H_{\min} = 2$.

Fig. 8.7. *Step 2 of the Omega algorithm.* Cut points are eliminated by the Omega algorithm using $H_{min} = 2$

1.1	1.2	1.3	1.4	1.5	1.6	1.7	1.8
C1	C2	C2	C1	C1	C2	C1	C1

cut points eliminated

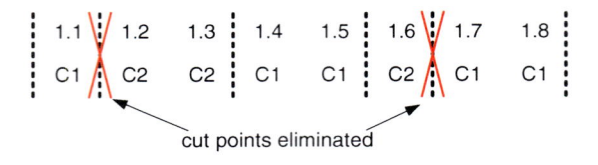

1.1	1.2	1.3	1.4	1.5	1.6	1.7	1.8
C1	C2	C2	C1	C1	C2	C1	C1

Fig. 8.8. *Step 3 of the Omega algorithm.* A cut point is eliminated by Omega using $\zeta_{max} = 0.35$

3. Omega fuses consecutive intervals, measuring the inconsistency rate to determine which intervals should be merged. Let M_{T_k} be the majority class of an interval T_k. The inconsistency rate ζ_{T_k} of an interval T_k as given by

$$\zeta_{T_k} = \frac{|T_k| - |M_{T_k}|}{|T_k|} \qquad (8.14)$$

where $|T_k|$ is the number of instances in the interval T_k, and $|M_{T_k}|$ is the number of instances of the majority class in the interval T_k. The Omega algorithm fuses consecutive intervals that have the same majority class and also have inconsistency rates below or equal to an input threshold ζ_{max} ($0 \leq \zeta_{max} \leq 0.5$). Figure 8.8 shows an example of a cut point (Fig. 8.7) that is eliminated using $\zeta_{max} = 0.35$. The inconsistency rates ζ_{T_k} of the second and third intervals shown in Fig. 8.8 are respectively $\zeta_{T_2} = 0/2 = 0$ and $\zeta_{T_3} = 1/3 = 0.33$. Since T_2 and T_3 have the same majority class, i.e. $M_{T_2} = M_{T_3} =$ "C1" and $\zeta_{T_2} \leq \zeta_{max}$ and $\zeta_{T_3} \leq \zeta_{max}$, the second and third intervals are fused. The cut points remaining in Step 3 are the final cut points returned by the algorithm.
4. Omega performs feature selection. Let T be the set of intervals in which a feature is discretized. For each feature, Omega computes the global inconsistency

$$\zeta_G = \frac{\sum_{T_k \in T}(|T_k| - |M_{T_k}|)}{\sum_{T_k \in T} |T_k|} \qquad (8.15)$$

The feature selection criterion employed by Omega removes from the feature vector every feature whose global inconsistency value is greater than an input threshold $\zeta_{G_{max}}$ ($0 \leq \zeta_{G_{max}} \leq 0.5$). Since the number of inconsistencies of a feature is the factor that most contribute to disturb the learning algorithm, discarding the most inconsistent features contributes to improve the accuracy as well as to speed up the learning algorithm.

8.5 Association Rule Mining

Feature selection can also be seen as a mining process to access the most meaningful features to classify images. A mining task that has been recently employed to perform feature selection is the association rule mining. Recently, the mining of association rules has been successfully employed to perform feature selection.

A mining process corresponds to the application of a data analysis algorithm to the raw data to produce interesting and useful knowledge about it. For some image mining process, after the feature extraction process, feature selection and feature discretization are employed in order to make the mining process more feasible and accurate.

8.5.1 Definition

Association rule mining is a popular and well researched method in data mining for discovering interesting relations between variables in large databases. It has been extensively studied and applied for market basket analysis.

The problem of mining association rules was firstly stated in [27], as follows. Let $I = \{i_1, \ldots, i_n\}$ be a set of literals called items. A set $X \subseteq I$ is called an itemset. Let R be a table with transactions t involving elements that are subsets of I. An association rule is an expression $X \rightarrow Y$, where X and Y are itemsets, X is the body or antecedent of the rule and Y is the head or consequent of the rule:

- *Support* is the ratio between the number of transactions of R containing the itemset $X \cup Y$ and the total number of transactions of R.
- *Confidence* is the fraction of the number of transactions containing X that also contain Y.

The problem of mining association rules, as it was firstly stated, consists of finding association rules that satisfy the restrictions of minimum support (*minsup*) and confidence (*minconf*) specified by the user.

However, traditional association rule mining is adequate when dealing with categorical (nominal) data items. However, image features consist of continuous attributes, so a type of association rule that considers continuous values is necessary. A recent type of continuous association rules is the statistical association rule, where the rules are generated using statistical measurements from images [20]. In this case, the mined association rules are used to weight the features according to their relevance, making a new and enhanced representation of the images.

8.5.2 Case Study: Improving Computer-Aided Diagnosis by Association Rule Mining

In this section, we present the Image Diagnosis Enhancement through Associations (IDEA) method [28], a method based on association rules to support the diagnosis of medical images. The method suggests a set of keywords to compose the diagnosis of a given image and uses a measure of certainty to rank the keywords according to their probability of occurring in the final diagnosis of the image prepared by the radiologist. It is an example of application of feature extraction and feature selection techniques together with image mining approaches to improve diagnosis of medical images, i.e., building a CAD method. The main mining technique employed in this case study is association rule mining. It is important to state that association rules were also successfully employed to classify mammograms [29] and to analyze brain tumors [30].

The IDEA Method

IDEA is a supervised method that mines association rules relating the features automatically extracted from images to the reports given by radiologists about the training images. It aims at identifying the set of keywords that have a high probability of being in the report given by the specialist, based on the features extracted. Figure 8.9 shows the pipeline execution of IDEA and Algorithm 1 summarizes its steps.

Algorithm 1 The steps of the IDEA Method

Require: Training images and Test images datasets
Ensure: Report (Set of keywords).
 1: Extract features of the training images
 2: Perform supervised discretization of training image features (Omega)
 3: Mine association rules
 4: Extract features of the test image
 5: Execute the associative classifier engine (ACE)
 6: Return the suggested report (set of keywords)

The IDEA method has two phases: training and testing. In the training phase, features are extracted from the images to create the feature vectors that are used to represent the images (Algorithm 1, Step 1). The feature vector and the *class* of each training image are submitted to the Omega module (Sect. 8.4), which removes irrelevant features from the feature vector and discretizes the remaining features (Step 2). The *class* is the most important keyword chosen by the specialist to describe the image. In the training phase, a processed feature vector is merged with the diagnosis keywords about the training images, producing the transaction representation of each image. The transaction representing all the training images are submitted to the Apriori algorithm [31] to perform the association rule mining (Step 3), setting the minimum confidence to high values. In the test phase, the feature vector of the test image is extracted and submitted to the Associative Classifier Engine

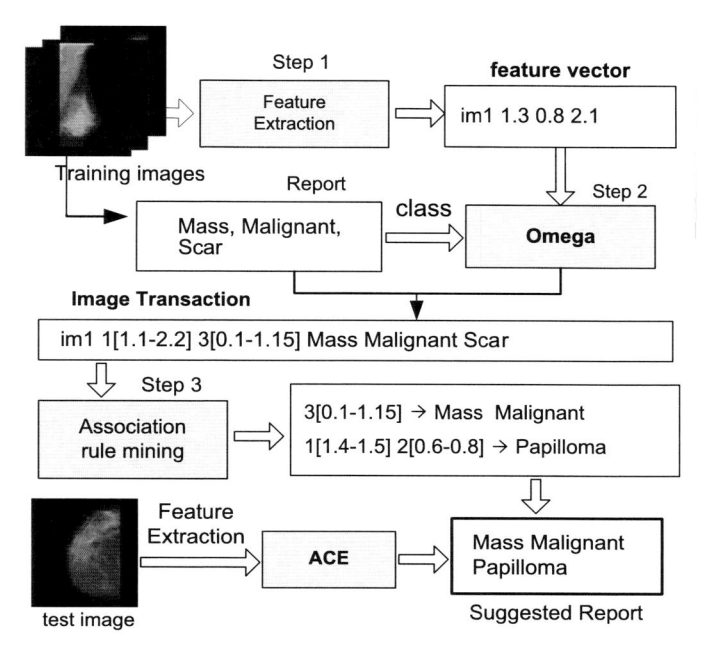

Fig. 8.9. *The IDEA method.* In the training phase, the image features are submitted to Omega algorithm, whose output is aggregated with high level information about the images to compose the image records. The image records are submitted to an association rule mining algorithm to produce a set of training association rules that is employed to classify new images

(ACE), which uses the association rules to suggest keywords to compose the diagnosis of the test image (see next section).

The IDEA method employs the Apriori algorithm [31] to mine association rules. The output of the Omega algorithm and the keywords of the report of the training images are submitted to the Apriori algorithm. A constraint is added to the mining process to restrict the diagnosis keywords to the head of the rules. The body of the rules is composed of indexes of the features and their intervals. The minimum confidence values are set high (usually greater than 97 %). The mined rules are used as input to the ACE algorithm.

Associative Classifier Engine

In the test phase, IDEA employs the associative classifier engine (ACE) to classify new test images. It is said that an image (a) matches, (b) partially matches or (c) does not matches a rule, if the image features, respectively, satisfy the whole body, satisfy part of the rule's body or does not satisfy any part of it. The ACE algorithm stores all sets of keywords (*itemsets*) belonging to the head of the rules in a data structure. An itemset h is returned by ACE in the suggested diagnosis whenever the following conditions are satisfied:

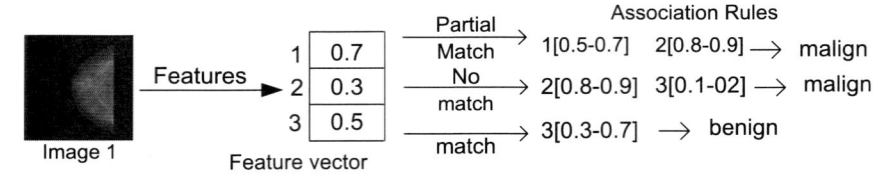

Fig. 8.10. *Example of ACE working.* In this example, $h = \{benign\}$ is returned if $\frac{4}{5} \geq w_{\min}$

$$M(h) \geq 1 \wedge w = \frac{3M(h)+P(h)}{3M(h)+P(h)+N(h)} \geq w_{\min}$$

where $M(h)$ is the number of matches of the itemset h, $P(h)$ is the number of partial matches, and $N(h)$ is the number of no matches automatically computed. Variable w is the weight of the itemset. The weight indicates the certainty level that an itemset h will belong to the final image diagnosis given by a specialist. The higher the weight, the stronger the confidence that h belongs to the image diagnosis. A threshold for the minimum weight $w_{\min}$ ($0 \leq w_{\min} \leq 1$) is employed to limit the weight of an itemset in the suggested diagnosis. If $w_{\min} = 0$, all itemsets matching at least one rule are returned.

Figure 8.10 shows an example of ACE working. In this example, $M(h) = 1$, $P(h) = 1$ and $N(h) = 1$ for the itemset $h = \{benign\}$. Therefore, if $\frac{4}{5} \geq w_{\min}$, the itemset $h = \{benign\}$ is returned, otherwise it is discarded.

Study Design

Following, an application of the IDEA method to mammography is presented. The parameters of the Omega algorithm were set to $H_{\min} = 2$, $\zeta_{\max} = 0.2$ and $\zeta_{G_{\max}} = 0.3$, which are tunning parameters of the algorithm. The values of minimum support minsup = 0.005 and confidence minconf = 1.0 were used as the Apriori input parameters. The value $w_{\min} = 0$, which maximizes the ACE accuracy, was employed as the ACE input parameter.

We employed in the experiments the Region of Interest (ROI) dataset, which consists of 446 images of ROIs comprising tumoral tissues, taken from mammograms collected from the Breast Imaging Reporting and Data System (BI-RADS) Tutorium[1] of the Department of Radiology of University of Vienna. The IDEA method was applied over the ROI dataset employing 10 % of the images from the dataset for testing and the remaining images for training. Each image has a diagnosis composed of three main parts:

1. Morphology (mass or calcification).
2. The Breast Imaging Reporting and Data System (BI-RADS) level.
3. Histology.

---

[1] http://www.birads.at

Feature	Position
Average intensity, contrast	1–2
Smoothness, skewness	3–4
Uniformity, entropy	5–6
Invariant moments	7–13
Histogram mean, standard deviation	14–15

Table 8.3. *Mammography signature.* The table shows the features extracted from the *ROI* dataset and their positions in the feature vector

Measure	Performance
Accuracy	96.7%
Sensitivity	91.3%
Specificity	71.4%

Table 8.4. *Results.* The results were achieved by IDEA in detecting the BI-RADS levels over the ROI dataset

In the feature extraction step, the images were segmented and the features of texture, shape and color were extracted from the segmented regions. The segmentation process were performed eliminating from each image the regions with gray-level smaller than 0.14 (in a gray-scale range of [0–1]) and applying the well-known Otsu's technique over the resultant image. The features shown in Table 8.3 were extracted from the segmented regions and used to compose the feature vector representation of the images.

The Omega algorithm was applied to the image features, removing the 13th feature, because the 13th feature is the least differentiating feature for the *ROI* dataset. The output from Omega was submitted to a priori, which mined 662 rules. The association rules generated and the test images were submitted to the ACE algorithm, which produced diagnosis suggestions for each test image.

Results

The experiments with the IDEA system were performed in a batch execution. The accuracy obtained considering the main parts of the diagnosis were 91.3% and 96.7% for morphology and BI-RADS value, respectively. Since BI-RADS categorization has a fuzzy separation among consecutive levels, even for a human being, we considered an answer correct if the BI-RADS level suggested was the same or an adjacent level of the one annotated by the radiologist in the image report. The result indicates that the features employed better represented the BI-RADS level of the lesion than the morphological properties of the images. Table 8.4 shows the values of accuracy, sensitivity and specificity achieved by IDEA in detecting the BI-RADS levels.

Figure 8.11 shows a screenshot of the system when analyzing the image shown on the left of the window. The values in parenthesis to the right of the suggested keywords are the degree of confidence that the keyword would be included in the diagnosis generated by the specialist.

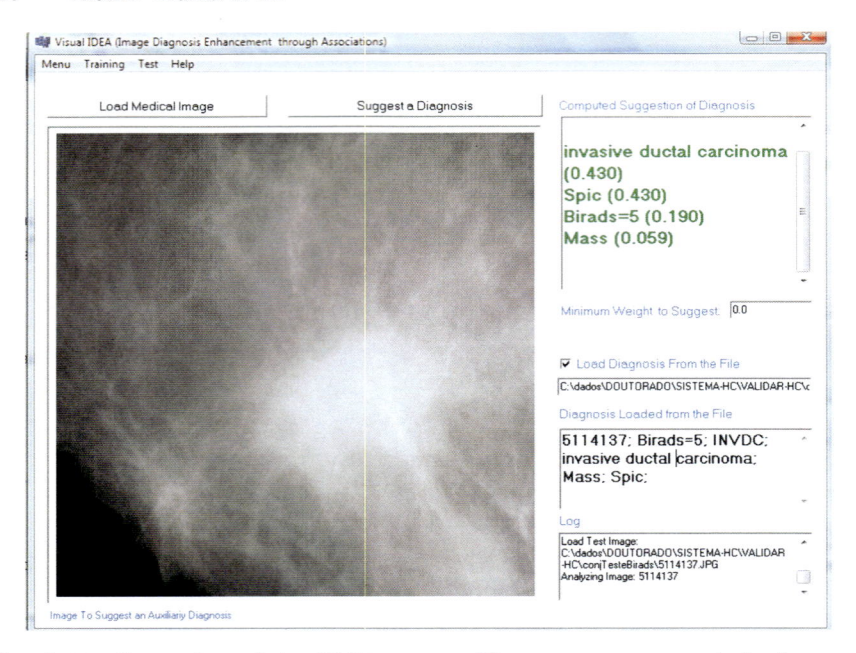

Fig. 8.11. *Screenshot of the IDEA system.* The mammogram and the keywords suggested by the system are shown on the left and on the top right, respectively. The values in parenthesis are the degree of confidence that the keyword would be included in the diagnosis given by the specialist

Resumee

The results obtained are very promising, mainly due to the fact that they show a very small error rate regarding the main part of the diagnosis (BI-RADS level), making the system reliable. In further studies, the performance of physicians with and without the aid of the IDEA-based system must be compared to finally prove the impact of association rule mining.

8.6 Conclusions

This chapter presented and discussed the use of image features to help constructing computational systems to aid in decision making over medical images. The main points highlighted in this chapter are:

- Present the main feature extractors that compose the image descriptors, based on color, shape and texture.
- Present the most common distance functions and discuss their relationship with the extracted features when processing similarity queries.
- Show that keeping low the dimensionality of the feature vector helps answering similarity queries in a more efficient and effective way.

The relationship among the extracted features and the distance function employed to compare them is relevant to improve precision when answering queries. It was also shown that, depending on the distance function employed, the precision can be improved up to 100 %, what certainly results in better answers when applying image retrieval techniques to support decision making when analyzing images.

Section 8.5 was dedicated to discuss a case study on the IDEA framework. It employs the concepts presented in the two previous sections and integrates association rule discovery techniques to mine patterns from the images that are related to the report provided by the specialist to train images used in the learning phase of the framework. By doing so, IDEA learns how to provide meaningful keywords that should be part of the report being generated to the new image under analysis.

From the results presented in this chapter, it can be seen that integrating image feature extraction, comparison selection and indexing, allied to image mining techniques provides strong support to CAD and are powerful allies to the medical specialist during the decision making process of analyzing sets of medical images and exams to provide diagnosis. These techniques are also valuable to help teaching students and radiologists.

Acknowledgments

This work has been supported by FAPESP (the State of Sao Paulo Research Foundation), CNPq (the Brazilian Council for Scientific and Technological Development), and CAPES (the Brazilian Federal Funding Agency for Graduate Education Improvement).

References

1. Muller H, Michoux N, Bandon D, et al. A review of content-based image retrieval systems in medical applications-clinical benefits and future directions. Int J Med Inform. 2004;73(1):1–23.
2. Bugatti PH, Traina AJM, Traina C Jr. Assessing the best integration between distance-function and image-feature to answer similarity queries. In: ACM Symp Appl Comput; 2008. p. 1225–1230.
3. Dempster AP, Laird NM, Rubin DB. Maximum likelihood for incomplete data via em algorithm. J R Stat Soc [Ser B]. 1977;39(1):1–38.
4. Comer ML, Delp EJ. The EM/MPM algorithm for segmentation of textured images: analysis and further experimental results. IEEE Trans Image Process. 2000;9(10):1731–44.
5. Balan AGR, Traina AJM, Marques PMDA. Fractal analysis of image textures for indexing and retrieval by content. Proc IEEE CBMS. 2005; p. 56–63.
6. Pagel BU, Korn F, Faloutsos C. Deflating the dimensionality curse using multiple fractal dimensions. In: Proc IEEE ICDE; 2000. p. 589–598.

7. Traina AJM, Traina C Jr, Bueno JM, et al. Efficient content-based image retrieval through metric histograms. World Wide Web J. 2003;6(2):157–85.
8. Gonzalez RC, Woods RE. Digital Image Processing. NJ: Pearson Prentice Hall; 2008.
9. Haralick RM, Shanmugam K, Dinstein I. Textural features for image classification. IEEE Trans Syst Man Cybern. 1973;3:610–621.
10. Alto H, Rangayyan RM, Desautels JEL. Content-based retrieval and analysis of mammographic masses. J Electron Imaging. 2005;14(2):023016 (1–17).
11. Zahn CT, Roskies RZ. Fourier descriptors for plane closed curves. IEEE Trans Comput. 1972;21:269–281.
12. Khotanzad A, Hua Hong Y. Invariant image recognition by Zernike moments. IEEE Trans Pattern Anal Mach Intell. 1990;12(5):489–497.
13. Chávez E, Navarro G, Baeza-Yates RA, et al. Searching in metric spaces. ACM Comput Surv. 2001;33(3):273–321.
14. Zezula P, Amato G, Dohnal V, et al. Similarity search: the metric space approach. vol. 32 of Advances in Database Systems. New York: Springer; 2006.
15. Wilson DR, Martinez TR. Improved heterogeneous distance functions. J Art Intell Res. 1997;6:1–34.
16. Rubner Y, Tomasi C. Perceptual metrics for image database navigation. The Kluwer International Series in Engineering and Computer Science. Norwell, MA: Kluwer; 2001.
17. Traina J Caetano, Traina AJM, Faloutsos C, et al. Fast indexing and visualization of metric datasets using slim-trees. IEEE Trans Knowl Data Eng. 2002;14(2):244–60.
18. Baeza-Yates RA, Ribeiro-Neto B. Modern Information Retrieval. Boston, MA, USA: Addison-Wesley Longman Publishing Co., Inc.; 1999.
19. Korn F, Pagel BU, Faloutsos C. On the dimensionality curse and the self-similarity blessing. IEEE Trans Knowl Data Eng. 2001;13(1):96–111.
20. Ribeiro MX, Balan AGR, Felipe JC, et al. Mining statistical association rules to select the most relevant medical image features. In: Mining Complex Data. Vol. 165 of Studies in Computational Intelligence. Hershey, PA, USA: Springer; 2009. p. 113–131.
21. Kira K, Rendell LA. A practical approach for feature selection. In: Int Conf Mach Learn; 1992. p. 249–256.
22. Kononenko I. Estimating attributes: analysis and extension of relief. In: Europ Conf Mach Learn; 1994. p. 171–182.
23. Cardie C. Using decision trees to improve case-based learning. In: Int Conf Mach Learn; 1993. p. 25–32.
24. Quinlan R. C4.5: Programs for Machine Learning. San Mateo, CA: Morgan Kaufmann; 1993.
25. Quinlan JR. Simplifying decision trees. Int J Man-Mach Stud. 1987;27(3):221–34.
26. Ribeiro MX, Ferreira MRP, Traina JC, et al. Data pre-processing: a new algorithm for feature selection and data discretization. In: Proc ACM/IEEE CSTST; 2008. p. 252–257.
27. Agrawal R, Imielinski T, Swami AN. Mining association rules between sets of items in large databases. In: ACM SIGMOD Int Conf Manage Data. vol. 1; 1993. p. 207–216.

28. Ribeiro MX, Bugatti PH, Traina C Jr, et al. Supporting content-based image retrieval and computer-aided diagnosis systems with association rule-based techniques. Data Knowl Eng. 2009;68(12):1370–82.
29. Dua S, Singh H, Thompson HW. Associative classification of mammograms using weighted rules. Expert Syst Appl. 2009;36(5):9250–59.
30. Pan H, Li J, Wei Z. Mining interesting association rules in medical images. In: Proc ADMA; 2005. p. 598–609.
31. Agrawal R, Srikant R. Fast algorithms for mining association rules. In: Proc VLDB; 1994. p. 487–499.

Part IV

Segmentation

9

Parametric and Non-Parametric Clustering for Segmentation

Hayit Greenspan and Tanveer Syeda-Mahmood

Summary. In this chapter, we contrast the medical image segmentation problem with general image segmentation and introduce several state-of-the-art segmentation techniques based on clustering. Specifically, we will consider two types of clustering, one parametric, and the other non-parametric, to group pixels into contiguous regions. In the first approach which is a statistical clustering scheme based on parametric Gaussian Mixture Models (GMMs), we develop the basic formalism and add variations and extensions to include a priori knowledge or context of the task at hand. In this formalism, each cluster is modeled as a Gaussian in the feature space. Each model component (Gaussian) can be assigned a semantic meaning; its automated extraction can be translated to the detection of an important image region, its segmentation as well as its tracking in time. We will demonstrate the GMM approach for segmentation of MR brain images. This will illustrate how the use of statistical modeling tools, in particular unsupervised clustering using Expectation-Maximization (EM) and modeling the image content via GMM, provides for robust tissue segmentation as well as brain lesion detection, segmentation and tracking in time. In the second approach, we take a non-parameterized graph-theoretic clustering approach to segmentation, and demonstrate how spatio-temporal features could be used to improve graphical clustering. In this approach, the image information is represented as a graph and the image segmentation task is positioned as a graph partitioning problem. A global criterion for graph partitioning based on normalized cuts is used. However, the weights of the edges now reflect spatio-temporal similarity between pixels. We derive a robust way of estimating temporal (motion) information in such imagery using a variant of Demon's algorithm. This approach will be illustrated in the domain of cardiac echo videos as an example of moving medical imagery.

9.1 Introduction

Image segmentation has been a long-standing problem in computer vision [1]. It is a very difficult problem for general images, which may contain effects such as highlights, shadows, transparency, and object occlusion. Segmentation in the domain of medical imaging has some characteristics that make the segmentation task easier and difficult at the same time. On the one hand, the imaging is narrowly focused on an anatomic region. The imaging context is also well-defined. While context may be present to some extent in segmenting general images (e.g., indoor vs. outdoor, city vs. nature, people vs. animals), it is much more precise in a medical imaging task, where the imaging modality, imaging conditions, and the organ identity is known. In addition, the pose variations are limited, and there is usually prior knowledge of the number of tissues and the Region of Interest (ROI). On the other hand, the images produced in this field are one of the most challenging due to the poor quality of imaging making the anatomical region segmentation from the background very difficult. Often the intensity variations alone are not sufficient to distinguish the foreground from the background, and additional cues are required to isolate ROIs. Finally, segmentation is often a means to an end in medical imaging. It could be part of a detection process such as tissue detection, or for the purpose of quantification of measures important for diagnosis, such as for example, lesion burden which is the number of pixels/voxels within the lesion regions in the brain.

To illustrate these subtleties, consider brain MRI imaging, as an example. We can select to use a single modality, such as T_1, or multi-modalities; we know that there are three main tissues of interest: white matter, gray matter, and cerebro-spinal fluid. We also know that in a pathological situation there may be one additional class such as lesions or tumors. Additional information may be on the relative sizes of the regions, their expected intensity distributions, and their expected geometric location (spatial layout). When treating a specific disease, e.g., Multiple Sclerosis (MS), we often also have context regarding the relevant regions in the brain. For example, we may be interested in specifically finding the relapsing-remitting lesions, which are actively changing in size, or alternatively, we may be interested in the static lesions.

The core operation in all these applications is the division of the image into a finite set of regions, which are smooth and homogeneous in their content and their representation. When posed in this way, segmentation can be regarded as a problem of finding clusters in a selected feature space. In this chapter, therefore, we will elaborate on approaches to segmentation based on clustering. Specifically, we will consider two types of clustering, one parametric, and the other non-parametric to group pixels into contiguous regions. In the first approach, which is a statistical clustering scheme based on parametric Gaussian Mixture Models (GMMs), we develop the basic formalism and add a variety of variations and extensions to include a priori knowledge or context of the task at hand. We will demonstrate the GMM approach for segmentation of brain MRI. In the second approach, we take a non-parameterized graph-theoretic clustering approach to segmentation, and demonstrate how spatio-temporal features could be used to improve graphical clustering. This approach will be illustrated in the domain of cardiac echo videos as an example of moving medical imagery. We begin by motivating the clustering approaches to segmentation.

9.2 Image Modeling and Segmentation

In general, the information contained in an image can be modeled in several ways. A simple approach is to record the intensity distribution within an image via a One-dimensional (1D) histogram and use simple thresholding to obtain the various segments. Based on the example images displayed in Fig. 9.1. Figure 9.2a shows such a histogram of a T_1-weighted brain MRI. As it can be seen, there are three intensity peaks each corresponding to the three main tissue types: white matter, gray matter, and cerebro-spinal fluid, which is indicated with high, mid-rande and low intensity values, respectively. By choosing appropriate thresholds, we can separate the desired classes. Two potential thresholds are shown in Fig. 9.2 as bars in the valleys of the histogram. The segmentation is then achieved by grouping all pixels of similar intensity ranges into a class.

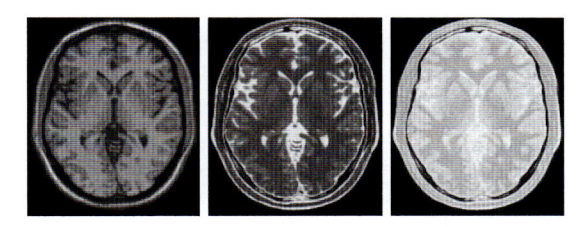

Fig. 9.1. *Brain MRI.* The brain slice from multiple acquisition sequences was taken from BrainWEB (http://www.bic.mni.mcgill. ca/brainweb). From left to right: T_1-, T_2-, and proton (P)-weighted image

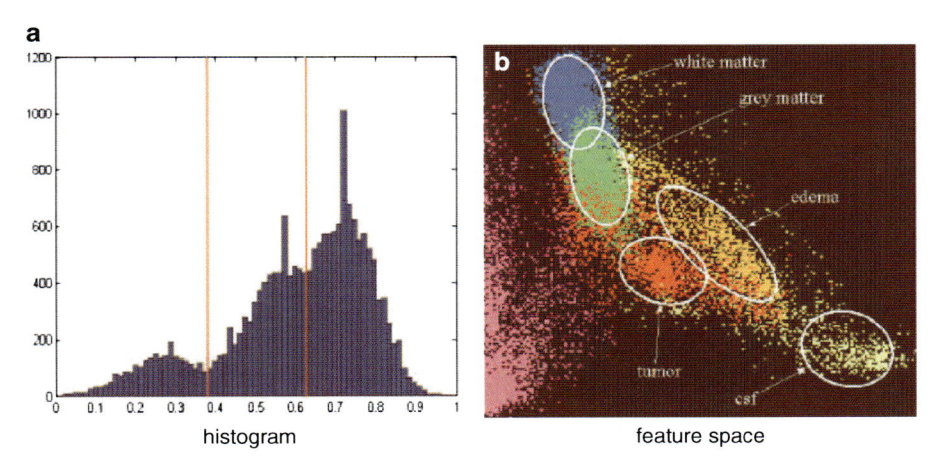

Fig. 9.2. *Brain MRI image characteristics.* The histogram of the T_1-weighted brain MRI shows three apparent classes (**a**). The 2D feature space shows five apparent classes (**b**)

Such a simple segmentation approach, however, is often insufficient for medical images where the imaging protocol can lead to variations in regional contrast making the task of segmentation difficult. This is illustrated in Fig. 9.1 which shows a sample MR brain slice, acquired with multiple acquisition protocols. Each protocol presents the image information in a unique range of contrasts. T_1-weighting, for example, causes fiber tracts (nerve connections) to appear white, congregations of neurons to appear gray, and cerebro-spinal fluid to appear dark. The contrast of white matter, gray matter, and cerebro-spinal fluid is reversed using T_2-weighted imaging, whereas proton-weighted imaging provides little contrast in normal subjects.

9.2.1 Image Modeling

Several variations on classical histogram thresholding have been proposed for medical image segmentation that incorporate extended image representation schemes as well as advanced information modeling. These include:

- *Multi-modal or multi-sequence data*: Multi-dimensional are histograms formed from the intensity values produced by each of the imaging protocols. It is often the case that several acquisitions are available for the same image. In MRI of the brain, for example, common imaging sequences include T_1, T_2, P and more. Each input provides different intensity valued information regarding the image to be analyzed. Thus, the intensity feature can be a single one $I(T_1)$ or a multi-dimensional one: $[I(T_1), I(T_2), I(P)]$. Figure 9.2b shows a scatter plot which is the intensity distribution of pixels in the image in two dimensions, with the x-axis and y-axis representing the intensity in the T_1 and T_2 image, respectively. With the multi-dimensional feature representation, other features besides intensity can be added, such as for example, texture around pixels in the image.
- *Spatial information*: Since intensity histograms do not preserve spatial contiguity of pixels, one variation is to add spatial position (x, y) or (x, y, z) to form a multi-dimensional feature vector incorporating spatial layout.
- *Temporal information*: If the medical images are in a time sequence (e.g., moving medical imagery), then time can be added as an additional feature in the representation space. If certain behaviors can be identified in time, and clustering in time is informative, the representation space can have time as one of its dimensions.

9.2.2 Segmentation

Thus, these approaches represent each image pixel as a feature vector in a defined multi-dimensional feature space. The segmentation task can be seen as a combination of two main processes:

1. *Modeling*: the generation of a representation over a selected feature space. This can be termed the modeling stage. The model components are often viewed as groups, or clusters in the high-dimensional space.

2. *Assignment*: the assignment of pixels to one of the model components or segments. In order to be directly relevant for a segmentation task, the clusters in the model should represent homogeneous regions of the image.

In general, the better the image modeling, the better the segmentation produced. Since the number of clusters in the feature space are often unknown, segmentation can be regarded as an unsupervised clustering task in the high-dimensional feature space.

9.2.3 State of the Art

There is a large body of work on clustering algorithms. For our purposes, we can categorize them into three broad classes: deterministic clustering, probabilistic clustering (model-based clustering), and graph-theoretic clustering. The simplest of these are the deterministic algorithms such as K-means [2], mean-shift [3], and agglomerative methods [4]. For certain data distributions, i.e. distributions of pixel feature vectors in a feature space, such algorithms perform well. For example, K-means provides good results when the data is convex or blob-like and the agglomerative approach succeeds when clusters are dense and there is no noise. These algorithms, however, have a difficult time handling more complex structures in the data. Further, they are sensitive to initialization (e.g. choice of initial cluster centroids).

The probabilistic algorithms, on the other hand, model the distribution in the data using parametric models, such as auto-regressive model, GMM, MRF, and conditional random field. Efficient ways of estimating these models are available using maximum likelihood algorithms such as the Expectation Maximization (EM) algorithm. While probabilistic models offer a principled way to explain the structures present in the data, they could be restrictive when more complex structures are present (e.g., data is a manifold). The last type of clustering algorithms we consider are non-parametric in that they impose no prior shape or structure on the data. Examples of these are graph-theoretic algorithms based on spectral factorization [5,6]. Here, the image data is modeled as a graph. The entire image data along with a global cost function is used to partition the graph, with each partition now becoming an image segment. In this approach, global overall considerations determine localized decisions. Moreover, such optimization procedures are often compute-intensive. Better data modeling by refining the cost of an edge using domain knowledge could improve the effectiveness of such clustering approaches as we will show in Sect. 9.5.

9.3 Probabilistic Modeling of Feature Space

We now consider an approach to segmentation that falls into the second class of clustering algorithms mentioned above, the probabilistic approach. Here, we will focus on the GMM-EM framework for parameterized Modeling of the feature space.

9.3.1 Gaussian Mixture Models

The feature space is generated from image pixels by a mixture of Gaussians. Each Gaussian can be assigned a semantic meaning, such as a tissue region. If such Gaussians could be automatically extracted, we can segment and track important image regions.

A single mode histogram can be modeled by a Gaussian. A mixture of Gaussians can be used to represent any general, multi-modal histogram. Using a Maximum Likelihood (ML) formalism, we assume that the pixel intensities (or more generally, the corresponding feature vectors) are independent samples from a mixture of probability distributions, usually Gaussian. This mixture, called a finite mixture model, is given by the probability density function

$$f(v_t|\Theta, \alpha) = \sum_{i=1}^{n} \alpha_i f_i(v_t|\theta_i) \tag{9.1}$$

where v_t is the intensity of pixel t; f_i is a component probability density function parameterized by θ_i, where $\Theta = [\theta_1 \ldots \theta_n]$ and the variables α_i are mixing coefficients that weigh the contribution of each density function, where $\alpha = [\alpha_1 \ldots \alpha_n]$.

9.3.2 Expectation Maximization

The EM algorithm [7] is often used to learn the model parameters, using iterative optimization. It obtains the maximum likelihood estimation by iterating two steps: the Expectation Step (E-step) and the Maximization Step (M-step). Based on the current estimation of the parameter set, the E-step produces the probabilistic classification values, $w_{it}, t = 1, \ldots, T$, which indicate the probabilistic affiliation of pixel v_t to Gaussian $i, i = 1, \ldots, n$

$$w_{it} = p(i|v_t) = \frac{\alpha_i f_i(v_t|\mu_i, \Sigma_i)}{\sum_{l=1}^{n} \alpha_l f_l(v_t|\mu_l, \Sigma_l)} \tag{9.2}$$

In the M-step, the model parameters are re-estimated using the classification values of the E-step

$$\alpha_i = \frac{n_i}{n}, \quad n_i = \sum_{t=1}^{T} w_{it} \tag{9.3}$$

$$\mu_i = \frac{1}{n_i} \sum_{t=1}^{T} w_{it} v_t, \quad \Sigma_i = \frac{1}{n_i} \sum_{t=1}^{T} w_{it} (v_t - \mu_i)^2$$

In each iteration of the EM algorithm, the log-likelihood of the data is guaranteed to be non-decreasing. In convergence, the parameter values found are therefore at a local maximum of the likelihood function. The iterative updating process above is repeated until a predefined condition is fulfilled. Possible conditions include a threshold applied to the increase of the log-likelihood and a limit on the number of iterations. In order to ensure a meaningful maximization point at the outcome, data-driven initialization schemes, such as the frequently used K-means algorithm [2], are appropriate.

9.3.3 Visualization

An immediate transition is possible between the extracted image representation and probabilistic segmentation. In the representation phase, a transition is made from voxels to clusters (Gaussians) in feature space. The segmentation process can be thought of as forming a linkage back from the feature space to the raw input domain. A segmentation map can be generated by assigning each pixel to the most probable Gaussian cluster, i.e., to the component i of the model that maximizes the a posteriori probability

$$\text{Label}\{v_t\} = \arg\max_i \{\alpha_i f(v_t|\mu_i, \Sigma_i)\} \tag{9.4}$$

We illustrate the GMM modeling of the feature space using an example x-ray image (Fig. 9.3). To produce this visualization, we projected the GMM models formed from a Three-dimensional (3D) feature space of intensity and pixel coordinates (I, x, y) onto the image plane (x, y). The resulting shapes of such regions are therefore ellipsoidal as shown in this figure. Figure 9.3 also shows the visual effect of varying the number n of Gaussians in the GMM. A small n provides a very crude description. As we increase the number of Gaussians, finer detail can be seen in the blob representation. Larger n provides a more localized description, including finer detail such as the fingers. This seems more representative to the human eye and definitely closer to the original image. Thus the GMM models can be an effective way to partition an input image into a collection of regions. This approach can also be used to group regions into semantically meaningful entities such as tissue regions and lesions.

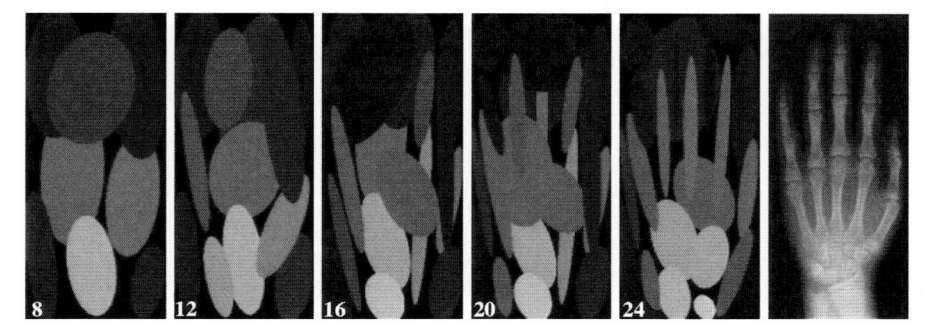

Fig. 9.3. *Level of granularity in the representation.* Different number n of Gaussians per image model

9.4 Using GMMs for Brain Tissue and Lesion Segmentation

The initial segmentation of pixels directly from the Gaussian clusters is sufficient for the search and retrieval function. In fact, GMMs were shown to be effective in image matching for general (e.g., web-based) as well as medical image search and retrieval [8,9]. However, if the goal of segmentation is also to identify tissue regions and lesions or tumors, then such fragmented regions in the image plane must be further grouped to form these structures. This can be guided by a priori knowledge, which is formulated algorithmically by constraints.

9.4.1 Application Domain

The tissue and lesion segmentation problems has been attempted earlier by different approaches for brain MRI images [10]. In such images, there is interest in mainly three tissue types: white matter, gray matter and cerebro-spinal fluid. The volumetric analysis of such tissue types in various part of the brain is useful in assessing the progress or remission of various diseases, such as Alzheimer's desease, epilepsy, sclerosis and schizophrenia.

The predominant approach to tissue segmentation in clinical settings is still manual. Manual partitioning of large amounts of low contrast and low Signal to Noise Ratio (SNR) brain data is strenuous work and is prone to large intra- and inter-observer variability. Automatic approaches to tissue segmentation have also become recently available, using intensity information and/or spatial layout.

Algorithms for tissue segmentation using intensity alone often exhibit high sensitivity to various noise artifacts, such as intra-tissue noise, inter-tissue contrast reduction, partial volume effects and others [11]. Reviews on methods for brain image segmentation (e.g., [10]) present the degradation in the quality of segmentation algorithms due to such noise, and recent publications can be found addressing various aspects of these concerns (e.g. partial volume effect quantification [12]). Due to the artifacts present, classical voxel-wise intensity-based classification methods may give unrealistic results, with tissue class regions appearing granular, fragmented, or violating anatomical constraints [13].

9.4.2 Spatial Constraints

One way to address the smoothness issue is to add spatial constraints. This is often done during a pre-processing phase by using a statistical atlas, or as a post-processing step via MRF models. A statistical atlas provides the prior probability for each pixel to originate from a particular tissue class [14–16].

Algorithms exist that use the Maximum A-Posteriori (MAP) criterion to augment intensity information with the atlas. However, registration between a given image and the atlas is required, which can be computationally prohibitive [17]. Further, the quality of the registration result is strongly dependent on the physiological variability of the subject and may converge to an erroneous result in the case of a diseased or severely damaged brain. Finally, the registration process is applicable only to complete volumes. A single slice cannot be registered to the atlas, thus, cannot be segmented using these state-of-the-art algorithms.

Segmentation can also be improved using a post-processing phase in which smoothness and immunity to noise can be achieved by modeling the neighboring voxel interactions using a MRF [15, 18, 19]. Smoother structures are obtained in the presence of moderate noise as long as the MRF parameters controlling the strength of the spatial interactions are properly selected. Too high a setting can result in an excessively smooth segmentation and a loss of important structural details [20]. In addition, MRF-based algorithms are computationally intractable unless some approximation is used. Finally, there are algorithms that use deformable models to incorporate tissue boundary information [21]. They often imply inherent smoothness but require careful initialization and precisely calibrated model parameters in order to provide consistent results in the presence of a noisy environment.

9.4.3 Modeling Spatial Constraints Through GMM

For complex tissue patterns or small lesions, a very refined segmentation is needed to achieve robust segmentation in the presence of strong thermal noise, without using an atlas as prior information. Ideally, since spatial information can be included in the feature space, coherent clusters in feature space can lead to coherent spatial localized regions in the image space. For regions of complex shapes in the image plane, for which a single convex hull is not sufficient (will cover two or more different segments of the image), a plausible approach is to utilize very small spatial supports per Gaussian. This in turn implies the use of a large number of Gaussians.

Using Gaussians that are very localized in space, and have a small spatial support, can result in a relatively small number of (affiliated) pixels affecting the Gaussian characteristics. This may be problematic in regions that have a substantial noise component, as well as in regions located close to tissue-borders. In order to provide a more global perspective, the over-segmented (localized) regions need to be grouped per tissue.

Figure 9.4 illustrates how a large set of Gaussians can be used to represent an image. Here again we show projection of the 3D feature space into the image plane. Different shades of gray represent the three distinct tissues present. All Gaussians of a particular tissue influence the global intensity modeling of that tissue (and are pseudo-colored in a constant shade in the Figure).

Our approach to tissue modeling using GMMs is called Constrained GMM (CGMM) [22]. CGMM model each tissue region by a separate collection of a large number of Gaussians. The overall approach is as follows. We first use

Fig. 9.4. *Image modeling via GMMs* [22]. Each Gaussian is colored with the mean gray level of its pixels and projected onto the (x,y)-plane

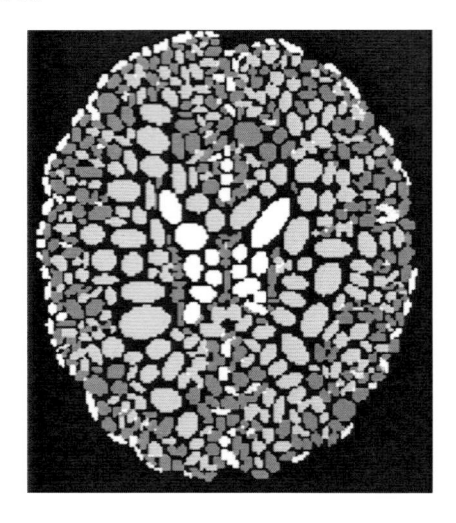

K-means algorithm based on intensity alone to result in an initial segmentation of the image into tissue groups (i.e., white matter, gray matter, cerebro-spinal fluid). Next, each tissue is modeled with many small locally convex Gaussians based on intensity and spatial features (I, x, y, z). Since each Gaussan supports a small location region, it can be assumed to be convex. Further, the intensity is roughly uniform within the Gaussian support region. This means, the spatial and intensity features are separable and can be independently estimated using the EM algorithm. Thus, a modified M-step is used in which the Gaussians for intensity and pixel locations are estimated separately. This modification ensures that smooth boundary segmentations are possible.

Constrained GMM Formulation

We now describe the CGMM approach in detail. Let the intensity component of the feature vector representing the voxel be denoted by v^I. In order to include spatial information, the (X, Y, Z) position is appended to the feature vector. The notation $v^{XYZ} = (v^X, v^Y, v^Z)$ is used for the three spatial features. The set of feature vectors extracted per volume is denoted by $\{v_t \,|\, t = 1, \ldots, T\}$ where T is the number of voxels. The image is modeled as a mixture of *many* Gaussians (9.1).

The spatial shape of the tissues is highly non-convex. However, since we use a mixture of many components, each Gaussian component models a small local region. Hence, the implicit convexity assumption induced by the Gaussian distribution is reasonable. The high complexity of the spatial structure is an inherent part of the brain image. The intra variability of the intensity feature within a tissue (bias) is mainly due to artifacts of the MRI imaging process and once eliminated (via bias-correction schemes) is significantly less than the inter-variability among different tissues. It is therefore sufficient to model the intensity variability within a tissue by a small number, or a single Gaussian (in the intensity feature). To incorporate this insight into the model, we further assume that each Gaussian is linked to a single tissue and all the Gaussians related to the same tissue share the same intensity parameters.

Technically, this linkage is defined via a grouping function. In addition to the GMM parameter set Θ, we define a grouping function $\pi : \{1,\ldots,n\} \to \{1,\ldots,k\}$ from the set of Gaussians to the set of tissues. We assume that the number of tissues is known and the grouping function is learned in the initialization step. The intensity feature should be roughly uniform in the support region of each Gaussian component, thus, each Gaussian spatial and intensity features are assumed uncorrelated. The above assumptions impose the following structure on the mean and variance of the Gaussian components

$$\mu_i = \begin{pmatrix} \mu_i^{\text{XYZ}} \\ \mu_{\pi(i)}^{\text{I}} \end{pmatrix}, \quad \Sigma_i = \begin{pmatrix} \Sigma_i^{\text{XYZ}} & 0 \\ 0 & \Sigma_{\pi(i)}^{\text{I}} \end{pmatrix} \quad (9.5)$$

where $\pi(i)$ is the tissue linked to the i-th Gaussian component and μ_j^I and Σ_j^I are the mean and variance parameters of all the Gaussian components that are linked to the j-th tissue.

The main advantage of the CGMM framework is the ability to combine, in a tractable way, a local description of the spatial layout of a tissue with a global description of the tissue's intensity.

Learning CGMM Parameters

The EM algorithm is utilized to learn the model parameters. In the proposed framework, Gaussians with the same tissue-label are constrained to have the same intensity parameters throughout. A modification of the standard EM algorithm for learning GMM is required, as shown in the following equations. The E-step of the EM algorithm for the CGMM model is the same as of the unconstrained version; denoting $\frac{n_i}{n} = \alpha_i$ yields

$$w_{it} = p(i|v_t) = \frac{\alpha_i f_i(v_t|\mu_i, \Sigma_i)}{\sum_{l=1}^{n} \alpha_l f_l(v_t|\mu_l, \Sigma_l)}, \quad n_i = \sum_{t=1}^{T} w_{it}, \quad k_j = \sum_{i \in \pi^{-1}(j)} n_i \quad (9.6)$$

such that n_i is the expected number of voxels that are related to the i-th Gaussian component and k_j is the expected number of voxels that are related to the j-th tissue. The maximization in the M-step is done given the constraint on the intensity parameters

$$\mu_i^{\text{XYZ}} = \frac{1}{n_i} \sum_{t=1}^{T} w_{it} v_t^{\text{XYZ}}, \quad \Sigma_i^{\text{XYZ}} = \frac{1}{n_i} \sum_{t=1}^{T} w_{it} \left(v_t^{\text{XYZ}} - \mu_i^{\text{XYZ}}\right)\left(v_t^{\text{XYZ}} - \mu_i^{\text{XYZ}}\right)^T$$

$$\mu_j^{\text{I}} = \frac{1}{k_j} \sum_{i \in \pi^{-1}(j)} \sum_{t=1}^{T} w_{it} v_t^{\text{I}}, \quad \Sigma_j^{\text{I}} = \frac{1}{k_j} \sum_{i \in \pi^{-1}(j)} \sum_{t=1}^{T} w_{it} \left(v_t^{\text{I}} - \mu_j^{\text{I}}\right)^2 \quad (9.7)$$

The grouping function π that links between the Gaussian components and the tissues is not altered by the EM iterations. Therefore, the affiliation of a Gaussian component to a tissue remains unchanged. However, since the learning is performed simultaneously on all the tissues, voxels can move between tissues during the iterations.

In the first step, the K-means clustering is done based only on the intensity features (T_1, T_2 and P), which gives a good initial segmentation into the three major tissue classes. We utilize T_1 to give tissue labels to the three groups. Then, each tissue is modeled with many small locally convex Gaussians. To ensure that small isolated areas are explicitly represented by local Gaussians, we first apply a 3D connected component process to each tissue. If a connected component contained less than three voxels, it was ignored and disregarded as noise. For each connected component (of voxels all from the same tissue) a subset of voxels is randomly chosen and a new Gaussian is formed at each of the chosen voxels. To compute initial covariance values for each Gaussian, we assign each voxel to the nearest center with the same tissue label. The Gaussian covariance (both intensity and spatial) is computed based on all the voxels that are assigned to the Gaussian center. As a result of the initialization process each Gaussian is linked to one of the tissues and each voxel is affiliated with a selected Gaussian.

9.4.4 Tissue Segmentation

The CGMM model uses multiple Gaussians for each tissue. Thus, we need to sum over the posterior probabilities of all the identical tissue Gaussians to get the posterior probability of each voxel to originate from a specific tissue

$$\text{Label}\{v_t\} = \arg\max_{j \in \{1,\ldots,k\}} \left\{ \sum_{i \in \pi^{-1}(j)} \alpha_i f_i(v_t | \mu_i, \Sigma_i) \right\} \quad (9.8)$$

such that $\text{Label}\{v_t\} \in \{1, \ldots, k\}$ is one of the tissues. The linkage of each voxel to a tissue label provides the final segmentation map.

Figure 9.5 illustrates the segmentation induced from the CGMM model and shows how the EM iterations improve the segmentation quality. Since much of the algorithmic effort is spent on finding a good initialization, the EM needs only few iterations to converge. Thus, the CGMM is an effective approach to segment tissues in images.

9.4.5 Lesion Segmentation

The CGMM approach can also be used effectively to segment lesions. Lesion segmentation is important for diagnosis of many brain disorders including MS. The most common quantitative parameter of interest in such cases is the lesion burden (*load*) of the disease expressed in terms of the number and volume of the brain lesions. The MRI-measured lesion burden is highly correlated with clinical finding [23–27].

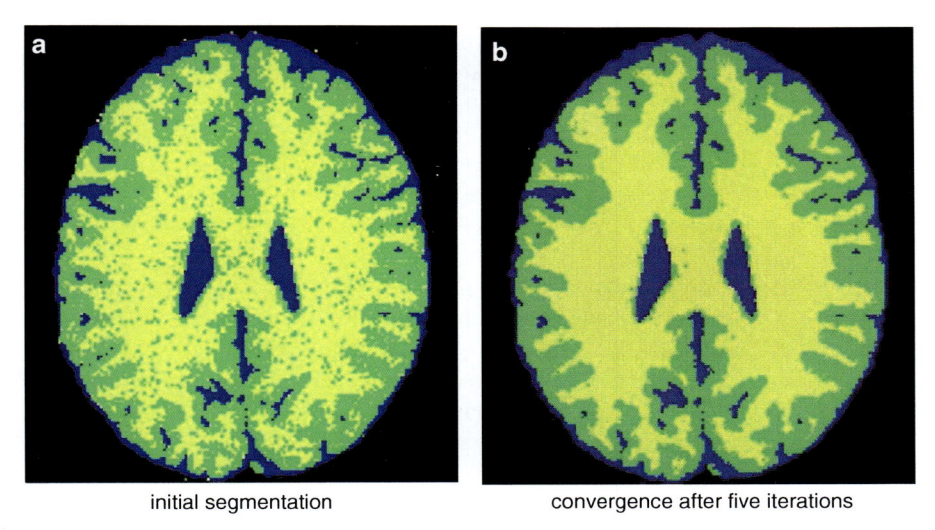

Fig. 9.5. *Tissue segmentation in brain MRI.* As the parameter estimation improves with the iterations, the segmentation induced from the model becomes more and more robust to noise. The colors *green, yellow* and *blue* indicate *gray matter, white matter*, and cerebro-spinal fluid, respectively

Segmentation of abnormal structures like MS lesions, is a difficult a task. Lesions differ from each other in size, location and shape. Also, as each MR modality reflects different physical properties, the exact lesion boundaries may vary between different MR channels. To complicate matters, the number of voxels that are associated with MS lesions is significantly smaller than the number of voxels that are associated with healthy tissues. Thus, simple clustering algorithms (e.g. K-means), fail to discern MS-lesion voxels from the rest of the image.

In a model-based approach, the lesion category can be modeled explicitly, as an additional category, or alternatively, the lesions may be defined as outliers to the tissue-extracted models. While several authors (e.g. [28, 29]) model the MS lesions as a distinguished class in addition to the healthy tissue classes (CSF, GM, WM), another approach is to model the lesions as outliers. For example, Van Leemput et al. treat voxels that are not well explained by either of the healthy tissue models) as candidates to be classified as lesions [30]. Among these candidates, the separation into lesion and non-lesion voxels is made according to contextual information and a set of rules. It is also possible to treat lesions as outliers only in an intermediate stage, and then to build an explicit model for them [31].

Using the CGMM framework, we exemplarily address MS-lesion segmentation. Instead of using a voxel-based lesion characterization, we use a Gaussian-based (or "blob"-based) approach, whereby lesions are detected as Gaussian outliers [32]. For this, the model is initialized with three tissue classes, and its parameters are learned as described above. Due to the fact that lesion voxels are significantly outnumbered by the healthy tissue voxels, the clusters still succeed in representing the healthy tissues. The lesion voxels are of course misclassified at this point as one or more of the healthy tissues. Moreover, at this stage there are misclassified Gaussians, i.e., Gaussians labeled as healthy tissues that are supposed to be labeled as lesions. The purpose of the current stage is to identify these Gaussians, and change their labels accordingly. In other words, lesion detection is performed on the Gaussian rather than the voxel level.

For each Gaussian a decision is made, based on its features, whether it should in fact be labeled as MS, or remain labeled as one of the healthy tissues. Both supervised and unsupervised approaches can be used to deal with this task. For example, a rule-based system can be used to distinguish the lesion Gaussians from the normal tissue Gaussians. A Gaussian for which all these conditions hold, is then labeled as a lesion. An example of such a rule set for Gaussians labeled as GM is the following:

1. T_2 mean-intensity of the Gaussian $> T_2$ mean-intensity of the GM tissue $+ \epsilon_{GM,T_2}$.
2. T_2 mean-intensity of the Gaussian $> P$ mean-intensity of the GM tissue $+ \epsilon_{GM,P}$.
3. A large Mahalanobis distance between the mean-intensity of the Gaussian and the mean-intensity of the three healthy tissue classes.
4. The majority of the Gaussian's k-nearest Gaussians are labeled as WM.

where ϵ_{GM,T_2} and $\epsilon_{GM,P}$ are thresholds that can be tuned and optimized. The first two rules reflect the general appearance of the MS lesions. The rules that rely on Mahalanobis distance imply that only Gaussians that are not well explained by the healthy tissue models are suspected as lesion Gaussians. The last rule incorporates contextual information by reflecting our expectation to find lesions within the WM tissue. These rules are similar to rules used by Van Leemput et al. [30]. However, note that at this point decisions are made at the Gaussian level rather than the voxel level.

Following the Gaussian-based MS lesion detection stage, all the re-labeled Gaussians now form a fourth class (MS lesion), with its own global intensity parameters. The EM is now applied to CGMM with four tissue types and segmentation of the tissues and the lesions is obtained (Fig. 9.6). Thus the CGMM can be seen as a versatile framework for image segmentation, tissue detection as well as lesion characterization.

9.5 Non-Parametric Clustering Approaches to Segmentation

So far, we have discussed model-based clustering approaches to segmentation. We now turn to a second segmentation algorithm that is a non-parametric approach to clustering regions. As mentioned in Sect. 9.1, intensity information is often insufficient in segmenting medical images. This is

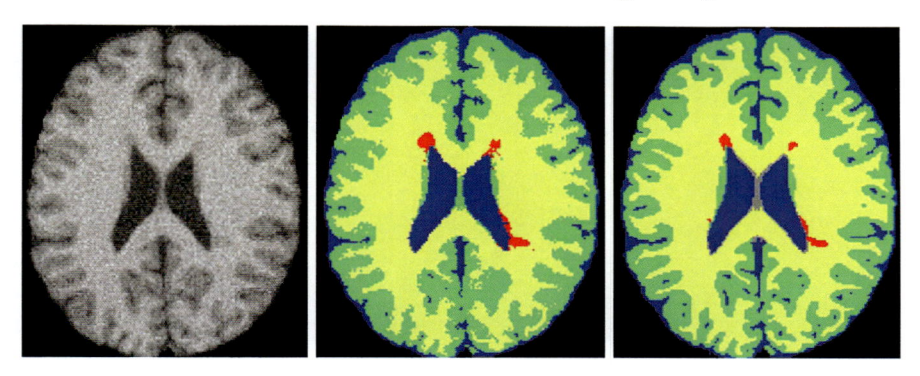

Fig. 9.6. *MS lesion segmentation.* The MS lesion is color-coded red. *Left*: Input T_2 images with 9% noise; *Middle*: CGMM segmentation; *Right*: ground truth segmentation

particularly true in challenging medical videos generated during an echocardiographic or a cardiac MRI study, where spatio-temporal information is captured by moving medical imagery.

We now describe an approach to segmentation that caters to these types of data. Due to the temporal nature of these videos, we exploit motion in addition to intensity and location information. Specifically, each pixel in the moving medical image is represented by its location, intensity and motion vectors. The relation between adjacent pixels is captured by the difference in their location, intensity and motion. Using the graph-theoretic formulation for clustering, each pixel represents a node of the graph and the distance between pixels is captured through the cost of the edge. The normalized cut algorithm is then used to partition the graph into segments.

9.5.1 Description of the Feature Space

Since motion is critical to segmenting objects in these videos, we first develop a method of reliably estimating inter-frame motion. Ideally, a simple way of estimating spatio-temporal motion is to use optical flow-based methods. These are suitable for small deformations in temporal sequences of images. In fact, Electrocardiography (ECG)-gated cardiac MRI was segmented using motion information derived from optical flow [33]. The basic idea of optical flow is as follows.

Optical Flow

The intensity between corresponding pixels is captured via the brightness change constraint equation which states that the corresponding points at successive time instants have equal intensity. Thus, if $I(x, y, t)$ is the pixel intensity at a given pixel location (x, y) at time t, then its corresponding pixel at time $t + \delta t$ has the same intensity as given by

$$I(x, y, t) = I(x + \delta x, y + \delta y, t + \delta t) \tag{9.9}$$

Assuming the relative change between successive images is small, the intensity can be approximated by the first few terms in a Taylor series expansion as

$$I(x+\delta x, y+\delta y, t+\delta t) = I(x, y, t) + \frac{\partial I}{\partial x}\partial x + \frac{\partial I}{\partial y}\partial y + \frac{\partial I}{\partial t}\partial t + \text{higher order terms} \tag{9.10}$$

Ignoring the higher order terms, this implies

$$\frac{\partial I}{\partial x}\partial x + \frac{\partial I}{\partial y}\partial y + \frac{\partial I}{\partial t}\partial t = 0 \tag{9.11}$$

or in matrix notation

$$\nabla I^T \cdot \boldsymbol{V} = -I_t \tag{9.12}$$

Thus the spatio-temporal motion estimation problem is to recover the velocity vectors from the above equation. Clearly, one equation alone is not sufficient to solve for the unknown velocity vectors $\boldsymbol{V}$ [34]. All optical flow methods, therefore, introduce additional constraints to estimate the velocity including a regularization term for smoothing[35]. One solution is to consider that the end point of velocity vectors are the closest point of a high-dimensional surface m with respect to spatial (x,y,z) translations which leads to a regularization term as

$$\boldsymbol{V} = \frac{-I_t \nabla I}{\nabla I^2} \tag{9.13}$$

This equation is unstable for small values of gradient, leading to infinite velocity values. In fact, optical flow estimates of motion in moving medical imagery are often noisy and incoherent. As such they have not been able to adequately model the complex motion of the heart that manifests as different directional motion at different locations.

Demon's Algorithm

To make (9.13) stable for small values of gradient, Thirion introduced the notion of regularization (or stabilization) by applying a smooth deformation field [36]. That is, we can estimate motion by treating each successive pairs of intensity image frames as surfaces $(x, y, I(x, y))$ and finding a deformable (changeable) surface model that warps one frame into the next [36]. It produces a displacement field between two successive frames which indicates the transform that should be applied to pixels of one of the images so that it can be aligned with the other. The resulting deformation field gives a consistent set of directional velocity vectors, sampling motion densely in both space and

9 Parametric and Non-Parametric Clustering for Segmentation

Table 9.1. *Final algorithm*

Step	Action
1	Compute initial gradient estimate for each pixel ∇I
2	Set initial velocity estimates $V = 0$
3	For each pixel i, repeat steps 4–6
4	Compute intensity in deformed image using (9.11)
5	Compute the additional displacement vector using (9.14)
6	Update the displacement vector $V = V_{\text{prev}} + V$ from above
7	Regularize the velocity field by applying Gaussian smoothing
8	Repeat steps 3–7 until convergence is reached

time. Thus this algorithm called the Demon's algorithm, treats the motion estimation as a surface registration problem. More specifically, to make (9.13) stable for small values of gradient, the Demon's method multiplies (9.13) by a term $\frac{\nabla I^2}{\nabla I^2 + (I_t)^2}$ to give the flow estimate as

$$V = \frac{-I_t \nabla I}{\nabla I^2 + (I_t)^2} \quad (9.14)$$

With this expression, the motion estimates can be computed in two steps. We first compute the instantaneous optical flow from every point in the target images using (9.12), and then regularize the deformation field as in (9.14). Thus the Demon's algorithm is inspired from the optical flow equations but is renormalized to prevent instabilities for small image gradient values. To further ensure unique and consistent motion in a local region, an elastic-like behavior can be ensured by smoothing with a Gaussian. Table 9.1 summarizes the algorithm.

While there is a large body of literature on surface registration algorithms [37], the Demon's algorithm has been particularly popular in medical imaging due to its proved effectiveness, simplicity, and computational efficiency. Hence we adopt this approach to estimate the spatio-temporal motion in moving medical imagery.

9.5.2 Clustering Intensity, Geometry, and Motion

Once a description of the spatio-temporal content of moving medical imagery is obtained, segmentation can proceed using one of the many available clustering approaches. In [33] in fact, spatio-temporal segmentation was based on using K-means clustering of intensity and motion features derived from optical flow. However, K-means clustering is sensitive to the selection of initial centroids for region growing, which together with noisy motion estimates from optical flow methods will give rise to inaccurate segmentation of such imagery.

Graph-Based Clustering

In our method, we take a non-parametric approach to clustering based on graph theory. Using the graph formalism, the segmentation problem for a set of moving medical images can be modeled by a graph $G = (V, E, W)$ where V are the pixels, and edges E connect pairs of neighboring pixels, and W denotes the strength of the connectivity between pixels. Typically, W captures the similarity between pixels using an exponential weighting function. The overall quality of segmentation depends on the pairwise pixel affinity graph. Two simple but effective local grouping cues are: intensity and proximity. Close-by pixels with similar intensity value are likely to belong to one object. But because of texture clutter, the intensity cue alone often gives poor segmentations, hence we add local neighborhood information. Using intensity and proximity, the affinity matrix can be given by

$$W_I(i,j) = e^{-||X_i - X_j||/\delta_x - ||I_i - I_j||/\delta_I} \tag{9.15}$$

where X_i denotes pixel location, and I_i is the image intensity at pixel location i.

Once the affinity between pairs of pixels is defined, there are several segmentation algorithms based on graph theory [5, 38]. Here, we used a graph clustering algorithm called the normalized cut algorithm [5] that was reported earlier to partition a single image pixels into self-similar regions, and adapt it for use on moving medical imagery.

Normalized Cut Algorithm

To understand this algorithm, consider the goal of segmentation which is to partition the image into regions. In terms of the graph formulation, this means dividing the graph by removing key edges such that the individual connected regions becomes disconnected. To ensure that such a removal does not remove close nodes, the set of edges (called a cut) to be removed should have the minimum weight. In order to ensure that the graph does not split into small isolated components, this weight must be normalized by the weight of the edges incident on a node. Thus, normalized cut edge cost between a pair of nodes is defined as

$$W(A, B) = \frac{w(A, B)}{w(A, V)} + \frac{w(B, A)}{w(B, V)} \tag{9.16}$$

where $w(A, B)$ is the cost of the affinity edge between nodes A and B.

If we assemble all the edge weights into a weight matrix W, let D be the diagonal matrix with entries $D(i, i) = W(i, j)$, then the normalized cut cost can be written as

$$\frac{y^T(D - W)y}{y^T D y} \tag{9.17}$$

where y is an indicator vector, $y = 1$ if the i-th feature belongs to A and -1 otherwise. While finding the exact minimum of the normalized cut is NP-complete, if we relax y to take on continuous values instead of binary values, we can minimize the cost by solving the generalized eigenvalue problem $(D - W)y = \lambda D y$. The solution y is given by the eigenvector corresponding to the second smallest eigenvalue from the Raleigh quotient. Using the edges indicated by the eigenvector, we can now split the graph into connected regions at the next level. This segmentation step recursively proceeds until convergence is reached.

Recently, Cour et al. improved the efficiency of the normalized cuts algorithm using a scale-space implementation [39]. In the multi-scale adaptation of normalized cuts, the graph links in W are separated into different scales according to their underlying spatial separation. That is

$$W = W_1 + W_2 + ...W_s \qquad (9.18)$$

where W_s contains affinity between pixels with certain spatial separation range: $W_s(i,j) \neq 0$ only if $G_{r,s-1} \leq r_{ij} \leq G_{r,s}$, where $G_{r,s}$ is the distance between pixels. This decomposition allows different graph affinities at different spatial separations to be modeled.

We adapt the above multi-scale version of normalized cuts to our segmentation of moving medical imagery problem as follows. First, we augment the affinity matrix with a motion term, since motion boundaries provide a natural way to group objects as well. Specifically, we define a new affinity matrix as

$$W_I(i,j) = e^{-||X_i-X_j||/\delta_x - ||I_i-I_j||/\delta_I - ||V_i-V_j||/\delta_V} \qquad (9.19)$$

Thus, the weighting function combines image intensity, proximity and motion information so that nearby pixels with similar intensity values and motion are likely to belong to one object. Using the revised affinity matrix, we use the multi-scale adaptation of normalized cuts to find the various clusters.

9.6 Using Non-Parametric Clustering for Cardiac Ultrasound

9.6.1 Application Domain

In an echocardiography study, an ultrasound-based diagnostic procedure is used for morphological and functional assessment of the heart. In particular, it is used to study the heart chambers called the atria and ventricles (Fig. 9.7). Bad blood enters the right atrium, and through the right ventricle is sent to the lungs for cleaning. The cleaned blood enters the left atrium and is supplied to the rest of the body through the aorta originating from the left ventricle. Of these chambers, the performance of the left ventricle is crucial to assessing the heart performance. The flow of blood is gated by the valves between the atria

Fig. 9.7. *Illustration of an echocardiogram.* A case of left ventricular hypertrophy is shown where the left ventricle has significantly altered shape due to muscle growth

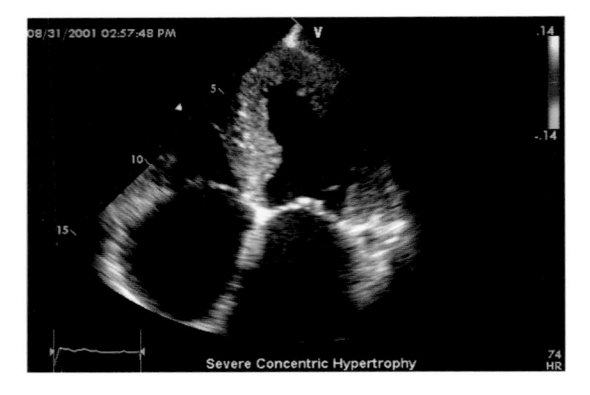

and ventricles, such as the mitral valve between left atrium and left ventricle, and the aortic valve in the entrance to the aorta. The opening and closing motion of these valves needs to be properly timed within the heart cycle. The echocardiogram is therefore used to study valvular motion abnormalities as well as septal and regional wall motion abnormalities in addition to estimating the performance of the left ventricular contraction.

The end result of this procedure is data in the form of echocardiogram videos. The objects of interest in echocardiogram videos are, therefore, the cardiac chambers (atria and ventricles), valves and septal walls as shown in Fig. 9.7. As can be seen, the various cardiac regions are difficult to segment based on image information alone. The interesting chambers are essentially dark feature-less regions. Further, the complex non-rigid motion alters their shape through the heart cycle, making it difficult to model them using parametric models such as Gaussian mixtures we described earlier. In such cases, non-parametric approaches such as graph-theoretic clustering are better suited as they keep alive multiple interpretation possibilities allowing both local and global constraints to be incorporated.

9.6.2 Cardiac Motion Estimation

We now illustrate the type of motion estimates produced by using the Demon's algorithm for echocardiographic sequences. Figure 9.8a,d show successive frames of an echocardiographic sequence. Using the Demon's algorithm, the displacement fields produced are shown in Fig. 9.8e. In comparison, the result of using straightforward optical flow is shown in Fig. 9.8b. The smooth and coherent estimation of motion in comparison to optical flow can be clearly seen in the enlarged portions of cardiac regions (Fig. 9.8c,f).

9.6.3 Segmentation of Meaningful Regions

We now describe the overall segmentation algorithm for moving medical imagery. The algorithm consists of two major steps: (a) pre-processing to

9 Parametric and Non-Parametric Clustering for Segmentation 247

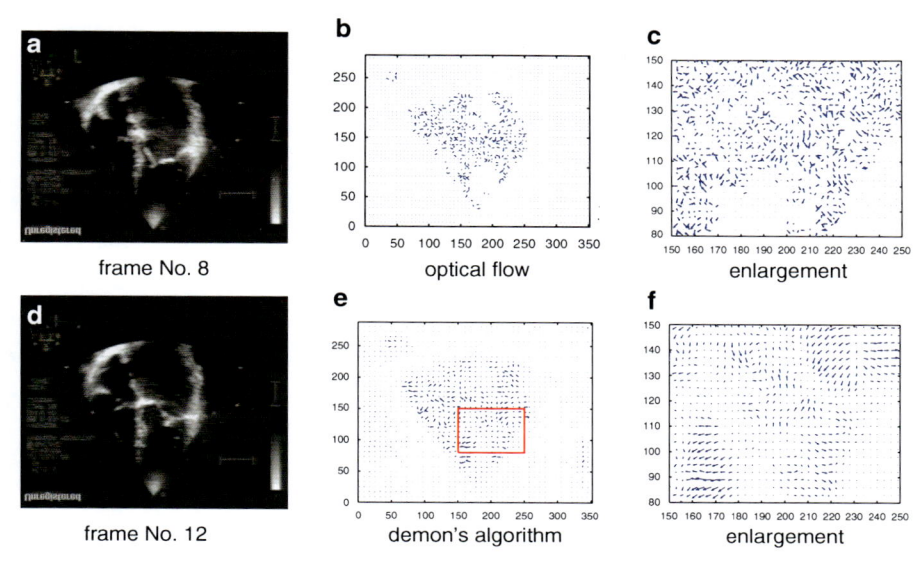

Fig. 9.8. *Motion estimation in echocardiograms.* The ultrasound images in (**a**) and (**d**) are three frames apart. Motion estimation using plain optical flow (**b, c**) is improved applying demon's algorithm (**e, f**)

extract affinity information between pixels and (b) multi-scale graph-theoretic clustering. To extract the motion field, we pre-process the videos to remove background text (recording information from instrument) through simple frame subtraction. An edge-enhancing diffusive smoothing algorithm (as described in Weickert [40]) is then applied to each frame to remove high-frequency noise from the tissue while retaining both tissue edges and intra-tissue intensity band structure. Demon's algorithm is run between frames that are less than three frames apart as described in the earlier section. Finally, the affinity values using (9.19) is used for multi-scale normalized cut-based segmentation.

Figure 9.9 illustrates cardiac region segmentation using the modified multi-level normalized cut combining intensity and motion information. Figure 9.9a,b show adjacent frames of an echo video. Figure 9.9c shows the results of using normalized cut-based segmentation based on intensity alone. Finally, Fig. 9.9d shows the result using both intensity and motion recovered from deformable registration. As can be seen, the combined use of intensity and motion information has resulted in improved delineation of relevant medical regions.

ultrasound frame No. 0   ultrasound frame No. 4

intensity-based graph cut result   intensity- and motion-based result

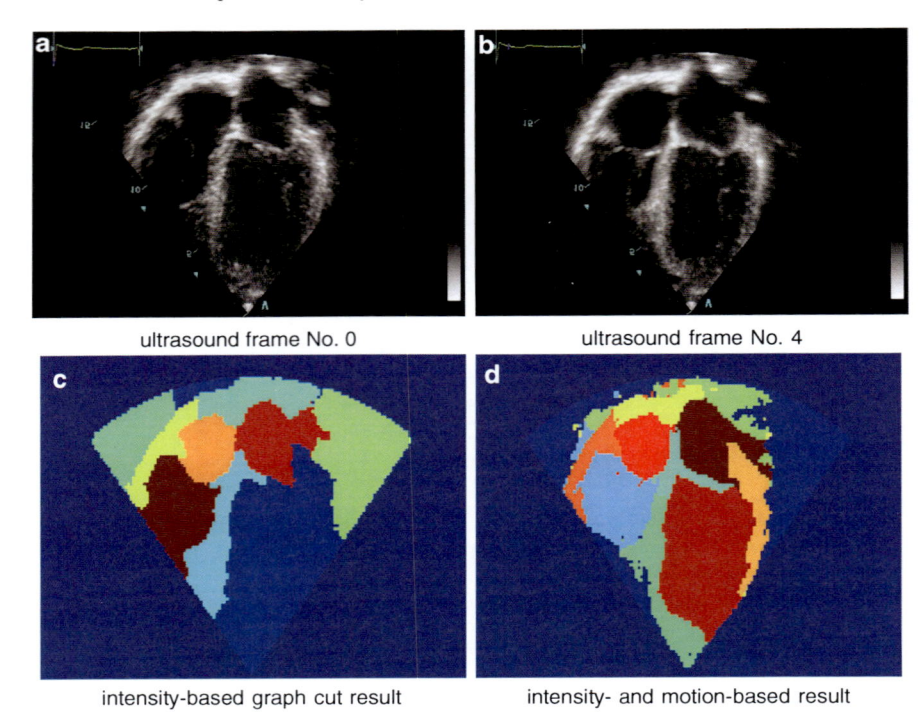

Fig. 9.9. *Segmentation of echocardiograms.* The combined use of intensity and motion information has improved the delineation of chamber and septal wall boundaries

9.7 Discussion

In this chapter, we have introduced two approaches to medical image segmentation based on parametric and non-parametric clustering. The GMM approach has been shown to be useful in not only segmentation but also in quantification of lesions in Multiple-sclerosis images. In the second approach, we showed how spatio-temporal information can be used to guide the normalized cut algorithm in segmentation of moving medical imagery based on graph-cut clustering. Future work will develop additional methodologies of segmentation, detection and quantification for improving quantitative disease evaluation, the prediction of disease outcome, and clinical knowledge.

References

1. Haralick RH, Shapiro LG. Computer and Robot Vision, Vol. I. Philadelphia: Addison-Wesley; 1993.
2. Bishop CM. Neural Networks for Pattern Recognition. University Press, England: Oxford; 1995.

3. Comaniciu D, Meer P. Mean shift: a robust approach toward feature space analysis. IEEE Trans Pattern Anal Mach Intell. 2002;24(5):603–19.
4. Duda RO, Hart PE, Stork DG. Pattern Classiffication. 2nd ed. New York: Wiley-Interscience; 2000.
5. Shi J, Malik J. Normalized cuts and image segmentation. IEEE Trans Pattern Anal Mach Intell. 2000;22(8):888–905.
6. Ng A, Jordan MI, Weiss Y. On spectral clustering: analysis and an algorithm. Proc Adv Neural Inf Process Syst. 2001.
7. Dempster AP, Laird NM, Rubin DB. Maximum likelihood from incomplete data via the EM algorithm. J R Stat Soc Ser B (Methodological). 1977;39(1):1–38.
8. Carson C, Belongie S, Greenspan H, et al. Blobworld: image segmentation using expectation-maximization and its application to image querying. IEEE Trans Pattern Anal Mach Intell. 2002;24(8):1026–38.
9. Greenspan H, Pinhas A. Medical image categorization and retrieval for PACS using the GMM-KL framework. IEEE Trans Inf Technol Biomed. 2007;11(2):190–202.
10. Pham DL, Xu C, Prince JL. Current methods in medical image segmentation. Annu Rev Biomed Eng. 2000;2:315–37.
11. Macovski A. Noise in MRI. Magn Reson Med. 1996;36(3):494–7.
12. Dugas-Phocion G, Ballester MÁG, Malandain G, et al. Improved EM-based tissue segmentation and partial volume effect quantification in multi-sequence brain MRI. Lect Notes Comput Sci. 2004;3216:26–33.
13. Kapur T, Grimson WE, Wells WM, et al. Segmentation of brain tissue from magnetic resonance images. Med Image Anal. 1996;1(2):109–27.
14. Prastawa M, Gilmore J, Lin W, et al. Automatic segmentation of neonatal brain MRI. Lect Notes Comput Sci. 2004;3216:10–7.
15. Leemput KV, Maes F, Vandermeulen D, et al. Automated model-based tissue classification of MR images of the brain. IEEE Trans Med Imaging. 1999;18(10):897–908.
16. Marroquin JL, Vemuri BC, Botello S, et al. An accurate and efficient bayesian method for automatic segmentation of brain MRI. IEEE Trans Med Imaging. 2002;21(8):934–45.
17. Rohlfing T, Jr CRM. Nonrigid image registration in shared-memory multiprocessor environments with application to brains, breasts, and bees. IEEE Trans Inf Technol Biomed. 2003;7(1):16–25.
18. Zhang Y, Brady M, Smith S. Segmentation of brain MR images through a hidden markov random field model and the expectation-maximization algorithm. IEEE Trans Med Imaging. 2001;20(1):45–57.
19. Held K, Kops ER, Krause BJ, et al. Markov random field segmentation of brain MR images. IEEE Trans Med Imaging. 1997;16(6):878–86.
20. Li SZ. Markov Random Field Modeling in Computer Vision. New York: Springer; 1995.
21. McInerney T, Terzopoulos D. Medical image segmentation using topologically adaptable surfaces. Lect Notes Comput Sci. 1997;1205:23–32.
22. Greenspan H, Ruf A, Goldberger J. Constrained Gaussian mixture model framework for automatic segmentation of mr brain images. IEEE Trans Med Imaging. 2006;25(9):1233–45.
23. Huber SJ, Paulson GW, Shuttleworth EC, et al. Magnetic resonance imaging correlates of dementia in multiple sclerosis. Arch Neurol. 1987;44(7):732–36.

24. Medaer R, Nelissen E, Appel B, et al. Magnetic resonance imaging and cognitive functioning in multiple sclerosis. J Neurol. 1987;235(2):86–9.
25. Paty DW, Li DK. Interferon beta-1b is effective in relapsing-remitting multiple sclerosis. Neurology 1993;43(4):662.
26. Rovaris M, Filippi M, Falautano M, et al. Relation between MR abnormalities and patterns of cognitive impairment in multiple sclerosis. Neurology 1998;50(6):1601–8.
27. Sperling RA, Guttmann CR, Hohol MJ, et al. Regional magnetic resonance imaging lesion burden and cognitive function in multiple sclerosis: a longitudinal study. Arch Neurol. 2001;58(1):115–21.
28. Sajja BR, Datta S, He R, et al. Unified approach for multiple sclerosis lesion segmentation on brain MRI. Ann Biomed Eng. 2006;34(1):142–51.
29. Warfield SK, Kaus M, Jolesz FA, et al. Adaptive, template moderated, spatially varying statistical classification. Med Image Anal. 2000;4(1):43–55.
30. Van Leemput K, Maes F, Vandermeulen D, et al. Automated segmentation of multiple sclerosis lesions by model outlier detection. IEEE Trans Med Imaging. 2001;20(8):677–88.
31. Wolff Y, Miron S, Achiron A, et al. Improved CSF classification and lesion detection in MR brain images with multiple sclerosis. Proc SPIE. 2007;6512:2P.
32. Greenspan H, Ruf A, Goldberger J. Constrained Gaussian mixture model framework for automatic segmentation of MR brain images. IEEE Trans Med Imaging. 2006;25(9):1233–45.
33. Galic S, Loncaric S, Tesla EN. Spatio-temporal image segmentation using optical flow and clustering algorithm. Proc Image and Signal Processing and Analysis (IWISPA) 2000; p. 63–8.
34. Simoncelli E, Adelson EA, Heeger DJ. Probability distributions of optical flow. In: IEEE Conf Comput Vis Pattern Recognit. 1991. p. 312–5.
35. Horn BKP, Schunck. Determining optical flow. Artiff Intell. 1981; 185–203.
36. Thirion JP. Image matching as a diffusion process: An analogy with Maxwell's demons. Med Image Anal. 1998;(3):243–60.
37. Audette MA, Ferrie FP, Peters TM. An algorithmic overview of surface registration. Med Image Anal. 2000;4:201–17.
38. Wu Z, Leahy R. An optimal graph-theoretic approach to data clustering: theory and its application to image segmentation. IEEE Trans Pattern Anal Mach Intell. 1993;11:1101–13.
39. Cour T, et al. Spectral segmentation with multiscale graph decomposition. In: Proc IEEE Comput Vis Pattern Recognit. 2005.
40. Informatik R, Weickert J, Scharr H. A scheme for coherence-enhancing diffusion filtering with optimized rotation invariance. 2000.

10

Region-Based Segmentation: Fuzzy Connectedness, Graph Cut and Related Algorithms

Krzysztof Chris Ciesielski and Jayaram K. Udupa

Summary. In this chapter, we will review the current state of knowledge on region-based digital image segmentation methods. More precisely, we will concentrate on the four families of such algorithms: (a) The leading theme here will be the framework of fuzzy connectedness (FC) methods. (b) We will also discuss in detail the family of graph cut (GC) methods and their relations to the FC family of algorithms. The GC methodology will be of special importance to our presentation, since we will emphasize the fact that the methods discussed here can be formalized in the language of graphs and GCs. The other two families of segmentation algorithms we will discuss consist of (c) watershed (WS) and (d) the region growing level set (LS) methods. Examples from medical image segmentation applications with different FC algorithms are also included.

10.1 Introduction and Overview

In this chapter, we will review the current state of knowledge in region-based digital image segmentation methods, with a special emphasis on the fuzzy connectedness (FC) family of algorithms. The other image segmentation methods are discussed in the other chapters of this book and we will refer to them only marginally. We will put a special emphasis on the delineation algorithms, that is, the segmentation procedures returning only one Object Of Interest (OOI) at a time rather than multiple objects simultaneously. This will make the presentation clearer, even for the methods that can be easily extended to the multi-object versions.

We will discuss only the region-growing-type delineation algorithms, which in Chap. 1 are referred to as agglomerative or bottom-up algorithms. More precisely, we will concentrate on the four families of such algorithms. The leading theme will be the framework of FC methods developed since 1996 [1–6], including a slightly different approach to this methodology, as presented in papers [7–9]. For some applications of FC, see also e.g. [10, 11]. We will also discuss the family of Graph Cut (GC) methods [12–20] and their relations to the FC family of algorithms. The GC methodology will be of special importance to

T.M. Deserno (ed.), *Biomedical Image Processing*, Biological and Medical Physics, Biomedical Engineering, DOI: 10.1007/978-3-642-15816-2_10,
© Springer-Verlag Berlin Heidelberg 2011

our presentation, since we will formalize the FC framework in the language of graphs and graph cuts. The other two families of segmentation algorithms we will discuss consist of Watershed (WS) [21–23] and region growing Level Set (LS) methods from [24, 25].

The common feature of all the presented algorithms is that the object to be segmented by them is indicated (by user, or automatically) by one or more space elements (*spels*) referred to as *seeds*. In addition, if P is an object returned by such an algorithm, then any spel belonging to P is connected to at least one of the seeds indicating this object. The word "connected" indicates, that the topological properties of the image scene play important role in this class of segmentation processes. So, we will proceed with explaining what we mean by the image scene, its topology, as well as the notion of connectedness in this context.

For the rest of this chapter, $\mathfrak{n} \geq 2$ will stand for the dimension of the image we consider. In most medically relevant cases, $\mathfrak{n}$ is either 2 or 3, but a time sequence of 3D images is often considered as a 4D image.

10.1.1 Digital Image Scene

A digital image scene C can be identified with any finite subset of the $\mathfrak{n}$-dimensional Euclidean space $\mathbb{R}^\mathfrak{n}$. However, we will concentrate here only on the case most relevant for medical imaging, in which C is of the rectangular form $C_1 \times \cdots \times C_\mathfrak{n}$ and each C_i is identified[1] with the set of integers $\{1, \ldots, m_i\}$.

A topology on a scene $\boldsymbol{C} = \langle C, \alpha \rangle$ will be given in terms of adjacency relation α, which intuitively determines which pair of spels $c, d \in C$ is "close enough" to be considered connected. Formally, an adjacency relation α is a binary relation on C, which will be identified with a subset of $C \times C$, that is, spels $c, d \in C$ are α-adjacent, if and only if, $\langle c, d \rangle \in \alpha$. From the theoretical point of view, we need only to assume that the adjacency relation is symmetric (i.e., if c is adjacent to d, then also d is adjacent to c).[2] However, in most medical applications, it is enough to assume that c is adjacent to d when the distance[3] $||c-d||$ between c and d does not exceed some fixed number. In most applications, we use adjacencies like 4-adjacency (for $\mathfrak{n} = 2$) or 6-adjacency (in the Three-dimensional (3D) case), defined as $||c - d|| \leq 1$. Similarly, the 8-adjacency (for $\mathfrak{n} = 2$) and 26-adjacency (in 3D) relations can be defined as $||c - d|| \leq \sqrt{3}$.

---

[1] This identification of the coordinates of spels with the integer numbers is relevant only for the computer implementations. For theoretical algorithmic discussion, especially for anisotropic images, we will assume that C_i's are the real numbers of appropriate distances

[2] Usually, it is also assumed that α is reflexive (i.e., any spel c is adjacent to itself, $\langle c, c \rangle \in \alpha$), but this assumption is not essential for most considerations

[3] In the examples, we use the Euclidean distance $|| \cdot ||$. But any other distance notion can be also used here

The adjacency relation on C translates to the notion of connectivity as follows. A (connected) path p in a subset A of C is any finite sequence $\langle c_1, \ldots, c_k \rangle$ of spels in A such that any consecutive spels c_i, c_{i+1} in p are adjacent. The family of all paths in A is denoted by $\mathbb{P}^A$. Spels c and s are connected in A provided there exists a path $p = \langle c_1, \ldots, c_k \rangle$ in A from c to s, that is, such that $c_1 = c$ and $c_k = s$. The family of all paths in A from c to d is denoted by $\mathbb{P}^A_{cd}$.

10.1.2 Topological and Graph-Theoretical Scene Representations

The topological interpretation of the scene given above is routinely used in the description of many segmentation algorithms. In particular, this is the case for FC, WS, and most of the LS methods. On the other hand, the algorithms like GC use the interpretation of the scene as a directed graph $G = \langle V, E \rangle$, where $V = C$ is the set of vertices (sometimes extended by two additional vertices) and E is the set of edges, which are identified with the set of pairs $\langle c, d \rangle$ from $V = C$ for which c and d are joined by an edge.

Note that if we define E as the set of all adjacent pairs $\langle c, d \rangle$ from C (i.e., when $E = \alpha$), then the graph $G = \langle V, E \rangle$ and the scene $\boldsymbol{C} = \langle C, \alpha \rangle$ are the identical structures (i.e., $G = \boldsymbol{C}$), despite their different interpretations. This forms the basis of the duality between the topological and graph-theoretical view of this structure: any topological scene $\boldsymbol{C} = \langle C, \alpha \rangle$ can be treated as a graph $G = \langle C, \alpha \rangle$ and, conversely any graph $G = \langle V, E \rangle$ can be treated as topological scene $\boldsymbol{C} = \langle V, E \rangle$.

Under this duality, the standard topological and graph theoretical notions fully agree. A path p in C is connected in $\boldsymbol{C} = G$ in a topological sense, if and only if, it is connected in the graph $G = \boldsymbol{C}$. A subset P of C is connected, in a topological sense, in $\boldsymbol{C} = G$, if and only if, it is connected in the graph $G = \boldsymbol{C}$. The symmetry of α translates into the symmetry of the graph $G = \langle C, \alpha \rangle$, and since any edge $\langle c, d \rangle$ in G can be reversed (i.e., if $\langle c, d \rangle$ is in $E = \alpha$, then so is $\langle d, c \rangle$), G can be treated as an undirected graph.

10.1.3 Digital Image

All of the above notions depend only on the geometry of the image scene and are independent of the image intensity function. Here, the image intensity function will be a function f from C into $\mathbb{R}^k$, that is, $f \colon C \to \mathbb{R}^k$. The value $f(c)$ of f at c is a k-dimensional vector of image intensities at spel c. A digital image will be treated as a pair $\langle \boldsymbol{C}, f \rangle$, where $\boldsymbol{C}$ is its scene (treated either as a topological scene or as a related graph) and f is the image intensity. We will often identify the image with its intensity function, that is, without explicitly specifying associated scene adjacency. In case when $k = 1$ we will say that the image is scalar; for $k > 1$ we talk about vectorial images. Mostly, when giving examples, we will confine ourselves to scalar images.

10.1.4 Delineated Objects

Assume that with an image $\langle C, f \rangle$ we have associated an energy function e, which for every set $P \subset C$ associates its energy value $e(P) \in \mathbb{R}$. Assume also that we have a fixed energy threshold value θ and a non-empty set $S \subset C$ of seeds indicating our OOI. Let $\mathcal{P}(S, \theta)$ be a family of all objects $P \subset C$, associated with e, S, and θ, such that $e(P) \leq \theta$, $S \subset P$, and every $c \in P$ is connected in P to some seed $s \in S$. Threshold θ will be always chosen so that the family $\mathcal{P}(S, \theta)$ is non-empty. Any of the region-based algorithms we consider here will return, as a delineated object, a set $P(S, \theta) \in \mathcal{P}(S, \theta)$. Usually (but not always) $P(S, \theta)$ is the smallest element of $\mathcal{P}(S, \theta)$.

In the case of any of the four methods FC, GC, WS, and LS, the value $e(P)$ of the energy function is defined in terms of the boundary $\mathrm{bd}(P)$ of P, that is, the set $K = \mathrm{bd}(P)$ of all edges $\langle c, d \rangle$ of a graph $C = \langle C, E \rangle$ with $c \in P$ and d not in P. We often refer to this boundary set K as a graph cut, since removing these edges from C disconnects P from its complement $C \setminus P$. The actual definition of e depends on the particular segmentation method.

Let $\kappa \colon E \to \mathbb{R}$ be a local cost function. For $\langle c, d \rangle \in E$ the value $\kappa(c, d)$ depends on the value of the image intensity function f on c, d, and (sometimes) nearby spels. Usually, the bigger is the difference between the values of $f(c)$ and $f(d)$, the smaller is the cost value $\kappa(c, d)$. This agrees with the intuition that the bigger the magnitude of the difference $f(c) - f(d)$ is, the greater is the chance that the "real" boundary of the object we seek is between these spels. In the FC algorithms, κ is called the affinity function. In the GC algorithms κ is treated as a weight function of the edges and is referred to as local cost function. For the classical GC algorithms, the energy function $e(P)$ is defined as the sum of the weights of all edges in $K = \mathrm{bd}(P)$, that is, as $\sum_{\langle c,d \rangle \in K} \kappa(c, d)$. The delineations for the FC family of algorithms are obtained with the energy function $e(P)$ defined as the maximum of the weights of all edges in $K = \mathrm{bd}(P)$, that is, as $\max_{\langle c,d \rangle \in K} \kappa(c, d)$. The same maximum function works also for the WS family with an appropriately chosen κ. The energy function for LS is more complicated, as it depends also on the geometry of the boundary, specifically its curvature.

10.2 Threshold-Indicated Fuzzy Connected Objects

Let $I = \langle C, f \rangle$ be a digital image, with the scene $C = \langle C, E \rangle$ being identified with a graph. As indicated above, the FC segmentations require a local measure of connectivity κ associated with I, known as affinity function, where for a graph edge $\langle c, d \rangle \in E$ (i.e., for adjacent c and d) the number $\kappa(c, d)$ (edge weight) represents a measure of how strongly spels c and d are connected to each other in a local sense. The most prominent affinities used so far are as follows [26], where $\sigma > 0$ is a fixed constant. The homogeneity-based affinity

$$\psi_\sigma(c, d) = \mathrm{e}^{-\|f(c)-f(d)\|^2/\sigma^2} \quad \text{where} \quad \langle c, d \rangle \in E \tag{10.1}$$

with its value being close to 1 (meaning that c and d are well connected) when the spels have very similar intensity values; ψ_σ is related to the notion of directional derivative.

The object feature-based affinity (single object case, with an expected intensity vector $m \in \mathbb{R}^k$ for the object)

$$\phi_\sigma(c,d) = e^{-\max\{||f(c)-m||,||f(d)-m||\}^2/\sigma^2} \quad \text{where} \quad \langle c,d \rangle \in E \quad (10.2)$$

with its value being close to one when both adjacent spels have intensity values close to m. The weighted averages of these two forms of affinity functions – either additive or multiplicative – have also been used. The values of these affinity functions, used in the presented algorithms, are in the interval $[0,1]$.

10.2.1 Absolute Fuzzy Connectedness Objects

Let κ be an affinity associated with a digital image I. As stated in Sect. 10.1, an FC delineated object $P_{\max}(S, \theta)$, indicated by a set S of seeds and an appropriate threshold θ, can be defined as

$$P_{\max}(S, \theta) \text{ is the smallest set belonging to the family } \mathcal{P}_{\text{FC}}(S, \theta), \quad (10.3)$$

where $\mathcal{P}_{\text{FC}}(S, \theta)$ is the family of all sets $P \subset C$ such that: (a) $S \subset P$; (b) every $c \in P$ is connected in P to some $s \in S$; (c) $\kappa(c,d) \leq \theta$ for all boundary edges $\langle c,d \rangle$ of P (i.e., $e(P) = \max_{\langle c,d \rangle \in \text{bd}(P)} \kappa(c,d) \leq \theta$). This definition of the object is very convenient for the comparison of FC with GC and with the other two methods. Nevertheless, for the actual implementation of the FC algorithm, it is more convenient to use another definition, standard in the FC literature. The equivalence of both approaches is given by Theorem 1.

A path strength of a path $p = \langle c_1, \ldots, c_k \rangle$, $k > 1$, is defined as $\mu(p) \stackrel{\text{def}}{=} \min\{\kappa(c_{i-1}, c_i) \colon 1 < i \leq k\}$, that is, the strength of the κ-weakest link of p. For $k = 1$ (i.e., when p has length 1) we associate with p the strongest possible value: $\mu(p) \stackrel{\text{def}}{=} 1$.[4] For $c, d \in A \subseteq C$, the (global) κ-connectedness strength in A between c and d is defined as the strength of a strongest path in A between c and d; that is,

$$\mu^A(c,d) \stackrel{\text{def}}{=} \max\left\{\mu(p) \colon p \in \mathbb{P}_{cd}^A\right\}. \quad (10.4)$$

Notice that $\mu^A(c,c) = \mu(\langle c \rangle) = 1$. We will often refer to the function μ^A as a connectivity measure (on A) induced by κ. For $c \in A \subseteq C$ and a non-empty $D \subset A$, we also define $\mu^A(c,D) \stackrel{\text{def}}{=} \max_{d \in D} \mu^A(c,d)$. The standard definition of an FC delineated object, indicated by a set S of seeds and an appropriate

---

[4] For $k = 1$ the set $\{\kappa(c_{i-1}, c_i) \colon 1 < i \leq k\}$ is empty, so the first part of the definition leads to equation $\mu(\langle c_1 \rangle) = \min \emptyset$. This agrees with our definition of $\mu(\langle c_1 \rangle) = 1$ if we define $\min \emptyset$ as equal to 1, the highest possible value for κ. Thus, we will assume that $\min \emptyset = 1$

threshold $\theta < 1$ and referred to as Absolute Fuzzy Connectedness (AFC) object, is given as $P_{S\theta} = \{c \in C : \theta < \mu^C(c, S)\}$.[5]

Theorem 1. $P_{S\theta} = P_{\max}(S, \theta)$ *for all* $S \subset C$ *and* $\theta < 1$.

10.2.2 Robustness of Objects

If a set of seeds S contains only one seed s, then we will write $P_{s\theta}$ for the object $P_{S\theta} = P_{\{s\}\theta}$. It is easy to see that $P_{S\theta}$ is a union of all objects $P_{s\theta}$ for $s \in S$, that is, $P_{S\theta} = \bigcup_{s \in S} P_{s\theta}$. Actually, if $G_\theta = \langle C, E_\theta \rangle$ is a graph with E_θ consisting of the scene graph edges $\langle c, d \rangle$ with weight $\kappa(c, d)$ greater than θ, then $P_{s\theta}$ is a connected component of G_θ containing s, and $P_{S\theta}$ is a union of all components of G_θ intersecting S.

One of the most important properties of the AFC objects is known as robustness. Intuitively, this property states that the FC delineation results do not change if the seeds S indicating an object are replaced by another nearby set T of seeds. Formally, it reads as follows.

Theorem 2. (Robustness) *For every digital image I on a scene $\mathbf{C} = \langle C, E \rangle$, every $s \in C$ and $\theta < 1$, if $P_{s\theta}$ is an associated FC object, then $P_{T\theta} = P_{s\theta}$ for every non-empty $T \subset P_{s\theta}$. More generally, if $S \subset C$ and $T \subset P_{S\theta}$ intersects every connected component of G_θ intersecting $P_{S\theta}$ (i.e., $T \cap P_{s\theta} \neq \emptyset$ for every $s \in S$), then $P_{T\theta} = P_{S\theta}$.*

The proof of this result follows easily from our graph interpretation of the object, as indicated above. The proof based only on the topological description of the scene can be found in [2, 5]. The robustness property constitutes the strongest argument for defining the objects in the FC fashion. Note, that none of the other algorithms discussed here have this property.

10.2.3 Algorithm for Delineating Objects

The algorithm presented below comes from [1].

Algorithm $\kappa\theta FOEMS$
Input: Scene $\mathbf{C} = \langle C, E \rangle$, affinity κ defined on an image $I = \langle \mathbf{C}, f \rangle$, a set $S \subset C$ of seeds indicating the object and a threshold $\theta < 1$.
Output: AFC object $P_{S\theta}$ for the image I.
Auxiliary Data Structures: A characteristic function $g \colon C \to \{0, 1\}$ of $P_{S\theta}$ and a queue Q of spels.

---

[5] In the literature, an AFC object is usually arrived at (see [3, 5]) as $P_{S\theta}^{\leq} = \{c \in C : \theta \leq \mu^C(c, S)\}$. However, if θ^+ denotes the smallest number greater than θ of the form $\kappa(c, d)$, with $\langle c, d \rangle \in E$, then $P_{S\theta} = P_{S\theta^+}^{\leq}$. Thus, our definition of AFC object can be also expressed in the standard form, with just slightly different threshold. On the other hand, the following presentation is considerably easier expressible with the AFC object defined with the strict inequality.

begin
    1.   set $g(s) = 1$ for all $s \in S$ and $g(c) = 0$ for all $c \in C \setminus S$;
    2.   push to Q all spels $c \in C$ for which $\kappa(c,s) > \theta$ for some $s \in S$;
    3.   *while* Q is not empty *do*
    4.   remove a spel c from Q;
    5.   *if* $g(c) = 0$ *then*
    6.   set $g(c) = 1$;
    7.   push to Q all spels $d \in C$ for which $\kappa(d,c) > \theta$;
    8.   *endif*;
    9.   *endwhile*;
   10.   create $P_{S\theta}$ as a set of all spels c with $g(c) = 1$;
end

It is easy to see that $\kappa\theta FOEMS$ runs in linear time with respect to the size n of the scene C. This is the case, since any spel can be pushed into the queue Q (Line 7) at most Δ-many times, where Δ is the degree of the graph $\boldsymbol{C}$ (i.e., the largest number of spels that can be adjacent to a single spel; e.g., $\Delta = 26$ for the 26-adjacency). Specifically, $\kappa\theta FOEMS$ runs in time of order $O(\Delta n)$.

10.3 Optimization in Foreground-Background Case

So far, we discussed algorithms delineating an object, P, indicated by some seeds S belonging to P. Since we had no direct information on the spatial extent of the desired object, the actual extent of the delineated object P was regulated only by a mysterious parameter: a threshold θ setting the upper limit on the energy function value $\boldsymbol{e}(P)$. The difficulty of choosing this threshold is overcome by setting up and solving an appropriate optimization problem for an energy function $\boldsymbol{e}$. The setting part is done as follows.

First, we choose a proper initial condition, which, in the case of FC and GC algorithms, consists of indicating not only the foreground object (i.e., the OOI) by a set S of seeds, but also a background (i.e., everything except the OOI) by another set T of seeds. The stipulation is that S is contained in the delineated P, while T is disjoint with P. This ensures that we will consider only non-trivial sets P as possible choices for the object.

Let $\mathcal{P}(S,T)$ be the family of all sets $P \subset C$ such that $S \subset P$ and $T \cap P = \emptyset$. We like the desired object P to minimize the energy $\boldsymbol{e}(P)$ over all $P \in \mathcal{P}(S,T)$, that is, sets P satisfying the initial condition indicated by seeds S and T. In other words, if we define $\boldsymbol{e}_{\min} = \min\{\boldsymbol{e}(P): P \in \mathcal{P}(S,T)\}$, then the OOI $P_{S,T}$ will be chosen, by an algorithm, as an element of the family $\mathcal{P}_{\min} = \{P \in \mathcal{P}(S,T): \boldsymbol{e}(P) = \boldsymbol{e}_{\min}\}$. This is a typical setup for the energy optimization image delineation algorithms.

Notice that, although the minimal energy $\boldsymbol{e}_{\min}$ is always uniquely defined, the family $\mathcal{P}_{\min}$ may have more than one element, so our solution $P_{S,T} \in \mathcal{P}_{\min}$

still may not be uniquely determined. In the case of GC framework, the family $\mathcal{P}_{\min}$ has always the smallest element (smallest in terms of inclusion) and this element is taken as $P_{S,T}$. The situation is the same in the FC framework, when S is a singleton. In the case when S has more seeds, the family $\mathcal{P}_{\min}$ is refined to a smaller family $\mathcal{P}^*_{\min}$ and $P_{S,T}$ is the smallest element of $\mathcal{P}^*_{\min}$. All of this is discussed in more detail below.

10.3.1 Relative Fuzzy Connectedness

In the FC framework, the optimization technique indicated above is called Relative Fuzzy Connectedness (RFC). Once again, the actual definition of the RFC object $P_{S,T}$ (see [2]) is in a slightly different format from the one indicated above – it emphasizes the competition of seed sets S and T for attracting a given spel c to their realms. The attraction is expressed in terms of the strength of global connectedness $\mu^C(c,S)$ and $\mu^C(c,T)$: $P_{S,T}$ claims a spel c when $\mu^C(c,S)$ exceeds $\mu^C(c,T)$, that is,

$$P_{S,T} = \{c \in C \colon \mu^C(c,S) > \mu^C(c,T)\}$$

Notice that, $P_{S,T} = \{c \in C \colon (\exists s \in S) \mu^C(c,s) > \mu^C(c,T)\} = \bigcup_{s \in S} P_{\{s\},T}$, as $\mu^c(c,S) = \max_{s \in S} \mu^C(c,s)$. Below, we will show that, if the number $\mu^C(S,T) = \max_{s \in S} \mu^C(s,T)$ is less than 1, then $P_{S,T} \in \mathcal{P}(S,T)$, that is, that $P_{S,T}$ contains S and is disjoint with T. (If $\mu^C(S,T) = 1$, then sets S and T need not be disjoint. In this case the set $P_{S,T}$ is empty.) It is also important that

$$\text{if } P \in \mathcal{P}(S,T), \text{ then } \boldsymbol{e}(P) \geq \mu^C(S,T) \tag{10.5}$$

Indeed, choose a path $p = \langle c_1, \ldots, c_k \rangle$ from $s \in S$ to a $t \in T$ such that $\mu(p) = \mu^C(S,T)$. Since $c_1 = s \in P$ and $c_k = t \notin P$, there exists a $j \in \{2, \ldots, k\}$ with $c_{j-1} \in P$ while $c_j \notin P$. This means that $\langle c_{j-1}, c_j \rangle \in \mathrm{bd}(P)$. Hence, $\boldsymbol{e}(P) = \max_{\langle c,d \rangle \in \mathrm{bd}(P)} \kappa(c,d) \geq \kappa(c_{j-1}, c_j) \geq \min\{\kappa(c_{i-1}, c_i) \colon 1 < i \leq k\} = \mu(p) = \mu^C(S,T)$.

Next note that each object $P_{\{s\},T}$ is indeed a result of the optimization, as stated above.

Lemma 1. *Assume that* $\theta_s = \mu^C(s,T) < 1$. *Then* $P_{\{s\},T} = P_{s\theta_s}$. *Moreover, θ_s equals $\boldsymbol{e}_{\min} = \min\{\boldsymbol{e}(P) \colon P \in \mathcal{P}(\{s\},T)\}$ and $P_{s\theta_s}$ is the smallest set in the family* $\mathcal{P}_{\min} = \{P \in \mathcal{P}(\{s\},T) \colon \boldsymbol{e}(P) = \boldsymbol{e}_{\min}\}$.

The description of the RFC object $P_{S,T}$ when S has more than one seed is given in the following theorem. Intuitively, it says that each seed $s \in S$ generates separately its own part $P_{\{s\},T} \in \mathcal{P}(\{s\},T)$ and although their union, $P_{S,T}$, minimizes only its own lower bound $\theta_S = \mu^C(S,T)$, each component $P_{\{s\},T}$ minimizes its own version of the minimum, $\theta_s = \mu^C(s,T)$, which may be (and often is) smaller than the global minimizer $\theta_S = \mu^C(S,T)$. In other words, the object $P_{S,T}$ can be viewed as a result of minimization procedure used separately for each $s \in S$, which gives a sharper result than a simple minimization of global energy for the entire object $P_{S,T}$.

Theorem 3. *Assume that $\theta_S = \mu^C(S,T) < 1$. Then $e(P_{S,T}) = \theta_S = e_{\min}$ and $P_{S,T} = \bigcup_{s \in S} P_{s\theta_s}$ is the smallest set in the family $\mathcal{P}^*_{\min}$ of the sets of the form $\bigcup_{s \in S} P^s$, where each P^s belongs to $\mathcal{P}^s_{\min} = \{P \in \mathcal{P}(\{s\}, T) \colon e(P) = \theta_s\}$. Moreover, $\mathcal{P}^*_{\min} \subset \mathcal{P}_{\min}$.*

10.3.2 Algorithm for Delineating Objects

The algorithm presented below is a multiseed version of the algorithm from [1]. It is a main step for defining the RFC object $P_{S,T}$.

Algorithm $\kappa FOEMS$
Input: Scene $\boldsymbol{C} = \langle C, E \rangle$, affinity κ defined on an image $I = \langle \boldsymbol{C}, f \rangle$, a set $T \subset C$.
Output: A connectivity function $h \colon C \to [0, 1]$, $h(c) = \mu^C(c, T)$.
Auxiliary Data Structures: A queue Q of spels.
begin
   1.   set $h(t) = 1$ for all $t \in T$ and $h(c) = 0$ for all $c \in C \setminus T$;
   2.   push to Q all spels $c \in C$ for which $\kappa(c,t) > 0$ for some $t \in T$;
   3.   *while* Q is not empty *do*
   4.   remove a spel c from Q;
   5.   find $M = \max\{\min\{h(d), \kappa(d,c)\} \colon \langle d, c \rangle \in E\}$
   6.   *if* $M > h(c)$ *then*
   7.   set $h(c) = M$;
   8.   push to Q all $d \in C$ for which $\min\{M, \kappa(c,d)\} > h(d)$;
   9.   *endif*;
  10.   *endwhile*;
  11.   output connectivity function $h \colon C \to [0, 1]$, $h(c) = \mu^C(c, T)$;
end

The algorithm runs in quadratic time with respect to the size n of a scene $\boldsymbol{C}$. More precisely, the maximal number of possible values for the connectivity function h is the size of the range of κ, which does not exceed the size of the set of all edges E, that is, Δn. Therefore, each spel d may be pushed back to Q at most Δn many times: when the value $h(c)$ is changed (maximum Δn-many times) for each of Δ-many spels c adjacent to d. Since each instance of performing the *while* command operation is of time order $O(\Delta)$ and we have n spels, the $\kappa FOEMS$ ends, in the worst case, in time of order $O(\Delta^2 n^2)$.

If a connectivity function $h(c) = \mu^C(c, T)$ is calculated, then numbers $\theta_s = \mu^C(s, T) < 1$ are readily available, and object $P_{S,T} = \bigcup_{s \in S} P_{s\theta_s}$ can be delineated, in quadratic time of order $O(\Delta^2 n^2)$, by calling algorithm $\kappa\theta FOEMS$ for each $s \in S$.

10.3.3 Graph Cut Delineation

For the GC algorithms, a graph $G^I = \langle V, E \rangle$ associated with the image $I = \langle \boldsymbol{C}, f \rangle$, where $\boldsymbol{C} = \langle C, \alpha \rangle$, is a slight modification of the graph $\langle C, \alpha \rangle$ discussed above. Specifically, the set of vertices V is defined as $C \cup \{s,t\}$, that is, the standard set C of image vertices is expanded by two new additional vertices s and t called terminals. Individually, s is referred to as source and t as sink. The set of edges is defined as $E = \alpha \cup \{\langle b,d \rangle \colon$ one of b,d is in C, the other in $\{s,t\}\}$. In other words, the edges between vertices in C remains as in $\boldsymbol{C}$, while we connect each terminal vertex to each $c \in C$.

The simplest way to think about the terminals is that they serve as the seed indicators: s for seeds $S \subset C$ indicating the object; t for seeds $T \subset C$ indicating the background. The indication works as follows. For each edge connecting a terminal $r \in \{s,t\}$ with a $c \in C$ associate the weight: ∞ if either $r = s$ and $c \in S$, or $r = t$ and $c \in T$; and 0 otherwise. This means, that the source s has infinitely strong connection to any seed c in S, and the weakest possible to any other spel $c \in C$. (We assume that all weights are nonnegative, that is, in $[0,\infty]$.) Similarly, for the sink t and seeds c from T.

Now, assume that for every edge $\langle c, d \rangle \in \alpha$ we give a weight $\kappa(c,d)$ associated with the image $I = \langle \boldsymbol{C}, f \rangle$. Since the algorithm for delineating RFC object uses only the information on the associated graph (which includes the weights given by the affinity κ), we can delineate RFC object $P^*_{\{s\},\{t\}} \subset V$ associated with this graph G^I. It is easy to see that the RFC object $P_{S,T} \subset C$ associated with I is equal to $P^*_{\{s\},\{t\}} \cap C$. Similarly, for $\theta < 1$, if $P^*_{s\theta} \subset V$ is an AFC object associated with the graph G^I, then the AFC object $P_{S\theta} \subset C$ associated with I is equal to $P^*_{S\theta} \cap C$. All of this proves that, from the FC framework point of view, replacing the graph $G = \langle C, \alpha \rangle$ with G^I is only technical in nature and results in no delineation differences.

Historically, the rationale for using in GC frameworks graphs G^I, with distinctive terminals, is algorithmic in nature. More precisely, for a weighted graph $G = \langle V, E \rangle$ with positive weights and two distinct vertices s and t indicated in it, there is an algorithm returning the smallest set P_G in the family $\mathcal{P}_{\min} = \{P \in \mathcal{P}(s,t) \colon \boldsymbol{e}(P) = \boldsymbol{e}_{\min}\}$, where $\mathcal{P}(s,t) = \{P \subset V \setminus \{t\} \colon s \in P\}$, $\boldsymbol{e}_{\min} = \min\{\boldsymbol{e}_\Sigma(P) \colon P \in \mathcal{P}(s,t)\}$, $\boldsymbol{e}_\Sigma(P) = \sum_{e \in \mathrm{bd}(P)} w_e$, and w_e is the weight of the edge e in the graph.

Now, let $G^I = \langle C \cup \{s,t\}, E \rangle$ be the graph associated with an image I as described above, that is, weights of edges between spels from C are obtained from the image I (in a manner similar to the affinity numbers) and weights between the other edges by seed sets S and T indicating foreground and background. It is easy to see that the object $P^\Sigma_{S,T} = C \cap P_{G^I}$ contains S, is disjoint with T, and has the smallest cost $\boldsymbol{e}_\Sigma$ among all such sets. Thus, the format of definition of the GC object $P^\Sigma_{S,T}$ is the same as that for RFC object $P_{S,T}$, the difference being only the energy functions $\boldsymbol{e}$ they use.

In spite of similarities between the GC and RFC methodologies as indicated above, there are also considerable differences between them. There are several theoretical advantages of the RFC framework over GC in this setting:

- *Speed:* The FC algorithms run faster than those for GC. Theoretical estimation of FC algorithms worst scenario run time (for slower RFC) is $O(n^2)$ with respect to the scene size n (Sect. 10.3.2), while the best theoretical estimation of the run time for delineating $P_{S,T}^{\Sigma}$ is of order $O(n^3)$ (for the best known algorithms) or $O(n^{2.5})$ (for the fastest currently known), see [15]. This is also confirmed by experimental comparisons.
- *Robustness:* The outcome of FC algorithms is unaffected by small (within the objects) changes of the position of seeds (Theorems 2 and 4). On the other hand, the results of GC delineation may become sensitive for even small perturbations of the seeds.
- *Multiple objects:* The RFC framework handles easily the segmentation of multiple objects, retaining its running time estimate and robustness property (Sect. 10.4.1). The GC in the multiple object setting leads to an NP-hard problem [12], so all existing algorithms for performing the required precise delineation run in exponential time. However, there are algorithms that render approximate solutions for such GC problems in a practical time [12].
- *Shrinking problem:* In contrast to RFC methods, the GC algorithms have a tendency of choosing the objects with very small size of the boundary, even if the weights of the boundary edges is very high [16, 19]. This may easily lead to the segmented object being very close to either the foreground seed set S, or the complement of the background seed set T. Therefore, the object returned by GC may be far from desirable. This problem has been addressed by many authors, via modification of the GC method. Notice that RFC methods do not have any shrinking problem.
- *Iterative approach:* The FC framework allows an iterative refinement of its connectivity measure μ^A, which in turn makes it possible to redefine e as we go along. From the viewpoint of algorithm, this is a powerful strategy. No such methods exist for GC at present.

All of this said, it should be noticed that GC has also some nice properties that FC does not possess. First notice that the shrinking problem is the result of favoring shorter boundaries over the longer, that is, has a smoothing effect on the boundaries. This, in many (but not all) cases of medically important image delineations, is a desirable feature. There is no boundary smoothing factor built in to the FC basic framework and, if desirable, boundary smoothing must be done at the FC post processing stage.

Another nice feature of GC graph representation G^I of an image I is that the weights of edges to terminal vertices naturally represent the object feature-based affinity, see (10.2), while the weights of the edges with both vertices in C are naturally connected with the homogeneity type of affinity (10.1). This is the case, since homogeneity-based affinity (a derivative concept) is a binary relation in nature, while the object feature-based affinity is actually a unary relation. Such a clear cut distinction is difficult to achieve in FC framework, since it requires only one affinity relation in its setting.

10.4 Segmentation of Multiple Objects

Now, assume that we like to recognize $m>1$ separate objects, $P_1,\ldots,P_m$, in the image $I=\langle C,f\rangle$. What general properties the family $\mathcal{P}=\{P_1,\ldots,P_m\}$ should have? The term "segmentation" suggests that $\mathcal{P}$ should be a partition of a scene C, that is, that sets are *pairwise disjoint* (i.e., no two of then have common element) and that they *cover* C (i.e., $C=\bigcup_{i=1}^{m}P_i$). Unfortunately, insuring both of these properties is usually neither desirable not possible for the medical image segmentation problems. We believe, that the most reasonable compromise here is to assume that the objects P_i are pairwise disjoint, while they do not necessarily cover the entire image scene C. The motivation here is the delineation of major body organs (e.g., stomach, liver, pancreas, kidneys). Therefore, the term image segmentation refers to a family $\mathcal{P}=\{P_1,\ldots,P_m\}$ of pairwise disjoint objects for which the background set $B_\mathcal{P}=C\setminus\bigcup_{i=1}^{m}P_i$ might be nonempty.

It should be stressed, however, that some authors allow overlap of the objects, while ensuring that there is no nonempty background $B_\mathcal{P}$ [7,8]. Other methods (like classical WS algorithms) return a partition of a scene.

10.4.1 Relative Fuzzy Connectedness

Assume that for an image $I=\langle C,f\rangle$ we have a pairwise disjoint family $\mathcal{S}=\{S_1,\ldots,S_m\}$ of sets of seeds, each S_i indicating an associated object P_i. If for each i we put $T_i=\left(\bigcup_{j=1}^{m}S_j\right)\setminus S_i$, then the RFC segmentation is defined as a family $\mathcal{P}=\{P_{S_i\mathcal{S}}\colon i=1,\ldots,m\}$, where each object $P_{S_i\mathcal{S}}$ is equal to $P_{S_i,T_i}=\{c\in C\colon \mu^C(c,S_i)>\mu^C(c,T_i)\}$.

Since, by Lemma 1, each P_{S_i,T_i} equals $\bigcup_{s\in S_i}P_{\{s\},T_i}=\bigcup_{s\in S}P_{s\theta_s}$, where $\theta_s=\mu^C(s,T_i)$, using the algorithms from Sect. 10.3.2, the partition $\mathcal{P}$ can be found in $O(n^2)$ time. Also, the robustness Theorem 2 can be modified to this setting as follows.

Theorem 4. (Robustness for RFC) *Let $\mathcal{S}=\{S_1,\ldots,S_m\}$ be a family of seeds in a digital image I and let $\mathcal{P}=\{P_{S_i\mathcal{S}}\colon i=1,\ldots,m\}$ be an associated RFC segmentation. For every i and $s\in S_i$ let $g(s)$ be in $P_{\{s\},T_i}$. If $\mathcal{S}'=\{S'_1,\ldots,S'_m\}$, where each $S'_i=\{g(s)\colon s\in S_i\}$, then $P_{S_i\mathcal{S}}=P_{S'_i\mathcal{S}'}$ for every i.*

In other words, if each seed s present in $\mathcal{S}$ is only "slightly" shifted to a new position $g(s)$, then the resulting RFC segmentation $\{P_{S'_i\mathcal{S}'}\colon i=1,\ldots,m\}$ is identical to the original one $\mathcal{P}$.

When an RFC object $P_{S_i\mathcal{S}}$ is indicated by a single seed, then, by Theorem 3, it is equal to the AFC object $P_{s_i\theta_i}$ for appropriate threshold θ_i. But even when all objects are in such forms, different threshold θ_i need not be equal, each being individually tailored.

This idea is best depicted schematically (Fig. 10.1). Figure 10.1a represents a schematic scene with a uniform background and four distinct areas denoted

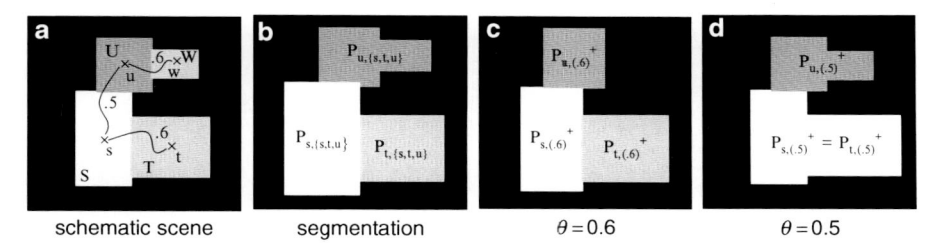

Fig. 10.1. *Relative fuzzy connectedness.* Each object is optimized separately. Panels (**c**) and (**d**) show delineations

by S, T, U, W, and indicated by seeds marked by ×. It is assumed that each of these areas is uniform in intensity and the connectivity strength within each of these areas has the maximal value of 1, the connectivity between the background and any other spel is ≤ 0.2, while the connectivity between the adjacent regions is as indicated in the figure: $\mu(s,t) = 0.6$, $\mu(s,u) = 0.5$, and $\mu(u,w) = 0.6$. (Part b): The RFC segmentation of three objects indicated by seeds s, t, and u, respectively. (Part c): Three AFC objects indicated by the seeds s, t, u and delineated with threshold $\theta = 0.6$. Notice that while $P_{s,\{s,t,u\}} = P_{s,.6}$ and $P_{t,\{s,t,u\}} = P_{t,.6}$, object $P_{u,.6}$ is smaller than RFC indicated $P_{u,\{s,t,u\}}$. (Part d): Same as in Part (c) but with $\theta = 0.5$. Note that while $P_{u,\{s,t,u\}} = P_{u,.5}$, objects $P_{s,.5}$ and $P_{t,.5}$ coincide and lead to an object bigger than $P_{s,\{s,t,u\}}$ and $P_{t,\{s,t,u\}}$.

10.4.2 Iterative Relative Fuzzy Connectedness

The RFC segmentation $\mathcal{P} = \{P_{S_i \mathcal{S}} \colon i = 1, \ldots, m\}$ of a scene can still leave quite a sizable "leftover" background set $B = B_{\mathcal{P}}$ of all spels c outside any of the objects wherein the strengths of connectedness are equal with respect to the seeds. The goal of the Iterative Relative Fuzzy Connectedness (IRFC) is to find a way to naturally redistribute some of the spels from $B_{\mathcal{P}}$ among the object regions in a new generation (iteration) of segmentation. Another motivation for IRFC is to overcome the problem of "path strength dilution" within the same object, of paths that reach the peripheral subtle and thin aspects of the object.

In the left part of Fig. 10.2, two object regions A and B, each with its core and peripheral subtle parts, are shown, a situation like the arteries and veins being juxtaposed. Owing to blur, partial volume effect and other shortcomings, the strongest paths from s_1 to t_1, s_1 to t_2, s_2 to t_1, and s_2 to t_2 are all likely to assume similar strengths. As a consequence, the spels in the dark areas may fall in $B_{\mathcal{P}}$, the unclaimed background set.

The idea of IRFC is to treat the RFC delineated objects $P_{S_i \mathcal{S}}$ as the first iteration $P^1_{S_i \mathcal{S}}$ approximation of the final segmentation, while the next step iteration is designed to redistribute some of the background spels $c \in B_{\mathcal{P}}$,

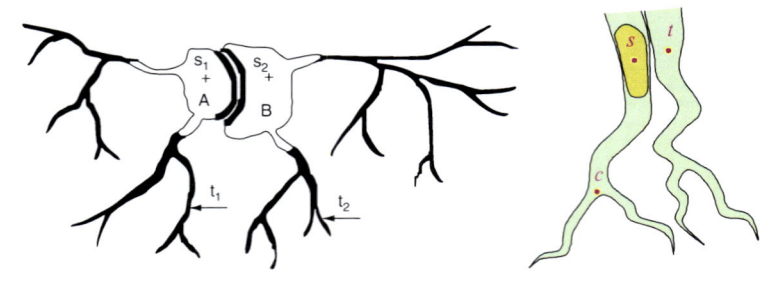

Fig. 10.2. *RFC vs. IRFC. Left*: The strongest paths from s_1 to t_1, s_1 to t_2, s_2 to t_1, and s_2 to t_2 are likely to have the same strength because of partial volume effects; *Right*: Pictorial illustration of IRFC advantages over RFC

for which $\mu^C(c, S_i) = \mu^C(c, T_i)$ for some i. Such a tie can be resolved if the strongest paths justifying $\mu^C(c, S_i)$ and $\mu^C(c, T_i)$ cannot pass through the spels already assigned to another object. In other words, we like to add spels from the set $P^* = \{c \in B \colon \mu^{B \cup P_{S_i S}}(c, S_i) > \mu^{B \cup P_{S_j S}}(c, S_j) \text{ for every } j \neq i\}$, to a new generation $P^2_{S_i S}$ of $P^1_{S_i S}$, that is, define $P^2_{S_i S}$ as $P^1_{S_i S} \cup P^*$. This formula can be taken as a definition. However, from the algorithmic point of view, it is more convenient to define $P^2_{S_i S}$ as

$$P^2_{S_i S} = P^1_{S_i S} \cup \left\{ c \in C \setminus P^1_{S_i S} \colon \mu^C(c, S_i) > \mu^{C \setminus P^1_{S_i S}}(c, T_i) \right\}$$

while the equation $P^2_{S_i S} = P^1_{S_i S} \cup P^*$ always holds, as proved in [5, thm. 3.7]. Thus, the IRFC object is defined as $P^\infty_{S_i S} = \bigcup_{k=1}^\infty P^k_{S_i S}$, where sets $P^k_{S_i S}$ are defined recursively by the formulas $P^1_{S_i S} = P_{S_i S}$ and

$$P^{k+1}_{S_i S} = P^k_{S_i S} \cup \left\{ c \in C \setminus P^k_{S_i S} \colon \mu^C(c, S_i) > \mu^{C \setminus P^k_{S_i S}}(c, T_i) \right\} \tag{10.6}$$

The right side of Fig. 10.2 illustrates these ideas pictorially. The initial segmentation is defined by RFC conservatively, so that $P_{S_i S}$ corresponds to the core aspects of the object identified by seed $s \in S_i$ (illustrated by the hatched area containing s). This leaves a large boundary set B where the strengths of connectedness with respect to the different seeds are equal (illustrated by the shaded area containing c). In the next iteration, the segmentation is improved incrementally by grabbing those spels of B that are connected more strongly to $P_{S_i S}$ than to sets $P_{S_j S}$. When considering the object associated with s, the "appropriate" path from s to any $c \in B$ is any path in C. However, all objects have to compete with the object associated with s by allowing paths from their respective seeds $t \in T_i$ to c not to go through $P_{S_i S}$ since this set has already been declared to be part of the object of s.

The IRFC segmentation is robust in the sense of Theorem 4, where in its statement the objects $P_{S_i S}$ are replaced by the first iteration $P^1_{S_i S}$ of $P^\infty_{S_i S}$. This follows easily from Theorem 4 [5]. It is also worth to notice that the witnessing strongest paths from $c \in P^\infty_{S_i S}$ to S_i can be found in $P^\infty_{S_i S}$ [5].

10.4.3 Algorithm for Iterative Relative Fuzzy Connectedness

The algorithm presented below comes from [5]. Note that we start the recursion with $P^0_{S_i\mathcal{S}} = \emptyset$. It is easy to see that with such definition $P^1_{S_i\mathcal{S}}$ obtained with (10.6) is indeed equal to $P_{S_i\mathcal{S}}$.

Algorithm $\kappa IRMOFC$

Input: Scene $\boldsymbol{C} = \langle C, E \rangle$, affinity κ defined on an image $I = \langle \boldsymbol{C}, f \rangle$, a family $\mathcal{S} = \{S_1, \ldots, S_m\}$ of pairwise disjoint set of seeds, a sequence $\langle T_1, \ldots, T_m \rangle$, with $T_i = \left(\bigcup_{j=1}^m S_j \right) \setminus S_i$ for every i.

Output: A sequence $\langle P^\infty_{S_1\mathcal{S}}, \ldots, P^\infty_{S_m\mathcal{S}} \rangle$ forming IRFC segmentation.

Auxiliary Data Structures: A sequence of characteristic functions $g_i \colon C \to \{0,1\}$ of objects $P^k_{S_i\mathcal{S}}$ and affinity κ_{g_i} equal to κ for pairs $\langle c, d \rangle$ with $g_i(c) = g_i(d) = 0$, and 0 otherwise. Note that $\mu^C(\cdot, T_i)$ for κ_{g_i} equals to $\mu^{C \setminus P_{S_i\mathcal{S}}}(\cdot, T_i)$ for $P_{S_i\mathcal{S}}$ indicated by g_i.

begin
1.   *for* $i = 1$ *to* m *do*
2.   invoke $\kappa FOEMS$ to find $h_0(\cdot) = \mu^C(\cdot, S_i)$;
3.   initiate $g_i(c) = 0$ for all $c \in C$;
4.   set $\kappa_{g_i} = \kappa$ and $flag = $ true;
5.   *while* $flag = $ true *do*;
6.   set $flag = $ false;
7.   invoke $\kappa FOEMS$ to find $h(\cdot) = \mu^C(\cdot, T_i)$ for κ_{g_i};
8.   *for* all $c \in C$ *do*
9.   *if* $g_i(c) = 0$ and $h_0(c) > h(c)$ *then*
10.   set $g_i(c) = 1$ and $flag = $ true;
11.   *for* every $d \in C$, $d \neq c$, adjacent to c *do*
12.   set $\kappa_{g_i}(c, d) = 0$ and $\kappa_{g_i}(d, c) = 0$;
13.   *endfor*;
14.   *endif*;
15.   *endfor*;
16.   *endwhile*;
17.   *endfor*;
18.   output sets $P^\infty_{S_i\mathcal{S}}$ indicated by characteristic functions g_i;

end

The proof that the algorithm stops and returns proper objects can be found in [5]. Since it can enter while loop at most once for each updated spel, it enters it $O(n)$ times, where n is the size of C. Since $\kappa FOEMS$ runs in time of order $O(\Delta^2 n^2)$, the worst scenario for $\kappa IRMOFC$ is that it runs in time of order $O(\Delta^2 n^3)$.

A slightly different approach to calculating IRFC objects comes from the Image Foresting Transform (IFT) [20, 27]. This approach distributes the spels unassigned by IRFC to different objects, according to some ad hoc algorithmic procedure.

10.4.4 Variants of IRFC

In the papers [7–9] (in particular, see [8, page 465]), the authors employ different affinity κ_i for each i-th object to be delineated, and apply the algorithm that returns the objects $\hat{P}^\infty_{S_i S} = \bigcup_{k=0}^\infty \hat{P}^k_{S_i S}$, with sets $\hat{P}^k_{S_i S}$ being defined recursively by the formulas $\hat{P}^0_{S_i S} = \emptyset$ and

$$\hat{P}^{k+1}_{S_i S} = \hat{P}^k_{S_i S} \cup \bigcup_{j \neq i} \left\{ c \in C \setminus \hat{P}^k_{S_i S} : \mu_i^C(c, S_i) \geq \mu_j^{C \setminus \hat{P}^k_{S_i S}}(c, S_j) \right\} \quad (10.7)$$

where μ_j is the global connectivity measure associated with the affinity κ_j.

In general, the segmentations defined with different affinities, in the format of (10.7) (even with just one step iteration, that is, in the RFC mode), are neither robust nor have path connectedness property mentioned at the end of Sect. 10.4.2 (See [2]). Although, the lack of path connectedness property may seem to be of little consequence, it undermines the entire philosophy that stands behind IRFC definitions. Nevertheless, it solves some problems with dealing with the object-feature based affinity in single affinity mode, which was discussed in [28].

10.5 Scale-Based and Vectorial Fuzzy Connectedness

In our discussion so far, when formulating affinities κ, we considered $\kappa(c, d)$ to depend only (besides the spatial relation of c and d) on the (vectorial or scalar) intensities $f(c)$ and $f(d)$ at c and d, cf. (10.1) and (10.2). This restriction can be relaxed, yielding us scale-based and vectorial affinity.

In scale-based FC [26], instead of considering just c and d, a "local scale region" around each of c and d is considered in scene C for defining κ. In the ball scale approach, this local region around c is the largest ball b_c, centered at c, which is such that the image intensities at spels within b_c are homogeneous. For defining $\kappa(c, d)$ then, the intensities within b_c and b_d are considered. Typically a filtered value $f'(x)$ is estimated for each $x \in \{c, d\}$ from all intensities within b_x by taking their weighted average, the weight determined by a k-variate Gaussian function centered at $f(x)$. The filtered values $f'(c)$ and $f'(d)$ are then used in defining $\kappa(c, d)$ instead of the original intensities $f(c)$ and $f(d)$. In place of the ball, an ellipsoid has also been proposed for the scale region, which leads to the tensor scale approach [29]. The underlying idea in these approaches is to reduce the sensitivity of FC algorithms to spel-level random noise. Note that when local scales are used in this manner, none of the theoretical constructs of FC needs change. Actually, the scale-based approach can be seen as a preprocessing step: replace the original intensity function f with its scale-based filtered version f', and then proceed with the regular FC algorithm applied to the image $I' = \langle C, f' \rangle$ in place of $I = \langle C, f \rangle$.

In vectorial FC [6], the vectorial intensity function $f(x) \in \mathbb{R}^k$ is used in defining κ. For example, in such a case, (10.1) and (10.2) become k-variate

Gaussian functions (i.e., we apply k-variate Gaussian to a vector, like $f(c) - f(d)$, instead of simple Gaussian function to its length $||f(c) - f(d)||$). Obviously, the scale concept can be combined with the vectorial idea [6]. In fact, these two concepts can be individually or jointly combined with the principles underlying AFC, RFC, and IRFC.

10.6 Affinity Functions in Fuzzy Connectedness

An affinity function for an image $I = \langle \boldsymbol{C}, f \rangle$, with $\boldsymbol{C} = \langle C, \alpha \rangle$, is a function, say κ, defined on a set $C \times C$. More precisely, it is of importance only for the adjacent pairs $\langle c, d \rangle$, that is, from $\alpha \subset C \times C$. The affinity functions defined in (10.1) and (10.2) have the values in the interval $[0, 1]$, are symmetric (i.e., $\kappa(c, d) = \kappa(d, c)$ for all $c, d \in C$) and have the property that $\kappa(c, c) = 1$ for all $c \in C$. We will refer to any such affinity as a standard affinity.

In general, any linearly ordered set $\langle L, \preceq \rangle$ can serve as a range (value set) of an affinity[30]: a function $\kappa \colon C \times C \to L$ is an affinity function (into $\langle L, \preceq \rangle$) provided κ is symmetric and $\kappa(a, b) \preceq \kappa(c, c)$ for every $a, b, c \in C$. Note that $\kappa(d, d) \preceq \kappa(c, c)$ for every $c, d \in C$. So, there exists an element in L, which we denote by a symbol $\mathbf{1}_\kappa$, such that $\kappa(c, c) = \mathbf{1}_\kappa$ for every $c \in C$. Notice that $\mathbf{1}_\kappa$ is the largest element of $L_\kappa = \{\kappa(a, b) \colon a, b \in C\}$, although it does not need to be the largest element of L. Clearly, any standard affinity κ is an affinity function with $\langle L, \preceq \rangle = \langle [0, 1], \leq \rangle$ and $\mathbf{1}_\kappa = 1$. In the discussion below, $\langle L, \preceq \rangle$ will be either the standard range $\langle [0, 1], \leq \rangle$ or $\langle [0, \infty], \geq \rangle$. Note that, in this second case, the order relation $\preceq$ is the *reversed* standard order relation $\geq$.

10.6.1 Equivalent Affinities

We say that the affinities $\kappa_1 \colon C \times C \to \langle L_1, \preceq_1 \rangle$ and $\kappa_2 \colon C \times C \to \langle L_2, \preceq_2 \rangle$ are *equivalent* (in the FC sense) provided, for every $a, b, c, d \in C$

$$\kappa_1(a, b) \preceq_1 \kappa_1(c, d) \quad \text{if and only if} \quad \kappa_2(a, b) \preceq_2 \kappa_2(c, d).$$

For example, it can be easily seen that for any constants $\sigma, \tau > 0$ the homogeneity-based affinities ψ_σ and ψ_τ, see (10.1), are equivalent, since for any pairs $\langle a, b \rangle$ and $\langle c, d \rangle$ of adjacent spels: $\psi_\sigma(a, b) < \psi_\sigma(c, d) \iff ||f(a) - f(b)|| > ||f(c) - f(d)|| \iff \psi_\tau(a, b) < \psi_\tau(c, d)$. (Symbol $\iff$ means "if and only if.") Equivalent affinities can be characterized as follows, where $\circ$ stands for the composition of functions, that is, $(g \circ \kappa_1)(a, b) = g(\kappa_1(a, b))$ [31].

Theorem 5. *Affinities $\kappa_1 \colon C \times C \to \langle L_1, \preceq_1 \rangle$ and $\kappa_2 \colon C \times C \to \langle L_2, \preceq_2 \rangle$ are equivalent if and only if there exists a strictly increasing function g from $\langle L_{\kappa_1}, \preceq_1 \rangle$ onto $\langle L_{\kappa_2}, \preceq_2 \rangle$ such that $\kappa_2 = g \circ \kappa_1$.*

The FC objects, defined in the previous sections, have the same definition with the general notion of affinity, when the standard inequality '$\leq$' is replaced by '$\preceq$.' The implications of and our interest in equivalent affinities are well encapsulated by the next theorem, which says that any FC segmentation (AFC, RFC, or IRFC) of a scene C remains unchanged if an affinity on C used to get the segmentation is replaced by an equivalent affinity.

Theorem 6 ([31]). *Let $\kappa_1 \colon C \times C \to \langle L_1, \preceq_1 \rangle$ and $\kappa_2 \colon C \times C \to \langle L_2, \preceq_2 \rangle$ be equivalent affinity functions and let $\mathcal{S}$ be a family of non-empty pairwise disjoint subsets of C. Then for every $\theta_1 \prec_1 \mathbf{1}_{\kappa_1}$ in L_1, there exists a $\theta_2 \prec_2 \mathbf{1}_{\kappa_2}$ in L_2 such that, for every $S \in \mathcal{S}$ and $i \in \{0,1,2,\ldots\}$, we have $P^{\kappa_1}_{S\theta_1} = P^{\kappa_2}_{S\theta_2}$, $P^{\kappa_1}_{S\mathcal{S}} = P^{\kappa_2}_{S\mathcal{S}}$, and $P^{i,\kappa_1}_{S\mathcal{S}} = P^{i,\kappa_2}_{S\mathcal{S}}$.*
Moreover, if $g \colon C \to C$ is a strictly monotone function such that $\kappa_2 = g \circ \kappa_1$ (which exists by Theorem 5), then we can take $\theta_2 = g(\theta_1)$.

Keeping this in mind, it makes sense to find for each affinity function an equivalent affinity in a nice form:

Theorem 7 ([31]). *Every affinity function is equivalent (in the FC sense) to a standard affinity.*

Once we agree that equivalent affinities lead to the same segmentations, we can restrict our attention to standard affinities without losing any generality of our method.

Next, we like to describe the natural FC equivalent representations of the homogeneity-based ψ_σ (10.1) and object feature-based ϕ_σ (10.2) affinities. The first of them, $\psi_\sigma(c,d)$, is equivalent to an approximation of the magnitude of the directional derivative $|D_{\overrightarrow{cd}}f(c)| = \left|\frac{f(c)-f(d)}{||c-d||}\right|$ of f in the direction of the vector $\overrightarrow{cd}$. If spels c and d are adjacent when $||c-d|| \leq 1$, then for adjacent $c,d \in C$ we have $\psi(c,d) \stackrel{\text{def}}{=} |D_{\overrightarrow{cd}}f(c)| = |f(c) - f(d)|$. Such defined ψ is an affinity with the range $\langle L, \preceq \rangle = \langle [0,\infty], \geq \rangle$. The equivalence of ψ with ψ_σ is justified by Theorem 5 and the Gaussian function $g_\sigma(x) = e^{-x^2/\sigma^2}$, as $\psi_\sigma(c,d) = (g_\sigma \circ \psi)(c,d)$ for any adjacent $c,d \in C$.

The natural form of the object feature-based ϕ_σ affinity (for one object) and a spel c is the number $||f(c) - m||$, a distance of the image intensity $f(c)$ at c from the expected object intensity m. For two adjacent distinct spels, this leads to the definition $\phi(c,d) = \max\{||f(c)-m||, ||f(d)-m||\}$. We also put $\phi(c,c) = 0$, to insure that ϕ is an affinity function, with the range $\langle L, \preceq \rangle = \langle [0,\infty], \geq \rangle$. Once again, ϕ is equivalent with ϕ_σ, as $\phi_\sigma = g_\sigma \circ \phi$.

The homogeneity-based connectivity measure, $\mu_\psi = \mu_\psi^C$, can be elegantly interpreted if the scene $\mathcal{C} = \langle C, f \rangle$ is considered as a topographical map in which $f(c)$ represents an elevation at the location $c \in C$. Then, $\mu_\psi(c,d)$ is the highest possible step (a slope of f) that one must make in order to get from c to d with each step on a location (spel) from C and of unit length. In particular, the object $P^\psi_{s\theta} = \{c \in C \colon \theta > \mu_\psi(s,c)\}$ represents those spels

$c \in C$ which can be reached from s with all steps lower than θ. Note that all we measure in this setting is the actual change of the altitude while making the step. Thus, this value can be small, even if the step is made on a very steep slope, as long as the path approximately follows the altitude contour lines – this is why on steep hills the roads zigzag, allowing for a small incline of the motion. On the other hand, the measure of the same step would be large, if measured with some form of gradient induced homogeneity-based affinity! (Compare Sect. 10.7.2.)

The object feature-based connectivity measure of one object has also a nice topographical map interpretation. For understanding this, consider a modified scene $\bar{C} = \langle C, |f(\cdot) - m| \rangle$ (called membership scene in [1]) as a topographical map. Then the number $\mu_\phi(c,d)$ represents the lowest possible elevation (in $\bar{C}$) which one must reach (a mountain pass) in order to get from c to d, where each step is on a location from C and is of unit length. Notice that $\mu_\phi(c,d)$ is precisely the degree of connectivity as defined by Rosenfeld [32–34]. By the above analysis, we brought Rosenfeld's connectivity also into the affinity framework introduced by [1], particularly as another object feature component of affinity.

10.6.2 Essential Parameters in Affinity Functions

Next, let us turn our attention to the determination of the number of parameters essential in defining the affinities:

- *Homogeneity-based affinity* ψ_σ has no essential parameter, that is, the parameter σ in its definition is redundant, as $\psi_\sigma = g_\sigma \circ \psi$ is equivalent to ψ, which is independent of σ. This beautiful characteristic says that FC partitioning of a scene utilizing homogeneity-based affinity is an inherent property of the scene and is independent of any parameters, besides a threshold in case of AFC.
- *Object feature-based affinity* ϕ_σ *for one object* has two explicit parameters, m and σ, of which only parameter m is essential. Parameter σ is redundant, since $\phi_\sigma = g_\sigma \circ \phi$ is equivalent to ϕ defined above.
- *Object feature-based affinity* $\bar{\phi}_{\bar{\sigma}}$ *for $n > 1$ objects* is usually defined as $\bar{\phi}(c,d) = \max_{i=1,\ldots,n} \phi_{\sigma_i}(c,d)$ [28], where each ϕ_{σ_i} is defined by (10.2), with the parameter m replaced with the ith object average intensity m_i. Here $\bar{\sigma} = \langle \sigma_1, \ldots, \sigma_n \rangle$. This affinity has $2n$ explicit parameters, but only $2n - 1$ are essential. Indeed, if $\bar{\delta} = \langle 1, \delta_2, \ldots, \delta_n \rangle$, where $\delta_i = \sigma_i/\sigma_1$, then $\bar{\phi}_{\bar{\sigma}}$ and $\bar{\phi}_{\bar{\delta}}$ are equivalent, since $\bar{\phi}_{\bar{\delta}} = h_{\sigma_1} \circ \bar{\phi}_{\bar{\sigma}}$, where $h_\sigma(x) = x^{\sigma^2}$.

Similar results for the averages, additive and multiplicative, of ψ and ϕ, as well as their lexicographical order combination, can be found in [28].

10.7 Other Delineation Algorithms

We have already discussed deep similarities between FC and GC methods. In both cases, the image scene can be represented as weighted graphs (with different ways of assigning these weights) and the segmentations consist of different subsets P's of the graph vertices. In both cases, we associate with each object P in the graph its energy (cost) value $\boldsymbol{e}(P)$ represented in terms of the weights of edges in the boundary of P, that is, with one spel in P, another in its complement. The difference is the format of the energy cost function: in GC it is a sum of the weights of the boundary edges, while in FC it is the maximum among all these numbers.

10.7.1 Generalized Graph Cut

Despite the similarities, the segmentations resulting from FC and GC have different properties. For example, the FC segmentations are robust with respect to seed choice, but GC delineations are not. On the other hand, GC output smoothes the boundary of the resulting object (via penalizing long boundaries) – which is sometimes desirable – while FC have no such properties. An interesting problem was considered in [35]:

> "For what classes of graph energy cost functions $\boldsymbol{e}(P)$ (not necessarily defined in terms of the edge weights) can we find graph weights such that the GC optimization of the resulting graph is identical to the optimization of the original function $\boldsymbol{e}$?"

The necessary condition given there implies, in particular, that the maximum energy of FC cannot be represented that way. This also follows from the fact that FC and GC segmentations have different properties, like robustness.

It should be clear that, if one uses in FC an object feature-based affinity, then, under an interpretation of μ as Rosenfeld's degree of connectivity, the resulting segmented object is the water basin, as in the WS segmentations. If one desires more than one basin/object, then RFC results agree with the WS basin interpretation, as long as one "stops pumping the water" when a spill to another basin occurs.

At that point, we face a problem discussed in Sect. 10.4.1: should we leave the spels where competition breaks unassigned to any object, or should we find a way to redistribute them among the objects. In RFC, we opt for the first of these choices. In standard WS, the second option is followed by "building the dams" at the "mountain passes" where conflict occurs, and then continuing "land submerging" process until every spel is assigned. In other words, the outcome of WS can be viewed as the outcome of RFC used with proper object feature-based affinity, if we opt for leaving unassigned the spels where "ties" occur.

In summary, the FC, GC, and WS methods, to which we will refer here as Generalized Graph (GG) methods, can be viewed as the same class of segmentation methods, with their outcomes – resulting from the optimization of appropriate energy functions – obtained as segmentations of appropriate weighted graphs. This was clearly demonstrated above, if one chooses treating segmentation as an assignment of disjoint regions, when some spels belong to no object. In the other extreme, when the "spels with ties" are assigned according a proper (slightly ad hoc) procedures typical for each method, the GG algorithms are also equivalent. They all can be expressed in the IFT framework [20, 27, 36].

10.7.2 Level Set vs. Generalized Graph Cut

The relation of GG to LS is not straightforward. First of all, we will understand that the name relates to the image segmentation methods that have the following properties:

1. Set the segmentation problem in the continuous setting (i.e., images are defined on the regions Ω in the Euclidean space $\mathbb{R}^{\mathfrak{n}}$, usually with $\mathfrak{n} = 2$ or $\mathfrak{n} = 3$), solve it as such, and only at the last stage of method development, use discretization (i.e., finite approximation) of the continuous case to the digital image case
2. In the problem set-up, use an energy function e associating with image segmentation $\mathcal{P}$ its energy value $e(\mathcal{P})$
3. Usually (but not always) considered as a problem solution a segmentation $\mathcal{P}$ that minimizes e in an appropriate class of segmentations
4. Usually (but not always) the minimization is achieved by variational methods, which leads to a differential equation and returns a local minimum for e

Some optimization methods, like active contour (snake) [37] satisfy all these properties, but are not region-based methods, since they concentrate on finding only parts of a region boundary at a time. Some others actually do not explicitly optimize an energy (i.e., there is no clear Step 3), but it can be viewed as a solution for a variational problem (i.e., Step 4), that is, a solution for an implicitly given optimization problem [24]. Perhaps the most influential and prominent LS delineation method is that of Mumford and Shah [38], and its special case, due to Chan and Vese [39].

The biggest difference between such described LS methods and GG methods is the property (1) of LS: it makes a precise theoretical comparison between the methods difficult, and, at the purely discrete level, actually impossible. This is the case, since the precise outcome of LS is a segmentation of Ω, while the other methods return segmentations on a discrete scene C. If we try to compare LS and GG segmentations of a discrete scene C, then the comparison is between a precisely defined GG output and an unspecified *approximation* of the continuous LS segmentation, and any conclusion of such effort will be only approximate. Therefore, the only precise theoretical comparison between LS and GG segmentation methods can be made at the continuous level, that is, on the images defined on an Euclidean scene Ω. A natural approach how to relate the GG with the continuous output is described in [40].

For a continuous image $F\colon \Omega \to \mathbb{R}^k$ and a digital scene $C \subset \Omega$ let $F \restriction C$ be a digital image on C approximating F. We think about it here as a restriction of F (i.e., $(F \restriction C)(c) = f(c)$ for all $c \in C$). For a segmentation algorithm $\boldsymbol{A}$, let $\boldsymbol{A}(F \restriction C, \boldsymbol{p})$ be the output of $\boldsymbol{A}$ applied to the image $F \restriction C$ and some parameters $\boldsymbol{p}$, like seeds and cost function. We like to think of an $\boldsymbol{A}$-segmentation of the entire scene Ω of F as a result of application of $\boldsymbol{A}$ to the "image $F \restriction C$ obtained with infinite resolution." More formally, it will be understood as a limit $\boldsymbol{A}^*(F, \boldsymbol{p}) = \lim_{C \to \Omega} \boldsymbol{A}(F \restriction C, \boldsymbol{p})$ over all appropriate finite sets $C \subset \Omega$ [40]. In this set-up, we can say that a discrete segmentation algorithm $\boldsymbol{A}$ agrees (or is equivalent) at infinite resolution with a continuous (say, level set) segmentation model $\boldsymbol{M}$ in the class $\boldsymbol{F}$ of images $F\colon \Omega \to \mathbb{R}^k$ provided for every $F \in \boldsymbol{F}$ and appropriate parameter vector $\boldsymbol{p}$, the limit $\boldsymbol{A}^*(F, \boldsymbol{p})$ exists and is equal to a segmentation $\boldsymbol{M}(F, \boldsymbol{p})$ of Ω predicted by $\boldsymbol{M}$. In this sense, we have proved

Theorem 8. [40] *The FC delineation algorithm $\boldsymbol{A}_\nabla$ used with the gradient based affinity is equivalent, at infinite resolution, with a level set delineation model $\boldsymbol{M}_{\mathrm{LS}}$ from Malladi, Sethian, and Vemuri paper [24].*

Here, the gradient based affinity is a natural discretization of a notion of gradient (see [40]) similar in spirit to the homogeneity based affinity. We should stress that there are a few hidden elements in this theorem. First of all, we consider, after the authors of [24], the outcome of the model as the viscosity solution of the propagation problem, in which the curvature parameter used in the model goes to zero. In other words, the actual outcome of the model $\boldsymbol{M}_\nabla$ does not guarantees smoothness (in a curvature sense) of the boundary. This is the only way the equivalence with GG algorithms can be achieved (at least, with the energy functions we consider), as the graphs associated with the images consist only of the first order image intensity information (the weights of edges are based on the intensities of at most two adjacent spels, which can be viewed as an approximation of the first derivative of the intensity function), while the curvature is the second order (based on the second derivative) notion, which requires information of at least three spels to be defined [16].)

The strange twist of Theorem 8 is that, in fact, it tells us nothing on the level set algorithm $\boldsymbol{A}_{\mathrm{LS}}$, which is obtained as a discretization of the model $\boldsymbol{M}_{\mathrm{LS}}$. Although we proved [40] that the limit $\lim_{C \to \Omega} \boldsymbol{A}_\nabla(F \restriction C, \boldsymbol{p})$ exists and is equal to $\boldsymbol{M}_{\mathrm{LS}}(F, \boldsymbol{p})$, there is no formal prove in the literature that, for appropriate functions F, the limit $\lim_{C \to \Omega} \boldsymbol{A}_{\mathrm{LS}}(F \restriction C, \boldsymbol{p})$ exists or that it is equal to $\boldsymbol{M}_{\mathrm{LS}}(F, \boldsymbol{p})$. Although there are general results in the theory of differential equations indicating when a discretization of a differential equation converges to its continuous solution (in the $\boldsymbol{M}_{\mathrm{LS}}$ case, the discrete approximation of the level set function, property (iv), can indeed converge to the continuous level set function), such convergence implies neither the existence of the limit $\lim_{C \to \Omega} \mathbf{A}_{\mathrm{LS}}(F \upharpoonright C, \mathbf{p})$ nor, even if it exists, that it is equal to the continuous object indicated by the limiting surface. The story of other level sets algorithms is similar – there is a great ravine between their continuous, mathematical models and the associated discrete approximation algorithms, which should approximate the continuous models, but that are unknown (at least, theoretically) when they do so.

10.8 Medical Image Examples

The FC algorithms have been employed in segmenting medical Computed Tomography (CT), Magnetic Resonance Imaging (MRI), and ultrasound under various medical applications. They have also been used in non-medical image segmentation tasks. Our own application focus has been medical. These include:

- Delineation of gray matter, white matter, Cerebrospinal Fluid (CSF), lesions, diseased parts of white matter, and parcellations of these entities in different anatomic and functional regions of the brain via multi-protocol MRI for studying the multiple sclerosis disease (Fig. 10.3) and in elderly subjects to study aging-related depression and dementia
- Delineation of bone and soft tissue structures in CT images for craniomaxillofacial surgery planning (Fig. 10.4)
- Separation of arteries and veins in Magnetic Resonance Angiography (MRA) images (Fig. 10.5)
- Delineation of brain tumors in multi-protocol MRI (Fig. 10.6)
- Delineation of upper airway and surrounding structures in MRI for studying pediatric Obstructive Sleep Apnea (OSA) (Fig. 10.7)

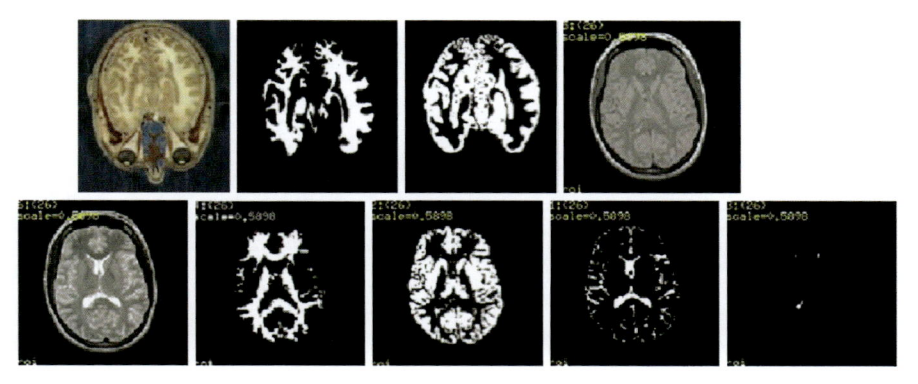

Fig. 10.3. *FC and AFC segmentation of brain images. Top*: Cross-sectional brain image from visible woman data set, white and gray matter segmentations via vectorial scale-based FC, and proton density-weighted MRI; *Bottom*: T2-weighted MRI, white matter, gray matter, CSF and lesion segmentations via AFC

Fig. 10.4. *Skin peeling via AFC segmentation. Left*: volume rendering from CT image of a patient with mid facial clefts; *Right*: Result after skin peeling

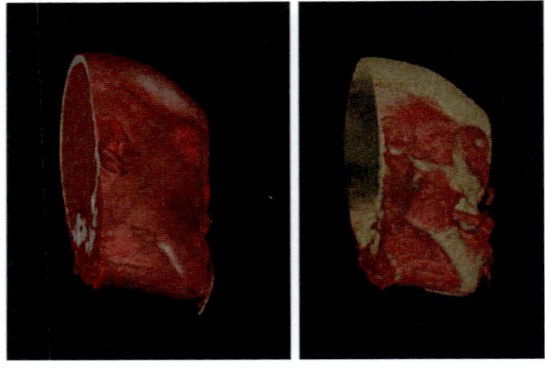

Fig. 10.5. *Vessel separation via IRFC. Left*: A segment of the peripheral vascular tree from MRA; *Right*: arteries and veins separated via IRFC

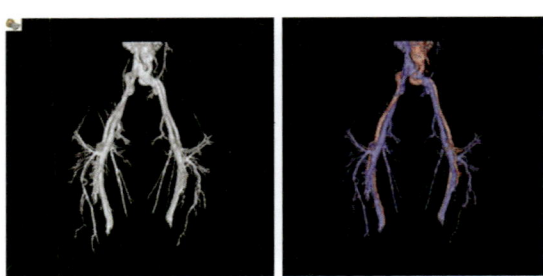

The need for image segmentation in medical applications arises from our desire to:

1. Characterize and quantify a disease process
2. Understand the natural course of a disease
3. Study the effects of a treatment regimen for a disease
4. Guide therapeutic procedures

In our applications, the motivation came from 1 to 3. The performance of the different FC algorithms has been evaluated in these applications quite extensively; please refer to the specific application related papers cited in [41]. The reasons for choosing FC in these applications are mainly three-fold:

- We are intimately familiar with the FC technology, have the full resources of its implementation, and have the expertise for optimally utilizing them in medical applications. These are crucial requirements for the optimal use of any segmentation algorithm.
- Among comparable other families of algorithms (such as graph cuts, watershed, level sets), FC constitutes one of the fastest groups of algorithms.
- The FC formulation is entirely digital starting from first principles, and so there are no ad hoc/unspecified continuous-to-digital conversion issues.

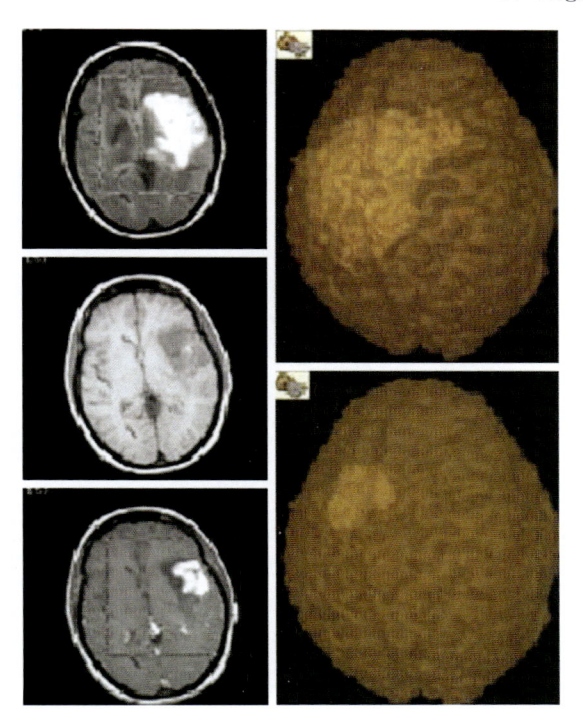

Fig. 10.6. *Tumor segmentation. Left column*: FLAIR and T1 weighted MRI without and with contrast agent; *Right column*: The edematous tumor region segmented via AFC from the FLAIR image and from the subtracted (post from pre-contrast) image showing enhancing aspects of the tumor

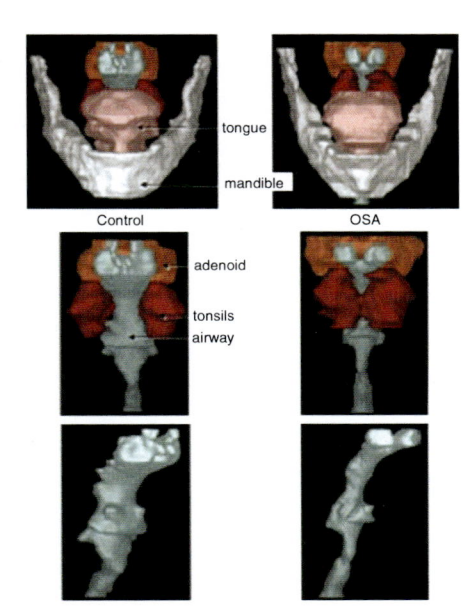

Fig. 10.7. *Surface rendering of AFC-segmented MRI. Left*: Upper airway and other surrounding structures (mandible, adenoid, tonsils, tongue) of a normal child; *Right*: a child with OSA

10.9 Concluding Remarks

Focusing mainly on FC methods, we have presented a unified mathematical theory wherein four currently predominant, purely image-based approaches – GC, WS, LS, and FC – are described in a single framework as energy optimization methods in image segmentation. Among these, LS has a continuous formulation and poses some challenges, unenunciated in the literature, on how to reconcile it with the eventual computational/algorithmic requirements of discretization. The remaining – GC, WS, FC – have an inherently discrete formulation and lend themselves naturally to combinatorial optimization solutions. The unifying treatment has helped us in delineating the similarities and differences among these methods and in pinpointing their strengths and weaknesses.

All segmentation methods rely on a (local) attribute functional of some sort – we called them affinity for FC and edge cost in general – for transforming intensity information into contributions to the energy functional. The notion of equivalent affinities is useful in characterizing the distinct and unique aspects of this function that have a real impact on the energy functional. Such an analysis can also be carried out for the attribute functionals of GC, WS, and LS, and of any other segmentation methods, although this does not seem to have been done (cf., [35]). Its consequence on nailing down the real independent parameters of a segmentation algorithm has implications in setting the segmentation algorithm optimally for a given application domain and in evaluating its robustness to parameter settings.

In all FC developments so far, for theoretical and algorithmic simplicity, only 2-ary fuzzy relations have been considered (meaning, affinity and connectedness have been considered only between two spels). Further, in the composition of fuzzy relations such as ψ_σ and ϕ_σ (for a given object and for all objects), only union and max-min constructs have been employed for the same reasons. Relaxing these restrictions may lead to new, more powerful and effective algorithms. For example, m-ary relations can be defined by considering all spels in the local scale region. Further, considering fuzzy relations as both fuzzy subsets of the scene and as m-ary relations ($m \geq 2$), various fuzzy subset operations (e.g., algebraic union, product) and compositing operations (e.g., max-star, sum-min, sum-product, algebraic sum-min) can also be used. Prior object shape and appearance fuzzy models can also be brought into this realm. These require considerable theoretical, algorithmic, and application related work.

References

1. Udupa JK, Samarasekera S. Fuzzy connectedness and object definition: theory, algorithms, and applications in image segmentation. Graph Models Image Process. 1996;58(3):246–61.

2. Saha PK, Udupa JK. Relative fuzzy connectedness among multiple objects: theory, algorithms, and applications in image segmentation. Comput Vis Image Underst. 2001;82(1):42–56.
3. Udupa JK, Saha PK, Lotufo RA. Relative fuzzy connectedness and object definition: theory, algorithms, and applications in image segmentation. IEEE Trans Pattern Anal Mach Intell. 2002;24:1485–1500.
4. Saha PK, Udupa JK. Iterative relative fuzzy connectedness and object definition: theory, algorithms, and applications in image segmentation. Proc IEEE Workshop Math Methods Biomed Image Anal. 2002; p. 28–35.
5. Ciesielski KC, Udupa JK, Saha PK, et al. Iterative relative fuzzy connectedness for multiple objects, allowing multiple seeds. Comput Vis Image Underst. 2007;107(3):160–182.
6. Zhuge Y, Udupa JK, Saha PK. Vectorial scale-based fuzzy connected image segmentation. Comput Vis Image Underst. 2006;101:177–93.
7. Carvalho BM, Gau CJ, Herman GT, et al. Algorithms for fuzzy segmentation. Pattern Anal Appl. 1999;2:73–81.
8. Herman GT, Carvalho BM. Multiseeded segmentation using fuzzy connectedness. IEEE Trans Pattern Anal Mach Intell. 2001;23:460–74.
9. Carvalho BM, Herman GT, Kong YT. Simultaneous fuzzy segmentation of multiple objects. Discrete Appl Math. 2005;151:65–77.
10. Pednekar A, Kakadiaris IA. Image segmentation based on fuzzy connectedness using dynamic weights. IEEE Trans Image Process. 2006;15(6):1555–62.
11. Fan X, Yang J, Cheng L. A novel segmentation method for MR brain images based on fuzzy connectedness and FCM. Lect Notes Comput Sci. 2005;3613:505–13.
12. Boykov Y, Veksler O, Zabih R. Fast approximate energy minimization via graph cuts. IEEE Trans Pattern Anal Mach Intell. 2001;23(11):1222–39.
13. Boykov Y, Jolly M. Interactive graph cuts for optimal boundary & region segmentation of objects in N-D images. Proc ICCV. 2001;I:105–12.
14. Boykov Y, Kolmogorov V. Computing geodesics and minimal surfaces via graph cuts. Proc ICCV. 2003;I:26–33.
15. Boykov Y, Kolmogorov V. An experimental comparison of min-cut/max-flow algorithms for energy minimization in vision. IEEE Trans Pattern Anal Mach Intell. 2004;26:1124–37.
16. Boykov Y, Funka-Lea G. Graph cuts and efficient N-D image segmentation. Int J Comput Vis. 2006;70:109–31.
17. Boykov Y, Kolmogorov V, Cremers D, et al. An integral solution to surface evolution PDEs via geo-cuts. Lect Notes Comput Sci. 2006;3953:409–22.
18. Boykov Y, Veksler O. Graph cuts in vision and graphics: theories and applications. In: Paragios N, Chen Y, Faugeras O, editors. Handbook of Mathematical Models and Computer Vision. Berlin: Springer; 2006. p. 79–96.
19. Shi J, Malik J. Normalized cuts and image segmentation. IEEE Trans Pattern Anal Mach Intell. 2000;22:888–905.
20. Miranda PAV, Falcao AX. Links between image segmentation based on optimum-path forest and minimum cut in graph. J Math Imaging Vis. 2009;35:128–42.
21. Beucher S. The watershed transformation applied to image segmentation. Proc 10th Pfefferkorn Conf Signal Image Process Microsc Microanal. 1992; p. 299–314.

22. Shafarenko L, Petrou M, Kittler J. Automatic watershed segmentation of randomly textured color images. IEEE Trans Image Process. 1997;6:1530–44.
23. Park J, Keller J. Snakes on the watershed. IEEE Trans Pattern Anal Mach Intell. 2001;23:1201–05.
24. Malladi R, Sethian J, Vemuri B. Shape modeling with front propagation: a level set approach. IEEE Trans Pattern Anal Mach Intell. 1995;17:158–75.
25. Sethian JA. Fast marching methods and level sets methods. Evolving Interfaces in Computational Geometry, Fluid Mechanics, Computer Vision, and Materials Science. Cambridge: Cambridge University Press; 1999.
26. Saha PK, Udupa JK, Odhner D. Scale-based fuzzy connectedness image segmentation: theory, algorithms, and validation. Computer Vis Image Underst. 2000;77:145–74.
27. Falcao AX, Stolp J, Lotufo RA. The image foresting transform: theory, algorithms, and applications. IEEE Trans Pattern Anal Mach Intell. 2004;26(1):19–29.
28. Ciesielski KC, Udupa JK. Affinity functions in fuzzy connectedness based image segmentation II: defining and recognizing truly novel affinities. Comput Vis Image Underst. 2010;114:155–66.
29. Saha PK. Tensor scale: a local morphometric parameter with applications to computer vision and image processing. Comput Vis Image Underst. 2005;99:384–413.
30. Ciesielski K. Set theory for the working mathematician. No. 39 in London Mathematical Society Students Texts. Cambridge: Cambridge University Press; 1997.
31. Ciesielski KC, Udupa JK. Affinity functions in fuzzy connectedness based image segmentation I: equivalence of affinities. Comput Vis Image Underst. 2010;114:146—54.
32. Rosenfeld A. Fuzzy digital topology. Inf Control. 1979;40:76–87.
33. Rosenfeld A. On connectivity properties of grayscale pictures. Pattern Recognit. 1983;16:47–50.
34. Rosenfeld A. The fuzzy geometry of image subsets. Pattern Recognit Lett. 1984;2:311–17.
35. Kolmogorov V, Zabih R. What energy functions can be minimized via graph. IEEE Trans Pattern Anal Mach Intell. 2004;26(2):147–59.
36. Audigier R, Lotufo RA. Duality between the watershed by image foresting transform and the fuzzy connectedness segmentation approaches. Proc 19th Brazilian Symposium Comput Graph Image Process. 2006.
37. Kass M, Witkin A, Terzopoulos D. Snakes: active contour models. Int J Comput Vis. 1987;1:321–31.
38. Mumford D, Shah J. Optimal approximations by piecewise smooth functions and associated variational problems. Commun Pure Appl Math. 1989;42:577–685.
39. Chan TF, Vese LA. Active contours without edges. IEEE Trans Image Process. 2001;10:266–77.
40. Ciesielski KC, Udupa JK. A general theory of image segmentation: level set segmentation in the fuzzy connectedness framework. Proc SPIE. 2007;6512.
41. Udupa JK, Saha PK. Fuzzy connectedness in image segmentation. Proc IEEE. 2003;91:1649–69.

11
Model-Based Segmentation

Tobias Heimann and Hervé Delingette

Summary. This chapter starts with a brief introduction into model-based segmentation, explaining the basic concepts and different approaches. Subsequently, two segmentation approaches are presented in more detail: First, the method of deformable simplex meshes is described, explaining the special properties of the simplex mesh and the formulation of the internal forces. Common choices for image forces are presented, and how to evolve the mesh to adapt to certain structures. Second, the method of point-based statistical shape models (SSMs) is described. The model construction process is explained and the point correspondence problem is treated in detail. Several approaches of how gray level appearance can be modeled are presented, and search algorithms that use this knowledge to segment the modeled structures in new images are described.

11.1 Introduction

Automatic medical image segmentation is such a challenging task because it involves highly variable objects that have to be extracted from images of very low quality. Often, lack of contrast or artifacts lead to missing data in the image to be segmented, and the boundary of the Object of Interest (OOI) cannot be determined from local information as edges or region homogeneity alone. Medical experts are still able to delineate the object because they know what it is supposed to look like: they have a model of the object in their mind, they have an a priori knowledge about its shape and appearance. Model-based segmentation methods strive to translate this knowledge into smart algorithms that have a prior knowledge about the structures of interest. Those methods can be qualified as *top down* and usually consists of two stages. The former initializes the location and appearance of the model. It is based on user input (mouse clicks in an image), the registration of an atlas or machine learning approaches. The latter optimizes the shape and appearance of the model such that it closely matches the ones measured in the images.

As the shape of a model is optimized during the segmentation process, one important property of model-based segmentation lies in the choice of

T.M. Deserno (ed.), *Biomedical Image Processing*, Biological and Medical Physics, Biomedical Engineering, DOI: 10.1007/978-3-642-15816-2_11,
© Springer-Verlag Berlin Heidelberg 2011

the deformation strategy. There are two basic ways of deforming a shape: by deforming the embedding space of a shape and by modifying their parameters or Degree of Freedom (DOF). For instance, applying a rigid transform (i.e., rotation and translation) lies in the first category while modifying the vertex position of a mesh lies in the second category. Of course, those two mechanisms are often combined but their formalism and approaches are fairly different. Deforming a shape by deforming the embedding space can be qualified as a *registration approach* whereas modifying the parameters of a shape can be qualified as a *deformable model approach*. The geometric representation of the models is specifically important in the latter case. In Fig. 11.1, we show a coarse taxonomy of possible geometric representations. Each representation having pros and cons, the choice of a representation is often problem dependent. In Sect. 11.2, we detail the deformable model formulation based on simplex meshes. Among others, common representations are level-sets, triangulated meshes and spline curves or surfaces.

If many segmentations of a specific object are available, there are further possibilities to make models more specific. In addition to knowledge about an average template shape, Statistical Shape Models (SSMs) also include some knowledge about the principal modes of variation. The SSMs presented in Sect. 11.3 are based on the Point Distribution Model (PDM). To perform statistical analysis on these shapes, point correspondences have to be known: a challenging problem, for which just recently automatic methods have been proposed. The visual appearance of the object can also be captured by statistics on training samples, using techniques from computer vision and machine learning. Finally, there exist specific algorithms to employ the statistical information to drive a model towards an instance of the object in an image.

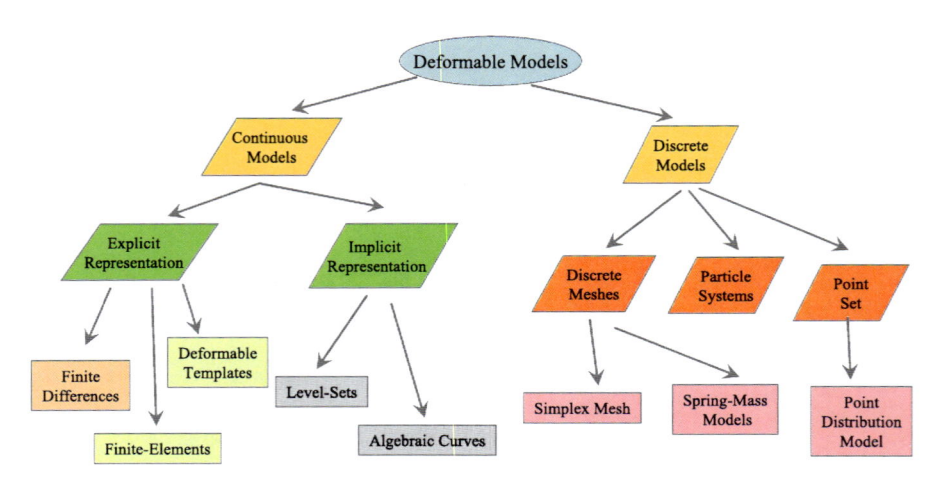

Fig. 11.1. *Taxonomy of deformable models.* The classification scheme of deformable models for medical image segmentation is based on their geometric representation

To conclude the chapter, we compare the concepts of a deformable simplex mesh and the SSM and give hints about which technique to use under which circumstances.

11.2 Deformable Simplex Meshes

In a deformable model approach, the shape of a model is optimized in order to match that of a structure of interest in an image. This technique has been pioneered in 1987 by Terzopoulos et al. [1] with the introduction of active contours or *snakes* [2]. This has been later generalized to active surfaces [3] but one difficulty arises when dealing with three-dimensional (3D) surfaces: the continuous parameterization of surfaces. Indeed, even for surfaces having the same topology as the sphere, it is not straightforward to define a C^1 parametric surface.

In order to avoid this issue, authors have proposed to work with implicit parameterization (e.g., using level-sets function) or with C^0 continuous meshes (e.g., using triangulated or simplex meshes). In the latter case, shape smoothing or regularization cannot be defined in the continuous domain (since the representation is not even C^1) but must be defined in a discrete fashion.

A simplex mesh is an original mesh representation that has been introduced in 1992 [4] as a model-based segmentation method for range data and volumetric images. A k-simplex meshes can represent a manifold surface of dimension k ($k = 2$ for 3D surfaces) and can be defined as a $k + 1$-cell. We provide below a recursive definition of a cell (Fig. 11.2).

Definition 1. *A 0-cell is defined as a vertex P and a 1-cell as an edge, i.e., an unordered pair of distinct vertices (P, M). We then recursively define a p-cell ($p \geq 2$) $\mathcal{C}$ as the union of c $(p-1)$-cells, $c \in \mathbb{N}$, such that:*

1. *Each vertex belonging to $\mathcal{C}$ also belongs to p distinct $(p-1)$-cells.*
2. *A $(p-2)$-cell belongs to 2 and only 2 $(p-1)$-cells.*
3. *The intersection of 2 $(p-1)$-cells is empty or is a $(p-2)$-cell.*
4. *A p-cell is simply connected, i.e. that given 2 vertices of that cell, there exists at least one set of edges that connect those 2 vertices.*

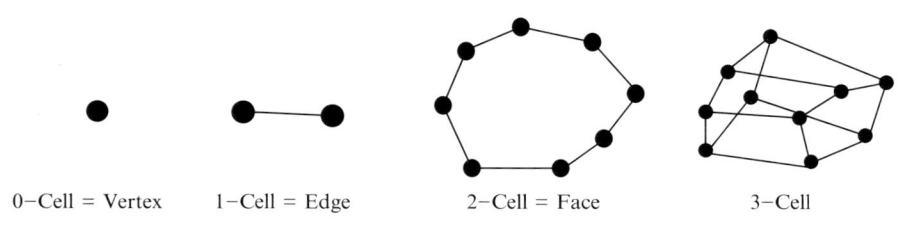

0−Cell = Vertex   1−Cell = Edge   2−Cell = Face   3−Cell

Fig. 11.2. *Examples of p-cells. A collection of p-cells is given for $0 \leq p \leq 3$. An edge is a 0-simplex mesh, a face is a 1-simplex mesh, and a 3-cell a 2-simplex mesh*

Two fundamental topological properties result from this definition. First, a k-simplex mesh has a fixed vertex connectivity: every vertex is adjacent to $k + 1$ distinct vertices. Second, a k-simplex mesh is a topological dual of a k-triangulation. In this duality relationship, a triangle is associated with a vertex of a 2-simplex mesh, a triangulation edge with an edge and a vertex with a face (a 2-cell).

In a deformable 2-simplex mesh for image segmentation, the vertex position $\mathbf{P}_i$ evolves over time under the influence of internal $\mathbf{f}_i^{\text{int}}$ and external forces $\mathbf{f}_i^{\text{ext}}$ capturing the influence of prior knowledge on shape and appearance, respectively

$$\mathbf{P}_i^{t+1} = \mathbf{P}_i^t + (1-\delta)(\mathbf{P}_i^t - \mathbf{P}_i^{t-1}) + \alpha \mathbf{f}_i^{\text{int}} + \beta \mathbf{f}_i^{\text{ext}} \qquad (11.1)$$

A description of those forces are the topic of the next two sections.

11.2.1 Internal Forces on Simplex Meshes

Local Internal Forces

There are two types of prior knowledge that can be applied. A *weak* shape prior consists in assuming that the structure of interest is smooth, which is a reasonable assumption for many anatomical structures (liver, heart, bones...) but not all of them (e.g., infiltrating tumor, brain gray matter). With a *strong* shape prior, it is assumed that the structure of interest has a typical shape, i.e. that its shape varies to a "small extent" around a reference shape.

Simplex meshes are well suited to enforce both types of priors. Indeed, one can define at each vertex of a 2-simplex mesh, discrete geometric quantities that locally control the shape of the mesh. At vertex $\mathbf{P}_i$ those quantities are the three metric parameters $(\epsilon_{1i}, \epsilon_{2i}, \epsilon_{3i})$ and the simplex angle ϕ_i (Fig. 11.3, left). Metric parameters control the relative spacing of vertices through a tangential internal force while the simplex angle controls the local surface smoothness through a normal internal force. By combining those two tangential and normal components of the internal force, it is possible to define weak and strong shape priors in a simple way.

For instance, Fig. 11.3 (right) shows the effect of a specific tangential force that leads to a concentration of vertices at parts of high curvatures. In Fig. 11.4, the tangential component impose metrics parameters equal to $1/3$, while the normal component of the internal forces smooths the surface with normal and mean curvature continuity. In some restricted cases, discrete mean curvature on a simplex mesh has been proved [5] to converge towards its continuous value.

The straightforward definition of forces imposing mean curvature continuity is a major advantage of simplex meshes. Indeed, with such regularizing constraint, no shrinking effect is observed unlike the mean curvature flow [6] widely used in the level-sets method [7]. With those flows minimizing surface

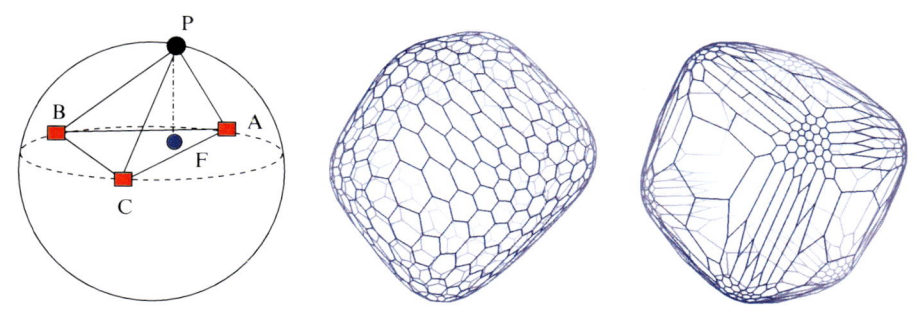

Fig. 11.3. *Metric parameters. Left*: point P and its three neighbors on a 2-simplex mesh define a circumscribed sphere. The simplex angle in P is related to the height $\|FP\|$ of point P with respect to triangle A, B, C such that the angle stays constant when P lies on the circumsphere. The metric parameters ($\epsilon_1, \epsilon_2, \epsilon_3$) are simply the barycentric coordinates of F in the triangle A, B, C; *Middle*: cube simplex mesh with metric parameters equal to $1/3$; *Right*: mesh with optimized metric parameters so that vertices are concentrated at parts of high curvature

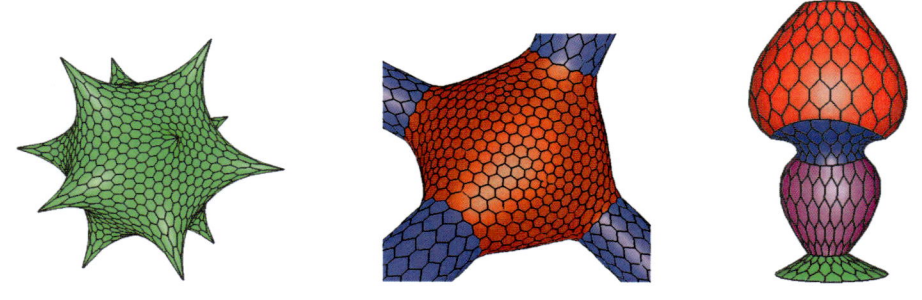

Fig. 11.4. *Regularization of a simplex mesh. Left*: Simplex mesh with zero mean curvature; *Middle*: The red part is smoothed with a mean curvature continuity to the four cylinders; *Right*: Normal continuity between the surface and 3D contours can also be enforced

areas, it is often necessary to limit the shrinking effect by adding a "balloon force", which may lead to additional problems [8].

To provide a strong shape prior, a specific internal force has been proposed that constrains the metrics parameters and simplex angle to be equal to some user-defined values corresponding to a reference shape. Under the influence of this shape-memory force, a simplex mesh is attracted towards a reference shape up to a translation, rotation and scale. This local representation of shape is also a unique feature of simplex meshes compared to other representations.

11.2.2 Image Forces

Image forces drive a deformable simplex mesh towards the apparent boundary of a structure of interest and rely on some prior knowledge about the appearance of that structure. In all cases, the external force is formulated as displacement in the normal direction towards the closest boundary point $\mathbf{C}_i$ of the structure: $\mathbf{f}_i^{\text{ext}} = ((\mathbf{C}_i - P_i) \cdot n_i)n_i$. Again one can rely on a weak appearance prior that assumes generic boundary features (such as gradient maxima) or on a strong appearance prior that relies on machine learning trained on segmented dataset. In fact, there is a continuum of priors between those two extremes:

- *Contour prior*: The closest boundary points are searched among the maxima of the gradient norm in the gradient direction. Additional criteria may be used to select the proper boundary point such as intensity values or the dot product between the surface normal n_i and the gradient vector at that point [9].
- *Contour and region prior*: The principle is to characterize the inner (or outer) region by a range of intensity. The boundary is then determined as the voxel located next to this region whose gradient norm is greater than a threshold [10].
- *Region histogram prior*: From a collection of segmented structure, one can build the probability density function that a pixel belongs to the structure from the region histogram (using Parzen windowing for instance). The closest point is then determined as the one which maximizes (resp. minimizes) the probability for the inner (resp. outer) voxels to belong to the inner region [11].
- *Intensity profile matching prior*: At each mesh vertex, an intensity profile along the normal direction is extracted. The external force is computed by searching along the normal direction the displacement which maximizes the correlation with a reference intensity profile. That reference profile may result from a Principal Component Analysis (Fig. 11.5) (PCA) as described in Sect. 11.3.4. Several criteria may be used such as the sum of the square differences, the linear correlation or the Mahalanobis distance.
- *Block matching prior*: This approach is similar to the intensity profile approach but based on the matching of image blocks rather than intensity profiles. The main difference is that the search for the closest point is no longer constrained to be along the normal direction but may be in all directions. This is of interest when homologous points can extracted from images (such as points of high curvatures or anatomically relevant features).
- *Texture classification prior*: The image force can be applied after performing some image classification based on machine learning methods. The output of the classification is a map of the probability that a voxel belongs to the structure of interest. The closest point along the normal direction whose probability is equal to 50 % may then be used to compute the image forces. Linear classifiers, support vector machines [12] or even neural networks [13] can be trained on segmented datasets to classify the image.

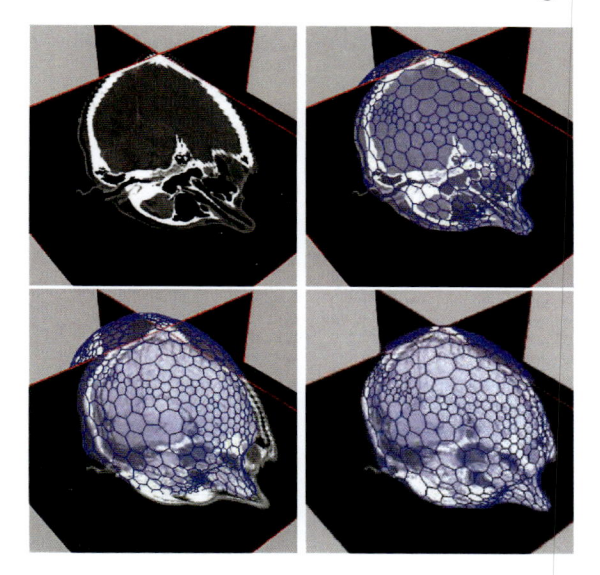

Fig. 11.5. *Deformable simplex meshes.* The meshes are driven by the matching of intensity profiles; *Upper row*: reference CT image scan (*left*) and head segmented with a simplex mesh (*right*). *Lower row*: an intensity profile is stored at each vertex as a reference and this mesh is used to register the head of the same patient on an MRI: mesh position before (*left*) and after rigid registration (*right*)

11.2.3 Globally Constrained Deformation

As mentioned previously, the first stage of model-based segmentation consists in positioning the deformable model in the image through various methods, manual or automatic. Due to the convergence to a local minimum of the functional, the second stage corresponding to the optimization of the model parameters may not give good segmentation results if the initialization is not close enough to the final shape. One common strategy to decrease the sensitivity of the optimal shape to the initialization position is to use a coarse-to-fine framework. In this approach, the complexity (in terms of DOF) of the model is slowly increased in order to start the optimization with a small number of local minima and then track the global optimum during the increase of complexity.

A first method to control the DOF consists in using a multi-resolution approach where a coarse mesh is deformed on a given image and replaced with a finer mesh when convergence is reached. This technique is fairly straightforward to implement but requires to create meshes of increasing resolution which is not trivial on unstructured meshes [14]. On simplex meshes, the $\sqrt{3}$ subdivision scheme is well suited to globally or even locally refine the mesh [15]. One limitation of multi-resolution schemes is that the coarser mesh

Table 11.1. *Geometric transforms and constrains.* The table shows geometric transforms that can constrain the deformation of a simplex mesh

Transform	Number of DOF	Transformation type
Rigid	6	Global
Similarity	7	Global
Affine	12	Global
B-spline	3 × Number of control points	Restricted to a bounding box
Axis-symmetric	3 × Number of points on the central line	Restricted to the mesh
PCA	6 + Number of modes	Restricted to the mesh

may still have too many DOF to prevent falling into a local mimimum of the functional.

An alternative is to use globally constrained deformations [16], where global spatial transforms are used in the early stages of segmentation in order to constrain the DOF. Table 11.1 lists some transforms that may be applied with a growing number of DOF. Starting from a rigid transform, it is possible to use only six DOF which greatly constrain the optimization problem.

However, is it often difficult to smoothly increase the number of DOFs, except when using the PCA (Sect. 11.3.3). To this end, the concept of globally constrained deformation has been introduced [16] that can tune the complexity of the deformation with a locality parameter λ. More precisely, the motion equation (11.1) is refined by adding a global force $\mathbf{f}_i^{\text{global}} = \mathcal{T}(\mathbf{P}_i^t) - \mathbf{P}_i^t$ computed from the application of a global transform $\mathcal{T}$

$$\mathbf{P}_i^{t+1} = \mathbf{P}_i^t + (1-\delta)(\mathbf{P}_i^t - \mathbf{P}_i^{t-1}) + \lambda \left(\alpha \mathbf{f}_i^{\text{int}} + \beta \mathbf{f}_i^{\text{ext}} \right) + (1-\lambda) \mathbf{f}_i^{\text{global}} \quad (11.2)$$

When $\lambda = 0$ the simplex mesh deforms according to the application of a global transformation (with few DOFs) whereas when $\lambda = 1$, the deformation is only driven by local internal and external forces. When $0 < \lambda < 1$ the deformation is a trade-off between global and local behaviors corresponding to an intermediate number of DOFs. The transformation $\mathcal{T}$ is computed such that it minimizes the discrepancy between the vertex position $\mathbf{P}_i^t$ and its closest point on the image $\mathbf{C}_i^t$.

In many cases, it is sufficient to apply global transforms from rigid to affine and then to slowly increase the locality parameter from 0 to 1 or to a smaller value (Fig. 11.6). The concept of globally constrained deformation may also be successfully applied to enforce to some degree the axis-symmetry around a deformable axis (for instance to segment vessels) or to take into account the statistical shape variation around a mean shape.

11.2.4 3D+t Deformable Simplex Meshes

For the analysis of time series of medical images, e.g., gated cardiac imaging, it is important to not only to segment a structure but also to track its

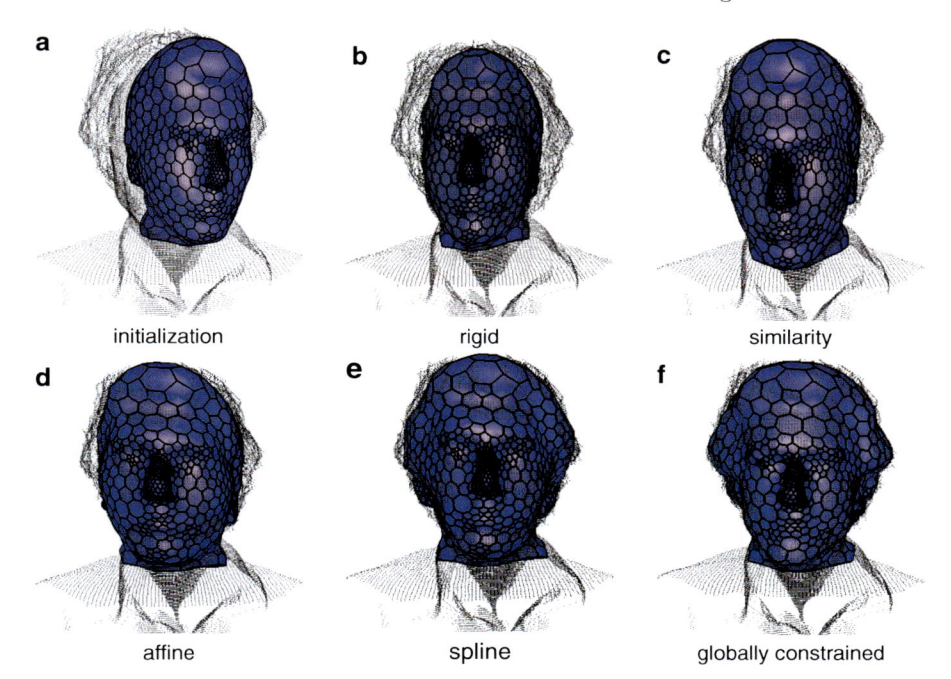

Fig. 11.6. *Coarse-to-fine strategy of 3D mesh reconstruction.* The coarse to fine strategy runs from clouds of points: the different stages of the deformations with increasing deformation complexity are shown ending with a globally constrained deformation ($\lambda = 0.3$ in panel (f))

deformation over time. A common approach consists in using the result of the segmentation at time t as the initialization of the segmentation at time $t + 1$. However this approach is prone to the propagation of errors if one image is especially challenging and tends to bias the result noticeably by underestimating the motion of the structure.

Rather than processing the images one after the other, 3D+T deformable models [17] can be used to process all images of the time series at once (if they can fit in the memory of the computer). In this framework, a family of simplex meshes having the same topology (Fig. 11.7) are deformed on each image of the time series, the mesh deformation at time t being coupled with that of time $t - 1$ and $t + 1$.

The temporal coupling forces act as additional internal forces and can be formulated in a weak or a strong way. The weak prior consists in minimizing the kinetic energy of the system and thus tries to make vertex $\mathbf{P}_i^t$ close to both points $\mathbf{P}_i^{t-1}$ and $\mathbf{P}_i^{t+1}$. With a strong motion prior, the shape of the trajectories of each vertex are constrained to be similar to some reference trajectory curves, described with geometric quantities that are invariant to translation, rotation and scale. Examples of segmentation of time-series of echocardiographic images are shown in Fig. 11.8.

Fig. 11.7. *3D+T simplex mesh.* The array of meshes has the same topology but different geometries

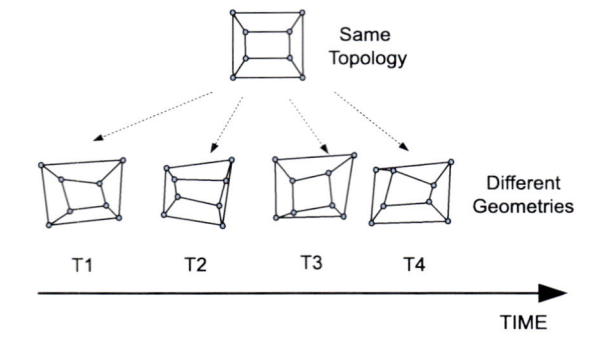

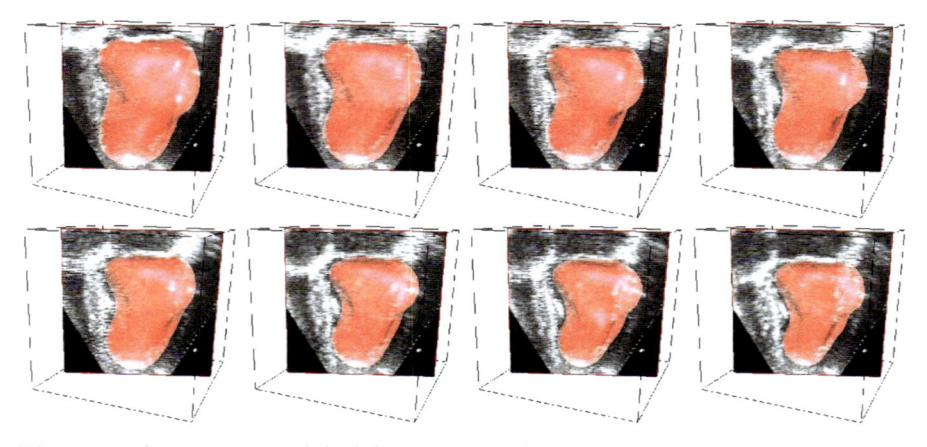

Fig. 11.8. *Segmentation of the left ventricle endocardium.* Segmentation from 3D+T echo-cardiographic images by 3D+T simplex meshes

11.2.5 Advanced Segmentation Strategies

The segmentation of anatomical structures from medical images is often a difficult problem due to their low contrast with surrounding structures, the inter-subject variability, the presence of pathologies, etc. Problem specific segmentation strategies must be defined in order to achieve a robust and accurate delineation.

Atlas-Based Initialization

The first stage of model-based segmentation consists in finding the location of the structure of interest in the image. This can be done manually by the user but the inter-user variability in providing this information may be very large and may impair the reproducibility of the method. An interesting alternative is to use non-rigid registration [12] to find a geometric transformation between a reference image (or an atlas) and the target image. In the reference image, meshes of the structures of interest can then be mapped by this transformation in the space of the target image and finally serve as initial shape for model-based segmentation. This approach has the advantage of being automatic (and therefore reproducible) and usually provides good model initializations. However, when the structure location varies greatly, it is required to provide a suitable initialization of the non-rigid registration algorithm, which can be achieved by performing some coarse exhaustive search [18].

Multi-Structure Segmentation

It is often a good strategy to first segment the most contrasted structures and then proceed with more challenging ones (least contrasted). Indeed, already delineated structures can restrict the possible shape of neighboring structures during the model-based segmentation of other structures. Furthermore, when boundaries between structures are hardly visible, it is best to segment both structures at the same time, the shape of the former providing inter-penetration constraints [19] for the latter. Those constraints are based on the computation of distance maps measuring the distance of a given points to the two surface meshes. Figure 11.9 shows an example of the joint segmentation of bladder and prostate.

Rule-Based Segmentation

Ideally, all model-based segmentation algorithms should proceed in two stages: initialization followed by an optimization stage. However, most segmentation algorithms are meant to be applied on patients with some pathologies (e.g., tumor, metastasis) or some rare anatomical configurations (e.g., abnormal branching of vessels, already ablated tissue). In such case, it is difficult to build a shape and appearance model that can cope with such variation of

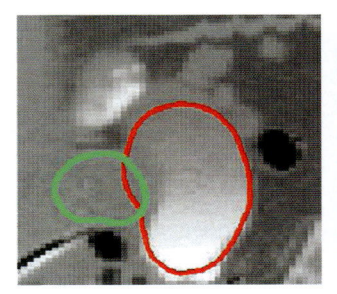

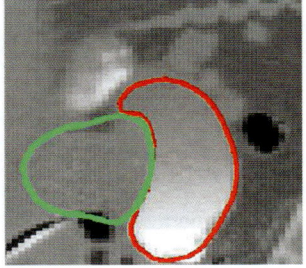

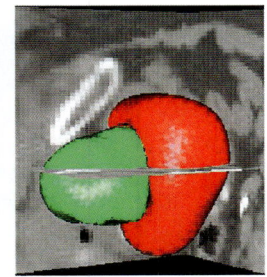

Fig. 11.9. *Joint segmentation of bladder and prostate.* The bladder (*red*) and prostate (*green*) are shown in a CT image. *Left*: No inter-penetration constraints are used which causes a leakage; *Middle*: With inter-penetration constraint and some statistical shape constraint on the prostate, the segmentation is improved despite the low contrast between structures; *Right*: 3D view of the two organs

cases. Therefore, to improve the robustness of the segmentation, some alert mechanisms should be included in order to detect that the segmentation is not performing well. For instance, for segmentation of hippocampi in brain MRI, a projection on the main PCA mode was used [20] to detect the leakage of a deformable model thus signaling that the locality parameter λ should be decreased. Additional rules and meta-rules can be used to test the validity of hypothesis and eventually reset the optimization stage with new parameters.

11.2.6 Geometric Representations for Model-Based Segmentation

The choice of a geometric representation of a model has a significant impact on the performance of a model-based segmentation algorithm. Many components of those models can be implemented independently of the representation, however some implementations may be far easier than others. For instance, PCA of the shape variability may be implemented for both point-set distributions [21] (Sect. 11.3.1) and level-set functions [22] but are far simpler for the former.

Simplex Meshes vs. Triangulations

Triangulations are widely used representations for computational purposes. They can represent manifold and non-manifold surfaces while simplex meshes are only suitable for manifold surfaces. Furthermore, it is straightforward to compute the closest distance between a point to a triangulation because of its planar faces (triangles). On simplex meshes, the faces are curved and therefore computing accurately the closest point is more complex (no closed form solution). However, due to its fixed vertex connectivity, it is much easier to control the spreading of vertices on simplex meshes for instance to concentrate vertices at parts of high curvature. Also one can define on simplex meshes shape memory regularization as well as curvature continuous smoothing without shrinkage. On triangulations, the most common way to smooth a shape is to use the mean-curvature flow [23] which entails shrinkage.

Simplex Meshes vs. Parametric Surfaces

Parametric surfaces, such as B-splines, the Non-Uniform Rational B-Spline (NURBS), or Finite Element Model (FEM), can provide C^k continuity ($k > 0$) everywhere while for simplex meshes only C^0 continuity is enforced. Therefore, they describe smoother surfaces which can be important for some applications (e.g., rapid prototyping, physical simulation). Also, differential quantities such as surface normals or curvatures can be described everywhere, which is important for shape analysis (e.g., extraction of feature points). However, using those parametric surfaces brings additional complexity especially when the surface cannot be parameterized by a single planar patch (i.e., surface of planar, cylindrical and toroidal topology). Indeed, even for surfaces of spherical topology, parametric surfaces may be complex to handle for instance with the creation of poles (deformable super-quadrics [3]). Simplex meshes as well as triangulated surfaces can cope with surfaces of any topology. Furthermore in many image segmentation problems, only the position, volume and distance of the segmented structure are of interest. Therefore, describing surfaces with C^0 continuity is a good trade-off between complexity and accuracy.

Simplex Meshes vs. Level-Sets

Level-sets rely on an implicit description of the geometry. They can describe manifolds of co-dimension one (i.e., hyper-surface) of any topology but cannot represent surfaces with borders (except if the border is on the edge of the image) such as a planar patch or a cylinder. Simplex meshes can represent a manifold of any dimension and co-dimension. In level-sets, differential quantities (normal, curvatures) may be computed everywhere by filtering the embedding image while in simplex meshes they are only available at vertices. Level-sets can adapt their topologies easily while it is much more difficult for simplex meshes [24] or triangulations [25]. Also, the coupled segmentation of structures is easier to implement with level-sets since the distance function is already computed to update their position. Since they are based on an Eulerian scheme, level-sets are not suited to transport a priori information from a model to the segmented structure. Also, the regularization of level-sets is usually local and based on the mean curvature flow which entails shrinkage.

11.3 Statistical Models of Shape and Appearance

As presented in Sect. 11.2.1, shape priors can be employed as stabilizing forces during the evolution of a deformable model. With strong shape priors, it is possible to enforce smooth shape changes around a defined template. Apart from the smoothness constraint, vertices generally move independently from one another, driven by their individual image forces. For many structures in the human body, it is possible to constrain the possible variations further, creating more specific models while still maintaining a good ability to adapt to individual instances. The higher the specificity of a model, i.e. the fewer non-plausible shapes it produces, the more robust it can segment images in the presence of noise and artifacts.

To determine plausible and non-plausible shapes for a certain structure requires either an artificial parametric model (which has to be hand-crafted each time), or a large amount of examples from which this information can be extracted automatically. The SSM employs the latter approach; it generally uses second order statistics to extract the principal modes of variation from a set of example shapes. These modes can be used instead of the internal forces of deformable models. In addition, it is also possible to employ statistics to generate more specific image forces. As for shape modeling, the required information is extracted from a set of example images. Using both techniques together allows to develop very specific and accurate segmentation algorithms for medical applications. A more detailed discussion of the techniques and approaches is given in the comprehensive review by Heimann and Meinzer [26].

11.3.1 Shape Representation

As mentioned in Sect. 11.2.6, there exist several different representations for shapes. In the following, we assume the most common and simplest representation: a cloud of points densely sampled on the surface of each shape. Thus, each training shape can be represented as a single vector **x** that concatenates coordinates from all k points on the surface

$$\mathbf{x} = (x_1, y_1, z_1, \ldots, x_k, y_k, z_k)^T \quad (11.3)$$

Connecting edges between neighboring points leads to a mesh structure. Although these connections are not required for a PDM [21], they are essential to define inside, outside, and normal vectors for segmentation applications.

11.3.2 Point Correspondence

To be able to perform statistical analysis on a set of training shapes, the individual points representing each training shape have to correspond over the entire set of examples (Fig. 11.10). This means that if a point is placed on

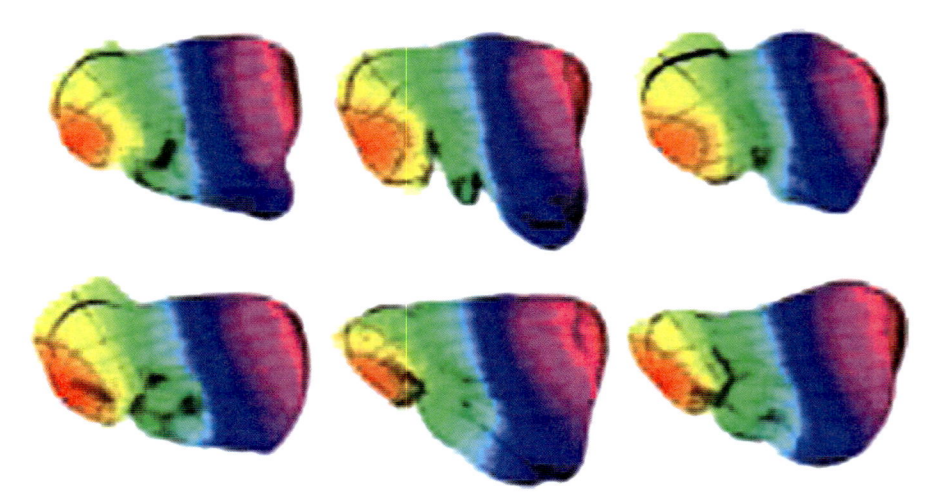

Fig. 11.10. *Correspondences for a set of liver shapes.* Corresponding areas are marked by colors and a sparse coordinate grid

a certain anatomic landmark in one example, it should be placed on the same landmark in all other shapes (and the number of sample points is necessarily equal for all training shapes). Thus, all points do represent a kind of landmark, and the two terms are often used synonymously in shape modeling.

Setting landmarks manually on 3D shapes is an extremely tedious and time-consuming process. Additionally, results are not reproducible, as different experts will label different landmarks, especially in areas without discernible features. For this reason, methods to automatically determine correspondences are of paramount importance to shape modeling, and the quality of extracted correspondences has a large impact on the quality of the resulting shape model. The variety of approaches to determine correspondences over a set of shapes can be classified according to the type of registration used to match one shape to the other. In the following sections, we briefly present the most important approaches.

Mesh-to-Mesh Registration

Assuming that the original training shapes are available as meshes, a straightforward solution to the point correspondence problem is to register the different surfaces to each other. One training shape is selected as template and matched to all others by a generic surface matching method as Iterative Closest Point (ICP) [27] or Softassign Procrustes [28]. For each point of the template, these algorithms deliver the closest point in the target mesh after both have been aligned by an optimal similarity transform. However, this solution has a number of draw-backs. Firstly, the selection of the template induces a bias in the process; different templates usually lead to different correspondences. Secondly, possible corresponding points are limited by the resolution of the target mesh: if this mesh is not sampled densely enough, the approach can introduce considerable artificial variation in the resulting SSM. Finally, the most serious problem is that the employed similarity transform may be insufficient to match the template to the other training shapes if the training set shows large geometric variations. In this case, the approach leads to invalid correspondences and the resulting statistical model is not representative of the training data.

An alternative approach is to use non-rigid registration to match the training surfaces. For this, a limited number of matching landmarks is determined in advance, either by manual labeling or by automatic feature matching. These sparse point sets can then be used to initialize a non-rigid registration as the thin-plate-spline deformation, which produces exact matches for the pre-defined landmarks. Correspondences for the remaining points can be determined simply by Euclidean distance. However, to avoid topological problems it is safer to regularize this matching, e.g., by mesh relaxation [29].

Mesh-to-Volume Registration

Most training shapes for medical imaging applications originate from volumetric image modalities as CT or MRI. In these cases, it is possible to acquire point correspondences by registering a deformable mesh to each volumetric training image. The basic idea is that the positions to which vertices of the deformable mesh converge are corresponding over the entire training set. Obviously, the deformable mesh has to adapt accurately to each shape, which is difficult when using one of the typical image forces (Sect. 11.2.2) on intensity images. Therefore, the deformable mesh is mostly used with the segmented, binary volumes, which provide a hard feature for the image force. The geometry for the deformable mesh can be extracted from an arbitrary instance of the training set. However, this procedure can introduce a bias in the corresponding detection. A technique to minimize this bias has been proposed in [30], where intermediate templates are generated for difficult cases.

Volume-to-Volume Registration

Similar to registering a deformable mesh to all training volumes, it is also possible to employ a volumetric atlas for that purpose. The atlas can either consist of a single volume or can be built from a collection of images. After registering the atlas to an instance of the training set, the resulting deformation field is used to propagate landmarks on the atlas to the respective training shape. As with mesh-to-volume registration, the atlas registration is usually conducted on the segmented volumes to increase accuracy. To provide the required flexibility, non-rigid transforms are essential for the registration; popular methods include B-splines or thin-plate-splines deformations. As usual, a multi-resolution approach increases robustness of the matching.

Parameterization-to-Parameterization Registration

A parameterization is a bijective mapping between a mesh and an appropriate base domain [31]. For closed 2D contours, this base domain is the circle, and determining correspondences by parameterization is equivalent to matching points by relative arc-length. Depending on the topology of training shapes the respective base domains can change, which makes registration between them impossible. This means that a prerequisite for using this approach is that all training shapes have to feature the same topology. Please note that this is also true for the previously presented registration approaches, with the difference that this constraint is not verified there. For 3D shapes, most parameterization approaches are geared to genus zero shapes, i.e. 2-manifolds without holes or self-intersections. The corresponding base-domain is the sphere. When using unconstrained parameterizations as spherical harmonics [32], the obtained correspondences are often arbitrary. Therefore, a common approach is to determine a limited set of matching landmarks in advance (as for the mesh-to-mesh registration) and employ parameterization to match the remaining regions of the shape.

Population-Based Optimization

In all the correspondence approaches presented above, matching points are determined in a pair-wise comparison or without a comparison at all (for the parameterization-based approaches). Population-based correspondence methods analyze the complete set of training shapes at the same time and select corresponding points based on a global cost function. Starting from initially arbitrary landmarks, an optimization algorithm iteratively modifies correspondences until the cost function converges at its minimum. Although this procedure sounds straight-forward and should lead to optimal results, there are a number of challenges involved in the design.

First, as it is difficult to define what *good correspondences* are, coming up with a practically usable and theoretically sound cost function is not evident. Practically all current approaches rely on a form of Minimum Description Length (MDL) as cost function, which was pioneered for correspondence optimization by Davies [33]. MDL implements the principle of Occam's razor that a simpler model is better than a complex one. Translated to SSMs, this means that a model is better when it features fewer and more compact modes of variation. As the cost function for the full MDL is computationally very expansive, a simplified version [34] is often used. Here, costs F are directly calculated from the eigenvalues λ of the shape model

$$F = \sum_m \mathcal{L}_m \quad \text{with} \quad \mathcal{L}_m = \begin{cases} 1 + \log(\lambda_m/\lambda_{\text{cut}}) & \text{for } \lambda_m \geq \lambda_{\text{cut}} \\ \lambda_m/\lambda_{\text{cut}} & \text{for } \lambda_m < \lambda_{\text{cut}} \end{cases} \quad (11.4)$$

where λ_{cut} describes the amount of noise in the dataset.

A second challenge is to adjust the landmark positions in the course of optimization. The difficulty lies in the fact that the optimization problem is heavily constrained, as individual landmarks cannot be moved independently from their neighbors (to avoid flipping triangles in the model mesh). Additionally, all 3D landmarks have to remain on the 2D manifolds of the surfaces of the training shapes. This is either accomplished by employing parameterizations to modify landmark positions [33, 35] or by particle approaches [36].

11.3.3 Construction of Statistical Shape Models

Once point correspondences are known, each training shape i can be represented as coordinate vector $\mathbf{x_i}$ in (11.3). Before applying the dimensionality reduction that delivers the predominant modes of variation, the individual shapes first have to be aligned into the same coordinate system.

Alignment

The commonly used procedure to align a set of training shapes is the Generalized Procrustes Analysis (GPA) [37]. The underlying Procrustes analysis determines the optimal similarity transform to minimize the sum of squared distances between two corresponding point sets. GPA employs this method iteratively to match a whole set of shapes to their unknown mean. An additional step for constructing a shape model is a rescaling of the thus aligned shapes. This is required since the minimization of squared distances introduces non-linearities in the shape space, which are difficult to capture by commonly used techniques. A straight-forward solution is to scale each training shape i with $1/(\mathbf{x_i} \cdot \bar{\mathbf{x}})$, where $\bar{\mathbf{x}}$ is the mean as determined by GPA.

Dimensionality Reduction

Statistical shape modeling is essentially a problem of dimensionality reduction: The input data of s training shapes $\mathbf{x_i}$ (with the dimension of $\mathbf{x_i}$ generally in the thousands) should be reduced to a limited number of modes of variation. This task is usually solved by PCA [38]. The first step is to calculate the mean shape $\bar{\mathbf{x}}$ of the training set

$$\bar{\mathbf{x}} = \frac{1}{s} \sum_{i=1}^{s} \mathbf{x_i} \tag{11.5}$$

Then, the covariance matrix S of the input data is computed by:

$$S = \frac{1}{s-1} \sum_{i=1}^{s} (\mathbf{x_i} - \bar{\mathbf{x}})(\mathbf{x_i} - \bar{\mathbf{x}})^T \tag{11.6}$$

An eigendecomposition of S yields the eigenvectors ϕ_m (representing the principal modes of variation) and the corresponding eigenvalues λ_m (indicating the variance per mode). Sorting all modes from largest to smallest variance, the first c modes are employed to model the distribution, while the remaining modes are discarded. Thus, all valid shapes can be approximated by a linear combination

$$\mathbf{x} = \bar{\mathbf{x}} + \sum_{m=1}^{c} b_m \phi_m \tag{11.7}$$

with **b** as the vector of shape parameters. To constrain variation of the model, **b** is usually bound to certain limits, either by constraining $|b_m| < 3\lambda_m$ for each mode individually or by constraining b globally to lie within an appropriate ellipsoid. Typically, c is chosen so that the accumulated variance $\sum_{m=1}^{c} \lambda_m$ reaches a certain percentage (e.g. 95 %) of the total variance. Figure 11.11 shows the first three principal modes of variation as extracted from a collection of liver shapes.

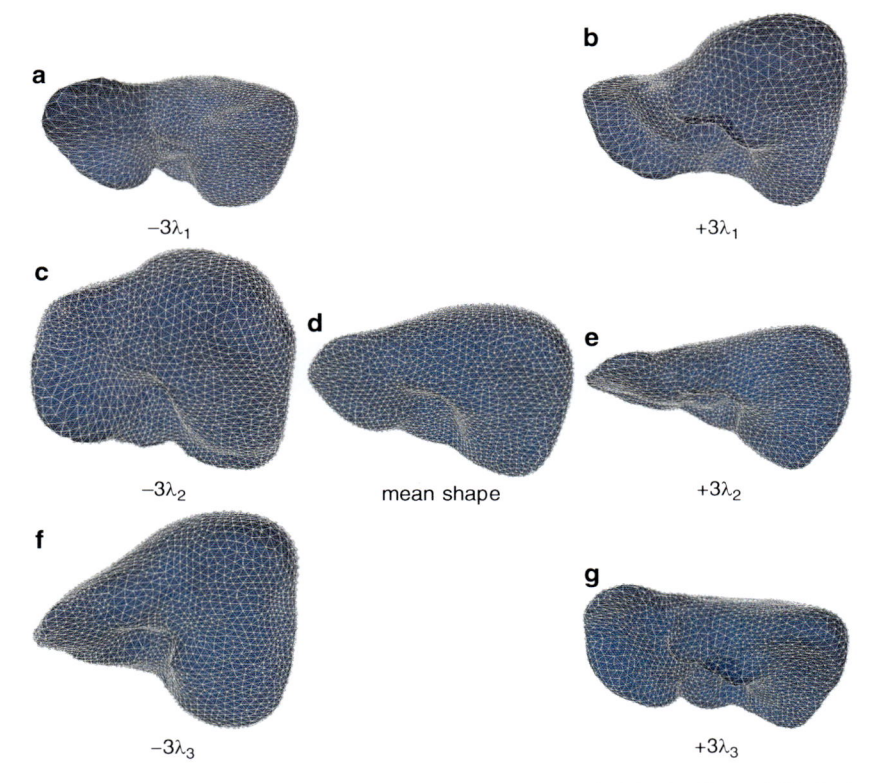

Fig. 11.11. *Statistical shape model of the liver.* Panels (**a,b**), (**c,d**), and (**f,g**) show the first three modes of variation for a SSM of the liver with mean shape (**d**)

11.3.4 Modeling Object Appearance

In Sect. 11.2.2, we already described possible image forces for deformable models. As we assume to have training data available for statistical modeling, we now focus on strong appearance priors.

Boundary-Based Features

The most popular appearance models for SSMs are related to the above-mentioned intensity profile. At each landmark of the shape model, a profile is sampled in normal direction and compared to a model built from training data at the same landmark. This implies that one shape model commonly features a large number of appearance models. The employed profiles are not limited to plain intensity alone; gradient profiles and normalized intensity or gradient profiles are also popular choices. Different features can also be combined in the same model. Each appearance model is built similar to the shape model, i.e. mean profile and covariance matrix are generated. Using the covariance matrix, the Mahalanobis distance to profiles sampled during the search can be determined. For Gaussian distributions, this represents a robust measure of the goodness of fit [39].

In cases where the distribution of profiles is essentially non-Gaussian (e.g., because the modeled object is adjacent to several tissues of different appearance), non-linear appearance model can deliver better results. A straight-forward method to employ a non-linear model is using a k-Nearest-Neighbor (k-NN) classifier [40]. To train it, both true and false examples of boundary profiles must be used. The required false examples can be generated by shifting profiles from their known positions towards the inside and outside of the OOI.

Region-Based Features

As mentioned in Sect. 11.2.2, region histogram priors can be used to evolve a deformable model in order to match a certain distribution of intensities. For statistical models, it is straight-forward to model this distribution from training data. However, there are a number of much more specific features that can be used in this case. The most consequent approach for learning object appearance uses the entire texture of the OOI [41]. To combine textures from all training images, the respective objects are morphed to the same shape (the mean shape of the SSM). Subsequently, the mean texture and its covariance matrix can be calculated as usual. A drawback of this approach is the high memory footprint of the appearance model, especially for 3D models where a volumetric texture is used. One solution is to model the texture only in certain parts of the object, e.g. around the boundary.

Modeling local regions inside and outside of the OOI opens up more possibilities regarding the search algorithm for model evolution (Sect. 11.3.5), as the goodness of fit can be determined locally (like when using profiles). In addition to the above mentioned techniques of histogram matching and texture matching, local appearance can also be modeled by multi-scale derivatives, Gabor filterbanks, wavelets, etc. In comparison to boundary-based features, region-based features are often more robust as they use more image data to determine the goodness of fit. At the same time and for the same reason, they are also computationally more demanding.

11.3.5 Local Search Algorithms

As mentioned in Sect. 11.2.5, the first step to employ a deformable model for segmentation is to initialize it for the respective image. This initialization places the model close to the OOI (e.g., by atlas registration), so that the model can iteratively evolve towards the correct contour. In the following, we present several techniques to conduct this subsequent evolution (the local search) for statistical models of shape and appearance.

Active Shape Model

The Active Shape Model (ASM) was introduced by Cootes et al. [21]. It is one of the most popular model-based approaches for segmentation of medical images. An instance **y** of the model in an image is defined by a similarity transform T and the shape parameter vector **b**

$$\mathbf{y} = T\left(\bar{\mathbf{x}} + \Phi \mathbf{b}\right) \tag{11.8}$$

where $\Phi = (\phi_1 \ldots \phi_c)$ is the matrix of eigenvectors (11.7). To evolve the shape over time, the optimal displacement $\mathbf{dy_p}$ is determined for each landmark point separately by querying the goodness of fit for the respective appearance model at various positions along the normal vector. The update then consists of two steps: First, transform T is updated by the result of the Procrustes match of the model to $\mathbf{y} + \mathbf{dy_p}$. This leads to new residual displacements $\mathbf{dy_s}$. Second, shape parameters **b** are updated by transforming $\mathbf{dy_s}$ into model space and then applying a projection into shape space

$$\mathbf{db} = \Phi^T \tilde{T}^{-1}\left(\mathbf{dy_s}\right) \tag{11.9}$$

where $\tilde{T}$ is equal to T without the translational part. The resulting adjustments **db** are added to **b**. Usually, **b** is restricted to certain limits to maintain a valid shape (Sect. 11.3.3). The presented two steps are repeated iteratively until the model converges at the best local fit in the image.

To increase robustness, especially against incorrect initialization, the search is commonly conducted in a multi-resolution fashion. Contrary to the multi-resolution approach presented for simplex meshes (Sect. 11.2.3), the DOF usually remains unchanged. Instead, appearance models are constructed for a number of down-sampled versions of all training images, and the search is started on a down-sampled version of the original image. Here the search radius is larger, since the goodness of fit is evaluated at points at a greater distance from the original contour. When the model evolution converges on the rough scale, appearance model and search radius are switched to the next (more detailed) versions, until the original resolution is reached.

Another approach to increase robustness is to decrease the effect of outliers, i.e. wrongly detected optimal displacements for specific landmarks. There are two basic techniques for this problem. Firstly, outliers can be detected (e.g. by comparing suggested landmark positions to collected statistics) and corrected before the shape is updated. Secondly, the update of transform and shape parameters can be weighted and each suggested landmark position receives a reliability weighting in each step.

Active Appearance Model

The Active Appearance Model (AAM) [41] belongs to the class of generative models, i.e., it can generate realistic images of the modeled data. This is accomplished by storing a complete texture model, consisting of the mean and the principal modes of variation, in addition to the shape model. However, AAMs are more than a shape model with region-based features, they employ a specific search method which is completely different from the ASM approach. Moreover, shape and appearance parameters are combined into one linear system, where shape $\mathbf{x}$ and appearance $\mathbf{g}$ are described by a common parameter vector $\mathbf{c}$

$$\begin{aligned} \mathbf{x} &= \bar{\mathbf{x}} + \Phi_s W_s Q_s \mathbf{c} \\ \mathbf{g} &= \bar{\mathbf{g}} + \Phi_g Q_g \mathbf{c} \end{aligned} \qquad (11.10)$$

Here, Φ_s and Φ_g are the independent eigenvector matrices of shape and appearance model, respectively, and W_s is a diagonal weight matrix for the shape parameters. $Q = \begin{pmatrix} Q_s \\ Q_g \end{pmatrix}$ is the eigenvector matrix of the combined shape and appearance parameters. It is the result of a PCA on the independent parameters $\mathbf{b} = \begin{pmatrix} W_s \mathbf{b_s} \\ \mathbf{b_g} \end{pmatrix}$. An instance of the model in an image is defined by a similarity transformation T and the combined shape-appearance parameters $\mathbf{c}$. In the following, we name the latter ones simply parameters $\mathbf{p}$. To evaluate the goodness of fit, the image texture is warped to the mean shape and normalized, resulting in $\mathbf{g_s}$. With the modeled appearance $\mathbf{g_m} = \mathbf{g}$ from (11.10), the residuals are given by $\mathbf{r}(\mathbf{p}) = \mathbf{g_s} - \mathbf{g_m}$, and the error by $E = \mathbf{r}^2$.

The key idea of AAM search is to assume a constant relationship between texture residuals $\mathbf{r}(\mathbf{p})$ and parameter updates $\mathbf{dp}$ over the entire search

$$\mathbf{dp} = -R\mathbf{r}(\mathbf{p}) \qquad (11.11)$$

The success of this optimization scheme largely depends on the derivative matrix R. In the first presentation of AAMs, R was computed using multivariate linear regression on a large number of simulated disturbances of the training images. Later, regression was replaced by numeric differentiation, claimed to be both faster and more reliable.

The main challenge when employing AAMs for medical image segmentation is the enormous amount of data that has to be captured by the model. For 3D models, appearance is generally modeled as a 3D texture, which quickly leads to prohibitively large equation systems. Therefore, texture resolution generally has to be scaled down. It it noteworthy that the AAM search procedure is not limited to region-based features. Indeed, the general idea of determining parameter updates from residuals can also be used with alternative appearance features as described in Sect. 11.3.4.

11.4 Conclusion

Model-based approaches to segmentation are arguably the most robust methods when image data is noisy or includes artifact. Therefore, they are a prime choice for many applications in medical image analysis. The deformable simplex meshes and point-based statistical models presented in this chapter represent state-of-the-art techniques for this purpose.

Comparing both methods, an obvious difference is that building a shape model requires a large collection of training images (typically 20–50 as a minimum for 3D models), while a simplex mesh with strong shape prior can be based on a single example. Another point is that the additional constraints of SSMs generally result in a higher robustness, but at the same time limit the accuracy of the final result. Statistical models are therefore often applied as part of a coarse-to-fine strategy that features an additional refinement step afterwards. As already hinted to in Sect. 11.2.3, it is also possible to combine both techniques in one framework. That way, the strict shape constraints can gradually be lifted and the final free deformation ensures an optimal fit to the data.

References

1. Terzopoulos D. Multiresolution Computation of Visible-Surface Representation. Cambridge, MA: MIT Press; 1984.
2. Kass M, Witkin A, Terzopoulos D. Snakes: active contour models. Int J Comput Vis. 1988;1:321–31.
3. Metaxas D, Terzopoulos D. Constrained deformable superquadrics and nonrigid motion tracking. In: Proc CVPR. Maui, Hawai; 1991. p. 337–43.
4. Delingette H, Hébert M, Ikeuchi K. Shape representation and image segmentation using deformable surfaces. Image Vis Comput. 1992;10(3):132–44.
5. Delingette H. Modélisation, Déformation et Reconnaissance D'objets Tridimensionnels a l'aide de Maillages Simplexes. Ecole Centrale de Paris; 1994.
6. Chopp DL, Sethian JA. Computing minimal surfaces via level set curvature flow. J Comput Phys. 1993;106:77–91.
7. Malladi R, Sethian JA, Vemuri BC. Shape modeling with front propagation: a level set approach. IEEE Trans Pattern Anal Mach Intell. 1995;17(2):158–75.
8. Cohen LD, Cohen I. Finite element methods for active contour models and balloons from 2-D to 3-D. IEEE Trans Pattern Anal Mach Intell. 1993;15(11).
9. Delingette H. General object reconstruction based on simplex meshes. Int J Comput Vis. 1999;32(2):111–46.
10. Montagnat J, Delingette H, Malandain G. Cylindrical echocardiographic images segmentation based on 3D deformable models. Lect Notes Comput Sci. 1999;1679:168–75.
11. Herbulot A, Besson SJ, Duffner S, et al. Segmentation of vectorial image features using shape gradients and information Measures. J Math Imaging Vis. 2006;25(3):365–86.
12. Pitiot A, Delingette H, Thompson PM, et al. Expert knowledge guided segmentation system for brain MRI. NeuroImage 2004;23(Suppl 1):S85–S96.
13. Pitiot A, Toga A, Ayache N, et al. Texture based MRI segmentation with a two-stage hybrid neural classifier. In: World Congr Comput Intell. 2002.
14. Ciampalini A, Cignoni P, Montani C, et al. Multiresolution decimation based on global error. Vis Comput. 1997;13(5):228–46.
15. Kobbelt L. Sqrt(3) subdivision. In: Proc SIGGRAPH. 2000. p. 103–12.

16. Montagnat J, Delingette H. Globally constrained deformable models for 3D object reconstruction. Signal Process. 1998;71(2):173–86.
17. Montagnat J, Delingette H. 4D Deformable models with temporal constraints: application to 4D cardiac image segmentation. Med Image Anal. 2005;9(1):87–100.
18. Montagnat J, Delingette H. Volumetric medical image segmentation using shape constrained deformable models. Lect Notes Comput Sci. 1997;1205:13–22.
19. Costa J, Delingette H, Novellas S, et al. Automatic segmentation of bladder and prostate using coupled 3D deformable models. Med Image Comput Assist Interv. 2007;10(Pt 1):252–60.
20. Pitiot A, Delingette H, Ayache N, et al. Expert-knowledge-guided segmentation system for brain MRI. Lect Notes Comput Sci. 2003;2879:644–52.
21. Cootes TF, Taylor CJ, Cooper DH, et al. Active shape models: their training and application. Comput Vis Image Underst. 1995;61(1):38–59.
22. Leventon ME, Grimson WEL, Faugeras OD. Statistical shape influence in geodesic active contours. In: CVPR; 2000. p. I: 316–23.
23. Chopp DL, Sethian JA. Flow under curvature: singularity formation, minimal surfaces, and geodesics. J Exp Math. 1993;2(4):235–55.
24. Delingette H, Montagnat J. New algorithms for controlling active contours shape and topology. Lect Notes Comput Sci. 2000;(1843):381–95.
25. McInerney T, Terzopoulos D. T-snakes: topology adaptive snakes. Med Image Anal. 2000;4:73–91.
26. Heimann T, Meinzer HP. Statistical shape models for 3D medical image segmentation: a review. Med Image Anal. 2009;13(4):543–63.
27. Besl PJ, McKay ND. A method for registration of 3-D shapes. IEEE Trans Pattern Anal Mach Intell. 1992;14(2):239–56.
28. Rangarajan A, Chui H, Bookstein FL. The softassign Procrustes matching algorithm. Lect Notes Comput Sci. 1997;1230:29–42.
29. Lorenz C, Krahnstöver N. Generation of point-based 3D statistical shape models for anatomical objects. Comput Vis Image Underst. 2000;77(2):175–91.
30. Zhao Z, Teoh EK. A novel framework for automated 3D PDM construction using deformable models. In: Proc SPIE, vol. 5747; 2005. p. 303–14.
31. Floater MS, Hormann K. Surface parameterization: a tutorial and survey. In: Dodgson NA, Floater MS, Sabin MA, editors. Advances in Multiresolution for Geometric Modelling. Berlin: Springer; 2005. p. 157–86.
32. Kelemen A, Székely G, Gerig G. Elastic model-based segmentation of 3-D neuroradiological data sets. IEEE Trans Med Imaging. 1999;18(10):828–39.
33. Davies RH, Twining CJ, Cootes TF, et al. 3D statistical shape models using direct optimization of description length. Lect Notes Comput Sci. 2002;2352:3–20.
34. Thodberg HH. Minimum description length shape and appearance models. Lect Notes Comput Sci. 2003;2732:51–62.
35. Heimann T, Wolf I, Williams TG, et al. 3D Active shape models using gradient descent optimization of description length. Lect Notes Comput Sci. 2005;3565:566–77.
36. Cates J, Fletcher PT, Styner MA, et al. Shape modeling and analysis with entropy-based particle systems. Lect Notes Comput Sci. 2007;4584:333–45.
37. Goodall C. Procrustes methods in the statistical analysis of shape. J R Stat Soc B. 1991;53(2):285–339.

38. Jolliffe IT. Principal Component Analysis, 2nd ed. New York: Springer; 2002.
39. Cootes TF, Taylor CJ. Using grey-level models to improve active shape model search. In: Proc ICPR. vol. 1; 1994. p. 63–7.
40. de Bruijne M, van Ginneken B, Viergever MA, et al. Adapting active shape models for 3D segmentation of tubular structures in medical images. Lect Notes Comput Sci. 2003;2732:136–47.
41. Cootes TF, Edwards GJ, Taylor CJ. Active appearance models. IEEE Trans Pattern Anal Mach Intell. 2001;23(6):681–85.

Part V

Classification and Measurements

12
Melanoma Diagnosis

Alexander Horsch

Summary. The chapter deals with the diagnosis of the malignant melanoma of the skin. This aggressive type of cancer with steadily growing incidence in white populations can hundred percent be cured if it is detected in an early stage. Imaging techniques, in particular dermoscopy, have contributed significantly to improvement of diagnostic accuracy in clinical settings, achieving sensitivities for melanoma experts of beyond 95% at specificities of 90% and more. Automatic computer analysis of dermoscopy images has, in preliminary studies, achieved classification rates comparable to those of experts. However, the diagnosis of melanoma requires a lot of training and experience, and at the time being, average numbers of lesions excised per histology-proven melanoma are around 30, a number which clearly is too high. Further improvements in computer dermoscopy systems and their competent use in clinical settings certainly have the potential to support efforts of improving this situation. In the chapter, medical basics, current state of melanoma diagnosis, image analysis methods, commercial dermoscopy systems, evaluation of systems, and methods and future directions are presented.

12.1 The Cutaneous Melanoma

Skin cancer develops in the upper layer of the skin, the epidermis (Fig. 12.1). The most common types are the basal cell carcinoma, the squamous cell carcinoma, and the Cutaneous Melanoma (CM). Of these three types, CM is the most aggressive one.

12.1.1 Medical Basics

The CM is a malignant type of a Pigmented Skin Lesion (PSL) or melanocytic skin lesion. A PSL typically is a dark *spot*, *mole* or *nevus* on the skin (Fig. 12.4) originating from an aggregation of the skin color pigment melanin. Special cells, the melanocytes, produce melanin in small granules called melanosomes. The melanin is then transported to cells in the outer skin keratocytes, where they appear as "color" of the skin.

T.M. Deserno (ed.), *Biomedical Image Processing*, Biological and Medical Physics,
Biomedical Engineering, DOI: 10.1007/978-3-642-15816-2_12,
© Springer-Verlag Berlin Heidelberg 2011

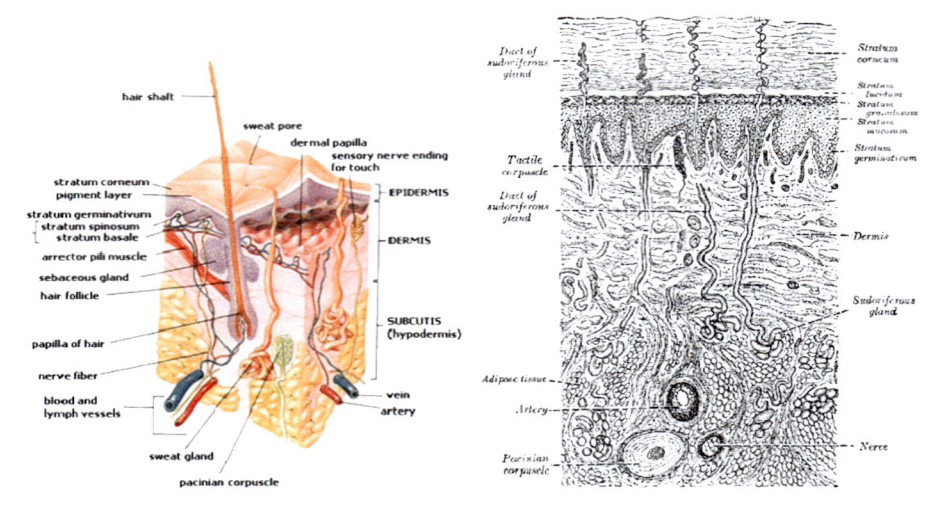

Fig. 12.1. *The human skin. Left*: Anatomy; *Right*: Histology

The CM, also referred to as malignant melanoma or simply melanoma, is a cancerous tumor that arises in the melanocytes. It accounts for about 4% of all skin cancers. The most common CM types are[1]:

- The Superficial Spreading Melanoma (SSM) accounts for about 70% of diagnosed CM). It usually spreads along the epidermis and then grows deeper to the dermis (Fig. 12.1). It is curable when it is removed before it invades the dermis.
- The Acral Lentiginous Melanoma (ALM) is the most common melanoma in dark-skinned races (50% in dark-skinned). It is frequently mistaken for a bruise or nail streak in its early stages and therefore often diagnosed in a later stages when it is very aggressive.
- The Lentigo Maligna Melanoma (LMM) accounts for about 10% of CMs diagnosed in the United States. It develops in a sun-induced freckle and typically occurs on sun-damaged skin in the elderly and may in its early stages be mistaken for a benign age spot or sun spot.
- The Nodular Melanoma (NM) accounts for approximately 15% of diagnosed melanomas. It is a very aggressive CM type that tends to grow downwards into deeper skin tissue rather than along the surface of the skin.

Cancer statistics prove increasing incidence of CM in many countries with white population over the last four decades [1]. Rates were highest in Australia. With an estimated increase of 3–7% it is the most rapidly increasing cancer in white populations, suggesting a doubling of rates every ten to twenty years. Incidence rates in central Europe are in the middle, with a north-south

---

[1] http://www.skincarephysicians.com/skincancernet/glossary.html

gradient due to darker skin type in Mediterranean compared to central and northern European populations. Risk factors include: the number of PSLs on the body; the presence of atypical PSLs; sun-exposure and sunburns; and hereditary factors [2].

12.1.2 Relevance of Early Diagnosis

Unlike other types of cancer, there is good chance for 100% cure for CM. Detected in an early, non-invasive stage, a CM can surgically be removed (excised), with an excellent prognosis for the patient. CM screening programs are therefore promoted [3]. In fact, currently in about 90% of cases a CM is diagnosed at the primary tumor stage. However, the rate of diagnostic excisions, i.e. the number of PSLs excised per CM, also referred to as Number Needed to Treat (NNT), is very high, with estimated averages of 10–90, depending on the age group of the patients and the experience of the examiner [4,5] (some expert centers may reach a NNT of around 5). CM diagnosis is very demanding and needs a lot of experience. Since the main goal is not to overlook any CM, in suspicious cases more often excision is decided (increasing sensitivity on cost of specificity). PSM excisions are performed with certain safety margins. Current recommendations are 0.5, 1, and 2 cm for in situ CM, CM with Breslow thickness ≤ 2 mm and CM with thickness >2 mm, respectively [2]. As surgical interventions, excisions implicate inherent risks and costs. Therefore, the average NNT should be reduced by suitable measures such as enhanced training and improvements of diagnostic imaging techniques and algorithms.

As digital image analysis is concerned, this article focuses on the diagnosis (i.e., malignancy assessment) of PSLs, not on detection. Both tasks, detection and diagnosis, can be supported by digital image analysis. Related systems are commonly referred to as Computer-Aided Detection (CADe), and Computer-Aided Diagnosis (CADx), respectively. For CADe in CM screening see the literature on mole mapping and total body mapping.

12.2 State of the Art in CM Diagnosis

Diagnosis of CM has been a very active research field over the past decades, both in terms of diagnostic algorithms to be applied by human examiners, and in terms of skin imaging techniques, including the support by digital image processing. However, the task of differing Cutaneous Melanomas (CMs) from benign Pigmented Skin Lesions (PSLs) still is challenging (Fig. 12.2).

12.2.1 Diagnostic Algorithms

The differential diagnosis for PSLs is challenging even for specialists. For this reason, a 2-step approach (Fig. 12.3) was suggested by a virtual consensus

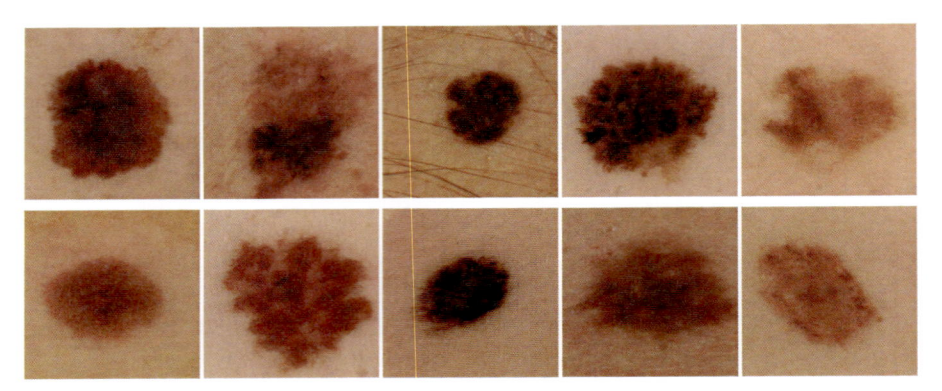

Fig. 12.2. *Pigmented skin lesions.* In the dermoscopic view, cutaneous melanomas (*upper row*) and benign pigmented skin lesions (*lower row*) appear very similar

Fig. 12.3. *Flow chart.* The differential diagnosis of pigmented skin lesions is performed with a 2-step approach

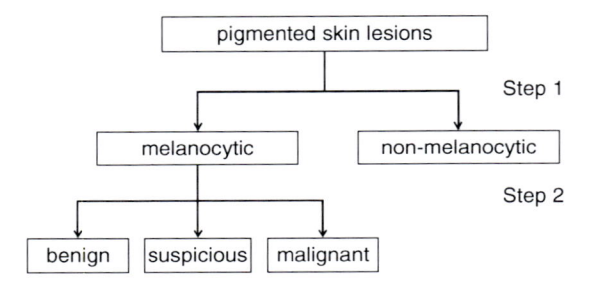

Table 12.1. *Evaluation of diagnostic algorithms.* The algorithms were evaluated by the CNMD regarding the inter- and intra-observer agreement

Algorithm	Inter-observer κ (95% CI)	Intra-observer κ (range)
First-step	0.63 (0.62–0.63)	1.00 (0.73–1.00)
Pattern analysis	0.55 (0.54–0.56)	0.85 (0.12–1.00)
ABCD rule	0.48 (0.47–0.48)	0.72 (0.11–1.00)
7-point checklist	0.47 (0.46–0.47)	0.72 (0.29–1.00)
Menzies method	0.52 (0.51–0.52)	0.75 (0.21–1.00)

meeting of 40 actively participating experienced clinicians known as the Consensus Net Meeting on Dermoscopy (CNMD). In the first step, a classification in melanocytic and non-melanocytic lesions is performed, following a 6-step algorithm. In the second step, melanocytic lesions are classified in benign and malignant. It is primarily the second step that is currently addressed by dermoscopy CADx.

Current diagnostic algorithms in dermoscopy evaluated by the CNMD (cf. [6]) are summarized in Table 12.1. *Pattern analysis* [7, 8] performed by physicians allows distinction between benign and malignant growth features. Typical patterns of some common PSLs are: dots, globules, streaks, blue-white veil, blotch, network, network borders:

- The Asymmetry, Border, Color, and Differential structures (ABCD) rule [9, 10] as applied by physicians scores the features:
  1. *Asymmetry*, 0–2 points: 0 = symmetric in two axes; 1 = symmetric in one axis; 2 = no symmetry at all;
  2. *Border*, 0–8 points: number of half-quadrants with diffuse border;
  3. *Color*, 1–6 points: number of colors appearing in the lesion (white, red, light brown, dark brown, blue-gray, black); and
  4. *Differential structures*, 1–5 points: number of differential structures in the lesion (pigment network, dots, globules, streaks, structureless areas)

  with particular weight factors and sums up to a total score ranging from 1.0 to 8.9. Depending on this Dermatoscopic Point Value (DPV), $DPV = 1.3A + 0.1B + 0.5C + 0.5D$, a lesion is considered benign ($DPV = 1.00 - 4.75$), suspicious ($DPV = 4.75 - 5.45$), or malignant ($DPV = 5.45 - 8.90$).
- The *7-point checklist* [11] distinguishes 3 major criteria, each with a score of two points (atypical pigment network; blue-white veil; atypical vascular pattern) and four minor criteria, each with a score of one point (irregular streaks; irregular pigmentation; irregular dots/globules; regression structures). Diagnosis of CM requires a minimum total score of three.
- For a diagnosis of melanoma according to the *Menzies method* [12] both of the two negative features (lesion shows only a single color; patterns are point or axial symmetric) must *not* be found, and at least one of 9 positive features (blue-white veil; multiple brown dots; pseudopods; radial streaming; scar-like depigmentation; peripheral black dots-globules; multiple colors; multiple blue/gray dots; broadened network) must be found.
- Two more diagnostic algorithms which were not evaluated by the CNMD are the *3-point score* and the Color, Architecture, Symmetry, and Homogeneity (CASH) system.
- According to the 2002 staging system of the American Joint Committee on Cancer (AJCC), melanomas classified in ten stages, from 0 (in situ) via II–III (primary tumor) to IV (tumor with distant metastases) [13].

12.2.2 Imaging Techniques

In order to improve diagnostic performance for PSL various imaging techniques have been explored [14–17]. Important imaging techniques in clinical settings are summarized with their main characteristics in Table 12.2:

- *Photography* is the most common form of skin imaging. Follow-up of lesions over time can help in early detection.
- *Dermoscopy* adds significant information compared to photography by making structures in the epidermis and the papillary dermis (Fig. 12.1) directly visible.

Table 12.2. *Imaging techniques.* Diagnosis of PSL in clinical settings

Technique	Physical principle	Penetration depth
Photography	Visible light	0.1 mm
Dermoscopy	Visible light	2 mm
Spectral imaging	Visible and near infrared	0.1–1 mm
LASER Doppler perfusion imaging	Laser light	N/A
Magnetic resonance imaging	Nuclear magnetic resonance	<7 mm
Infrared thermal imaging	Infrared spectrum	N/A

- *Spectral imaging* can map biochemical properties such as chromophore concentration.
- *LASER Doppler perfusion imaging* creates maps of the blood flow, which can be used for melanoma detection and lymph node assessment. Imaging techniques such as X-ray Computed Tomography (CT), Magnetic Resonance Imaging (MRI), or Positron Emission Tomography (PET) combined with CT (PET-CT) have a high spatial resolution and a high sensitivity for the detection of metastatic disease. However, they increase the number of false-positive findings [2].
- For research, a variety of other imaging techniques with different depth penetration are explored, including:
  - *Profilometry* (assessment of changes in skin surface topology);
  - *Optical Coherence Tomography (OCT)* (near infra-red low coherence laser, 1–2 mm);
  - *Confocal mode imaging* (near infra-red laser, 300 µm);
  - *Terahertz pulse imaging*;
  - *Video-microscopy.*

Currently, dermoscopy [6, 10, 18, 19], also referred to as dermatoscopy, skin surface microscopy, Epi-Luminescence Microscopy (ELM), or in vivo cutaneous surface microscopy, is clinically the most relevant in vivo imaging technique for CM diagnosis [20]. The history of this technique goes back to 1663, the beginning of skin surface microscopy [6]. The use of immersion oil in microscopy was introduced by Abbe (1878) and transferred to skin surface microscopy by Unna (1893). Saphier (1920) and Goldman (1950s) further developed the technique.

Modern dermoscopy devices create high-quality magnified images of the skin subsurface (typically 10–20 fold magnification [2]). Normally, around 4–7% of light is reflected from the dry skin surface [15], limiting the visualization of deeper structures. To reduce reflection, in dermoscopy either (a) the light source is directly coupled to the epidermis by immersion oil sprayed on the skin before the optics with its front-glass plate is pressed on the skin surface, or (b) polarized light is used. The optically magnified image of the skin surface and subsurface – flattened by the glass plate in case (a), non-flattened in case (b) – is then either visually inspected or captured by a computer for subsequent digital image analysis and examination at the computer screen [6].

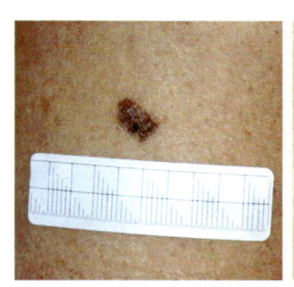

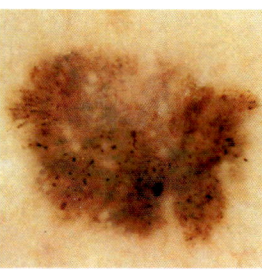

Fig. 12.4. *Example of a malignant melanoma. Left*: macroscopic ("clinical") view with the naked eye; *Right*: microscopic ("dermoscopic") view

Compared to photographic images, dermoscopic images are very rich of details (Fig. 12.4). To support the difficult interpretation of dermoscopic images, various computer diagnosis systems have been proposed in the published literature, commonly referred to as *computer dermoscopy* as a specific type of CADx for skin lesions. In analogy to the field of computer-aided breast cancer diagnosis [21], *dermoscopy CADx* is a suitable synonym.

12.2.3 Diagnostic Accuracies

Compared to purely visual examination, dermoscopy is clearly more accurate [20]. According to a meta-analysis by Vestergaard et al. [22] from 2008, the relative diagnostic odds ratio for CM in this comparison is 15.6 (95% CI 2.9–83.7, $p = 0.016$) for 9 eligible studies published from 1987. Similarly, CADx systems accuracies on dermoscopic images outperform CADx systems accuracies on clinical images (log odds ratios 4.2 vs. 3.4, $p = 0.08$) [23].

Annessi et al. [24] compared the performance of three diagnostic algorithms – pattern analysis, ABCD rule, 7-point checklist – on doubtful PSLs in 195 patients (89 male, 106 female, average 43 y): 198 consecutive atypical melanocytic lesions, of these 102 Atypical Melanocytic Nevi (AMN) (Clark's melanocytic nevus, benign), 96 Thin Melanoma (TM) (malignant, 24 in situ melanomas and 72 SSMs with an average tumor thickness of 0.3 mm). Two dermoscopy-experienced dermatologists classified the lesions in benign and malignant using the three algorithms. Surgical excision followed. The diagnostic accuracy (ratio of true positives to real positives plus false positives), sensitivity and specificity are: 70.8, 85.4, 79.4 for the pattern analysis, 67.8, 84.4, 74.5 for the ABCD rule, and 57.7, 78.1, 64.7 for the 7-point checklist, respectively.

There is evidence that dermoscopy CADx systems can reach the performance of CM experts and help improving early CM diagnosis [2]. Rajpara et al. [25] reviewed 765 articles and performed a meta-analysis on 30 eligible studies comparing human dermoscopy to computer dermoscopy. They found the pooled sensitivity for computer dermoscopy was slightly higher than for human dermoscopy (91% vs. 88%, $p = 0.076$), while pooled specificity was significantly better for human dermoscopy (86% vs. 79%, $p = 0.001$). In the meta-analysis of Rosado et al. [23] from 2003 on 10 eligible studies with a total of 6579 cases CADx systems diagnosis was statistically not different from human diagnosis (log odds rations 3.36 vs. 3.51, $p = 0.80$).

12.3 Dermoscopy Image Analysis

The analysis of digital dermoscopy images can follow different approaches, either driven by the medical expert perspective, or driven by a machine intelligence perspective. All approaches share basic steps in the processing pipeline, including: segmentation, feature extraction, and classification.

12.3.1 Image Analysis Approaches

Two major approaches can be distinguished in dermoscopy digital image analysis for melanoma diagnosis:

1. *Machine learning approach* or *black box approach*. Typically, a large number of (more or less) low-level features are computed, typically some hundreds, features discriminating best between melanoma and benign PSLs are automatically selected by a certain strategy, reducing the number of features to typically some tens, and finally a classifier based on these features is constructed. In general, it is not transparent to the human user what such features really measure and how the classifier comes to its conclusion.
2. *Mimic approach*. Typically, a small number of high-level diagnostic features used by medical experts for visual evaluation of the images are modeled mathematically. Semi-quantitative diagnostic algorithms such as the ABCD rule are then automated using the computed features instead of the human scores. Usually, the user can easily understand the meaning of the features and the way the proposed diagnosis is created by the classification algorithm.

In practice, both approaches seldom appear in a pure form. Rather, following a mimic approach, a system developer will slightly extend the number of features (e.g., by introducing variants) and adapt the classification algorithm, if this improves the performance of the system and pertains the comprehensibility by users. On the other hand, applying the black box approach, the number and type of features as well as the choice of classifiers can be steered to gain transparency.

An example for the black box approach is Blum et al. [26], where the researchers did not follow a preformed strategy (e.g., ABCD rule). Instead, the authors applied a large number of algorithms of vision algebra to the pictures. In particular, the authors used 64 analytical parameters to build their classifier.

An example for the mimic approach is Horsch et al. [27–29]. This research group used high-level features in line with the ABCD rule of dermatoscopy. Some methodological details of the approach will be given as examples in the next sections.

12.3.2 Segmentation of Skin Lesions

The first step in the analysis of a dermoscopic PSL image is segmentation, i.e. the automatic or semiautomatic separation of the lesion from the surrounding skin [30]. This can be done e.g. by transforming the original color image in Red, Green, and Blue (RGB) color model into a better suited color space. In [28], for example, a simplified YUV color space with luminance channel Y and chrominance channels U and V is used:

$$Y = \frac{1}{3}(R+G+B), \quad U = \frac{1}{2}\frac{R-B}{R+G+B}, \quad V = \frac{1}{2\sqrt{3}}\frac{2G-R-B}{R+G+B} \quad (12.1)$$

In order to emphasize the intensity difference between skin and lesion, subsequently a Mahalanobis transform $X \to X - E(X), X \to X/E(X^2)$ and an enhancement of Y by factor 4 are performed, resulting in a scaled YUV representation $f_{sYUV}(x,y)$ of the dermoscopic image. Next step is a reduction to the first principal component $f^1_{sYUV}(x,y) = z$, $0 \leq z \leq 255$. Due to the clear intensity difference between lesion and skin, f^1_{sYUV} usually shows a bimodal histogram, which is used to compute a segmentation threshold z_{seg} between the two modes. Simple thresholding and a subsequently applied cleaning operation Clean() for filling holes and detecting and masking out hair, delivers the lesion segment

$$L = \text{Clean}\big(\{(x,y) : f^1_{sYUV}(x,y) < z_{\text{seg}}\}\big) \quad (12.2)$$

Possible results of the automatic segmentation algorithm are illustrated in Fig. 12.5. In cases where the segmentation result is considered correct in shape, but slightly too big or too small, the dermatologist can adjust the threshold manually. In few cases, no reasonable segmentation is possible, especially if the lesion is very large. It is worth mentioning that this pragmatic and transparent segmentation procedure with on-demand, minimal machine-user interaction, fits well for systems following the mimic approach.

Another segmentation procedure, employing a hybrid method that combines statistical clustering of the color space and hierarchical region growing,

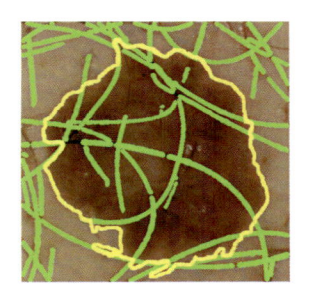

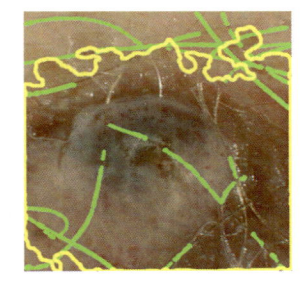

Fig. 12.5. *Examples of segmentation.* The results are obtained fully automatic (*left*), manually adjusted (*middle*), and the automatic segmentation fails (*right*)

is used in the Artificial Neural Network (ANN) classifier of the DANAOS expert system [31]. A validation and comparison of automatic segmentation with manual segmentation performed by the physician by drawing the lesion outline into the image can be found in [30].

12.3.3 Feature Extraction

Numerous features can be extracted from segmented PSLs. For example, in the case of a mimic system following the dermatoscopic ABCD rule, features modeling the diagnostic criteria asymmetry (A), border diffusiveness (B), color variety (C), and differential structures (D) are in the core of the approach. The more asymmetric, diffuse-bordered, diversely colored, and diversely structured a lesion, the higher its probability of being malignant (Sect. 12.2.1).

Asymmetry

Mathematically, asymmetry can be modeled, for instance, as *field color asymmetry* A_H or as *pixel color asymmetry* A_C [28] with symmetry axes computed by a principal component analysis (PCA). Absolute differences of color entropies on symmetric quadratic tiles q_{ij} (edge 50–100 pixels, depending on the lesion size) or single color values $f_H(x,y)$ are summed up in the hue channel H of the HSI color space (for the complicated color space transform see e.g. [32]), resulting in asymmetry scores

$$D_x^H = \sum |H_{i,j} - H_{i,-j})| \tag{12.3}$$

$$D_x = \frac{1}{N_L} \sum_{(x,y) \in L} |f_H(x,y) - f_H(x,-y)| \tag{12.4}$$

and D_y^H and D_y, analogously (D_x and D_y normalized by lesion size N_L). Then, the maxima

$$A_H = \max\{D_x^H, D_y^H\} \tag{12.5}$$
$$A_C = \max\{D_x, D_y\} \tag{12.6}$$

serve as quantifications of lesion asymmetry. Both features are of very high disciminative power, and they are correlated. It should be mentioned that these features are obviously also sensitive to irregular spatial distribution of differential structures in the lesion.

Border Diffusiveness

Border diffusiveness can be measured e.g. by quantifying the properties of a portion of pixels around the lesion border. In [28], for this purpose the Scaling Index Method (SIM) [33,34] has been used. It assigns a scaling index $\alpha(x,y)$ to each pixel (x,y). Pixels belonging to point-like structures (on a certain, definable scale) will have indexes α close to 0, those in linear structures around one, and those in area-like structures close to two. The higher the number of pixels with an $\alpha > 1$, the more diffuse is the lesion border. This leads to the feature

$$B_{\text{SIM}} = N_{\alpha>1}/N_B \tag{12.7}$$

as a possible quantification of the dermatoscopic B score. Since it is the ratio of pixels with diffuse local environment to the total number of pixels in the border area, the feature is independent of absolute lesion size and border length.

Color Variety

Color variety can be modeled as the entropy on the hue (H) channel of the HSI color space (compare asymmetry, above), considering the relative portions $p_i = p(c_i) = N(c_i)/N_L$ of color tones c_i, and computing from these the color entropy

$$C_H = -\frac{1}{\ln 256} \sum_{i=1}^{256} (p_i \ln p_i) \tag{12.8}$$

of the lesion area. It reflects the diversity of colors in the PSL. The higher C_H, the more distinguishable colors occur in the lesion.

Differential Structures

For the SIM-based modeling of differential structures it is referred to [29]. Instead, the lesion size expressed by the number of lesion pixels

$$S_L = N_L \tag{12.9}$$

shall be included, here. The size of a PSL is known to be an easily comprehensible and strong feature commonly used in the so-called clinical ABCD rule (naked eye inspection of the skin), with D for diameter as simple size measure. At this point, the strict modeling of the dermatoscopic ABCD rule is altered, without leaving the mimic approach, though.

12.3.4 Feature Visualization

As the finial diagnosis always lies in the responsibility of the physician, the visualization of computed features is an important part in supporting the diagnostic decision making process. Moreover, if computed features are made visible, e.g. as colored overlays on the original images, the criteria for the final decision become more transparent and documented. This, in turn, may improve the diagnostic skills of the physician.

While for the machine learning approach the visualization of features typically will cause problems because of the large number of features, and also because normally a considerable number of features hard to comprehend for a human being are employed (e.g. frequency-based features), the typically few high-level diagnostic criteria used in the mimic approach should be easy to visualize, especially because they usually follow well-known diagnostic procedures. Visualizations of three features introduced in the previous section shall illustrate some options.

Figure 12.6 shows the visualization of field color asymmetry A_H. Degrees of asymmetry are codified by a color scale, from green (low asymmetry) via blue to red (high asymmetry). Since entropy values on tiles are compared for computing the feature A_H, the visualization shows homogeneously colored tiles with the color representing the difference in entropy of tiles lying symmetrically to the first or second principal axis (maximum, cf. (12.5)). The axes are visualized as white lines. Only one half of the asymmetry coding overly along the first principal axis is shown, because the other half is symmetric and displaying it would not add further information, but rather hide information by covering the entire PSL. It seems worth noting that the spontaneous visual impression of asymmetry does, at least for the layman, not necessarily coincide with the complex asymmetry to be assessed in PSL diagnosis.

An example for the visualization of border diffusiveness measured by the feature B_{SIM} is given in Fig. 12.7. In this visualization, pixels belonging to the lesion border area are colored either blue or red. The color indicates whether the local environment of the pixel shows low (blue) or high (red) diffusiveness (scaling index $\alpha \leq 1$, or $\alpha > 1$, cf. (12.7)). In a lesion or lesion part sharply

Fig. 12.6. *Field color asymmetry.* Visualization of feature A_H for a CM (*left*) and a benign PSL (*right*)

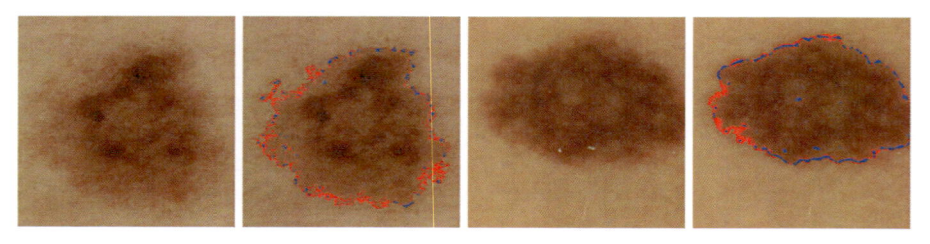

Fig. 12.7. *Border diffusiveness.* Visualization of feature B_{SIM} for a CM (*left*) and a benign PSL (*right*)

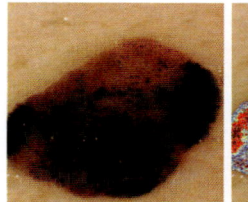

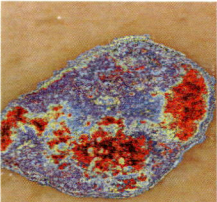

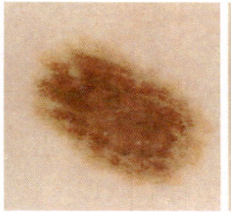

Fig. 12.8. *Color variety.* Visualization of feature C_H for a CM (*left*) and a benign PSL (*right*)

separated from the surrounding skin the border area appears (more or less) as a one or few pixels broad line. Dominating color is blue for such a lesion or lesion part. For diffuse sections of a lesion border, the border area is broad and the majority of pixels belonging to such sections are colored red.

The visualization of color variety modeled by the feature C_H in (12.8) is illustrated in Fig. 12.8. A color scale from blue via green to red is employed to codify the number of colors appearing in a PSL. In this visualization, a specific color code does not correspond to a specific lesion color. Rather than coding colors, the relative amount of colors appearing in a lesion are visualized, so that comparably homogeneous lesions appear in blue tones only, more inhomogeneous lesions will in addition show green tones, and lesions with a high number of different colors will in the feature visualization show red tones, as well. Note that for such a color variety visualization, in order to utilize the full range of color codes from blue to red, the spectrum of colors that practically can appear in dermoscopic skin images has to be used as reference, not the entire color spectrum.

12.3.5 Classification Methods

A type of classification methods frequently used in a machine learning approach is the ANN (Fig. 12.9). The ANN is trained by a set of features computed from images of nevi with known malignancy. Applying the trained ANN to a feature vector of an unknown PSL will result in a certain malignancy score (in the simplest case with only two possibilities, *benign* and *malignant*). Unless other methods are used in parallel for computation and visualization of diagnostic criteria, the way in which the ANN comes to its conclusion is normally not transparent to the user. This has often been criticized as one of the major drawbacks of ANN. An example of a system using an ANN classifier is the DANAOS expert system [31]. For this system it has been shown that the area under the Receiver Operating Characteristic (ROC) curve (AUC) increases rapidly with an increase of training dataset size from 50 images ($AUC = 0.67$) to 2,000 images ($AUC = 0.86$).

Classification methods appropriate for the mimic approach are e.g. logistic regression (Logreg) and Classification And Regression Tree (CART). In [28],

Fig. 12.9. *Principle of a 3-layer ANN*

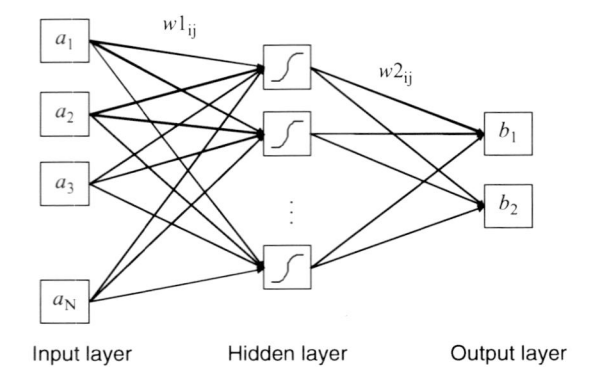

Table 12.3. *Classifier comparison on 466 PSLs (125 CM, 341 benign PSL)*

Classifier	Validation	ρ	κ	SN	SP
Logreg	10 × 10-fold cross-validations	0.895	0.737	0.862[a]	0.911[a]
CART (AACS)	10-fold cross-validation	0.89	–[b]	0.76[a]	0.94[a]
ANN (linear layer)	10 × 100/100 train/test	0.896	0.792	0.896	0.896

[a] at ρ-maximizing operating point; [b] not computed by S-Plus

both methods have been applied to a set of 466 PSLs (125 melanomas, 341 benign nevi), using the 5 mimic features A_H (field color asymmetry), A_C (pixel color asymmetry), B_{SIM} (border diffusiveness), C_H (color variety) and S_L (lesion size) introduced in Sect. 12.3.3.

To estimate the classification performance of logistic regression, a ten-fold cross-validation was computed 10 times, resulting in a mean value of 0.945 for the AUC (cf. Table 12.3). To build a best classifier from all data, logistic regression was computed for all 5 features. Significant features in the resulting regression model were A_H ($p < 0.001$), and C_H ($p < 0.001$) and S_L ($p < 0.05$). With these three features another logistic regression was computed resulting in an estimate for the melanoma probability of

$$P(f \in K_{CM}) = 1/(1 + e^{-2.25 A_H - 21.30 C_H - 0.0000128 S_L + 13.89}) \qquad (12.10)$$

and a set of threshold classifiers Logreg_s for cut points s:

$$\text{Logreg}(f) = \begin{cases} \text{benign} & \text{if } P(f \in K_{CM}) < s \\ \text{malignant} & \text{otherwise} \end{cases} \qquad (12.11)$$

with K_{CM} being the class of CMs. Figure 12.10 shows the ROC curve for this set of classifiers.

To create a reasonable CART classifier, a CART analysis was performed. In the course of this analysis, a large number of classification trees with binary splits have been computed, and for each of these trees, the ten-fold cross-validated discriminative power has been estimated. Based on these results, a tree with a reasonable number of splits has been chosen (Fig. 12.11), not too

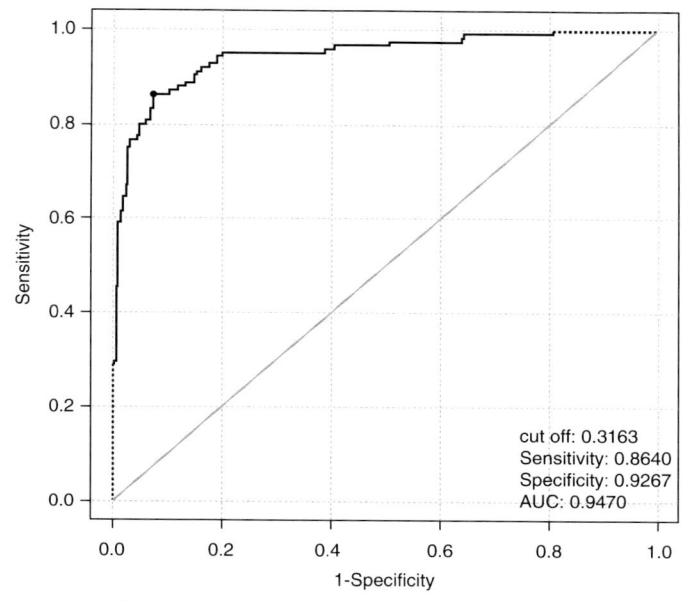

cut off: 0.3163
Sensitivity: 0.8640
Specificity: 0.9267
AUC: 0.9470

Cut-off that minimizes the distance between the curve and upper left corner.

Fig. 12.10. *ROC curve for the Logreg classifier in (12.11)*

big, in order to avoid over-fitting of the model, and not too small, in order to gain satisfactory accuracy. As in the logistic regression analysis, the accuracy is estimated by cross-validation, while the final tree is constructed from the entire set of images.

As a look at the splits shows, the two most powerful (i.e., top-level) ones are based on the two asymmetry features A_C and A_H, classifying $286 + 29 = 315$ of the benign PSLs correctly, while subsequently color variety C_H and lesion size S_L are involved in the final decision for CM in 95 cases.

For comparison, also an ANN has been trained and tested with a cross-validation scheme on the same set of 466 PSLs in [28]. Figure 12.9 shows the principle layout of the employed 3-layer ANN, a so-called linear-layer perceptron. The weight factors $w1_{ij}$ and $w2_{ij}$ reflect the knowledge about how to distinguish CMs from benign PSLs. These factors have to be trained by a suitable learning strategy such as error back propagation. The input layer receives the feature values, at the output layer the classification result is obtained, here: $b_1 =$ malignant and $b_2 =$ benign. Assuming five input nodes ("neurons") for the five features, four hidden layer nodes, and two output layer nodes, this network has $5 \times 4 + 4 \times 2 = 28$ weight factors that had to be learned from the training set in each of the cross-validation tests.

Classification results achieved with Logreg, CART and ANN classifiers are compared in Table 12.3. The cross-validated total correct-classification rate ρ is around 0.89 for all three types of classifiers. Logreg and CART show, at

Fig. 12.11. *CART decision tree.* The example shows a decision tree with four splits

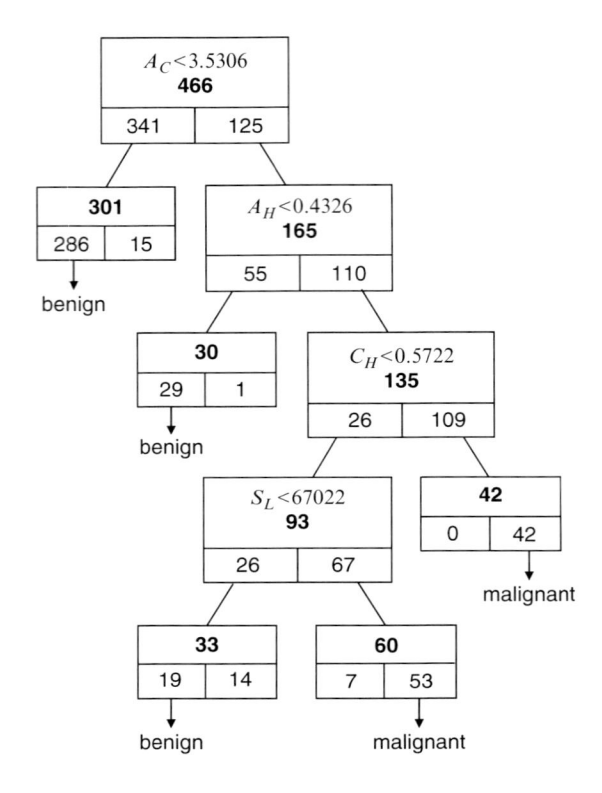

```
                    A_C < 3.5306
                        466
                    341 | 125
                   /         \
              301            A_H < 0.4326
           286 | 15              165
              ↓              55 | 110
           benign           /         \
                          30         C_H < 0.5722
                        29 | 1           135
                          ↓          26 | 109
                        benign      /         \
                              S_L < 67022      42
                                  93         0 | 42
                              26 | 67           ↓
                             /       \       malignant
                            33        60
                         19 | 14    7 | 53
                            ↓          ↓
                          benign    malignant
```

the operating point that minimizes the distance between the ROC curve and the ideal classifier point (sensitivity = specificity = 1), specificities superior to sensitivities due to the fact that the dataset contained less CMs (125) than benign PSLs (341) and therefore correct assessment of benign lesions is trained better than that of malignant lesions. In contrast to this, the ANN compensates different class sizes (sensitivity = specificity).

12.4 Commercial Systems

The current market offers a variety of dermoscopy devices and systems. Beyond common system design principles the products show more or less essential differences in terms of image acquisition devices, system functionalities, and the approach and degree of CADx support.

12.4.1 System Design Principles

The design of dermoscopy systems for the medical practice is commonly characterized by the goal of supporting the entire workflow of diagnostic decision making and patient management. Hereby, digital image analysis is one aspect, only. Required functions of a dermoscopy system include:

1. Electronic patient record;
2. Image acquisition and management;
3. Report creation and storage;
4. Image analysis and decision support;
5. Support of continued medical training;
6. Tele-consultation (optional).

Important features of a dermoscopy system built according to these requirements include: patient-, case- and lesion-based documentation; follow-up support by means of (qualitative and quantitative) comparison of the same lesion at different examination dates; automatic computation of scores derived from diagnostic algorithms (Sect. 12.2.1) such as the ABCD rule or the 7-point checklist; comparison of pathohistologic report with the corresponding dermoscopic image, so that the physician can validate his diagnostic skills; support of diagnostic skills training by offering similar cases from a reference database; explanations and visualizations of diagnostic scores and the features employed for a skin lesion under examination.

12.4.2 Image Capture Devices

Common image capture devices use calibrated Charge-Coupled Device (CCD) video camera technology or digital still image camera technology for dermoscopic image acquisition [14, 16]. The devices acquire either one image in the visible light spectrum (mono-spectral imaging), or a set of images in different wavelength bands (multi-spectral imaging). The latter is motivated by the fact that light of different wavelengths penetrates the skin to different depths. For example, MelaFind and SIAscope (Table 12.4) acquire a set of narrow band images between 400 and 1,000 nm (violet to near-infrared). Instead of using immersion oil, the MoleMax uses polarized light to suppress reflection from the skin surface.

Dermoscopic devices with image acquisition capability are, for example:

- The HEINE DELTA 20 Dermatoscope (HEINE Optotechnik GmbH & Co. KG, Herrsching, Germany) handheld device with digital camera adapter.
- The DermLite FOTO (3gen LLC, San Juan Capistrano, CA, USA), a dermoscope that is combined with a digital camera (polarized light).
- The DermLite II Multi-spectral (3gen LCC), a handheld device for multi-spectral imaging with polarized light (white, blue, yellow, red).
- The EasyGenius (BIOCAM GmbH, Regensburg, Germany) handheld device, combining the DermoGenius basic dermatoscope with a digital camera.
- The Fotofinder dermoscope (FotoFinder Systems GmbH, Bad Birnbach, Germany) camera. the SolarScan Sentry (Polartechnics Ltd, Sydney, Australia) high-resolution digital oil immersion dermoscopy camera.

Table 12.4. *Commercial dermoscopic systems with image analysis component*

System[a]	Company	Spect	Image analysis
DB-Mips	Biomips Engineering	mono	ANN, similarity
DermAssist	Romedix	mono	clin-ABCD
DermoGenius[b] ultra	BIOCAM	mono	ABCD, ref-DB
Fotofinder dermascope	FotoFinder Systems	mono	ref-DB
MelaFind[c]	Electro-Optical-Sciences	multi	ref-DB
MicroDerm	Visiomed AG	mono	ANN, ABCD
MoleMax II/3	Derma Medical Systems	mono	ABCD, 7-point
MoleView	Astron Clinica	multi	SIAscopy score[d], 7-point
Solarscan	Polartechnics	mono	ref-DB
VideoCap 100	DS Medica	mono	ref-DB

spect = spectrum (mono/multi); *ABCD* = dermatoscopic ABCD rule with D for differential structures; *clin-ABCD* = clinical ABCD rule with D for diameter; *ref-DB* = reference image database. [a]system names are registered trademarks of the companies; [b]registered trademark of LINOS AG, exclusive rights with BIOCAM; [c]under approval; [d]proprietary scoring scheme

- The MoleMate (Astron Clinica Ltd, Cambridge, UK) melanoma screening device with the SIAscope handheld scanner.
- The Nevoscope Ultra (TransLite, Sugar Land, TX, USA) handheld with adapter for digital camera or camcorder.
- The EasyScan Pico (Business Enterprise S.r.l., Trapani, Italy) handheld high-definition video dermatoscope system.
- The DG-3 digital microscope and the VL-7EXII video microscope (Scalar America, Sacramento, CA, USA).

12.4.3 Dermoscopy Computer Systems

Several commercial dermoscopy systems with integrated image analysis components are on the market (Table 12.4) [14–17]. Most of these systems support both microscopic (typically 5- to 15-fold magnification) and macroscopic imaging (e.g., EasyScan TS Pico, FotoFinder dermascope). In addition, some systems comprise room cameras capable of capturing whole body images (mole mapping) in a standardized way to aid the detection of new moles (e.g., MoleMax II, Molemax 3, DB-Mips). The systems are therefore equipped with one up to three digital video and still image cameras of specific types.

12.5 Evaluation Issues

High-quality, state-of-the-art evaluation of CADx systems is, without any doubt, a challenging task. Attempts to give reliable estimates of the performance of such systems are hampered by different types of biases caused by limitations in resources and study designs. This may, to a certain extent, explain why the methodological quality of studies is generally not good [15, 23, 25].

12.5.1 Case Databases

Particularly critical for evaluation is the composition of case databases used for evaluation. In order to guarantee that a case database reflects realistic clinical conditions, the choice of cases must meet certain requirements [23]:

- Random or consecutive selection of lesions (to avoid selection bias).
- Clear definition of inclusion and exclusion criteria.
- Inclusion of all lesions clinically diagnosed as PSLs.
- Inclusion of clearly benign lesions that were not excised (to avoid verification bias); diagnostic gold standard in these cases is short-term follow-up.

In order to gain sufficient accuracy in statistical terms, the size of the case database has to be large enough. Depending on the classification method under evaluation and the narrowness of the Confidence Interval (CI) aimed at for point estimates such as sensitivity and specificity, datasets will have to be of a size of some hundreds to some thousands of cases with biopsy-proven and clearly benign PSLs.

12.5.2 Evaluation Methods

In order to avoid methodological weaknesses, requirements have to be added to the list of database-related requirements of Sect. 12.5.1 ([23], extended):

- Clear definition of the study setting.
- Reporting of instrument calibration.
- Intra- and inter-instrumental repeatability.
- Classification on independent test set.
- Comparison of computer diagnosis with human diagnosis.
- Comparison of human diagnosis without and with CADx support.

Recent publications indicate that there is awareness for the problem, especially in the field of lung and breast cancer CADe and CADx (see e.g. [21, 35]). But the implementation of good practice addressing the above listed requirements is still a great challenge also in these fields.

12.6 Conclusion

In the published literature, the accuracy reached by current dermoscopy CADx systems for CM diagnosis is comparable to the accuracy achieved by human specialists. But this evidence is still weak. Large randomized controlled trials evaluating dermoscopy CADx systems under routine conditions are lacking. Many dermatologists therefore continue to be skeptical about the routine use of such systems [15]. Obviously, there is a need for large, high-quality case databases to support system validation. Future trials should not only evaluate the accuracy of the computer system and compare this accuracy to that of human experts, but should also compare the diagnostic performance of human readers alone with the performance of human readers using a dermoscopy system as second opinion.

New imaging techniques, such as multi-spectral imaging, could be a next step in the development of more accurate systems for the classification of PSMs. Intensified efforts towards an extension of the computer support to non-melanocytic skin lesions are another important step forward. To develop and validate sophisticated algorithms and systems supporting such new imaging techniques for an extended range of skin lesions and bring them into routine use, embedded in the physician's work flow as one of several components of computer support – electronic patient record, whole body imaging (CADe), tele-dermatology, and others – will demand a lot of research and innovative system development. But certainly there is the potential to further improve early recognition of CM and other threatening diseases of the human skin.

References

1. Garbe C, Leiter U. Melanoma epidemiology and trends. Clin Dermatol. 2009;27(1):3–9.
2. Garbe C, Eigentler TK. Diagnosis and treatment of cutaneous melanoma: state of the art 2006. Melanoma Res. 2007;17(2):117–27.
3. Losina E, Walensky RP, Geller A, et al. Visual screening for malignant melanoma: a cost-effectiveness analysis. Arch Dermatol. 2007;143(1):21–28.
4. Hansen C, Wilkinson D, Hansen M, et al. How good are skin cancer clinics at melanoma detection? Number needed to treat variability across a national clinic group in Australia. J Am Acad Dermatol. 2009;61(4):599–604.
5. English DR, Mar CD, Burton RC. Factors influencing the number needed to excise: excision rates of pigmented lesions by general practitioners. Med J Aust. 2004;180(1):16–19.
6. Braun RP, Rabinovitz HS, Oliviero M, et al. Dermoscopy of pigmented skin lesions. J Am Acad Dermatol. 2005;52(1):109–21.
7. Pehamberger H, Steiner A, Wolff K. In vivo epiluminescence microscopy of pigmented skin lesions. I. Pattern analysis of pigmented skin lesions. J Am Acad Dermatol. 1987;17(4):571–83.
8. Braun RP, Rabinovitz HS, Oliviero M, et al. Pattern analysis: a two-step procedure for the dermoscopic diagnosis of melanoma. Clin Dermatol. 2002;20(3):236–39.
9. Nachbar F, Stolz W, Merkle T, et al. The ABCD rule of dermatoscopy. High prospective value in the diagnosis of doubtful melanocytic skin lesions. J Am Acad Dermatol. 1994;30(4):551–59.
10. Stolz W, Braun-Falco O, Bilek P, et al. Color Atlas of Dermatoscopy, 2nd ed. London: Wiley-Blackwell; 2002.
11. Argenziano G, Fabbrocini G, Carli P, et al. Epiluminescence microscopy for the diagnosis of doubtful melanocytic skin lesions. Comparison of the ABCD rule of dermatoscopy and a new 7-point checklist based on pattern analysis. Arch Dermatol. 1998;134(12):1563–70.

12. Menzies SW, Ingvar C, McCarthy WH. A sensitivity and specificity analysis of the surface microscopy features of invasive melanoma. Melanoma Res. 1996;6(1):55–62.
13. Thompson JF, Shaw HM, Hersey P, et al. The history and future of melanoma staging. J Surg Oncol. 2004;86(4):224–35.
14. Psaty EL, Halpern AC. Current and emerging technologies in melanoma diagnosis: the state of the art. Clin Dermatol. 2009;27(1):35–45.
15. Rallan D, Harland CC. Skin imaging: is it clinically useful? Clin Exp Dermatol. 2004;29(5):453–59.
16. Wang SQ, Rabinovitz H, Kopf AW, et al. Current technologies for the in vivo diagnosis of cutaneous melanomas. Clin Dermatol. 2004;22(3):217–22.
17. Marghoob AA, Swindle LD, Moricz CZM, et al. Instruments and new technologies for the in vivo diagnosis of melanoma. J Am Acad Dermatol. 2003;49(5):777–97; quiz 798–9.
18. Marghoob AA, und Alfred W Kopf RB. An atlas of dermoscopy. In: Marghoob AA, und Alfred W, Kopf RB, editors. Encyclopedia of Visual Medicine Series. London: Taylor & Francis; 2004.
19. Menzies SW, Crotty KA, Ingvar C. An atlas of surface microscopy of pigmented skin lesions. In: Menzies SW, Crotty KA, Ingvar C, editors. Dermoscopy. Sydney: Mcgraw-Hill; 2003.
20. Argenziano G, Ferrara G, Francione S, et al. Dermoscopy: the ultimate tool for melanoma diagnosis. Semin Cutan Med Surg. 2009;28(3):142–48.
21. Elter M, Horsch A. CADx of mammographic masses and clustered microcalcifications: a review. Med Phys. 2009;36(6):2052–68.
22. Vestergaard ME, Macaskill P, Holt PE, et al. Dermoscopy compared with naked eye examination for the diagnosis of primary melanoma: a meta-analysis of studies performed in a clinical setting. Br J Dermatol. 2008;159(3):669–76.
23. Rosado B, Menzies S, Harbauer A, et al. Accuracy of computer diagnosis of melanoma: a quantitative meta-analysis. Arch Dermatol. 2003;139(3):361–7; discussion 366.
24. Annessi G, Bono R, Sampogna F, et al. Sensitivity, specificity, and diagnostic accuracy of three dermoscopic algorithmic methods in the diagnosis of doubtful melanocytic lesions: the importance of light brown structureless areas in differentiating atypical melanocytic nevi from thin melanomas. J Am Acad Dermatol. 2007;56(5):759–67.
25. Rajpara SM, Botello AP, Townend J, et al. Systematic review of dermoscopy and digital dermoscopy: artificial intelligence for the diagnosis of melanoma. Br J Dermatol. 2009;161(3):591–604.
26. Blum A, Luedtke H, Ellwanger U, et al. Digital image analysis for diagnosis of cutaneous melanoma: development of a highly effective computer algorithm based on analysis of 837 melanocytic lesions. Br J Dermatol. 2004;151(5):1029–38.
27. Horsch A, Stolz W, Neiss A, et al. Improving early recognition of malignant melanomas by digital image analysis in dermatoscopy. Stud Health Technol Inform. 1997;43 Pt B:531–35.
28. Horsch A. Computergestützte Diagnostik für Hautkrebsfrüherkennung, Ösophagustumorstaging und Gastroskopie. PhD thesis Technische Universität München; 1998.
29. Pompl R. Quantitative Bildverarbeitung und ihre Anwendung auf melanozytäre Hautveränderungen. PhD thesis Technische Universität München; 2000.

30. Joel G, Schmid-Saugeon P, Guggisberg D, et al. Validation of segmentation techniques for digital dermoscopy. Skin Res Technol. 2002;8(4):240–49.
31. Hoffmann K, Gambichler T, Rick A, et al. Diagnostic and neural analysis of skin cancer (DANAOS). A multicentre study for collection and computer-aided analysis of data from pigmented skin lesions using digital dermoscopy. Br J Dermatol. 2003;149(4):801–09.
32. Gonzalez RC, Woods RE. Digital Image Processing. Reading: Addison-Wesley; 1992.
33. Morfill G, Scheingraber H, Wiedemann G. Spatial filtering process and system. European Patent EP0700544, PCT No. PCT/EP94/01752 Sec. 371 Date Mar. 18, 1996 Sec. 102(e) Date Mar. 18, 1996 PCT Filed May 27, 1994 PCT Pub. No. WO94/28500 PCT Pub. Date Dec. 8, 1994.
34. Räth C, Morfill G. Texture detection and texture discrimination with anisotropic scaling indices. J Opt Soc Am. 1997;12:3208–15.
35. Horsch A, Blank R, Eigenmann D. The EFMI reference image database initiative: concept, state and related work. Proc CARS. 2005;447–52.

13
CADx Mammography

Lena Costaridou

Summary. Although a wide variety of Computer-Aided Diagnosis (CADx) schemes have been proposed across breast imaging modalities, and especially in mammography, research is still ongoing to meet the high performance CADx requirements. In this chapter, methodological contributions to CADx in mammography and adjunct breast imaging modalities are reviewed, as they play a major role in early detection, diagnosis and clinical management of breast cancer. At first, basic terms and definitions are provided. Then, emphasis is given to lesion content derivation, both anatomical and functional, considering only quantitative image features of micro-calcification clusters and masses across modalities. Additionally, two CADx application examples are provided. The first example investigates the effect of segmentation accuracy on micro-calcification cluster morphology derivation in X-ray mammography. The second one demonstrates the efficiency of texture analysis in quantification of enhancement kinetics, related to vascular heterogeneity, for mass classification in dynamic contrast-enhanced magnetic resonance imaging.

13.1 Introduction

Breast cancer is the most common cancer in women worldwide and the second leading cause of cancer deaths after lung cancer. Breast imaging modalities have a major role in early detection, diagnosis and clinical management of breast cancer, relying on observer interpretation. The accuracy of mammographic image interpretation depends on both image quality, provided by the various breast imaging modalities, and quality of observer interpretation, subjected to intra- and inter-observer performance variabilities.

Image analysis methods targeted to aid medical image interpretation have evolved in Computer-Aided Detection (CADe) and Computer-Aided Diagnosis (CADx) schemes [1–3]. CADe is targeted to identification of the location of suspect regions in a medical image. CADx is targeted to characterization (malignancy vs. benignity) of a Region of Interest (ROI) or a lesion region (presupposing a segmentation stage), initially located (or delineated) either by a CADe or an observer, by providing the probability of malignancy/benignity.

T.M. Deserno (ed.), *Biomedical Image Processing*, Biological and Medical Physics, Biomedical Engineering, DOI: 10.1007/978-3-642-15816-2_13,
© Springer-Verlag Berlin Heidelberg 2011

Image enhancement methods have also been proposed to aid detection and diagnosis, by improving image contrast and visibility of detail, often taking advantage of multi-scale image representations, such as wavelets to deal with varying mammographic image lesion sizes [4–6].

CADx, by definition, is targeted to increasing diagnostic specificity (i.e. fraction of benign lesions correctly characterized by the system), while maintaining high sensitivity (i.e. fraction of malignant lesions correctly characterized by the system), to aid patient management (follow-up vs. biopsy) and reduce intra- and inter-observer variability. The large variability of lesion appearance and low conspicuity challenges CADx research [7, 8].

In this chapter, lesion content derivation, by means of both anatomical and functional quantitative image features of Micro-Calcification (MC) clusters and masses across modalities is highlighted. Additionally, two CADx application examples are provided. The first example highlights the effect of segmentation algorithm accuracy on MC cluster content derivation and classification in X-ray mammography. The second example demonstrates the efficiency of texture analysis to quantify enhancement kinetics (vascular) heterogeneity for mass classification in Dynamic Contrast Enhanced (DCE) Magnetic Resonance Imaging (MRI).

13.2 Basic Terms and Definitions

In this section, we provide basic terms and definitions of CADx schemes, including a short mention of CADe schemes and their performance.

13.2.1 Breast Imaging Modalities

Screen-Film Mammography (SFM) provides structural (anatomical) detail of tissue properties (spatial resolution 30 μm). Currently, it is the most effective modality in detecting lesions, such as MCs, breast masses and architectural distortions, however challenged by the presence of dense breast parenchyma. Inherent limitations of SFM, such as the sigmoid response of the screen-film, have lead to the development of Full-Field Digital Mammography (FFDM). FFDM utilizes digital detectors to accurately and efficiently convert x-ray photons to digital signals, although with somewhat decreased spatial resolution (i.e. 45 μm), demonstrating improved detection accuracy in screening of dense breasts [9]. Currently, emerging Digital Breast Tomosynthesis (DBT) offers the capability of tomographic images at various breast depths, dealing with tissue superimposition along the breast depth axis, to improve detection of soft tissue abnormalities, lending itself also to functional imaging, by means of contrast-enhanced DBT.

When image interpretation is not supported by X-ray imaging of breast anatomy, additional structural and functional (i.e. related to physiology) tissue properties are solicited from breast imaging modalities, such as Ultrasonography (US) and MRI.

Technological developments in high frequency (≥ 10 MHz) transducer US technology combined with compounding and tissue harmonic imaging have provided improved contrast resolution and margin detection 2D/3D US anatomical imaging, with axial resolution of B-Mode scanning, representing the clinical standard, ranging from 130 to 300 μm. Contrast harmonic imaging and power Doppler US, providing information about blood flow, offer functional information to assess tissue vascularity (vessel architecture), related to angiogenesis, while strain (compression) US imaging provides the capability to characterize elastic breast tissue properties. US adjunctively used with X-ray mammography has been found to increase mass detection rates [10].

MRI provides 3D anatomy of the uncompressed breast through specially designed breast coils, with in plane spatial resolution of 1.5 T systems, a current clinical standard, at least 1 mm and z-axis resolution ≤ 2.5 mm, with emerging systems targeting sub-millimeter resolution. More importantly MRI may assess physiologic information indicative of increased vascular density and vascular permeability changes related to angiogenesis, by means of DCE-MRI, with temporal resolution of 60–120 s per volume acquisition, representing the current clinical standard. The functional character of MRI is further augmented by proton (^{1}H) Magnetic Resonance Spectroscopy Imaging (MRSI). MRI plays a major role as an adjunctive screening tool for high risk of hereditary cancer in women, detecting mammographically occult cancer, as well as a diagnostic tool for lesions undetermined by mammography and US, providing also extent and multi-centricity of the disease [11,12].

In addition to improvements within each modality, there is a trend towards hybrid breast imaging systems combining anatomical and functional breast tissue properties, providing co-registered breast images, including but not limited to the above mentioned modalities [13]. Specifically, prototype systems have been developed combining whole-breast 3D ultrasound with FFDM-DBT imaging and optical imaging (diffuse optical tomography) to MRI anatomical imaging.

13.2.2 Mammographic Lesions

The Breast Imaging Reporting and Data System (BI-RADS) of American College of Radiology (ACR) [14] lexicon defines masses, MC clusters, architectural distortion and bilateral asymmetry as the major breast cancer signs in X-ray mammography. A mass is a space occupying lesion seen at least in two different mammographic projections. If a mass is seen only in a single projection is called asymmetric density. When a focal area of breast tissue appears distorted with spiculations radiating from a common point and focal retraction at the edge of the parenchyma, while no central mass is definable, it is called architectural distortion.

Masses have different density (fat containing masses, low density, isodense, high density), different margins (circumscribed, micro-lobular, obscured, indistinct, spiculated) and different shape (round, oval, lobular, irregular). Round and oval shaped masses with smooth and circumscribed margins usually indicate benign changes. On the other hand, a malignant mass usually has a spiculated, rough and blurry boundary.

Architectural distortion of breast tissue can indicate malignant changes especially when integrated with visible lesions such as mass, asymmetry or calcifications. Architectural distortion can be classified as benign when including scar and soft-tissue damage due to trauma. Asymmetry in a pair of left and right mammograms is expressed as volume, breast density or duct prominence differences without a distinct mass.

MCs are deposits of calcium in breast tissue, associated to underlying biological disease processes. A number of MCs grouped together is termed a cluster and it may be a strong indication of cancer. A cluster is defined as at least three MCs within a $1\,cm^2$ area. Benign MCs are usually larger and coarser with round and smooth contours. Malignant MCs tend to be numerous, clustered, small, varying in size and shape, angular, irregularly shaped and branching in orientation.

13.2.3 CADe Schemes

CADe schemes in X-ray imaging account for a successful paradigm of application of image analysis to the clinical environment, resulting in several FDA approved systems. MC clusters and masses have been the main subject of CADe research [1–3], especially challenged by subtlety, while architectural distortions and bilateral asymmetry, also important early non-palpable breast cancer signs, are currently under-researched, however important subjects of future research [3].

CADe schemes for MCs are based on supervised pattern classification methods, usually involving an initial stage of candidate MC cluster detection (segmentation), followed by feature extraction and classification for removal of false positive detections. In addition, image enhancement and stochastic image modeling methods have been exploited to enhance MC local contrast to their surrounding tissue.

Most of CADe schemes for masses also follow the two stage pattern classification paradigm, involving detection (segmentation) of candidate mass regions and classification of candidate masses to true masses or normal tissue. ROI-based pixel-wise or mass region analysis, pre supposing a mass segmentation step, have both been utilized in the first stage, often in combination to mass filtering methods.

The advent of FFDM and emerging DBT systems has already motivated the development CADe schemes, taking advantage of successful film-based CADe schemes [15].

At the laboratory level, clinical evaluation studies of CADe systems have reported increased cancer detection rate (i.e. detection sensitivity). In case of MCs, they enable radiologists to detect more subtle cancers. In case of masses, decreased detection rates have been observed. Here, the higher false positive detection rates represent a major challenge, reducing radiologists confidence in CADe system output and increased recall rate. The impact of commercially available CADe systems, as assessed by large scale clinical studies, is somewhat mixed. There are studies reporting significant increase of breast cancers detected with an acceptable increase of recall rate [16], while no improvement was recently reported [17].

13.2.4 CADx Architectures

Classification is the main task of a CADx system. The majority of the reported systems formulates mammographic ROI or lesion characterisation in the context of quantitative image feature extraction and supervised feature pattern classification in two classes (i.e. benignity or malignancy) or more (e.g. cyst, benign and malignant solid masses in breast US). During training, classification rules are learned from training examples and subsequently applied to the classification of new unknown ROIs or lesion segment/segments.

The typical architecture of a single breast imaging modality ("unimodal") CADx is depicted in Fig. 13.1. Initialization of a CADx scheme requires identification of a lesion ROI provided either by a radiologist or by the output of a CADe scheme. This step benefits from denoising, contrast enhancement and/or segmentation techniques, the later introduced to restrict image analysis in the breast parenchyma region.

The first two stages deal with lesion segmentation and feature extraction, specifically designed to derive lesion ROI image content. The lesion segmentation stage is often optional, with subsequent classification relying on features extracted from the input ROI.

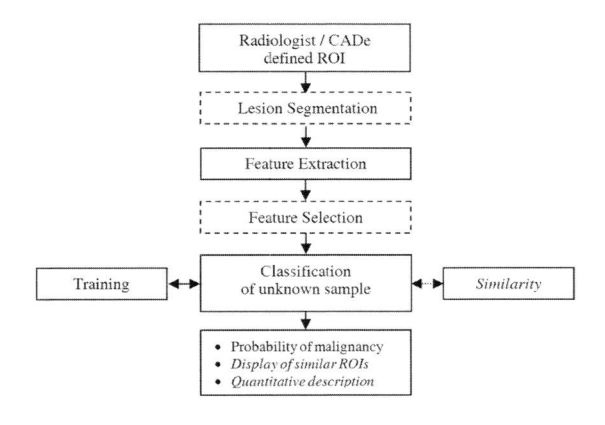

Fig. 13.1. *Flowchart of a CADx scheme.* Typical steps are indicated by *solid lines*, while *dashed* ones represent more elaborate options. Additional outputs of Content-Based Image Retrieval (CBIR) systems are indicated by italics

In case of a segmentation stage, pixel-wise accuracy should be quantitatively accessed utilizing pixel-wise ground truth. Segmentation performance metrics utilized include shape differentiation metrics, Area Overlap Measure (AOM), which is defined as the ratio of the intersection to the union between computer- and radiologist-based segmented areas, and the fraction of correctly segmented lesions as a function of area overlap threshold.

Image feature extraction methods aim to capture and quantify image appearance alterations, due to underlying biological processes reflected either as morphology or texture variations, mimicking or complementing radiologist interpretation. Recent advances of CADx architectures involve extraction of complementary image content derived from additional mammographic views [18] or breast imaging modalities [19, 20], following the multi-modality approach to breast imaging and leading to lesion intra- and inter-modality feature fusion or decision fusion systems. Quantification of temporal change involving serial analysis of lesion ROI in one modality is another promising architecture category [21]. When merging lesion content from multiple images, automated image registration methods play an important role [22].

Increasing dimensionality of the feature vector describing lesion content often decreases supervised classification performance, due to sparse distribution in the feature space. This, combined with the fact that feature discriminative power is varying, introduces the need for selection of the most discriminant features. As optimal feature selection by means of exhaustive search is not feasible in case of highly dimensional feature vectors, heuristic approaches, such as Linear Discriminant Analysis (LDA), sequential forward selection and backward elimination, as well as Genetic Algorithms (GA) are applied. Use of feature selection techniques has been reported to significantly influence classification performance [23].

The most widely used classifiers are the k-Nearest Neighbor (kNN), Artificial Neural Network (ANN) and Support Vector Machine (SVM), as well as regression methods such as LDA and Logistic Regression Analysis (LRA). Comparison of different classifiers performance has been reported for specific case samples analyzed [23, 24].

Recently, the Content-Based Image Retrieval (CBIR) architectures have enriched CADx architectures, by providing the classification outcome not in terms of a decision, but as a pictorial display of relevant image examples of known pathology, stored in an image database relevant (similar) to a pictorial query regarding an unknown ROI sample [25–27]. This type of output acts as a "visual" aid to support clinical decision making. The key factor of CBIR systems is selection of appropriate features to quantify image content in terms of perceptual similarity, ranging from image distance metrics to machine learning algorithms, trained to predict qualitative (observer/perceptual) image similarity ratings in terms of quantitative image lesion features, "learned similarity"[25]. Post-query user feedback (relevance feedback) indicating positive or negative relevance of retrieved images, due to differences between human and computer implemented similarity, is suggested as an additional alternative to previously "silent"(i.e. decision-only) CADx.

The performance of CADx schemes is evaluated by a Receiver Operating Characteristic (ROC) curve, expressing sensitivity as a function of 1− specificity, and utilizing the A_z metric and its Standard Error (SE), where A_z is denoting the area under the ROC curve. As the performance requirements of CADx systems are targeted to increase specificity without sacrificing sensitivity, partial ROC curve, such as $_{0.9}A_z$ defined as the area under the ROC curve with 0.9 indicating the target minimum sensitivity, is a more representative indicator of classification performance [28]. In case of CBIR performance, precision-recall curves are also utilized [25].

Following, CADx schemes for MCs and masses across three breast imaging modalities (X-ray mammography, US, MRI) are reviewed with emphasis on feature extraction methods across modalities.

13.3 CADx Schemes in X-ray Mammography

CADx schemes performances obtained report higher performances for masses, rendering MC cluster characterization more challenging [1–3] and reversing the trend observed in performance of corresponding CADe schemes. Computer extracted features are obtained on a single image lesion-basis from one mammographic view (left or right; CC or MLO) or on a case-basis from two or more views (e.g. CC, MLO, special diagnostic views or prior views).

13.3.1 Morphology Analysis of MC Clusters

In Fig. 13.2, the variability in morphology of individual MCs of a benign and a malignant cluster is depicted. The images are obtained from Digital Database for Screening Mammography (DDSM)[1]. Image content analysis includes morphology and location of the cluster, morphology of individual MCs and spatial distribution of MCs within the cluster. In particular, two major approaches are followed in deriving cluster content:

A: *Individual MCs* – Content is derived from statistics features of individual MC members of a cluster. In particular:

- Size (area, perimeter, length, effective thickness, and volume).
- Shape (circularity, elongation, compactness, eccentricity, concavity index, central and boundary moments, shape signature, boundary Fourier descriptors).
- Contrast.

---

[1] http://marathon.csee.usf.edu/Mammography/Database.html

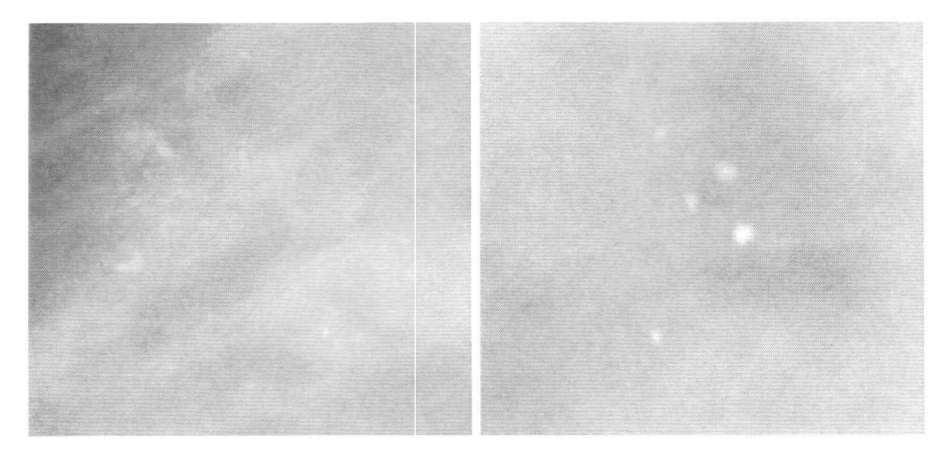

Fig. 13.2. *Morphology variability of a malignant and a benign MC cluster in X-ray mammography. Left*: Malignant MCs are numerous, more densely packed, small, varying in size, shape and orientation (B_3136_RIGHT_MLO); *Right*: Benign MCs are larger, more rounded, smaller in number, less densely packed and more homogenous in size and shape (C_0246_LEFT_CC)

B: *Cluster region* – The MC cluster is considered as region entity. Corresponding features are:

- Size (area, perimeter), number of MCs in a cluster.
- Distribution of location of MC in a cluster (proximity: mean number of nearest neighbors, distance to nearest neighbor; cluster density: mean distance between MCs), cluster location (distance to pectoral and breast edge).
- Shape (circularity, elongation, eccentricity, central moments).

In deriving cluster features from individual MCs, statistics such as mean, standard deviation, coefficient of variation, maximum, median, range of individual MC shape, contrast and spatial distribution are considered. Most of the statistics utilized highlight the variability of individual features as a strong indicator of malignancy.

In Table 13.1, discriminant MC cluster features of representative CADx schemes, which include a filter-based feature selection method enabling identification of discriminant features, are summarized.

The commonly adopted pixel size of 100 μm challenges shape estimation of small size MCs, with pixel sizes ≤50 μm suggested for accurate analysis. Even at 50 μm, only the three largest MCs in a cluster are considered reliable [29]. Furthermore, MC cluster shape analysis is also dependent on the accuracy of MC segmentation algorithms, challenged by MC shape variability, superimposition of dense or heterogeneously dense surrounding tissue and high frequency noise.

Table 13.1. *Morphology-based CADx schemes for MC clusters.* A and B refer to MC cluster feature categories. Image- and case-based performance considers one or two mammographic views, respectively

Study	Discriminant features	Performance ($A_z \pm$SE)
Betal et al. [29]	A: Percentage of irregular and round MCs, inter-quartile range of MC area B: Number of MCs	0.84 (case)
Chan et al. [30]	A: Coefficient of mean density variation, moment ratio variation and area variation, maximum moment ratio and area	0.79±0.04 (image)
Jiang et al. [31]	A: Mean area and effective volume, Standard Deviation (SD) of effective thickness and effective volume, second highest MC shape irregularity measure B: Number of MCs, circularity, area	0.92±0.04 (case) 0.83±0.03 (image)
Veldkamp et al. [32]	A: Mean and SD of individual MC area, orientation and contrast, cluster area B: Number of MCs, distance to pectoral edge and breast edge	0.83 (case) 0.73 (image)
Sklansky et al. [33]	A: Mean area, aspect ratio and irregularity B: Number of MCs	0.75 (image)
Leichter et al. [34]	A: Mean shape factor, SD of shape factor, brightness and area B: Mean number of neighbors, mean distance to the nearest MC	0.98 (image)
Buchbinder et al. [35]	A: Average of length extreme values	0.81 (image)
Paquerault et al. [36]	A: Mean area and effective volume, relative SD of effective thickness and effective volume, second highest MC shape irregularity B: Number of MCs, circularity, area	0.86 (case) 0.82 (image)
Arikidis et al. [37]	A: SD of length extreme values	0.86±0.05 (case) 0.81±0.04 (image)

The segmentation of individual MC of an MC cluster ROI is a critical pre-requisite for the extraction of features representing its morphological attributes. Early approaches were based on simple criteria, such as high absolute gray level or local contrast to label pixels belonging to the MC or the surrounding tissue. To deal with surrounding tissue, image enhancement methods relying on edge detection [29], the Laplacian of Gaussian (LoG) operator applied to the difference of two Gaussian smoothed images [38], fractal models [39], texture classification-based segmentation [40], and wavelet coefficient weighting and modeling [5,41], were utilized.

False positive segmentations, resulting from application of such criteria applied to the MC cluster ROI, compromise CADx performance [38], leading to semi-automatic segmentation methods, requiring manual seed annotation.

Those methods are based on region growing [31], radial gradient analysis [36] and multi-scale active rays [37].

In Sect. 13.6.1, an application paradigm of the effect of two recently proposed individual MC segmentation algorithms on MC cluster content derivation and classification accuracy is provided.

13.3.2 Texture Analysis of MC Clusters

Texture analysis of ROIs containing MC clusters is an alternative to morphology analysis, based on the hypothesis that malignancy, as indicated by MCs, would cause changes in the texture of tissue surrounding a cluster.

Another texture-based approach for classification of MC clusters focuses on analyzing texture of the tissue surrounding MC [42–44], rather than ROIs containing MCs. This approach takes into account the fact that the MC is a tiny deposit of calcium in breast tissue that can neither be malignant nor benign. This approach, taking account of the MC surrounding tissue only, is also aligned to tissue pathoanatomy and immunochemistry, subjecting only surrounding tissue to analysis.

A main advantage of texture-based schemes is that they overcome the increased accuracy demands of shape analysis CADx schemes on MC segmentation algorithms. In case of ROI analysis containing MCs, the MC segmentation step is completely omitted; while in the MC surrounding tissue approach only a coarse MC segmentation step is required.

Gray-Level Co-occurrence Matrices (GLCM) analysis provides image second order statistics characterizing the occurrence of pairs of pixel gray levels in an image at various pixel-pair distances and orientations is widely used in texture based MC clusters schemes. Grey level quantization is commonly adopted in co-occurrence analysis.

Features extracted from GLCMs provide texture heterogeneity/homogeneity and coarseness, not necessarily visually perceived. High spatial resolution (pixel size $\leq 50\,\mu\text{m}$) is required to capture fine texture.

The discriminating ability of GLCMs features, as extracted from original image ROIs containing MCs, has been demonstrated by most studies [30, 45, 46]. In addition, GLCM-based features have shown to be more effective than morphology analysis [30], while their combination can provide an even higher classification performance. Soltanian-Zadeh et al. demonstrated that GLCMs extracted from ROIs containing the MCs were superior to GLCMs extracted from segmented MCs and suggested that "there may be valuable texture information concerning the benignity or malignancy of the cluster in those areas that lie outside the MCs"[46].

To capture tissue texture alterations in multiple scales, First Order Statistics (FOS) (i.e. energy, entropy and Square Root of the Norm of coefficients (SRN)) were extracted from wavelet or multi-wavelet transform sub-images. Wavelet/multi-wavelet FOS have shown to be more effective than GLCMs features [47] and shape features [46], suggesting the advantages offered by

Table 13.2. *Texture-based CADx schemes for MC clusters.* Image- and case-based performance considers one or two mammographic views, respectively

Study	Discriminant features	Performance ($A_z \pm$SE)
Dhawan et al. [45]	• GLCM features • Entropy, energy (decomposition: wavelet packets; filters: Daubechies 6/20; levels: 0, 1) • Cluster features	0.86±0.05 (image)
Kocur et al. [47]	• SRN (decomposition: wavelet transform; filters: Daubechies 4 & Bi-orthogonal 9.7; levels: 0-5) • GLCM features (angular second moment) • Eigenimages (Karhunen-Loève coefficients)	Wavelet: 88% overall classification accuracy (image)
Chan et al. [30]	• GLCMs features • Cluster features (morphological)	0.89±0.03 (image) 0.93±0.03 (case)
Soltanian-Zadeh et al. [46]	• GLCMs features from segmented MCs and ROIs containing MCs • Entropy, energy (decomposition: wavelet packets; filters: Daubechies 6/10/12; levels: 1, 2) • Entropy, energy (decomposition: multi-wavelet (3 Filters); levels: 1, 2) • Cluster features (shape)	Multi-wavelet: 0.89 (image)
Karahaliou et al. [44] (analyzing tissue surrounding MCs)	• First order statistics • GLCMs features • Laws texture energy measures • Energy, entropy (decomposition: redundant wavelets; filter: B-spline; levels: 1-3) • Co-occurrence based (decomposition: redundant wavelets; filter: B-spline; levels: 1-3)	0.98±0.01 (image)

GLCM = Gray-Level Co-occurrence Matrices; SRN = Square Root of the Norm of coefficients

the multi-scale analysis. Table 13.2 summarizes representative feature spaces exploited for MC cluster characterization.

13.3.3 Morphology and Texture Analysis of Masses

Morphology and texture analysis also play an important role in mass content derivation in X-ray mammography. A mass ROI includes the mass region and its surrounding tissue. The mass region, resulting from a segmentation method, is further differentiated to the mass central region and the mass margin region, including pixels inner and outer to the mass contour. Features extracted from mass margin region are specific of masses, providing strong discriminant descriptors for mass classification. Mass features include:

- Intensity/contrast, extracted from the mass region or mass central region.
- Size, extracted from the mass region (e.g. perimeter, area, perimeter-to-area ratio, boundary Fourier descriptor).
- Shape, extracted from the mass region (e.g. convexity, rectangularity, circularity, normalized radial length).
- Mass margin region sharpness (e.g. radial distance metrics and scale space analysis of directional derivatives).
- Radial gradient analysis of the mass region or/and surrounding tissue (e.g. gradient convergence, gradient uniformity) in case of masses with spiculations.
- Texture analysis, including fractals, of the mass region or/and mass margin region.
- Micro-lobulation, extracted from the mass margin region based on a gradient analysis at different scales.
- MC likelihood, extracted from the mass region.

Examples of mammographic appearance of a benign mass with sharp margin and a spiculated malignant mass, originating from DDSM database, both challenged by superimposition with heterogeneously dense surrounding tissue, are provided in Fig. 13.3.

A number of mass segmentation methods have been proposed, such as region growing [48], active contour models [49] and dynamic programming-based algorithms [50]. Multistage methods, including the radial gradient index-based method [48] as an initialization step with an active contour model [51] and a level set method initialized by K-means clustering and morphological operations [52] have been proposed.

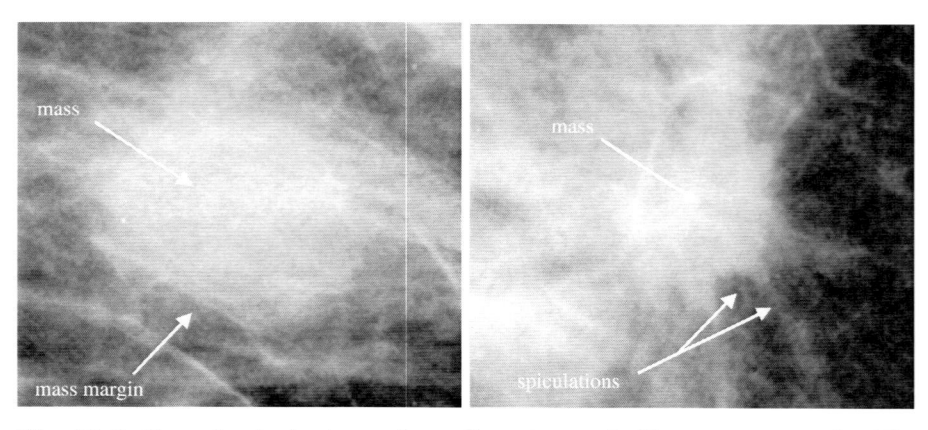

Fig. 13.3. *Example of a benign and a malignant mass in X ray-mammography.* The arrows indicate corresponding mass regions and margins. *Left*: The benign mass is of oval shape and circumscribed margin (A_1594_Left_CC). *Right*: The malignant mass is irregularly shaped and spiculated margin (B_3012_Right_CC)

Table 13.3. *CADx schemes for masses in X-ray mammography.* Image- and case-based performance considers one or two mammographic views, respectively

Study	Discriminant features	Performance ($A_z \pm$SE)
Mudigonda et al. [55]	• GLCM texture (mass region) • Radial gradient • GLCM texture (mass margin region)	0.85 (image) mass margin only
Lim and Er [56]	• FOS and GLCM texture (mass ROI)	0.868 ± 0.020 (image)
Varela et al. [50]	• Intensity/contrast (mass region) • Sharpness, micro-lobulation, GLCM and RLS texture (mass margin region) • Radial gradient (surrounding tissue)	0.81 ± 0.01 (image) 0.83 ± 0.02 (case)
Park et al. [54]	• Intensity, size, shape, radial gradient (mass region)	0.914 ± 0.012 (image)
Georgiou et al. [53]	• Sharpness of DFT, DWT (mass margin region)	0.993 (image)
Delogu et al. [57]	• Size, shape, intensity (mass region)	0.805 ± 0.030 (image)
Rangayyan and Nguyen [58]	• Fractal (mass margin region)	0.93 (image)
Shi et al. [52]	• Size, shape, MC likehood (mass region) • GLCM, RLS texture of RBST, sharpness (mass margin region)	0.83 ± 0.01 (image) 0.85 ± 0.01 (case)

GLCM = Grey-Level Co-occurrence Matrices; RLS = Run-Length Statistics; RBST = Rubber Band Straightening Transform; FOS = First-Order Statistics; DFT = Discrete Fourier Transform; DWT = Discrete Wavelet Transform

Pixel-wise segmentation accuracy, with ground truth provided by radiologists delineation, has been reported for some of the proposed mass segmentation algorithms utilized for mass classification performance [50–52].

In Table 13.3, representative recent publications are provided, including feature categories employed in feature analysis and best performance achieved. It is worth noting that features are extracted not only from spatial domain, but also from other transformed domains (rubber-band straightening, Fourier and wavelet) with rich information content, achieving improved classification performance [53]. This, in combination with more advanced classification schemes, such as SVM, resulted in further improvement of classification accuracy ($A_z = 0.993$) [53]. Several CADx schemes for masses in X-ray mammography have exploited the CBIR classification paradigm demonstrating encouraging results [26, 54].

13.4 CADx Schemes in Breast Ultrasound

CADx schemes in breast US have focused on cysts, benign and malignant solid masses, with the emerging high spatial resolution systems capable of detection of MCs. Most of the reported systems, employ user-defined most representative 2D conventional cross-sectional images, while emerging 3D systems have demonstrated obvious advantages in mass volumetry [59].

Ultrasonic imaging provides scattering characteristics of tissue, by capturing backscattered echoes averaged from contributions of tissue components (scatterers). Scatter statistics depend on the location, individual cross-sections and the number of the scatterers, rendering analysis of echogenic texture a powerful tool in benign from malignant lesion discrimination. Figure 13.4 provides examples of echogenic appearance of a benign and a malignant mass.

A characteristic of ultrasound images is the presence of speckle noise generated by the reflections on a number of randomly distributed scatterers, whose size is smaller than the wavelength of the ultrasound beam. Two approaches are followed regarding speckle in subsequent lesion content derivation, one considering de-speckling mandatory and the other utilizing speckle related image features, associated to the distribution pattern of noise (speckle emphasis) [59].

Texture is extracted from mass ROIs, mass margin and surrounding tissue ROIs, as well as mass posterior acoustic behavior ROIs, unique to ultrasound imaging. Autocorrelation, auto-covariance [60] and GLCM [61] analysis have been widely utilized to derive discriminant features. In addition, statistical models, such as the Nakagami and the K-distribution ones, result in model derived discriminant features [62].

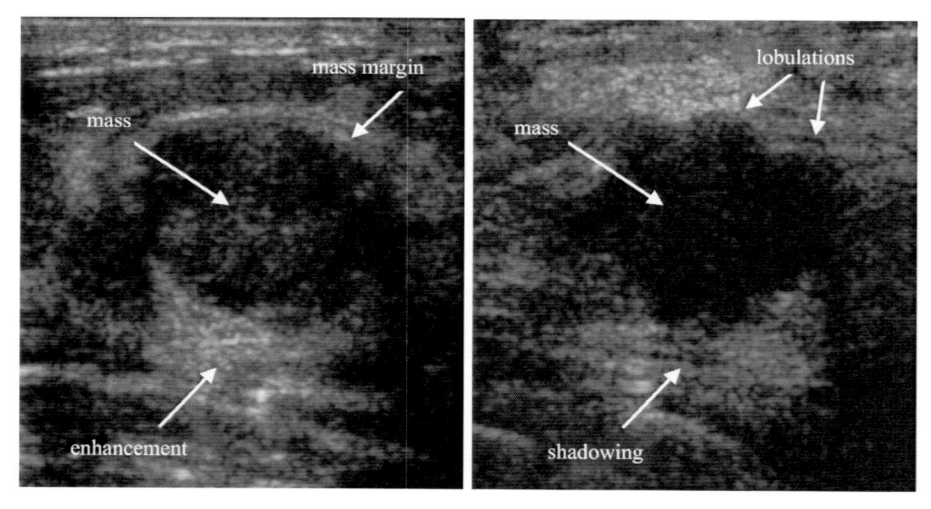

Fig. 13.4. *Example of a benign and a malignant mass in B-mode US.* The arrows indicate corresponding mass regions, mass margins, and the posterior acoustic behavior. *Left*: The benign mass has circumscribed margin and posterior enhancement. *Right*: The malignant mass has a lobulated margin and some posterior shadowing (absence of enhancement)

In addition to texture and in accordance to BI-RADS, mass morphology, mainly shape, size, orientation and margin analysis are also highly discriminant, however pre-supposing accurate mass segmentation methods as well. In Table 13.4, representative recent publications are provided, including feature categories employed in feature analysis and best performance achieved.

Accurate segmentation of the mass region, a prerequisite of morphology analysis, is a highly challenging task in breast US, due to the presence of speckle noise. Consequently, most CADx schemes employ either manual delineations for mass region and mass margin, while a few incorporate an automated mass segmentation stage. Specifically, grey level thresholding followed by morphological operations on pre-processed ROIs [65], radial gradient-based analysis [63], active contours [66] and level sets applied on pre-processed, by anisotropic diffusion filtering and the stick method [67] have been proposed.

Table 13.4. *CADx schemes for masses in US imaging.* Image- and case-based performance considers one or two mammographic views, respectively

Study	Discriminant features	Performance ($A_z \pm SE$)
Chen et al. [60]	Auto-correlation/auto-covariance texture of DWT (mass region and surrounding tissue ROI)	0.940 ± 0.018 (image)
Sivaramakrishna et al. [61]	GLCM texture (mass region), PAB (surrounding tissue ROIs)	0.954 cyst/non-cyst 0.886 benign/malignant (case)
Horsch et al. [63]	Shape (mass region), texture (mass ROI), PAB (surrounding tissue ROIs)	0.87 ± 0.02 all benign 0.82 ± 0.02 solid benign (case)
Shankar et al. [62]	K- and Nakagami distributions (ROIs inside mass), margin sharpness (mass margin ROI) spiculations, PAB (surrounding tissue ROIs)	0.959 ± 0.020 (case)
Chang et al. [64]	Auto-covariance texture in speckle-emphasis, non speckle emphasis and conventional (mass ROI)	0.952 speckle-emphasis (image)
Joo et al. [65]	Shape, texture, intensity (mass region)	0.95 (image)
Sahiner et al. [66]	Shape, PAB (mass region), GLCM texture (mass margin ROI)	0.92 ± 0.03 3D AC 0.87 ± 0.03 2D AC (image)
Chang et al. [67]	Shape, size (mass region)	0.947 (image)
Shen et al. [68]	Shape, intensity, orientation, gradient (mass region), sharpness (mass margin and surrounding tissue ROIs), PAB (surrounding tissue ROI)	0.97 (image)

PAB = Posterior Acoustic Behavior; DWT = Discrete Wavelet Transform; AC = Active Contour

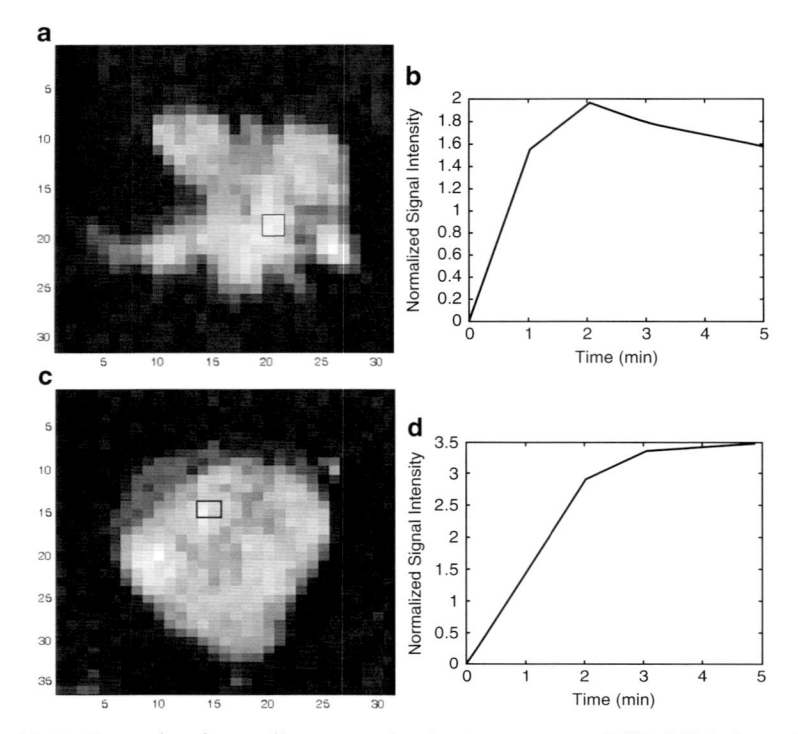

Fig. 13.5. *Example of a malignant and a benign mass in DCE-MRI.* A wash-out (**b**) and a persistent (**d**) type signal intensity curve generated from a rectangular ROI within a malignant (**a**) and a benign (**c**) mass. Masses are depicted in images subtracting the pre-contrast from the first post-contrast frame

13.5 CADx Schemes in Breast MRI

DCE-MRI is a significant complement to mammography, characterized by high sensitivity in detecting breast masses. However, specificity in distinguishing malignant from benign tissue is varying (ranging from 37% up to 90%), attributed to lack of standardized image acquisition protocols and interpretation schemes adopted in the clinical practice at present.

Diagnostic criteria in DCE-MRI of breast masses, according to the ACR BI-RADS MRI lexicon include assessment of morphologic features, from early post-contrast frames, such as lesion shape, margin and enhancement homogeneity (internal architecture) and time analysis of signal intensity curves, generated from manually selected ROIs, within the lesion area [14, 69]. In Fig. 13.5, examples of representative benign and malignant masses in DCE-MRI and corresponding signal intensity curves, are provided.

The main contribution of DCE-MRI to breast imaging and ultimately to CADx in mammography, is imaging of mass vascular heterogeneity (heterogeneity of micro-vascular structure) related to angiogenesis, by means of series acquisition of a time signal related to the distribution of intravenously administered contrast agent [70].

ROI-based analysis within a lesion, providing average enhancement kinetics, ignores the heterogeneity of tumor micro-vascular structure [71]. Pixel-wise analysis of enhancement kinetics enables visualization of lesion vascular heterogeneity, however, care has to be taken to compensate for pixel misregistration among time frames, referred to as noise or low SNR, due to respiratory, cardiac motion and patient movement artifacts.

Signal intensity-time curves can be analyzed quantitatively employing pharmaco-kinetic/physiological model fitting, or by means of semi-quantitative analysis employing empirical parameters of time curves, without fitting [72]. Meaningful fitting of pharmaco-kinetic/physiological models to time curves requires assumptions about mass tissue physiology, knowledge of contrast concentration in the blood as a function of time, which is difficult to obtain, as well as noise free measurements. Alternatively, empirical mathematical functions can be used to fit time curve data accurately, relaxing the requirements of the physiological models [73]. Empirical parameters of time curves, such as relative enhancement, time-to-peak enhancement, washout ratio have been proposed, however indirectly related to tumor physiology.

Texture analysis has been proposed as a method for quantifying lesion vascular heterogeneity, which is important for cancer diagnosis and evaluation of anticancer therapy [70].

CADx schemes in breast DCE-MRI, focusing on quantifying lesion vascular heterogeneity, have exploited FOS analysis applied on exchange rate parameter maps [74], on normalized maximum intensity-time ratio projection data [75] and on empirical enhancement kinetics parameters [76], as well as GLCM-based texture analysis applied on a particular post-contrast time frame [77–81] and on empirical enhancement kinetics parameter maps [82]. Table 13.5 summarizes studies analyzing texture properties of masses, including those integrating mass morphology (shape/size) and/or enhancement kinetics.

Furthermore, a 4D co-occurrence texture analysis approach (considering signal intensity variation over time) [83], and a multi-spectral co-occurrence analysis with three random variables (defined by three pharmaco-kinetic parameters) [84] were proposed for voxel classification-based segmentation of the malignant breast tissue yielding promising results.

Additional representative approaches towards lesion segmentation [85] account for thresholding on difference images and on images reflecting lesion enhancement properties, unsupervised (by means of fuzzy c-means clustering) or supervised classification of pixel/voxel intensity curves, dynamic programming and level-set methods on single post-contrast data and Gaussian mixture modeling combined with a watershed-based method applied on serial post-contrast data [86].

In Sect. 13.6.2, an application paradigm of texture analysis in quantification of enhancement kinetics, related to vascular heterogeneity of masses,

Table 13.5. *Texture analysis of masses in DCE-MRI.* 2D and 3D indicate two- and three-dimensional analysis, respectively. Performance is given either in terms of $A_z \pm SE$ or in terms of sensitivity (sens) and specificity (spec)

Study	Features	Time type analysis considered in texture feature extraction	Performance ($A_z \pm SE$)
Sinha et al. [77]	GLCMs, enhancement kinetics, morphology	Single post-contrast data (2D)	93% (spec) 95% (sens)
Issa et al. [74]	FOS	Serial post-contrast data (2D) (exchange rate parameter map)	88% (spec) 88% (sens)
Gibbs et al. [78]	GLCMs, enhancement kinetics, morphology	Single post-contrast data (2D)	0.92±0.05
Chen et al. [76]	Enhancement-variance dynamics, enhancement kinetics, morphology	Serial post-contrast data (3D)	0.86±0.04
Ertas et al. [75]	FOS	Serial post-contrast data (3D) (normalized maximum intensity-time ratio map)	0.97±0.03
Chen et al. [79]	GLCMs	Single post-contrast data (3D)	0.86±0.04
Nie et al. [80]	GLCMs, morphology	Single post-contrast data (3D)	0.86
McLaren et al. [81]	GLCMs, LTEMs, morphology	Single post-contrast data (3D)	0.82
Karahaliou et al. [82]	GLCMs	Serial post-contrast data (2D) (signal enhancement ratio map)	0.92±0.03

proposed in a recently reported CADx scheme in DCE-MRI breast cancer diagnosis, is provided.

13.6 Application Examples

13.6.1 Segmentation Accuracy on MC Cluster Content

In this application example, the effect of segmentation accuracy on MC content derivation is demonstrated. Specifically, two recently proposed semi-automatic segmentation methods are investigated by means of derivation of 'relevant' features and overall classification performance.

Segmentation accuracy is quantitatively assessed by means of AOM, utilizing manual segmentation of individual MCs as ground truth, provided by expert radiologists. A total of 1,073 individual MCs in a dataset of 128 MC clusters, digitized at 50 µm pixel resolution, originating from the DDSM database, were segmented manually (by one expert radiologist) and automatically (by two methods) [36, 37]. The dataset used consists of mainly pleomorphic MC clusters, characterized by increased size and shape variability. The first method [36] is a radial gradient-based method and the second one employs directional active contours implemented in a rich multi-scale framework [87]. In terms of the AOM measure, the method of multi-scale active contours achieved 0.61±0.15, outperforming the radial gradient method (0.42±0.16) statistically significantly (Fig. 13.6).

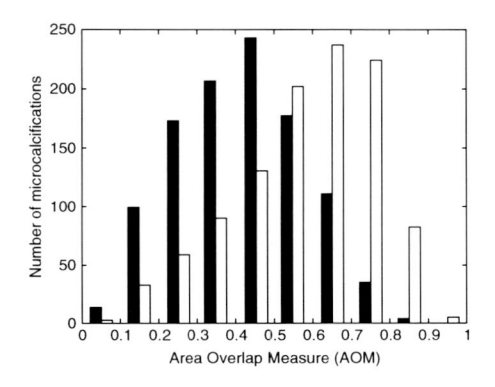

Fig. 13.6. *Method comparison.* Area Overlap Measure (AOM) for the radial gradient (black) and multi-scale active contours (white) segmentation method

Table 13.6. *Features types of individual MCs and MC clusters*

Type	Feature
Individual spot	Area, length, eccentricity, compactness, radial standard deviation, contrast, first regional 2D moment, second regional 2D moment, boundary 1D moment, boundary Fourier descriptor
MC cluster	Average of feature values of the three largest in size MCs. Range of feature values between the largest and the smallest, in size, MCs in a cluster

In Fig. 13.7, an application example of the performance of the two automated segmentation methods and expert manual delineations, applied on an MC cluster, originating from DDSM database, is provided. As observed, the radial gradient method cannot adapt to size and shape variations of individual MCs.

The cluster features utilized to analyze morphology are based on the average of individual morphology features of the 3 largest in size MC [29] and the range of individual morphology features of all MC in a cluster [35].

Individual morphology features are listed in Table 13.6. To capture the degree of variability within a cluster, range statistics are considered suitable, as a strong indicator of malignancy.

To assess the relevance of the features, Pearson correlation coefficient of features, extracted from expert manual delineations and the automated methods, were utilized. Only features demonstrating high correlation ($r > 0.70$) were considered within the feature spaces of the two automatic segmentation methods.

For features extracted from manual segmentations, the radial gradient and the multi-scale active contours segmentation methods resulted in two and eleven MC cluster features with high correlation. Stepwise LDA resulted in two (contrast average, contrast range) and three out of eleven (length average, contrast average, contrast range) uncorrelated MC cluster features, respectively. Accordingly, the corresponding four uncorrelated MC cluster features from manual delineations are compactness average, contrast average, contrast range, and boundary moment range.

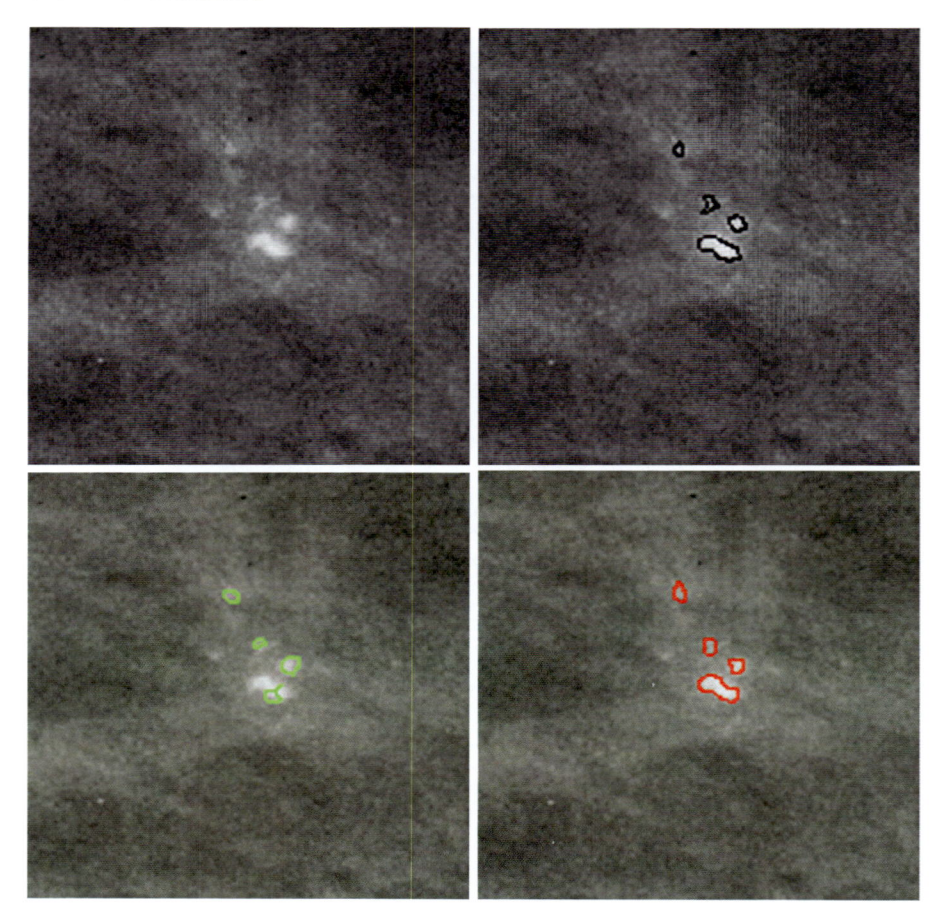

Fig. 13.7. *Automatic vs. manual delineation. Top*: Original ROI (128×128) of an MC cluster (*left*) and expert manual delineations (*right, gray*). *Bottom*: Segmentation using radial gradient (*left, green*) and multi-scale active contours (*right, red*)

Three CADx schemes were implemented, utilizing the selected features from the outcome of the two automated segmentation methods and the manual delineations. Their classification accuracy was evaluated on the same dataset of 128 MC clusters. Table 13.7 presents selected relevant features and performances achieved. Differences in performances achieved between the radial gradient and (a) multi-scale active contour and (b) the manual delineations were statistically significant ($p = 0.0495$ and $p = 0.0039$, respectively). However, the difference of the multi-scale active contour segmentation method to the manual one is not statistically significant, indicating similar performance.

Table 13.7. *Segmentation accuracy and classification performance*

Segmentation method	Selected features (stepwise LDA)	Performance ($A_z \pm$SE)
Radial gradient	Contrast average, contrast range	0.739±0.044
Multi-scale AC	Length average, contrast average, contrast range	0.802±0.041
Manual	Compactness average, contrast average, contrast range, range of boundary first moment	0.841±0.037

AC = Active Contour

13.6.2 Heterogeneity of Enhancement Kinetics in DCE-MRI

An approach toward the quantification of lesion enhancement kinetics heterogeneity for breast cancer diagnosis [82] is summarized in this section. The study is focused on quantifying heterogeneity of masses with respect to three enhancement kinetic parameters, commonly adopted in clinical practice for analysis of signal intensity-time curves, and on investigating its feasibility in discriminating malignant from benign breast masses.

The method is demonstrated on a dataset of 82 histologically verified breast masses (51 malignant, 31 benign) originating from 74 women subjected to MRI with a 1.5 T system. A coronal 3D T_1-weighted spoiled gradient echo sequence (TR 8.1 ms, TE 4 ms, flip angle 20°, matrix 256 × 256, FOV 320 mm, in-plane resolution 1.25×1.25 mm^2, slice thickness 2.5 mm, number of slices 64, acquisition time 1 min) was acquired before and five times after intravenous administration of 0.2 mmol/kg gadopentetate dimeglumine.

For each mass, a single most representative slice was selected in consensus by two experienced radiologists, the one containing the largest cross section of the mass. From the corresponding time series, three enhancement kinetic parameters, namely, initial enhancement, post-initial enhancement and Signal Enhancement Ratio (SER), were calculated in a pixel-wise fashion and used to create three parametric maps. The initial enhancement parameter describes the initial signal increase within the first 3 min after the administration of contrast medium. Both post-initial enhancement and SER describe the post-initial behavior of the signal curve, with the second one incorporating both the signal change in the initial and the post-initial phase relative to the pre-contrast signal measurement. The initial enhancement map was further used to delineate mass boundary by applying histogram thresholding on a rectangular ROI containing the mass, followed by morphological operations (Fig. 13.8). The delineated mass boundary was subsequently used to define corresponding mass areas in each parametric map (Fig. 13.8) providing the basis for texture analysis. Fourteen GLCM-based texture features were extracted from each mass parametric map and their ability in discriminating malignant from benign masses was investigated using a least squares minimum distance classifier. For comparison purposes, GLCM-based texture features were extracted from the 1st post-contrast frame mass area.

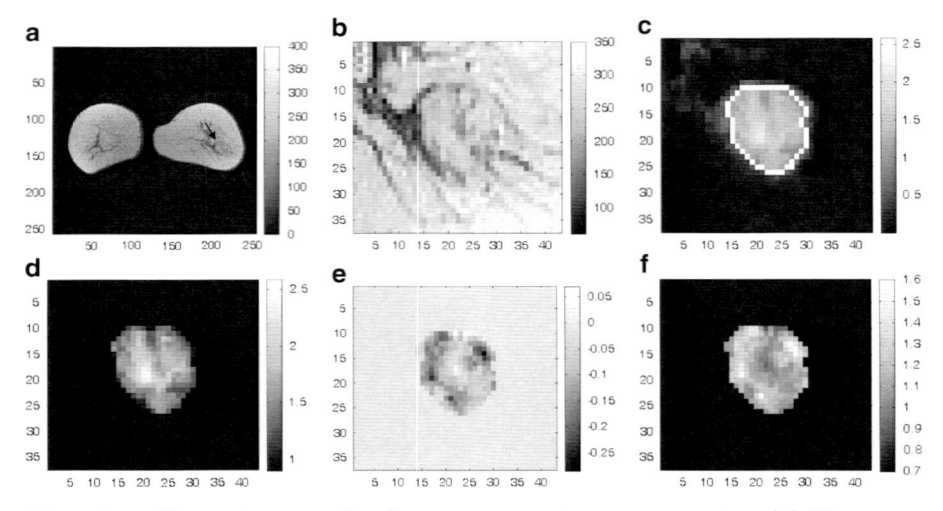

Fig. 13.8. *Illustrative example of mass parametric map generation.* (**a**) First post-contrast coronal slice with black arrow indicating a mass (invasive ductal carcinoma). (**b**) Magnified ROI containing the mass on 1st post-contrast slice. (**c**) Initial enhancement map ROI with delineated mass boundary (boundary pixels assigned to white gray level). (**d**) Initial enhancement mass map. (**e**) Post-initial enhancement mass map. (**f**) Signal enhancement ratio mass map

GLCM-based features were capable of capturing such heterogeneity properties and thus discriminate malignant from benign breast masses. GLCM-based features extracted from the SER and postinitial enhancement map demonstrated an increased discriminating ability, as compared to corresponding features extracted from the initial enhancement map and the 1st post-contrast frame mass area (Fig. 13.9). When considering classification based on selected feature subsets per parametric map (by means of stepwise LDA) a similar trend was observed (Table 13.8).

Results suggest that texture features extracted from parametric maps that reflect mass washout properties (postinitial enhancement and SER map) can discriminate malignant from benign masses more efficiently as compared to texture features extracted from either the 1st post-contrast frame mass area or from a parametric map that reflects mass initial uptake (initial enhancement map).

The approach of quantifying the heterogeneity of mass parametric maps for breast cancer diagnosis should be further exploited with respect to additional enhancement kinetic parameters, including those directly related to tumor physiology provided by pharmacokinetic modeling, and additional texture features.

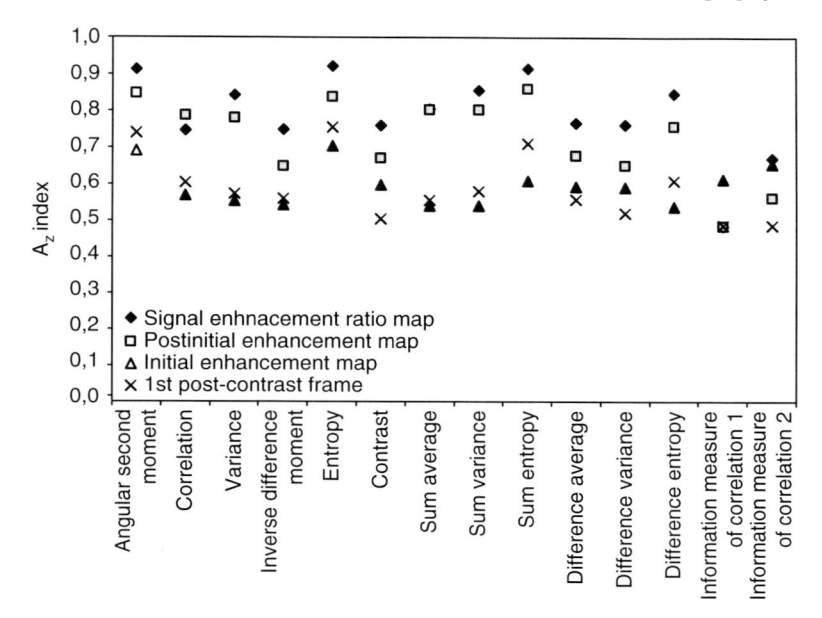

Fig. 13.9. *Classification performance.* The performance is given in terms of A_z index of individual texture features extracted from each parametric map and the first post-contrast frame mass area

Table 13.8. *Classification performance of selected feature subsets*

Type	Selected features	Performance ($A_z \pm$SE)
Initial enhancement map	Entropy, sum average	0.767±0.053
Post-initial enhancement map	Sum entropy, sum average	0.906±0.032
Signal Enhancement Ratio map	Entropy	0.922±0.029
First post-contrast frame	Entropy	0.756±0.060

13.7 Discussion and Conclusions

The potential of CADx schemes across breast imaging modalities in improving radiologists performance and reducing intra- and inter-radiologist performance variability has been well recognized.

A major limitation of reported CADx schemes is that achieved performances are reported at the laboratory level, i.e. very few large scale clinical studies have been carried out at present. As almost entirely the reported breast lesion CADx schemes are based on the supervised classification paradigm, use of heterogeneous data sets, in terms of origin, number and level of difficulty of the image datasets analyzed renders direct comparison of reported CADx methodologies not feasible.

A major step towards CADx methods inter-comparison is taken only in X-ray mammography, by means of publicly available data sets, such as the DDSM database, comprised entirely of digitized films. However, the reported

CADx schemes in X-ray mammography have also utilized varying DDSM subsets. Recently, a coded and annotated benchmark collection, aimed at supporting requirements of current and upcoming CADx schemes, has been introduced [88].

Additionally, pixel-wise accuracy evaluation of lesion segmentation algorithms is not supported by existing ground truth of available databases. To aid radiologists in the tedious task of pixel-wise ground truth derivation, efficient semi-automatic segmentation methods for initial contour delineation are needed.

A further limitation is lack of separate optimization of each of the prior to classification stages, i.e. lesion segmentation, feature spaces exploited and feature selection methods, capable of identifying useful or "relevant" lesion image features.

Imaging modalities adjunct to X-ray mammography, such as breast US and MRI contribute additional unique anatomical and highly promising functional breast lesion features, such as those related to time intensity curves.

Multi- and prior-image and multimodality CADx schemes are currently researched approaches, with the latter following the migration of breast imaging from X-ray mammography to multiple modalities and exploiting complementary tissue properties, as provided by these modalities. "Bimodal" CADx schemes (i.e. X-ray mammography and breast US) have been reported, based either on feature fusion or on decision fusion.

CBIR based CADx schemes, by providing the classification outcome not in terms of a decision, but as a pictorial display of image examples of known pathology, stored in an image database, relevant (similar) to a pictorial query, have the potential to make the CADx decision making transparent (visible) to the user, while providing interaction capabilities (e.g. relevance feedback). Such "interactive" CADx schemes have been proposed as alternatives to "silent" CADx schemes, which provide only a classification decision and are expected to increase radiologists confidence in use of CADx schemes.

Appropriately designed graphical user interfaces of CADx schemes, including accurate and efficient lesion segmentation and registration algorithms, have the potential to further enhance the role of CADx schemes into quantitative image tools of assessment of lesion response to various therapeutic schemes.

Acknowledgments

The author is grateful to professor G. Panayiotakis and research staff Dr. S. Skiadopoulos, Dr. N. Arikidis and Dr. A. Karahaliou of the Department of Medical Physics, University of Patras, Greece, for the many fruitful discussions. Part of this work was supported by the Caratheodory Programme (C.183) of the University of Patras, Greece.

References

1. Nishikawa RM. Current status and future directions of computer-aided diagnosis in mammography. Comput Med Imaging Graph. 2007;31:224–35.
2. Giger ML, Chan HP, Boone J. Anniversary paper: history and status of CAD and quantitative image analysis: the role of medical physics and AAPM. Med Phys. 2008;35(12):5799–20.
3. Tang J, Rangayyan RM, Xu J, et al. Computer-aided detection and diagnosis of breast cancer with mammography: recent advances. IEEE Trans Inf Technol Biomed. 2009;13(2):236–51.
4. Sakellaropoulos P, Costaridou L, Panayiotakis G. A wavelet-based spatially adaptive method for mammographic contrast enhancement. Phys Med Biol. 2003;48:787–803.
5. Heinlein P, Drexl J, Schneider W. Integrated wavelets for enhancement of microcalcifications in digital mammography. IEEE Trans Med Imaging. 2003;22(3):402–13.
6. Papadopoulos A, Fotiadis DI, Costaridou L. Improvement of microcalcification cluster detection in mammography utilizing image enhancement techniques. Comput Biol Med. 2008;38(10):1045–55.
7. Costaridou L, Skiadopoulos S, Karahaliou A, et al. Computer-aided diagnosis in breast imaging: trends and challenges. In: Exarchos TP, Papadopoulos A, Fotiadis D, editors. Handbook of Research on Advanced Techniques in Diagnostic Imaging and Biomedical Applications. IDEA Group Inc Global. 2009. vol. 10, p. 142–159.
8. Elter M, Horsch A. CADx of mammographic masses and clustered microcalcifications: a review. Med Phys. 2009;36(6):2052–68.
9. Pisano ED, Yaffe MY. Digital mammography. Radiology 2005;234(2):353–62.
10. Weismann CF, Datz L. Diagnostic algorithm: how to make use of new 2D, 3D and 4D ultrasound technologies in breast imaging. Eur J Radiol. 2007;64(2):250–57.
11. Kuhl C. The current status of breast MR imaging. Part 2: choice of technique image interpretation diagnostic accuracy and transfer to clinical practice. Radiology 2007;244(2):356–78.
12. Kuhl CK. Current status of breast MR imaging. Part 2: clinical applications. Radiology 2007;244(3):672–91.
13. Karellas A, Vedantham S. Breast cancer imaging: a perspective for the next decade. Med Phys. 2008;35(11):4878–97.
14. American College of Radiology (ACR) Breast Imaging Reporting and Data System Atlas (BI-RADS ® Atlas): Mammography, Ultrasound, and Magnetic Resonance Imaging, 4th ed. Reston, VA: ACR, 2003.
15. Chan HP, Wei J, Zhang Y, et al. Computer-aided detection of masses in digital tomosynthesis mammography: comparison of three approaches. Med Phys. 2008;35:4087–95.
16. Freerand TM, Ulissey MJ. Screening with computer-aided detection: prospective study of 12,860 patients in a community breast center. Radiology. 2001;220:781–86.
17. Fenton JJ, Taplin SH, Carney PA, et al. Influence of computer-aided detection on performance of screening mammography. N Engl J Med. 2007;356(14):1399–409.

18. Gupta S, Markey MK. Correspondence in texture features between two mammographic views. Med Phys. 2005;32(6):1598–606.
19. Drukker K, Horsch K, Giger ML. Multimodality computerized diagnosis of breast lesions using mammography and sonography. Acad Radiol. 2005;12(8):970–79.
20. Horsch K, Giger ML, Vyborny CJ, et al. Classification of breast lesions with multimodality computer aided diagnosis: observer study results on an independent clinical data set. Radiology 2006;240(2):357–68.
21. Timp S, Varela C, Karssemeijer N. Temporal change analysis for characterization of mass lesions in mammography. IEEE Trans Med Imaging. 2007;26(7):945–53.
22. Engeland S, Timp S, Karssemeijer N. Finding corresponding regions of interest in mediolateral oblique and craniocaudal mammographic views. Med Phys. 2006;33(9):3203–12.
23. Suri JS, Chandrasekhar R, Lanconelli N, et al. The current status and likely future of breast imaging CAD. In: Suri JS, Rangayyan RM, editors. Recent Advances in Breast Imaging, Mammography, and Computer-Aided Diagnosis of Breast Cancer. Bellingham: SPIE Press; 2006.
24. Wei L, Yang Y, Nishikawa RM, et al. A study on several machine-learning methods for classification of malignant and benign clustered microcalcifications. IEEE Trans Med Imaging. 2005;24(3):371–80.
25. El-Naqa I, Yang Y, Galatsanos NP, et al. A similarity learning approach to content-based image retrieval: application to digital mammography. IEEE Trans Med Imaging. 2004;23(10):1233–44.
26. Zheng B, Mello-Thoms C, Wang XH, et al. Interactive computer-aided diagnosis of breast masses: computerized selection of visually similar image sets from a reference library. Acad Radiol. 2007;14:917–27.
27. Mazurowski MA, Habas PA, Zurada JM, et al. Decision optimization of case-based computer-aided decision systems using genetic algorithms with application to mammography. Phys Med Biol. 2008;53:895–908.
28. Jiang Y, Metz CE, Nishikawa RM. A receiver operating characteristic partial area index for highly sensitive diagnostic tests. Radiology 1996;201(3):745–50.
29. Betal D, Roberts N, Whitehouse GH. Segmentation and numerical analysis of microcalcifications on mammograms using mathematical morphology. Br J Radiol. 1997;70(837):903–17.
30. Chan HP, Sahiner B, Lam KL, et al. Computerized analysis of mammographic microcalcifications in morphological and texture feature spaces. Med Phys. 1998;25(10):2007–19.
31. Jiang Y, Nishikawa RM, Schmidt RA, et al. Improving breast cancer diagnosis with computer-aided diagnosis. Acad Radiol. 1999;6(1):22–33.
32. Veldkamp WJH, Karssemeijer N, Otten JDM, et al. Automated classification of clustered microcalcifications into malignant and benign types. Med Phys. 2000;27(11):2600–8.
33. Sklansky J, Tao EY, Bazargan M, et al. Computer-aided, case-based diagnosis of mammographic regions of interest containing microcalcifications. Acad Radiol. 2000;7(6):395–405.
34. Leichter I, Lederman R, Buchbinder S, et al. Optimizing parameters for computer-aided diagnosis of microcalcifications at mammography. Acad Radiol. 2000;7(6):406–12.

35. Buchbinder SS, Leichter IS, Lederman R, et al. Can the size of microcalcifications predict malignancy of clusters at mammography? Acad Radiol. 2002;9(1):18–25.
36. Paquerault S, Yarusso LM, Papaioannou J, et al. Radial gradient-based segmentation of mammographic microcalcifications: observer evaluation and effect on CAD performance. Med Phys. 2004;31(9):2648–57.
37. Arikidis N, Skiadopoulos S, Karahaliou A, et al. B-spline active rays segmentation of microcalcifications in mammography. Med Phys. 2008;35(11):5161–71.
38. Salfity MF, Nishikawa RM, Jiang Y, et al. The use of a priori information in the detection of mammographic microcalcifications to improve their classification. Med Phys. 2003;30(5):823–31.
39. Bocchi L, Coppini G, Nori J, et al. Detection of single and clustered microcalcifications in mammograms using fractals models and neural networks. Med Eng Phys. 2004;26(4):303–12.
40. Fu JC, Lee SK, Wong ST, et al. Image segmentation feature selection and pattern classification for mammographic microcalcifications. Comput Med Imaging Graph. 2005;29(6):419–29.
41. Regentova E, Zhang L, Zheng J, et al. Microcalcification detection based on wavelet domain hidden markov tree model study for inclusion to computer aided diagnostic prompting system. Med Phys. 2007;34(6):2206–19.
42. Thiele DL, Kimme-Smith C, Johnson TD, et al. Using tissue texture surrounding calcification clusters to predict benign vs malignant outcomes. Med Phys. 1996;23(4):549–55.
43. Karahaliou A, Skiadopoulos S, Boniatis I, et al. Texture analysis of tissue surrounding microcalcifications on mammograms for breast cancer diagnosis. Br J Radiol. 2007;80(956):648–56.
44. Karahaliou A, Boniatis I, Skiadopoulos S, et al. Breast cancer diagnosis: analyzing texture of tissue surrounding microcalcifications. IEEE Trans Inf Technol Biomed. 2008;12(6):731–38.
45. Dhawan AP, Chitre Y, Kaiser-Bonasso C, et al. Analysis of mammographic microcalcifications using gray-level image structure features. IEEE Trans Med Imaging. 1996;15(3):246–59.
46. Soltanian-Zadeh H, Rafiee-Rad F, Pourabdollah-Nejad SD. Comparison of multiwavelet, wavelet, Haralick, and shape features for microcalcification classification in mammograms. Pattern Recognit. 2004;37(10):1973–86.
47. Kocur CM, Rogers SK, Myers LR, et al. Using neural networks to select wavelet features for breast cancer diagnosis. IEEE Eng Med Biol. 1996;15(3):95–102.
48. Kupinski MA, Giger ML. Automated seeded lesion segmentation on digital mammograms. IEEE Trans Med Imaging. 1998;17(4):510–17.
49. te Brake GM, Karssemeijer N. Segmentation of suspicious densities in digital mammograms. Med Phys. 2001;28(2):259–66.
50. Varela C, Timp S, Karssemeijer N. Use of border information in the classification of mammographic masses. Phys Med Biol. 2006;51(2):425–41.
51. Yuan Y, Giger ML, Li H, et al. A dual-stage method for lesion segmentation on digital mammograms. Med Phys. 2007;34(11):4180–93.
52. Shi J, Sahiner B, Chan HP, et al. Characterization of mammographic masses based on level set segmentation with new image features and patient information. Med Phys. 2008;35(1):280–90.

53. Georgiou H, Mavroforakis M, Dimitropoulos N, et al. Multi-scaled morphological features for the characterization of mammographic masses using statistical classification schemes. Artif Intell Med. 2007;41(1):39–55.
54. Park SC, Sukthankar R, Mummert L, et al. Optimization of reference library used in content-based medical image retrieval scheme. Med Phys. 2007;34(11):4331–39.
55. Mudigonda NR, Rangayyan RM, Desautels JE. Gradient and texture analysis for the classification of mammographic masses. IEEE Trans Med Imaging. 2000;19(10):1032–43.
56. Lim WK, Er MJ. Classification of mammographic masses using generalized dynamic fuzzy neural networks. Med Phys. 2004;31(5):1288–95.
57. Delogu P, Fantacci ME, Kasae P, et al. Characterization of mammographic masses using a gradient-based segmentation algorithm and a neural classifier. Comput Imaging Graph. 2007;37(10):1479–91.
58. Rangayyan RM, Nguyen TM. Fractal analysis of contours of breast masses in mammograms. J Digit Imaging. 2007;20(3):223–37.
59. Cheng HD, Shan J, Ju W, et al. Automated breast cancer detection and classification using ultrasound images: a survey. Pattern Recognit. 2010;43:299–317.
60. Chen DR, Chang RF, Kuo WJ, et al. Diagnosis of breast tumors with sonographic texture analysis using wavelet transform and neural networks. Ultrasound Med Biol. 2002;28(10):1301–10.
61. Sivaramakrishna R, Powell KA, Lieber ML, et al. Texture analysis of lesions in breast ultrasound images. Comput Med Imaging Graph. 2002;26:303–07.
62. Shankar P, Dumane V, Piccoli C, et al. Computer-aided classification of breast masses in ultrasonic B-scans using a multiparameter approach. IEEE Trans Ultrason Ferroelectr Freq Control. 2003;50(8):1002–9.
63. Horsch K, Giger ML, Venta LA, et al. Computerized diagnosis of breast lesions on ultrasound. Med Phys. 2002;29(2):157–64.
64. Chang RF, Wu WJ, Moon WK, et al. Improvement in breast tumor discrimination by support vector machines and speckle-emphasis texture analysis. Ultrasound Med Biol. 2003;29(5):679–86.
65. Joo S, Yang YS, Moon WK, et al. Computer-aided diagnosis of solid breast nodules: use of an artificial neural network based on multiple sonographic features. IEEE Trans Med Imaging. 2004;23(10):1292–1300.
66. Sahiner B, Chan HP, Roubidoux MA, et al. Computerized characterization of breast masses on three-dimensional ultrasound volumes. Med Phys. 2004;31(4):744–54.
67. Chang RF, Wu WJ, Moon WK, et al. Automatic ultrasound segmentation and morphology based diagnosis of solid breast tumors. Breast Cancer Res Treat. 2005;89:179–85.
68. Shen WC, Chang RF, Moon WK, et al. Breast ultrasound computer-aided diagnosis using BI-RADS features. Acad Radiol. 2007;14:928–39.
69. Mann RM, Kuhl CK, Kinkel K, et al. Breast MRI: guidelines from the European society of breast imaging. Eur Radiol. 2008;18:1307–18.
70. Jackson A, O'Connor JP, Parker GJ, et al. Imaging tumor vascular heterogeneity and angiogenesis using dynamic contrast-enhanced magnetic resonance imaging. Clin Cancer Res. 2007;13:3449–59.
71. Collins DJ, Padhani AR. Dynamic magnetic resonance imaging of tumor perfusion Approaches and biomedical challenges. IEEE Eng Med Biol Mag. 2004;23:65–83.

72. Eyal E, Degani H. Model-based and model-free parametric analysis of breast dynamic-contrast-enhanced MRI. NMR Biomed. 2007;22:40–53.
73. Fan X, Medved M, Karczmar GS, et al. Diagnosis of suspicious breast lesions using an empirical mathematical model for dynamic contrast-enhanced MRI. Magn Reson Imaging. 2007;25(5):593–603.
74. Issa B, Buckley DL, Turnbull LW. Heterogeneity analysis of Gd-DTPA uptake: Improvement in breast lesion differentiation. J Comput Assist Tomogr. 1999;23:615–21.
75. Ertas G, Gülçür HO, Tunaci M. Improved lesion detection in MR mammography: three-dimensional segmentation, moving voxel sampling, and normalized maximum intensity-time ratio entropy. Acad Radiol. 2007;14:151–61.
76. Chen W, Giger ML, Lan L, et al. Computerized interpretation of breast MRI: investigation of enhancement-variance dynamics. Med Phys. 2004;31:1076–82.
77. Sinha S, Lucas-Quesada FA, DeBruhl ND, et al. Multifeature analysis of GD-enhanced MR images of breast lesions. J Magn Reson Imaging. 1997;7:1016–26.
78. Gibbs P, Turnbull LW. Textural analysis of contrast-enhanced MR images of the breast. Magn Reson Med. 2003;50:92–98.
79. Chen W, Giger ML, Li H, et al. Volumetric texture analysis of breast lesions on contrast-enhanced magnetic resonance images. Magn Reson Med. 2007;58:562–71.
80. Nie K, Chen JH, Yu HJ, et al. Quantitative analysis of lesion morphology and texture features for diagnostic prediction in breast MRI. Acad Radiol. 2008;15:1513–25.
81. McLaren CE, Chen WP, Nie K, et al. Prediction of malignant breast lesions from MRI features: a comparison of artificial neural network and logistic regression techniques. Acad Radiol. 2009;16:842–51.
82. Karahaliou A, Vassiou K, Arikidis NS, et al. Assessing heterogeneity of lesion enhancement kinetics in dynamic contrast enhanced MRI for breast cancer diagnosis. Br J Radiol. 2010;83:296–309.
83. Woods BJ, Clymer BD, Kurc T, et al. Malignant-lesion segmentation using 4D co-occurrence texture analysis applied to dynamic contrast-enhanced magnetic resonance breast image data. J Magn Reson Imaging. 2007;25:495–501.
84. Kale MC, Clymer BD, Koch RM, et al. Multispectral co-occurrence with three random variables in dynamic contrast enhanced magnetic resonance imaging of breast cancer. IEEE Trans Med Imaging. 2008;27:1425–31.
85. Behrens S, Laue H, Althaus M, et al. Computer assistance for MR based diagnosis of breast cancer: present and future challenges. Comput Med Imaging Graph. 2007;31:236–47.
86. Cui Y, Tan Y, Zhao B, et al. Malignant lesion segmentation in contrast-enhanced breast MR images based on the marker-controlled watershed. Med Phys. 2009;36:4359–69.
87. Arikidis NS, Karahaliou A, Skiadopoulos S, et al. Size-adapted microcalcification segmentation in mammography utilizing scale-space signatures. Comput Med Imaging Graph. 2010;34(6):487–93.
88. Oliveira JEE, Gueld MO, Araújo AA, et al. Towards a standard reference database for computer-aided mammography. Procs SPIE. 2008;6915(2):1Y-1–9.

14

Quantitative Medical Image Analysis for Clinical Development of Therapeutics

Mostafa Analoui

Summary. There has been significant progress in development of therapeutics for prevention and management of several disease areas in recent years, leading to increased average life expectancy, as well as of quality of life, globally. However, due to complexity of addressing a number of medical needs and financial burden of development of new class of therapeutics, there is a need for better tools for decision making and validation of efficacy and safety of new compounds. Numerous biological markers (biomarkers) have been proposed either as adjunct to current clinical endpoints or as surrogates. Imaging biomarkers are among rapidly increasing biomarkers, being examined to expedite effective and rational drug development. Clinical imaging often involves a complex set of multi-modality data sets that require rapid and objective analysis, independent of reviewer's bias and training. In this chapter, an overview of imaging biomarkers for drug development is offered, along with challenges that necessitate quantitative and objective image analysis. Examples of automated and semi-automated analysis approaches are provided, along with technical review of such methods. These examples include the use of 3D MRI for osteoarthritis, ultrasound vascular imaging, and dynamic contrast enhanced MRI for oncology. Additionally, a brief overview of regulatory requirements is discussed. In conclusion, this chapter highlights key challenges and future directions in this area.

14.1 Introduction

Advances in development of new and novel therapeutics, medical devices and associated procedures have significantly contributed to improved global health and well being [1]. The impact of such development is directly noticeable in increased quality of life and economic productivity and prosperity. Research has been a backbone for discovery, development, and delivery of novel therapeutics. However, due to scientific complexity and regulatory processes involved, introduction of a new drug could take between 11–15 years [2]. This long process is associated with ever-increasing cost, which is estimated to be around $ 1 Billion on the average per drug [3]. Key reason for such a high

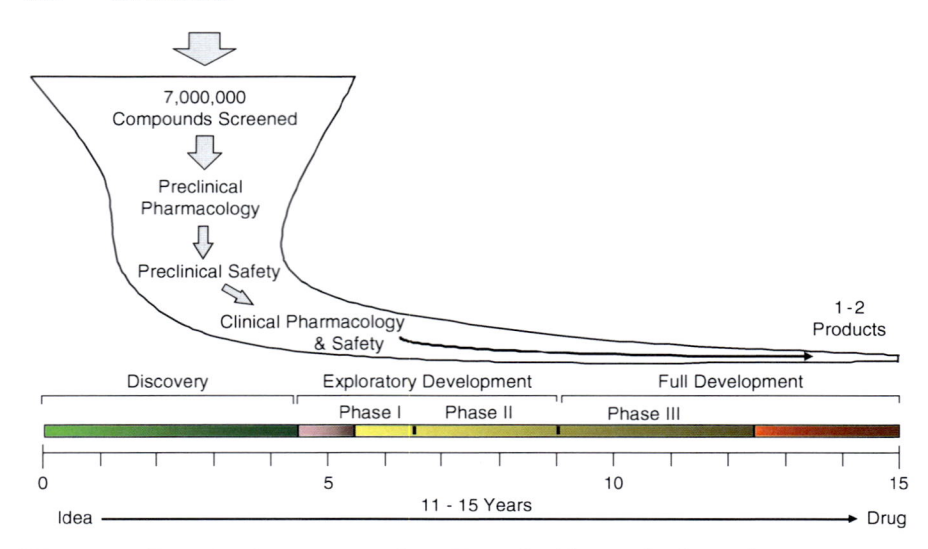

Fig. 14.1. *Drug development time line.* From the idea to the drug, there are various stages in drug discovery and development. The time line is calibrated in years

cost and long duration is attrition through various stages of drug Research and Development (R&D). Figure 14.1 depicts various stages associated with drug discovery and development. Of millions of chemical entities enter drug discovery pipeline, perhaps one or two become approved drug available for patients.

Despite consistent increase in global R&D budget, the pace of approval of new chemical and molecular entities has not kept pace with expenditure [3]. It is evident that with the current failure rate and ever-increasing cost, such a model for drug development is not sustainable. In response, drug development industry, along with their academic and regulatory partners, has been looking for alternative approaches to develop new therapeutics much faster and more efficiently. Such approaches include a wide variety of business models, investment strategies and scientific approaches. Examples include rational drug design, use of computational modeling/simulations and utilization of biologocal markers (short: biomarker) for timely and cost-effective decision making.

In this chapter, we specifically focus on imaging biomarkers and how these are helping resolving key challenges in drug R&D. Next section provides a brief overview of biomarkers, followed by examples of imaging biomarkers currently being used. The main focus of this chapter is quantitative methods for medical imaging in drug R&D. Several challenges and examples for semi- and fully-automated imaging analysis will be presented. Among other issues that readers need to be aware are steps involved in technical and clinical validation of such imaging tools, as well as regulatory acceptance for these biomarkers, which will be covered briefly. In closing, this chapter offers key challenges that are focus of future directions in this field.

14.2 Key Issues in Drug Research and Clinical Development

14.2.1 Biological Marker

Among approaches to achieve better and more cost-effective decision making is use of precise pre-clinical and clinical measurements tools for assessment of impact of therapeutics. These tools are ultimately intended to generate biological and clinical markers. To assure clarity for topics in this chapter we will use definition of biomarkers put forward by the National Institutes of Health (NIH) Biomarker Definitions Working Group (BDWG) [4]. The following definitions and characteristics are intended to describe biological measurements in therapeutic development and assessment:

- *Biological marker*: A characteristics that is objectively measured and evaluated as an indicator of normal biological processes, pathogenic processes, or pharmacologic response to a therapeutic intervention.
- *Clinical endpoint*: A characteristic or variable that reflects how a patient feels, functions, or survives.
- *Surrogate endpoint*: A biomarker that is intended to substitute for a clinical endpoint. A surrogate endpoint is expected to predict clinical benefit (or harm or lack of benefit/harm) based on epidemiologic, therapeutics, pathophysiologic, or other scientific evidence.

Biomarkers have broad applications in early efficacy and safety assessment, in vitro and in vivo animal studies to establish Proof of Mechanism (POM), and early stage clinical trial to establish Proof of Concept (POC). Other applications include diagnostics for patient selection, staging disease, prognosis, and patient monitoring.

Clinical endpoints are specifically intended for measurement and analysis of disease and response to treatment in the clinical trials. These endpoints are often considered credible observation and measurement for clinical evaluation of risk and benefit.

The surrogate endpoints, a subset of biomarkers, are key clinical measurements that expected to provide key decision points in lieu of specific clinical outcome. Achieving such status for a clinical measurement requires significant and large-scale evidence, which is source of decision making for clinicians and regulators. Currently, there are numerous biomarkers that are being utilized from drug screening to preclinical and clinical assessments. Examples include genetic makers, blood and plasma assays, and imaging.

Imaging is one of biomarkers, perhaps with the broadest applications from in vitro assessment to clinical endpoints in late stage clinical trials. Choice of imaging modality, with and without contrast agent, and specific measurements are among key steps for design and execution of imaging biomarker strategy for a preclinical or clinical program. Rest of this chapter will focus on issues specifically around quantitative clinical imaging in drug development.

14.2.2 Imaging Modality

In recent years, several imaging modalities have been used in drug discovery and development. These modalities offer a broad range of options for in vitro, in vivo preclinical and clinical imaging. A comprehensive coverage of such a broad range of applications will require extensive length and it is certainly beyond a single chapter. To narrow down the focus, this chapter will cover applications of imaging for clinical trials, thus in vivo imaging for healthy and diseased populations. Figure 14.2 depicts most commonly used modalities in clinical imaging, i.e. Ultrasound (US), Computed Tomography (CT), Magnetic Resonance Imaging (MRI), Single Photon Emission Computed Tomography (SPECT), and Positron Emission Tomography (PET). Optical imaging modalities (not shown in Fig. 14.2) offer a broad family of tools with rapidly expanding utilization for clinical trials.

Key factors for selecting an imaging modality in a clinical trial for a specific disease area depends on multiple factors, including (i) the physical performance of imaging modality, (ii) specific presentation of disease, (iii) impact of therapeutics, and (iv) dynamics of change (or lack of it) as a result of therapy. For example, spatial and temporal resolutions, as well as molecular affinity of contrast agent (if one is involved) and sensitivity of a modality are among physical performance criteria for selection of appropriate modality:

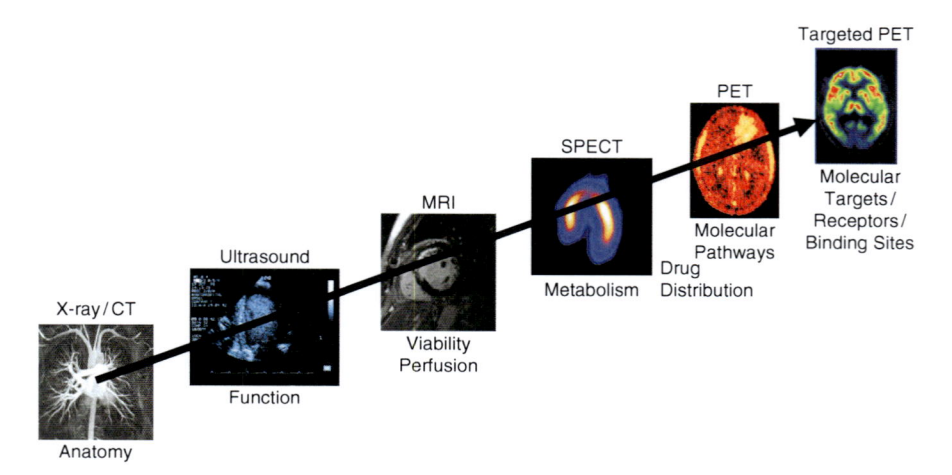

Fig. 14.2. *Imaging modalities.* Common medical imaging modalities used for in vivo clinical assessment for drug development (Courtesy: GE Healthcare)

- *Radiographic imaging* also referred to as radiography (projection and quantitative imaging), has been used extensively for osteoporosis studies quite frequently [5, 6]. In its simplest form, these imaging modalities provide a crude measure of bone quantity for a selected Region of Interest (ROI);
- *Computed tomography* has been a dominant tool for tumor detection and assessment in oncology. These measures often include one-dimensional (1D) and two-dimensional (2D) slice-based analysis of tumor morphology and its response to treatment;
- *Magnetic resonance imaging* as well as functional MRI (fMRI), has found a very strong traction for a number of disease areas, such as Central Nervous System (CNS), arthritis, and pulmonary disease. Recent progress in MRI, with and without contrast agents, has positioned it as an indispensable tool for structural and molecular imaging.
- *Nuclear medicine imaging* also referred to as diagnostic nuclear medicine imaging, provide modalities such as PET and SPECT are among critical tools for clinical molecular imaging for a broad range of applications in oncology and psychotherapeutics. As resolution and SNR steadily improve in these modalities, they are finding their ways in a number of clinical protocols for patient selection and therapy assessment.
- *More sophisticated approaches* such as Peripheral Quantitative CT (pQCT) and high-resolution MRI, provide tree-dimensional (3D) view of various components of bone structure, as well as structure of new formed bone [7].

While there is fundamental physical and utility difference in the modalities, one key question is repeatability of imaging and measurements between and within such modalities. More importantly, there are several acquisition parameters involved for a given modality, which also attribute greatly to quality differences from images acquired with the same modality.

Another key complicating factor is subjectivity involved in definition, identification and analysis of imaging content, which heavily impacts study design and analysis. To address these issues, there has been a shift in scientific community toward development of tools to assure: (i) repeatable acquisition of clinical images, independent of operator or manufacturer and (ii) objective and quantitative measurements for assessment of response to treatment.

There is a large body of work focusing on various techniques to assure repeatability of image acquisition. Thus, the focus of this chapter is to provide an overview of quantitative analytical approaches for handling such images.

14.3 Quantitative Image Analysis

There are numerous automated and semi-automated approaches of quantitative image analysis in clinical trials and a comprehensive coverage of such applications is certainly beyond scope of a single chapter. Thus, this section offers a selective set of examples that should provide the readers with some

of key approaches being used, as well as innovative tools under development
and validation. These examples also cover a range of modalities and disease
areas.

14.3.1 Assessment of Osteoarthritis

Osteoarthritis (OA) is generally viewed as clinical and pathological outcome
of range of disorders that cause structural degradation and functional failure of synovial joints [8]. Progressive joint failure associated with OA may
lead to pain and disability, which is ranked as the leading cause of disability in the elders [9]. Conventionally, measurement of osteoarthritis structural
changes is been made via radiographic examination. The standardized measurement of Joint Space Width (JSW) is a commonly accepted endpoint
for assessment of cartilage [10]. Due to inherent limitation associated with
radiographic assessment, MRI has been proposed as a non-invasive tool for
3D imaging of cartilage morphology [11]. Figure 14.3 shows saggital view,
marked bone-cartilage boundaries and 3D view of cartilage, based on high
resolution T1-weighted slices. As shown, 3D acquisition of signal, as well as
variation in signal intensity, could provide unique anatomic and content-based
measurements.

Although there is a large body of research on identifying correlation
between MRI-based measurements and OA, regulatory acceptance of such
end points is still an open question. Key MRI measurement are focused on
longitudinal change in cartilage morphology (e.g., thickness, volume, shape),
as well as bone-related anomalies, such as bone marrow edema.

While initial MRI image analysis was heavily done by manual tracing of
cartilage boundaries in consecutive slices, the most commonly used methods
are currently either based on semi-automated or fully automated approaches.

Semi-automated approaches generally start with manual boundary marking of selective slices, followed by algorithm-based method for extending
boundary identification to other slides. This step is followed by formation of

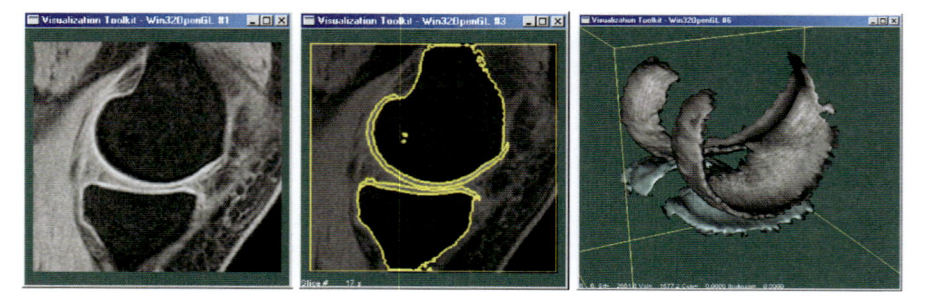

Fig. 14.3. *Cartilage assessment with MRI.* Based on saggital high resolution
T1-weighted slices (*left*), bone and cartilage boundaries marked with yellow lines
(*middle*), and femoral (*grey*) and tibial (*blue*) cartilages are highlighted (*right*)

volume by computing iso-voxel model, which also can accommodate the scenario where slice resolution differs that of slice thickness. While this method is relatively easy to implement, it is labor-intensive and heavily requires training and calibrations of analysts.

Alternatively, fully automated methods have been under development and validation. For example, the method developed by Tames-Pena et al. achieves 3D segmentation via two steps [12]:

1. initial segmentation using intensity-based, unsupervised 3D segmentation,
2. fine tuning of the initial 3D segmentation using a knee atlas and surface/volume computation.

Also, in this approach, rather than using voxel-based representation of volume, spline-based surfaces are computed for defining and calculating 3D volumes and its substructures. Figure 14.4 shows examples of initial input used in this approach, as well as segmented volume and labeled sub-regions. It must be noted, however, that labeling of sub-region often involves some subjective decision making and requires confirmation by a trained radiologist.

14.3.2 Assessment of Carotid Atherosclerosis

Atherosclerosis is a broad class of disease beginning by a process in which deposits of fatty substances, cholesterol, cellular waste products, calcium and other substances build up in the inner lining of an artery. This buildup, plaque, usually affects large and medium-sized arteries. Plaques can grow large enough to significantly reduce the blood's flow through an artery. But most of the damage occurs when they become fragile and rupture. Plaques that rupture cause blood clots to form that can block blood flow or break off and travel to

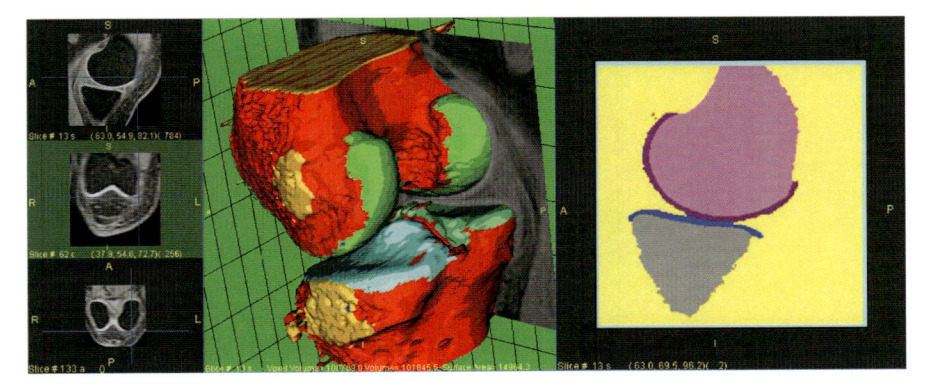

Fig. 14.4. *Automatic segmentation by Tames-Pena et al. [12]. Left*: selected saggital, coronal slices; *middle*: fully automated 3D segmentation of bone and cartilage; *right*: tibial (*gray*) and femoral (*pink*) bones and cartilage (*brown and blue*) labeled

Fig. 14.5. *Carotid intima-media thickness assessment.* In the US image of human carotid segment, the boundaries of lumen, intima and adventia are marked by technician

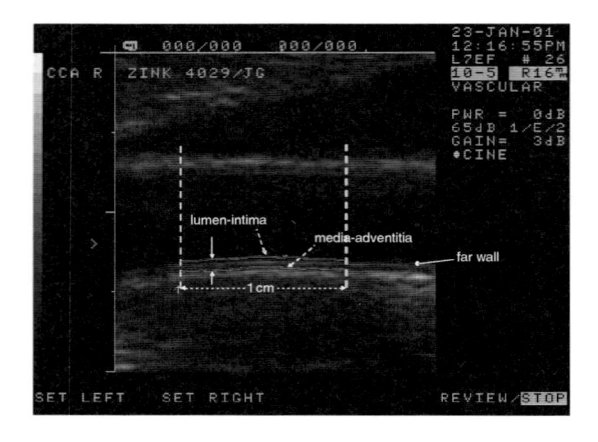

another part of the body. If either happens and blocks a blood vessel that feeds the heart, it causes a heart attack. If it blocks a blood vessel that feeds the brain, it causes a stroke. And if blood supply to the arms or legs is reduced, it can cause difficulty walking and eventually lead to gangrene[1].

There are a number of imaging approaches for detection and quantification of plaque. Examples include external and intra-vascular US, MRI and PET imaging. The external US approach is the most common for assessment of carotid imaging. In this approach, an external transducer is used to image carotid to measure thickness changes in intima-media segment. Thus, the method is referred to as Carotid Intima-Media Thickness (CIMT) assessment (Fig. 14.5).

In a typical procedure for CIMT assessment, US technician acquires a video sequence containing several frames. Then, one frame is selected via visual examination. Intima and media boundaries are either marked manually or via automatic edge detection methods. Manual edge detection clearly requires training and calibration of analysts. This in turn extends analysis time and potential variability due to use of multiple analysts. Automated edge detection has been shown to obtain accurate and precise measurement, without requiring manual operation [13].

Another key challenge is selection of the best frame for image analysis. For a video of duration of 8 s, with 20 fps, technician has 160 frames to examine, ultimately select one and discard 159 frames. The key question has been whether it would be possible to utilize the entire 160 frames for composing a much higher resolution image to be used in automated edge segmentation.

Super-Resolution (SR) is an approach by which a series of images from a fixed-resolution sensor array is used to compute an image at a resolution higher than original sensor array, with improved Signal-to-Noise Ratio (SNR) [14].

---

[1] see, for instance, the Web site of the American Heart Association (AHA) at http://www.americanheart.org

Using this concept, a technique called "pixel compounding" was proposed to achieve a sub-pixel accuracy reconstruction of US images [15]. Pixel compounding parallels the more common spatial and frequency compounding and uses the additional intensity information available from random movements of edges within the image. Images in the sequence should have slight, sub-pixel shifts due to either transducer or subject motion (which naturally occurs during course of image acquisition). Since the intensity at a particular pixel is given by the integral of the incident signal over the area of the pixel patch, the intensity weight center of a pixel for a feature that is smaller than a pixel should be a sub-pixel location within the pixel.

The technique operates on a sequence of US B-mode images acquired with random motion. Sub-pixel registration is estimated and a Maximum a Posteriori (MAP) approach with the shift information is used to reconstruct a high-resolution single image. A non-homogeneous anisotropic diffusion algorithm follows from the estimation process and is used to enhance the high-resolution edges.

Figure 14.6(top) shows a selected frame and rectangular ROI shown. Using nine consecutive frames, a new super-resolution image is computed (3 × resolution increase). Figure 14.6(left) compares a subset of ROI, representing CIMT to the conventional bicubic interpolation and SR. Pseudo-colored reconstructed ROIs are also shown on the right.

This approach provides a unique opportunity for achieving a resolution beyond inherent resolution of sensor, which in turns adds significant accuracy and precision in the clinical measurement. This increased accuracy and precision will be used to either conduct clinical trials with a smaller number of subjects or maintain the same number of subjects in each arm, and in turn increase accuracy in measurement of clinical response.

14.3.3 Assessment of Cancer

With increased effort for better understanding of underlying molecular events that lead to cancer, there is an increased demand for a wider range of structural, functional and molecular imaging modalities. Structural imaging, based on projection radiography and CT, is the most common modality of tumor response assessment [16]. In addition to CT, MRI also provides unique anatomical view for tumor size and structure, which is quite important for quantitative assessment of tumor response to treatment. Also, other modalities, such as PET, SPECT and Magnetic Resonance Spectroscopy (MRS) provide ability to characterize and measure biological processes at the molecular level [17, 18]. It must be noted that specific utility of such modalities is directly related to biological, functional and anatomical manifestation of the disease.

Among these imaging techniques, Dynamic Contrast Enhanced (DCE)-MRI is a unique modality that provides opportunity to examine microcirculation of tumor and normal tissues at high spatial resolution. Angiogenesis

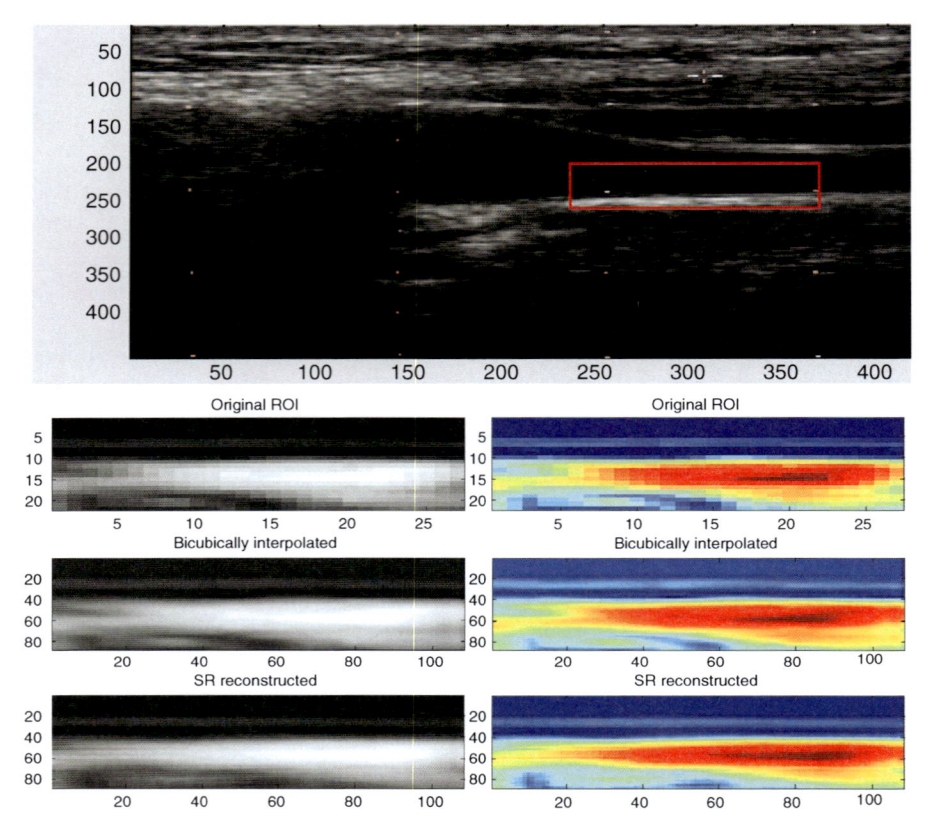

Fig. 14.6. *Super-resolution in ultrasound imaging. Top*: A selected frame from a series of clinical scans, with ROI marked; *left*: original ROI, bicubically interpolated ROI, and results of super-resolution reconstruction; *right*: pseudo-colored images corresponding to the gray scale images shown in the left

is a complex process critical to growth and metastasis of malignant tumors, with directly impacting micro-vasculature growth and behavior of tumors.

This process results in the development of vascular networks that are both structurally and functionally abnormal. A broad range of compounds are developed and are under development to disrupt new vessel formation (anti-angiogenic) or destroy existing vessels. There has been clear demand and significant development of imaging biomarkers for angiogenesis, which can serve as early indicators of drug activity in clinical trials and may facilitate early Pharmacodynamic (PD) assessment. DCE-MRI is intended to capture the Pharmacokinetic (PK) of injected low-molecular weight contrast agents as they pass through the tumor micro-vasculature [19]. This technique is sensitive to alterations in vascular permeability, extracellular and vascular volumes, and in blood flow.

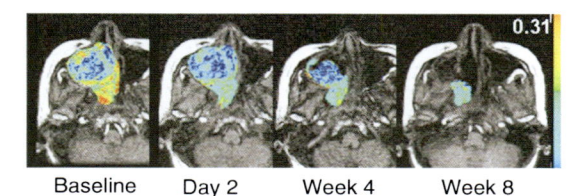

Fig. 14.7. *Treatment followup.* The sample DCE-MRI image from a patient with adenoid cyst carcinoma shows decline in tumor perfusion post treatment [21]

In T1-weighted DCE-MRI, an intravenous bolus of gadolinium contrast agent enters tumor arterioles, passes through capillary beds and then drains via tumor veins [20]. Gadolinium ions interact with nearby hydrogen nuclei to shorten T1-relaxation times in local tissue water. This causes increase in signal intensity on T1-weighted images to a variable extent within each voxel. The degree of signal enhancement is dependent on physiological and physical factors, including tissue perfusion and Arterial Input Function (AIF). T1-weighted DCE-MRI analysis generates parameters that represent one of, or combinations of these processes, and can be used to measure abnormalities in tumor vessel function such as flow, blood volume, and permeability.

There are various approaches for DCE-MRI imaging. However, commonly three types of data are acquired:

1. initial image for tumor localization and anatomic referencing,
2. sequences that allow calculation of baseline tissue T1-values before contrast agent administration,
3. dynamic data are acquired every few seconds in T1-weighted images over a period of around 5–10 min.

Using this data, a number of parameters, such as volume transfer function, are computed. These parameters are used for quantitative analysis of vascular response for a given therapeutic agent. Figure 14.7 shows sample DCE-MRI images from a patient with adenoid cystic carcinoma at baseline and after three cycles of treatment with an experimental anti-angiogentic compound [21].

While there is a consensus on utility of DCE-MRI for tumor imaging and it's response to therapy, there are a number of critical issues impacting quantitative analysis of such approach. Among these issues are choice of model for pharmacokinetic models, imaging protocol, estimation of arterial estimation function, and selection of ROI for analysis. These parameters are even more critical when imaging for clinical trials is conducted across multiple sites using different MRI machines, and there is heavy manual-analysis involved. Lack of control in these parameters is among key sources of error.

14.4 Managing Variability in Imaging Biomarkers

In every clinical and preclinical measurement, one needs to be aware of source of variations, how it can be controlled and accounted for. Perhaps the most critical question in value of an imaging endpoint is how repeatable it is.

Fig. 14.8. *Technical precision of measurements.* Various sources of variations related to image acquisition and analysis, impacting precision of image-based measurements

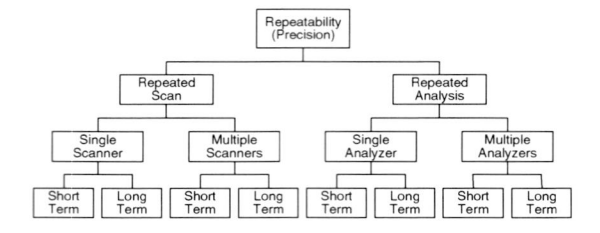

Regardless of whether imaging endpoint is intended to measure pharmacokinetics or -dynamics, target affinity, dose response, etc., it would essentially useless if one cannot reliably repeat the results of a given procedure. There is a broad range of sources of variability in imaging procedures. However, in this chapter we will classify them into broad categories:

- *Clinical variations* (and one can extend this to in vitro and preclinical variations) are related to biological/physiological characteristics that lead to between and within differences for patients under identical treatments. This is also referred to as biologic noise [22].
- *Technical variations* also contains a range of sources that one can divide them into modality/device related and operator-related sources.

While in managing variability we deal with accuracy and precision [23, 24], it is important to note that technical accuracy is directly linked to biologic noise and its impact on the measurement. It is common to assess technical repeatability of imaging measurement through repeated scans and repeated measurements. Figure 14.8, captures various possible combinations factors contributing to technical precision of measurement.

14.4.1 Technical Validation

To establish impact of variation within and between imaging devices, for a given modality, it is common to run repeated imaging sessions using a single device (scanner) or multiple devices. When possible, it is preferable to use the same brand and model of imaging device throughout clinical trial(s). However, since most clinical trials are conducted across multiple clinical centers, in multiple countries, it is not practical and/or cost effective to enforce this. Moreover, as technology evolves and clinical centers upgrade their imaging hardware/software, it is not possible to guarantee that in clinical trials of long duration, baseline and follow up imaging will be done by the same platform in a given clinical center. Quantitative understanding of device-related variability is extremely crucial, which directly impacts number of patients required, duration of clinical trials, and number of imaging observations throughout a given study. To establish and manage device performance, it is important to have a clear protocol for calibration and monitoring imaging devices, as well as operator/technician performing patient preparation and image acquisition.

The other source of variability in this graph is related to post-acquisition image analysis. Image analysis approaches that heavily rely on subjective identification of imaging content and manual segmentation are often prone to high degree of variability. The most critical step in initiating such analysis is assuring that there is well-defined and well-communicated set of rules for defining anatomical landmark, lesions, etc.

Another key element in analysis approaches that heavily rely on manual processing is assuring that the image analysis operators are well calibrated throughout analysis period. This would require assessment and calibration for short and long term experiments. To avoid calibration drift, some tend to conduct image analysis for entire clinical trial at the end of trials, as opposed to analyzing images as they arrive. While this could address long-term drift and variation between analysts, it tends to miss any imaging error that could have been detected by interim analysis.

In general, to minimize technical variations in the clinical trials is highly advisable to use automated methods that require minimal post-analysis adjustment by human operator. While there has been significant progress in development of automation for image analysis, the need for quality control remains with us.

14.4.2 Standard Operation Procedures

Every good quality management system is based on its Standard Operating Policies (SOP) or Standard Operating Procedures (SOP), which represents the afore mentioned set of well-defined and well-communicated set of rules. Hence, SOPs can act as effective catalysts to drive performance- and reliability-improvement results.

For example, in a comparative image analysis, a set of MRI images of patient's knee were provided to two independent groups. Each group was asked to conduct and report measurement of knee cartilage volume. No further instruction was provided. Figure 14.9 compares the results initially reported from these two analyses. While both groups agreed that images are of high quality, measurements show very low correlation. Detailed examinations identified two sources for such a low agreement: (i) use of different criteria for defining cartilage boundaries and (ii) use of direct voxel counting vs. spline-based method to measure volumes. The other contributing factor was heavy use of manual segmentation by one group and semi-automated approach by the second group.

Collectively, a clear definition for selection of cartilage boundary was identified and provided to both image analysis teams. Figure 14.10 shows results of new analysis, which clearly confirms improvement in measurement correlation.

Fig. 14.9. *Variation without SOPs.* Comparison of cartilage volume measurement by two independent image analysis groups, using manual and semi-automated methods

Fig. 14.10. *Variation using SOPs.* Second set of independent measurements made by the same methods as shown in Fig. 14.9, after both groups were offered the same definition for identifying cartilage boundary. The technical variation is significantly reduced

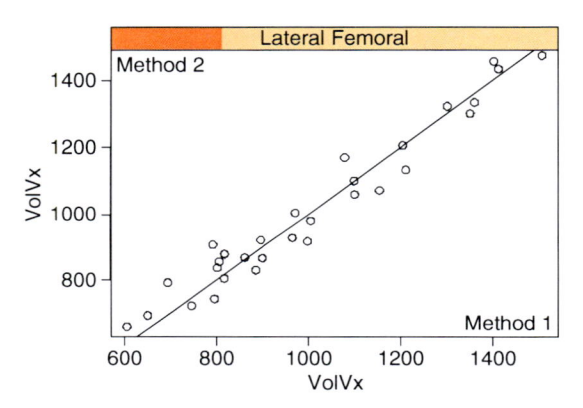

14.4.3 Regulatory Issues

Two critical regulatory questions for acceptance and deployment of imaging biomarkers can be summarized as:

1. What is utility of imaging endpoint (surrogate, auxiliary, etc)?
2. How reliable is such measurement (biologic and technical noises)?

These two questions are not unique for imaging endpoint from regulatory perspective, and they arise every time a new tool is proposed for existing or new endpoints. While the first question is very clear, establishing answer for it could be quite complex and often requiring very long and tedious validation studies. Examples of such studies include a number of completed and ongoing experimental and clinical studies for validation of imaging endpoints in oncology, CNS, cardiovascular, inflammation diseases.

Perhaps progress in technical validation of such imaging endpoints are further along mainly due to relative simplicity of addressing technical variations (compared to biologic noise), and significant progress in hardware availability and algorithm development.

In general, addressing regulatory concerns requires a well-coordinated collaboration between various players involved this area. This is to assure the framework for acceptance is clearly stated, validation studies are accepted by clinical and technical communities, and a timeline for achieving goals are committed to. Currently there are a number of such collaborative efforts ongoing, such as Alzheimer's Disease Neuroimaging Initiative (ADNI) and OsteoArthritis Initiative (OAI).

14.5 Future Directions

While there has been significant progress in enhancing current imaging modalities and introduction of new ones, the focus of this chapter has been on image analysis issues related to imaging biomarkers. In looking into current landscape of ongoing developments and projecting future demand, the author offers the following areas as the key issues to be addressed in future directions:

- *Moving toward quantitative and objective imaging*: As volume of medical data increases and additional modalities are introduced, reliance on visual and manual assessment of medical imaging will be becoming impractical and unreliable. Also, dealing with such complexity will increase time required to render final diagnosis that would be quite critical to patient care, as well as slowing rampant increase in cost of medical imaging in clinical practice and clinical trials. Thus key drivers are cost-effective and reliable tools for addressing ever-increasing imaging data.
- *Minimizing operator variability by developing fully-automated algorithms*: Considering global and distributed nature of clinical trials, it is inevitable that in the course of a clinical program, imaging data will be examined by a number of radiologists and technicians. Such operators are trained and licensed under a diverse set of curricula and re-certification rules. While such differences may not have noticeable impact on patient care, it will certainly have adverse effect on quality of analysis associated with clinical trials. Moving to fully-automated analytical tools faces two key challenges:
    1. Lack of availability of such tools for most of modalities and disease areas.
    2. Lack of regulatory acceptance for most of tools currently available.

  Addressing such challenges require further development of automated analytical tools and prospective, broad clinical validation. Success in these two steps would be critical for regulatory acceptance, hence, broad deployment of such tools.
- *Integration of analytical algorithm within pathophysiology of disease and imaging modalities*: While imaging modalities are capable of capturing tremendous amount of structural and functional data, full understanding

of underlying pathophysiology require additional non-imaging data. Non-imaging data includes:

1. general and population-based a priori knowledge about disease and/or organ under study, and
2. specific non-imaging data collected for a given patient in the course of clinical trial.

Examples of the latter include medical history, clinical evaluation, laboratory tests. To development an accurate and automated image analysis tool, it is critical to incorporate such non-image in the process.

- *Model-based approaches for adaptive and predictive measurements*: Historically, early approaches for medical image analysis deployed pixel/voxel information capture in an imaging dataset to offer measurement/diagnosis for what has been already observed. Accuracy of voxel-based approaches is heavily impacted by resolution and speed of imaging device, as well as overlapping and . Although increase in physical performance of imaging devices has been quite significant in past few years, and there is more on the near-term horizon, inclusion of mathematical model of underlying organs and system will provide complementary information for significant improvement in speed and accuracy of image analysis. Also, such models are quite crucial for development of tools to project forward from current observation; i.e. prognostic and quantitative tools.

References

1. Lichtenberg F. The effect of drug vintage on survival rates: evidence from Puerto Rico's ASES program. NBER Working Paper. 2004.
2. DiMasi J, Hansen R, Grabowski H. The price of innovation: new estimates of drug development costs. J Health Econ. 2003;22(2):151–85.
3. Munos B. Lessons from 60 years of pharmaceutical innovation. Nat Rev Drug Discov. 2009;8:959–68.
4. Munos B. Biomarkers definitions working group: biomarkers and surrogate endpoints: preferred definitions and conceptual framework. Clin Pharmacol Ther. 2001;69:89–95.
5. Boutroy S, Bouxsein ML, Munoz F, et al. In vivo assessment of trabecular bone microarchitecture by high-resolution peripheral quantitative computed tomography. J Clin Endocrinol Metab. 2005;90(12):6508–15.
6. Bachrach LK. Dual energy x-ray absorptiometry (DEXA) measurements of bone density and body composition: promise and pitfalls. J Pediatr Endocrinol Metab. 2000;13(2):983–8.
7. Link T, Majumdar S, Lin PAJ, et al. In vivo high resolution MRI of the calcaneus: differences in trabecular structure in osteoporosis patients. J Bone Mineral. 1998;13:1175–82.
8. Nuki G. Osteoarthritis: a problem of joint failure. Z Rheumatol. 1999;58:142–47.
9. McNeil JM, Binette J, Prevalence of disabilities and associated health conditions among adults. Morb Mortal Wkly Rep. 2001;50(7):120–5.

10. Hunter DJ, Graverand MPH, Eckstein F. Radiologic markers of osteoarthritis progression. Curr Opin Rheumatol. 2009;21:110–7.
11. Eckstein F, Guermazi A, Roemer FW. Quantitative MR imaging of cartilage and trabecular bone in osteoarthritis. Radiol Clin North Am. 2009;47(4):655–73.
12. Tamez-Pena GJ, Barbu-McInnis M, Totterman S. Knee cartilage extraction and bone-cartilage interface analysis from 3D MRI data sets. Proc SPIE. 2009;5370:1774–84.
13. Gustavsson T, Abu-Gharbieh R, Hamarneh G, et al. Implementation and comparison of four different boundary detection algorithms for quantitative ultrasonic measurements of the human carotid artery. IEEE Comp Cardiol. 1997;24:69.
14. Ng M, Bose NK. Mathematical analysis of super-resolution methodology. IEEE Signal Process Mag. 2009;20:49–61.
15. Yang Z, Tuthill TA, Raunig DL, et al. Pixel compounding: resolution-enhanced ultrasound imaging for quantitative analysis. Ultrasound Med Biol. 2007;33(8):1309–19.
16. Zhao B, Schwartz LH, Larson SM. Imaging surrogates of tumor response to therapy: anatomic and functional biomarkers. J Nucl Med. 2009;50:239–49.
17. Cai W, Mohamedali KCKA, Cao Q, et al. PET of vascular endothelial growth factor receptor expression. J Nucl Med. 2006;47:2048–56.
18. Hoffman JM, Gambhir SS. Molecular imaging: the vision and opportunity for radiology in the future. Radiology 2007;244(1):2048–56.
19. Padhani AR. Dynamic contrast-enhanced MRI in clinical oncology: current status and future directions. J Magn Reson Imaging. 2002;16(4):407–22.
20. O'Connor JPB, Jackson A, Parker GJM, et al. DCE-MRI biomarkers in the clinical evaluation of antiangiogenic and vascular disrupting agents. Br J Cancer. 2007;96:189–95.
21. Liu G, Rugo HS, Wilding G, et al. Dynamic contrast-enhanced magnetic resonance imaging as a pharmacodynamic measure of response after acute dosing of AG-013736, an oral angiogenesis inhibitor, in patients with advanced solid tumors: results from a phase I study. J Clin Oncol. 2005;23:5464–73.
22. Sullivan DC. Imaging as a quantitative science. Radiology 2008;248:328–32.
23. Gonen M, Panageas KS, Larson SK. Statistical issues in analysis of diagnostics imaging experiments with multiple observations per patient. Radiology 2001;221:763–67.
24. Richter WS. Imaging biomarkers as surrogate endpoints for drug development. Eur J Nucl Med Mol Imaging. 2006;33:S6–S10.

Part VI

Image Data Visualization

15

Visualization and Exploration of Segmented Anatomic Structures

Dirk Bartz[†] and Bernhard Preim

Summary. This chapter provides an introduction into the visualization of segmented anatomic structures using indirect and direct volume rendering methods. Indirect volume rendering typically generates a polygonal representation of an organ surface, whereas this surface may exhibit staircasing artifacts due to the segmentation. Since our visual perception is highly sensitive to discontinuities, it is important to provide adequate methods to remove or at least reduce these artifacts. One of the most frequently visualized anatomical structures are blood vessels. Their complex topology and geometric shape represent specific challenges. Therefore, we explore the use of model assumptions to improve the visual representation of blood vessels. Finally, virtual endoscopy as one of the novel exploration methods is discussed.

15.1 Introduction

The visualization of volumetric and multi-modal medical data is a common task in biomedical image processing and analysis. In particular after identifying anatomical structures of interest, cf. Chap. 11, page 279, and aligning multiple datasets, cf. Chap. 5, page 130, a Three-dimensional (3D) visual representation helps to explore and to understand the data.

Volume visualization aims at a visual representation of the full dataset, hence of all images at the same time. Therefore, the individual voxels of the dataset must be selected, weighted, combined, and projected onto the image plane. The image plane itself acts literally as a window to the data, representing the position and viewing direction of the observer who examines the dataset.

[†] On March 28, 2010, our colleague Dirk Bartz passed away unexpectedly. We will remember him, not only for his contributions to this field, but for his personal warmth and friendship.

T.M. Deserno (ed.), *Biomedical Image Processing*, Biological and Medical Physics, Biomedical Engineering, DOI: 10.1007/978-3-642-15816-2_15,
© Springer-Verlag Berlin Heidelberg 2011

15.2 Indirect and Direct Volume Rendering

In general, two different options are available to generate this visual representation; indirect and direct volume rendering. Indirect volume rendering, also frequently called surface – or isosurface – rendering, extracts an intermediate, polygonal representation from the volume dataset. This representation is then rendered using commonly available standard graphics hardware. In contrast, direct volume rendering generates the visual representation without an intermediate representation, by projecting the voxels onto the image plane [1, 2].

15.2.1 Indirect Volume Rendering

The standard approach of extracting an intermediate, polygonal representation of the volume dataset is the "marching cubes" algorithm [3], where the volume dataset is processed voxel cell by voxel cell. A voxel cell is represented by eight neighboring voxels in two neighboring volume slices, where these voxels form a cube shaped cell. The edges of this cube indicate the boundaries of the cell between the directly neighboring voxels. In that process, the respective isosurface is extracted for each voxel cell by first identifying the edges of the voxel cell that the isosurface intersects. Based on intersection points and a case table, the isosurface for this voxel cell is computed.

Typically, an isosurface indicates a material interface of an object (e.g., an organ). Technically, this isovalue is a binary opacity transfer functions, where the voxels below the isovalue mark the outside, the voxels above that isovalue mark the inside of the object. If the voxel is equal to the isovalue, it is located directly on the respective isosurface. While binary opacity transfer functions are equivalent to threshold-based segmentation (thresholding, cf. Sect. 1.6.1, page 27), they do not take into account any spatial information of the voxels' locations. To achieve a voxel specific isosurface extraction, we need to segment the respective object in the dataset first and to limit the isosurface extraction to the segmented region. If more than one object is segmented, the information is combined into a *Label volume*, where the label of each voxel determines its affiliation to an object or organ. The isosurface extraction is then performed consecutively one object at the time with the respective isovalue.

Note that the standard isosurface extraction method assumes that only one isosurface intersects an edge (up to one time). Hence, if more than one object or material intersects a volume cell, the algorithm leads to severe inconsistencies in the triangulation. A general solution to this problem was proposed by Hege et al.[4] and later by Banks and Linton [5].

15.2.2 Rendering of Multiple Objects

Multiple objects are typically rendered in polygonal graphics with different colors to enable a good differentiation between these objects (Fig. 15.1). If

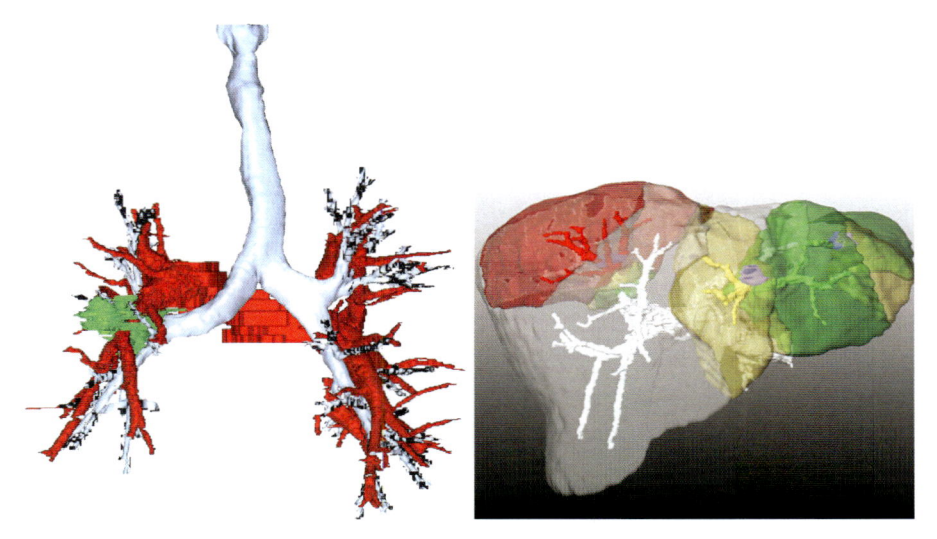

Fig. 15.1. *Segmented polygonal model. Left*: Model of the lungs, with blood vessels and tumor (*green*); *Right*: Model of the liver, where the colored regions indicate risk areas with respect to the portal vein (Courtesy: Milo Hindennach, Bremen)

one object is embedded into another object, the transparency of at least the outer object can be reduced to allow an unobscured view on the inner object (Fig. 15.1b). Furthermore, we need to address several additional aspects for a successful semitransparent rendering. Since the rendering of the individual objects are blended together, the sufficiently correct blending order is important to ensure a correct representation. Therefore, the objects must be ordered according to the depths of their barycenters, where the farthest objects are rendered first, the closest last. Nevertheless, highly nested objects should be avoided, since they lead to a convoluted visual representation.

One of the drawbacks of semitransparent polygonal rendering is the limited ability to represent depths, since the transparency value does not include any distance-related attenuation term. To address this significant perception issue, the transparency of every individual polygon can be modulated based on its orientation to the viewer. If the polygon is parallel to the user, it should be rendered with a low transparency, if it is oriented more orthogonal to the viewer, a high transparency should be used (Fig. 15.2). This approach goes back to an idea of Kay and Greenberg who assumed that a light ray will be more attenuated, if it traverses a medium in a more acute angle [6].

If objects from several volume datasets (multi-modal volume data) are combined, the user needs to ensure that the datasets are correctly aligned (registered) to each other, even if the different data volumes provide incomplete or even very limited overlap, cf. Chap. 5, page 130.

Fig. 15.2. *Transparency modulated objects. Left*: Oriented transparency modulation; *Right*: Constant transparency modulation (Courtesy: Detlev Stalling, Berlin)

Fig. 15.3. *Conceptual model of direct volume rendering.* A ray S from the view point accumulates contributions at samples s_k from entry point $k = n - 1$ to exit point $k = 0$

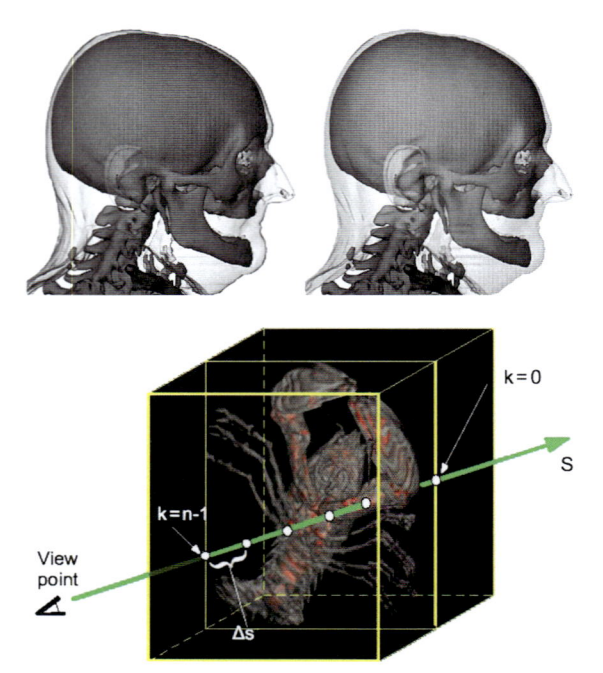

15.2.3 Direct Volume Rendering

In a direct volume rendering approach, no intermediate representation is computed. Instead, the contributions of the volume dataset are either directly projected on the image plane, or the contributes are accumulated by a collecting ray casted from the image plane through the volume (Fig. 15.3) [7]. Direct volume rendering (short: volume rendering) is based on the transport theory of light [8]. After several simplification steps, the basic optical model of volume rendering is the Density Emitter Model (DES) of Sabella [9], where every voxel is considered as a tiny light source (emitter) that is attenuated or absorbed when the light rays travel through a dense volume (density). Hence, this model only considers emission and absorption as the basic physical components [10].

Since volume data is only defined on voxel positions, each approach must specify how to reconstruct the contributions from the whole voxel space between the voxel positions:

1. *Reconstruction*: This is typically specified by the used reconstruction kernel. For image space algorithms like ray casting [11], the standard kernel is the trilinear interpolation in the voxel cell, which essentially is a cubic function. The standard reconstruction kernel (or filter) for object space algorithms like splatting [12, 13] is a three-dimensional Gaussian kernel with a specific cut-off range that limit the support of the filter.

2. *Sampling*: The approaches also need to specify the sampling, or how often a contribution ("sample") is taken from the volume dataset. Since all voxels are classified by transfer functions into a 4-tupel of red, green, blue, and opacity (r, g, b, α) based on their voxel value, their individual color and opacity contribution must be accumulated appropriately.
3. *Order of operations*: The standard order first computes the samples, classifies the samples by applying the transfer functions, and computes the lighting based on the individual sample. This order is called post-shading, since the lighting (and shading) is performed after sampling. In contrast, the pre-shading approaches first classify and shade the voxels, and the samples are computed based on the colored and opacity-weighted voxels. While pre-shading enables several performance advantages, its quality is far inferior to post-shading, since the reconstruction leads to a blurred representation in most cases.

Today, the most popular implementations of volume rendering use an hardware accelerated Graphics Processing Unit (GPU)-based or Central Processing Unit (CPU)-optimized ray casting with trilinear interpolation and post-shading [14].

One of the key elements of volume rendering is the specification of the transfer functions, since it defines how the individual samples are weighted in the final image [15]. While most approaches focus on the identification of object boundaries [16], this topic is still subject of active research. Rendering multiple objects, color blending [17] and opacity adaption [18] become relevant.

15.2.4 Rendering of Segmented Data

The integration of transfer functions and segmentation does not easily arise in direct volume rendering. Early approaches pre-classified the volume dataset based on a segmentation [19]. This pre-shading approach, however, leads to excessive blurring and henceforth to a reduced image quality. Furthermore, color bleeding effects might occur, if no precaution have been taken [20]. Alternatively, separate transfer functions can be provided for every label of a segmentation (Fig. 15.4). These *tagged* voxel sets are then volume rendered locally, and afterwards combined in a global compositing step. Due to the local and global compositing, this approach is also called two-level volume rendering [21].

Alternatively, Hastreiter et al. suggested to handle the voxel set specific transfer functions through hardware supported lookup tables that switch between the specific transfer functions, which are also implemented through hardware supported Look-Up Tables (LUTs) [22].

However, sampling on the boundaries between different compositing regions may lead to an abrupt material change or an abrupt end of the sampling, which in turn leads to color bleeding and staircasing artifacts. A solution

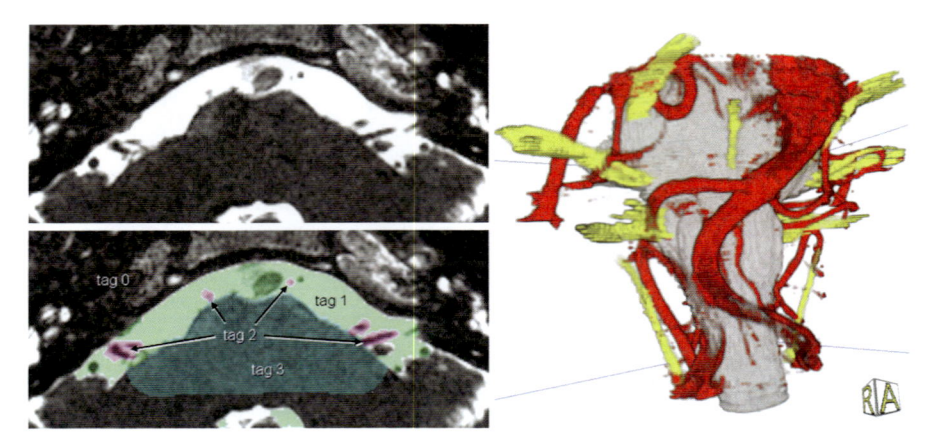

Fig. 15.4. *Direct volume rendering of neurovascular compression syndrome. Left*: Important anatomical structures (backround/tag 0, cerebrospinal fluid and blood vessels/tag 1, cranial nerves/tag 2 and brain stem/tag 3) near the brain stem based on MRI data. *Right*: Visualization of the segmented structures using individual transfer functions (Courtesy: Peter Hastreiter, Erlangen)

to this problem was presented by Tiede et al., who suggested to use a trilinear interpolation that takes into account only the current (closest) segment and which computes a sample closer to the segments surface than the regular sampling pattern [23], thus approximating the actual segment surface. Beyer et al. have improved Tiede's approach by extending the considered area along the gradient direction of the sample [24]. The new sample is then re-classified to determine a closer segment surface (Fig. 15.5).

15.2.5 Discussion

Overall, both indirect and direct volume rendering are viable options to visualize segmented medical data, although each option provides a different set of advantages or disadvantages. The biggest advantage for indirect volume rendering is the significantly easier specification of color and transparency in a visual representation of multiple objects and datasets. Because of its high flexibility, direct volume rendering is advantageous when small changes of the material interfaces have to be addressed. A small shift of a feature in the opacity and color transfer functions can be done quite fast and the effect can be seen instantaneously. With indirect volume rendering, new polygonal isosurfaces have to be extracted, which typically requires more time than generating a new direct volume rendering image. Hence, the exploration of unknown datasets is typically easier with direct volume rendering.

Another remarkable advantage of direct volume rendering is image quality. Due to superior reconstruction filter, the standard direct volume rendering algorithms provide a significantly better visual quality than indirect volume

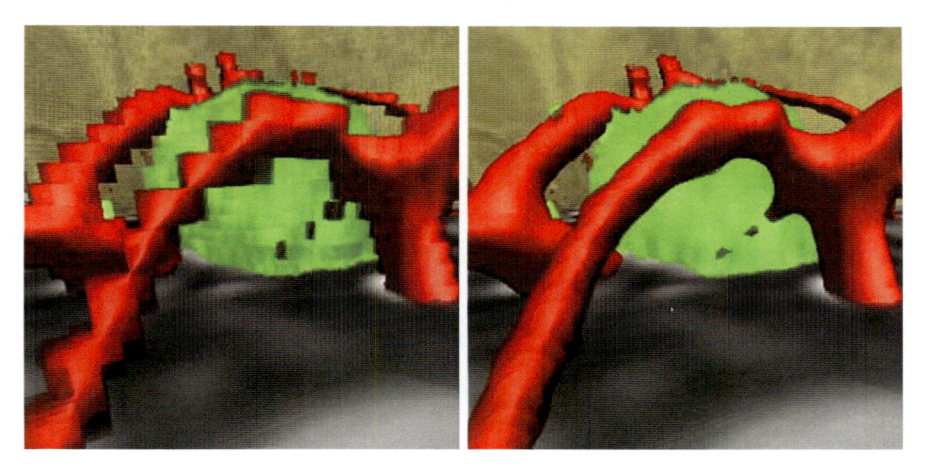

Fig. 15.5. *Direct volume rendering of segmented data.* Tumor and blood vessel segments are rendered using direct volume rendering. The samples are re-classified along the gradient direction (*right*) for a better surface reconstruction, whereas trilinear interpolation of the labels exposes staircasing artifacts (Courtesy: Johanna Beyer, Vienna)

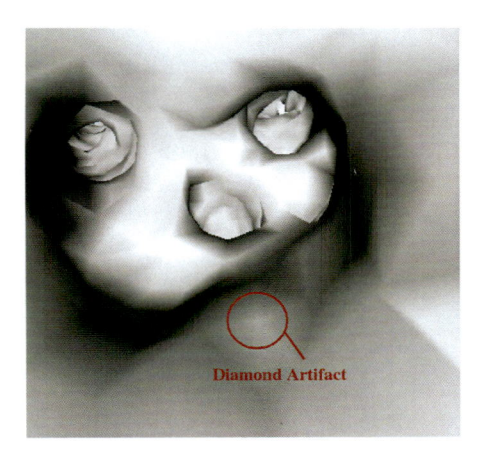

Fig. 15.6. *Diamond artifact of marching cubes.* The image shows a close-up of a virtual bronchoscopy

rendering approaches. In particular, the marching cubes algorithm provides only a linear interpolation for vertex computation and a bilinear interpolation for the triangle surface (Gouraud shading). This frequently leads to *diamond* interpolation artifacts (Fig. 15.6). This effect can be compensated with sub-voxel decomposition methods that induce higher computational costs [25] or avoided applying implicit surface methods [26].

15.3 Generation of Smooth and Accurate Surface Models

Patient-specific medical surface models are used to convey the morphology of anatomic and pathologic structures as well as spatial relations between them. Moreover, surface models are essential for generating volume models for simulations.

An inherent problem of medical image data is the limited resolution and the anisotropic character of the voxels (slice thickness is usually considerably larger than the in-plane resolution). Thus, the extracted surface meshes may contain several artifacts such as staircases, holes, and noise (Fig. 15.7).

Due to these artifacts, the reconstructed vessel may differ significantly from the real anatomic structures, which are usually smooth, and influence the visual and numerical evaluation of spatial relationships. Especially for surgical planning, it is essential to employ accurate models, e.g. to ensure the correct computation and visualization of safety margins and potential infiltrations.

15.3.1 Mesh Smoothing with Fairing

An essential class of smoothing approaches addresses the fairing of meshes. Since traditional surface fairing methods from geometric modeling are computationally too expensive to be considered, a variety of local fairing methods are frequently used. Most local fairing operators take into account the weighted average of the direct neighbors of the current vertex (the "umbrella region" to re-position the current vertex (Fig. 15.8).

However, it is not trivial for an application to select appropriate smoothing algorithms and parameters (neighborhood, number of iterations, weighting factor) from the class of umbrella operators. Bade et al. [27] compared several

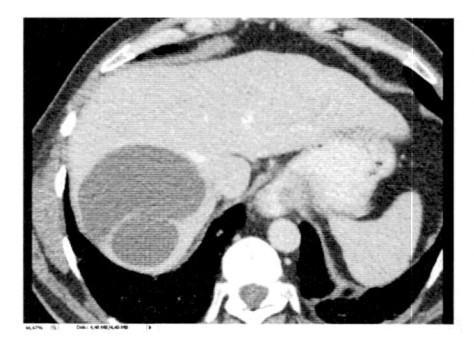

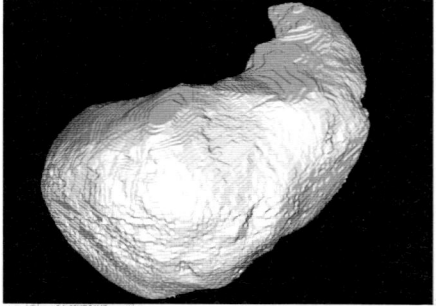

Fig. 15.7. *3D reconstruction of binary segmentation.* The liver was accurately segmented from CT data. However, it appears noisy (Courtesy: Tobias Mönch, Magdeburg)

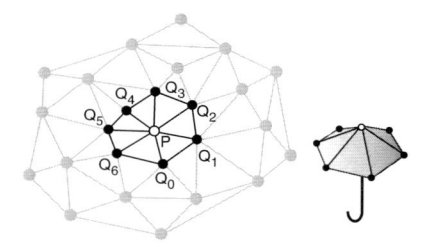

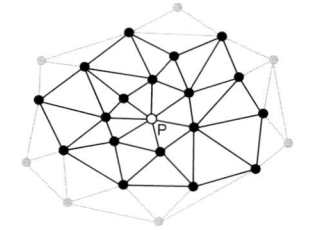

Fig. 15.8. *Part of a triangle mesh. Left*: the bold lines represent the neighborhood. The small umbrella illustrates why this region is called umbrella; *Right*: the topological neighborhood of two is displayed

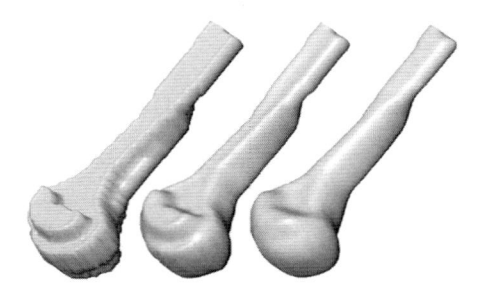

Fig. 15.9. *Visual smoothing. Left*: A surface model of bones extracted from strongly anisotropic MRI data; *Middle*: Smoothed mesh based on Laplacian smoothing; *Right*: Laplacian smoothing with extended neighborhood. The smoothing results in strongly reduced curvature of the surface (Courtesy: Jens Haase, Magdeburg)

variations of umbrella operators for different categories of segmented medical structures, such as organs, skeletal structures, small compact structures, such as lymph nodes and elongated structures. They also included extended neighborhoods, where instead of direct neighbors their second order neighbors are also considered (a topological neighborhood of two, where two represents the maximum path length in a local graph from the current vertex v_i to a neighbor v_j. Figure 15.8, right). In its simplest form, all vertices have the same influence. In a more elaborate version, the influence of a vertex represents the distance to the central vertex v_i.

$$v'_i = (1 - \alpha)v_i + \frac{\alpha}{n}\sum(v_j - v_i) \quad (15.1)$$

The simplest approach is to apply the well known Laplacian function, which smoothes every vertex in the mesh according to (15.1). With this equation, the vertex v_i is modified according to its previous value and the vertices v_j in the neighborhood of v_i, where α is the smoothing factor. Laplacian smoothing is applied iteratively, usually with 10 to 20 iterations (Fig. 15.9).

Unfortunately, the Laplacian filter tends to shrink the smoothed object. Hence, it is usually considered unsuitable for the smoothing of medical segmentations [27]. An extension of Laplacian fairing was proposed in [28]. In this approach, the smoothed umbrella regions are corrected in a second stage by moving the current vertex back towards its original position, thus maintaining the overall volume of the object. Another approach to surface

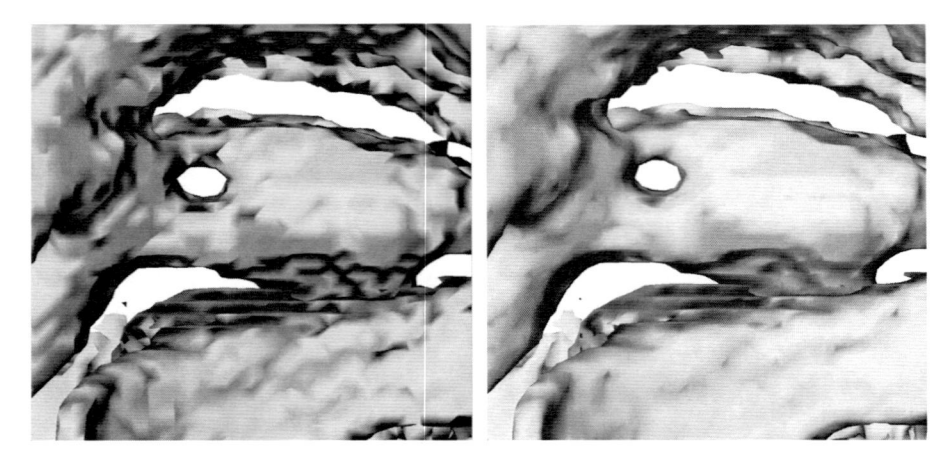

Fig. 15.10. *Smoothed dual marching cubes mesh.* The mesh shows a close-up of a segmented isosurface of the third cerebral ventricle. The hole corresponds to the adhesio interthalamica. *Left*: Original marching cubes mesh with staircasing artifacts; *Right*: Smoothed dual marching cubes surface

smoothing is based on signal theory, where the discrete Fourier theory is used to provide low pass filtering to (two dimensional discrete) surface signals by interpreting the eigenvalues of the Laplacian matrix as frequencies [29]. Model shrinking is controlled by alternating low pass filtering with different filter sizes.

As a final smoothing approach, we discuss a variation of marching cubes itself, the dual marching cubes algorithm [30]. After computing a quadrilateral patch structure from an original marching cubes generated mesh, the dual mesh of the patch mesh is generated. A dual of a mesh replaces a patch surface cell by a vertex and connects every neighboring dual vertex with an edge. Thus, every dual edge crosses an edge of the patch mesh, and every dual mesh cell contains a patch vertex. This also means that in contrast to the original triangular mesh, the dual mesh is composed of quadrilaterals. By iteratively applying this dualing operator, the original marching cubes mesh is successively smoothed (Fig. 15.10).

15.3.2 Improving Mesh Quality

While the methods mentioned above ensure a high visual quality of a surface mesh, the triangle quality with respect to the ratio between triangle sizes may still be too low for deriving simulation meshes. There are a variety of basic operations improving this quality significantly. As an example, very small edges and triangles may be removed or edges flipped (Fig. 15.11). The most common criterion for mesh quality is the *equi angle skewness* (Fig. 15.12). Cebral et al. discuss a pipeline of algorithms to yield smooth high quality meshes for visualization and simulation of blood flow [31].

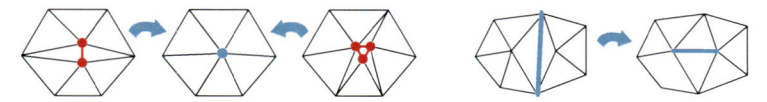

Fig. 15.11. *Improving mesh quality. Left*: Simple geometric operations: collapsing of small edges and triangles; *Right:* Edge flipping (Courtesy: Ragnar Bade, Magdeburg)

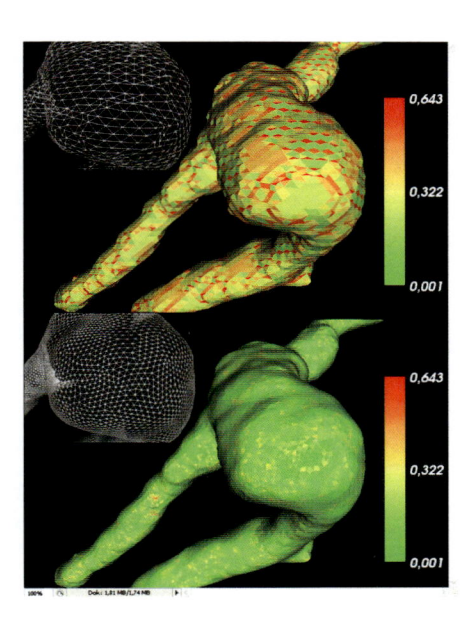

Fig. 15.12. *Surface model of a cerebral aneurysm. Top*: Initial mesh derived from marching cubes; *Bottom*: Smoothed mesh (Taubin's $\lambda|\mu$ filter) after subdivision and optimization. With these modifications, the equi angle skewness is significantly reduced (Courtesy: Tobias Mönch, Magdeburg)

15.4 Visualization of Vascular Structures

The visualization of vascular structures is an important and established topic within the broader field of visualizing medical volume data. General visualization techniques, such as slice-based viewing, direct volume rendering, and surface rendering (Sect. 15.2), are applicable in principle to display vascular structures from contrast-enhanced CT or MRI data. However, to recognize the complex spatial relations of a vascular tree and its surrounding structures more clearly, dedicated techniques are required. Furthermore, the required accuracy of the visualizations is different for diagnostic purposes, such as the search and analysis of vascular abnormalities, and therapy planning scenarios, where the vascular structures, themselves are not pathologic.

In general, there are two classes of surface visualization methods:

- Strictly adhering to the underlying data (model-free visualization).
- Relying on certain model assumptions and forcing the resulting visualization to adhere to these assumptions at the expense of accuracy (model-based visualization).

15.4.1 Surface-based Vessel Visualization

The most common surface reconstruction technique is marching cubes with an appropriate threshold. Unfortunately, the quality of the resulting visualizations is relatively low, due to inhomogeneities in the contrast agent distribution and due to the underlying interpolation (linear interpolation along the edges compared to trilinear interpolation in most direct volume rendering realizations). Textbooks on radiology warn their readers on the strong sensitivity of the selected isovalue on the visualization illustrating that a small change of 1 or 2 Hounsfield Unit (HU) may lead to a different diagnosis (e.g., presence or absence of a severe stenosis – a region where the vessel diameter is strongly reduced).

Recently, a surface visualization based on segmentation results was developed, which provides a superior quality compared to constrained elastic surface nets [26]. This method is based on Multi-level Partition of Unity (MPU) implicits [32], a technique originally developed for visualizing point clouds with implicit surfaces. The points are generated for the border voxels of the segmentation result. At thin and elongated structures, additional subvoxels are included and more points are generated (Fig. 15.13). The point set is locally sampled and approximated with quadric surfaces which nicely blend together. Implicit surfaces, in general, are able to represent a given geometry smoothly without explicitly constructing the geometry (Fig. 15.14).

15.4.2 Model-based Surface Visualization of Vascular Structures

For therapy planning (in case that not the vessels themselves are affected) and for educational purposes, model-based techniques are appropriate. The essential model assumption is usually that the cross-section is always circular. Cylinders [33] and truncated cones [34] have also been employed.

Model-based techniques require another preprocessing step beyond noise reduction and segmentation: vessel centerline and local vessel diameter have to be determined. For this purpose, skeletonization algorithms are used [35, 36].

To achieve smooth transitions at branchings, a variety of methods have been investigated. B-spline surfaces have been proposed to approximate small vascular structures and nerves, which could not be completely segmented [37].

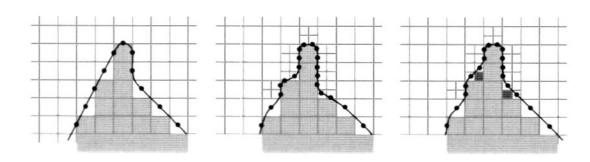

Fig. 15.13. *Point generation with MPU implicits [26]. Left*: Points resulting from segmentation; *Middle*: Additional points avoiding artifacts at corners; *Right*: Inclusion of additional subvoxels improves visualization

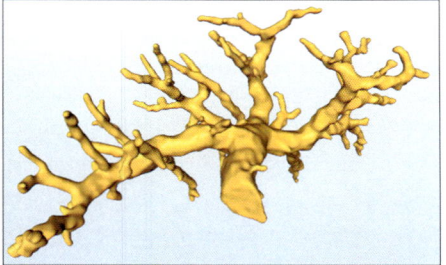

Fig. 15.14. *Visualization based on MUP implicits [26].* A segmented liver tree is visualized with marching cubes (*left*) and with MPU Implicits (*right*). Maintaining accuracy, the appearance is strongly improved

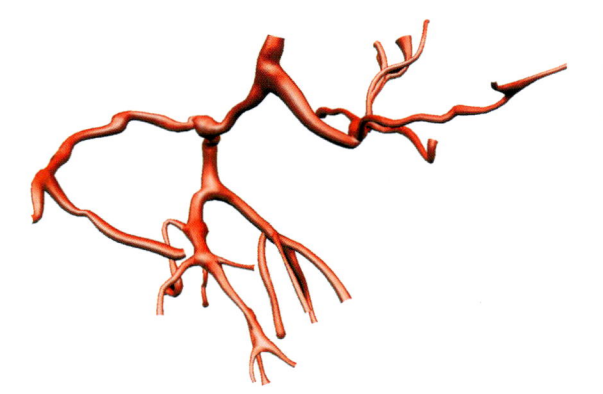

Fig. 15.15. *Subdivision surfaces.* The vessel tree is visualized applying subdivision surfaces

Felkl et al. [38] describe a method based on subdivision of an initial coarse base mesh. The advantage of their method is the adaptivity with respect to branchings, where more polygons are created (Fig. 15.15).

Also for model-free visualization of vascular structures, implicit surfaces may be used. Convolution surfaces, developed by Bloomenthal [39], allow to construct a scalar field along skeletal structures. With an appropriate filter for the convolution, the scalar field can be polygonized in such a way that the vessel diameter is precisely represented. The filter selection must also avoid so-called "unwanted effects" [40], such as blending and bulging, which are typical for implicit surfaces (Fig. 15.16).

A general problem of the high-quality visualization methods is that they are typically slower. For reasonably large datasets, the visualization with convolution surfaces takes 20–50 s [40], compared to 3–5 s with truncated cones. Another problem is that the accuracy of most of these methods has not been carefully investigated. Comparisons of different methods with respect to the resulting surfaces and their distances are necessary to state how reliable the results really are [41]. However, it must be noted that the good results (mean deviation: 0.5 times the diagonal size of a voxel) relate to datasets without any pathologies.

Fig. 15.16. *Convolution surfaces [26]*. Visualization of the portal vein derived from CT with 136 edges

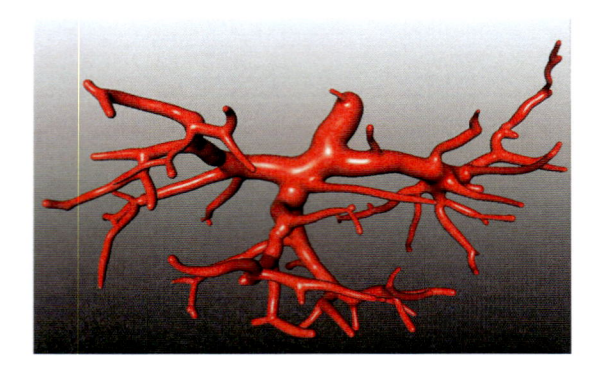

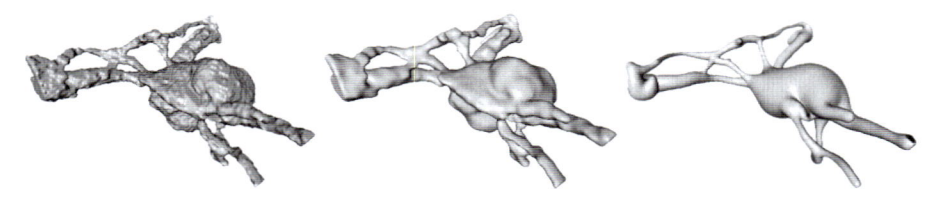

Fig. 15.17. *Visualization of a segmented cerebral aneurysm [26]*. *Left*: Marching cubes; *Middle*: MPU implicits; *Right*: Convolution surfaces

Figure 15.17 compares visualizations of an aneurysm with marching cubes, MPU implicits and convolution surfaces. Despite the high visual quality, the convolution surface is less appropriate than MPU Implicits due to pathologic variation of the blood vessels (strong deviation from the circular cross-section).

15.4.3 Volume Rendering of Vascular Structures

The most commonly used volume rendering technique in a clinical environment is the Maximum Intensity Projection (MIP), which basically displays the voxel with the highest image intensity for every ray. Based on the enhancement with a contrast agent, vessels are usually the brightest structures and can thus be selectively visualized with this method. This basic strategy fails when contrast-enhanced vessels and skeletal structures, which also exhibit high intensity values, are close to each other [42]. More general, the 3D visualization of vascular structures benefits from an enhancement of elongated vessel-like structures, which may be accomplished with a shape analysis (Fig. 15.18).

Since static MIP images do not provide any depth perception they are either interactively explored or presented as animation sequences. This, however, does not account for another problem related to MIP: small vascular structures are often suppressed, since they are represented primarily by border voxels. Due to averaging, those voxels appear less intense than the inner voxels of large vascular structures (partial volume effect). In order to display

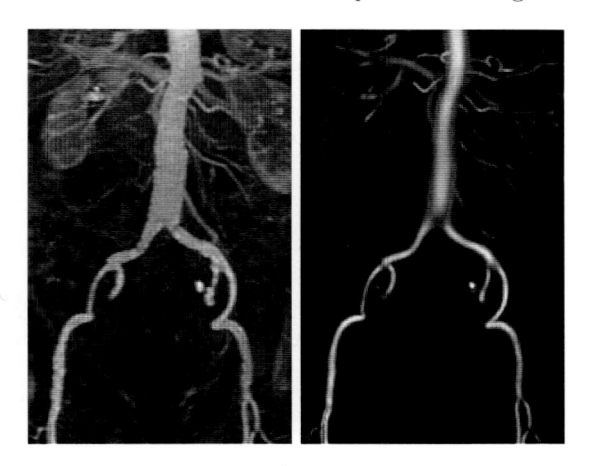

15 Visualization and Exploration of Segmented Anatomic Structures 393

Fig. 15.18. *Vesselness filter [43]. Left*: MIP image of MRA data; *Right*: Preprocessing with the "vesselness filter" suppresses other structures

small vessels in front of larger ones, the MIP method was modified by a threshold which specifies that the first local maximum above the threshold should be depicted instead of the global maximum. This method is known as either local MIP or Closest Vessel Projection (CVP).

In contrast to projection methods, direct volume rendering (e.g., ray casting, splatting) computes a weighted accumulation of different samples along viewing rays according to their spatial position. These rendering techniques provide realistic depth cues by blending data, since the samples are accumulated in depth order (front-to-back, back-to-front). Volume rendering of vascular structures usually involves only a small fraction of the overall data size. Frequently, the Transfer Function (TF) is adjusted, such that only 1%–2% of the voxels become visible [44]. Therefore, volume rendering may be strongly accelerated through techniques, such as empty space skipping.

The ability of 1D TFs to discriminate vascular structures from its surrounding is limited particularly in the neighborhood of skeletal structures which exhibit a similar range of image intensity values. Therefore, 2D TFs have been explored [45], where gradient magnitude has been used as second dimension in addition to image intensity (Fig. 15.19).

As a example of using direct volume rendering for diagnosis, we introduce a method for the diagnosis of the coronary artery disease based on multislice CT data with high spatial resolution. The major diagnostic task is to detect, characterize and quantify abnormalities of the vessel wall, so-called plaque which might be calcified (hard plaque) or fibrous (soft plaque). The structures to be emphasized are too small to yield a substantial footprint in a global histogram. Moreover, contrast agent distribution cannot be ensured to be uniform. Hence, a TF with fixed parameters will not be able to cope with the variability of the datasets. Based on a segmentation of the coronary vessels, a local histogram may be derived and analyzed with respect to a Gaussian distribution (mean and standard deviation) of the vessel wall voxels (Fig. 15.20).

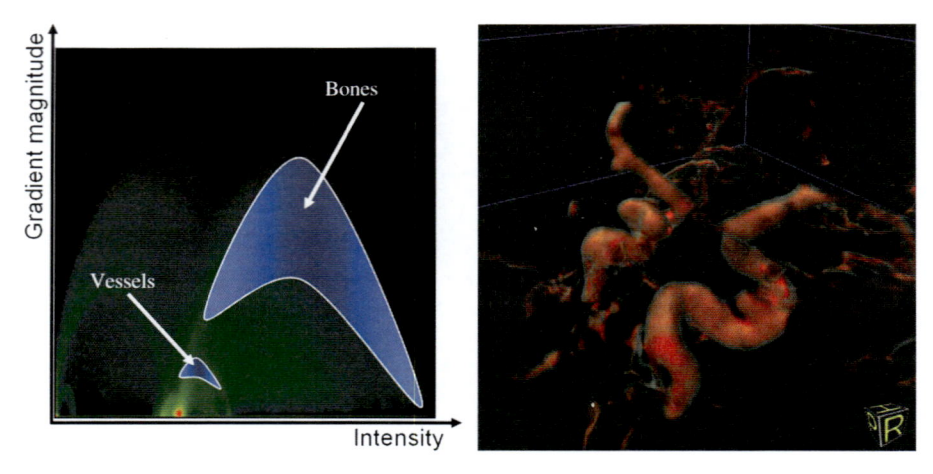

Fig. 15.19. *2D transfer function. Left*: Vessels and skeletal structures differ in the 2D histogram; *Right*: Visualization of skeletal structures is effectively suppressed using an appropriate 2D TF (Courtesy: Peter Hastreiter, Erlangen)

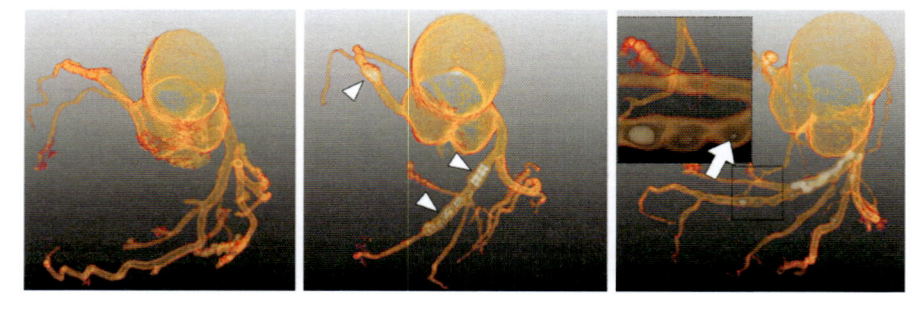

Fig. 15.20. *Direct volume rendering. Left*: Patient without plaque burden; *Middle*: Three stents (*arrowheads*); *Right*: Patient with many hard plaques. Even small hard plaques are recognizable (*arrow*) (Courtesy: Sylvia Glaßer, Magdeburg)

15.5 Virtual Endoscopy

The typical examination of a graphical representation of anatomical image data is from an exo- or geocentric perspective, i.e., the representation is viewed with a viewpoint that is located outside of the anatomical scene. This is analogous to a camera that captures a scene from the outside. To change the perspective to another image section, affine transformations (e.g., rotation, translation, scaling) are applied, which in turn transform the coordinate system of the graphical representation. Typical changes include zooming to an interesting detail, after moving that detail into the focus. While the coordinate system is transformed, the camera remains fixed at its position.

Specific diagnostic questions, however, cannot be sufficiently addressed with an overview of the anatomical visualization. Instead, details are of

15 Visualization and Exploration of Segmented Anatomic Structures   395

interest that cannot be viewed from the outside. In these cases, we speak of an egocentric perspective, since the camera is now moving with the observer through the respective anatomy. Due to the similarity to an endoscopic camera that is moved through a preformed cavity of the body, this visual representation is called virtual endoscopy [46]. It has four different application scenarios:

- Diagnosis and screening, e.g. [47].
- Intervention and therapy planning, e.g. [48, 49].
- Intra-operative support, e.g. [50].
- Education and training.

15.5.1 Graphical Representation

Once the Organ Of Interest (OOI) is identified, a graphical representation can be derived that represents the organ. However, the image acquisition stage must ensure a sufficient contrast between the preformed inside of the organ to the surrounding voxels. This is to ensure a clear separation between the inside and the outside of the organ and hence of the inspected organ cavity.

An important advantage of virtual endoscopy (compared to video endoscopy) is the possibility for semi-transparent rendering of organs. Hence, occluded structures (e.g., blood vessel, nerve tract, tumor) may become visible through the organ wall (Fig. 15.21). Nevertheless, the used visualization methods must provide for semi-transparent rendering to ensure the correct blending, and hence the correct rendering of the structures.

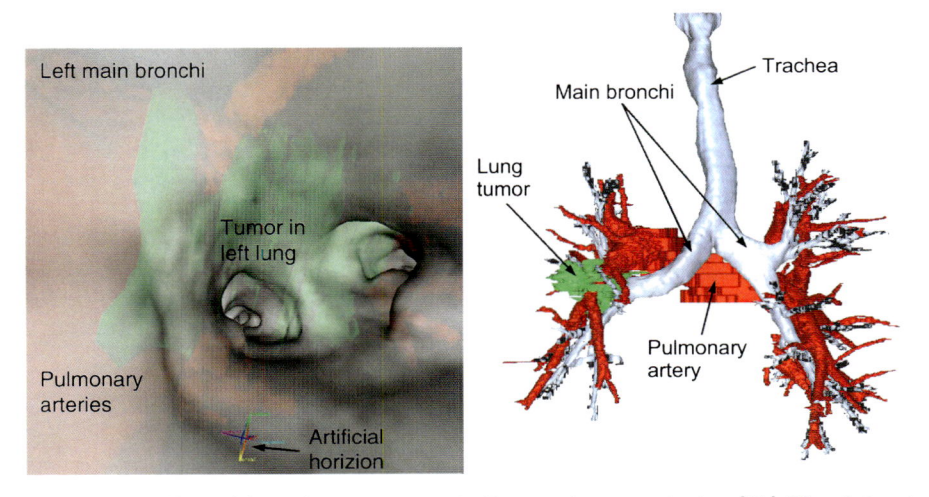

Fig. 15.21. *Virtual bronchoscopy using indirect volume rendering [51].* The dataset shows the lower airways of the lungs, the pulmonary arterioles (*red*), and a lung tumor (*green*). *Left*: Virtual endoscopic; *Right*: Lungs overview

15.5.2 Interaction Model

The probably most under-estimated technical aspect of virtual endoscopy is the interaction necessary to steer a virtual camera through the visual OOI representation. Depending on the quality of the underlying interaction model, camera control can be very intuitive and easy, or very cumbersome. While considering the interaction model, it is interesting to note that the actually used interaction device (e.g., computer mouse, graphic tablet, 3D mouse, joystick) plays only a minor role.

Since virtual endoscopy of an organ (e.g., colon) should cover the whole organ surface, the interaction model must ensure sufficient coverage. Experiments have shown that a regular fly-through of a tubular organ in one direction – probably one of the topological simple cases of virtual endoscopy – ensures only the inspection of 70% of the inner surface, while a fly-through in both directions covers up to 95% [52].

In general, we differentiate three interaction paradigms for virtual endoscopy [53]:

- *Planned or automatic interaction* limits the possible interaction of the camera to the reply along a pre-defined camera path. This paths is either pre-defined by a user or generated automatically. This interaction provides only a movie player-like functionality pause, forward, and backward.
- *Manual or free interaction* imposes no limitations to the movement of the virtual camera. Unfortunately, the often narrow and non regular structures of a patient's anatomy do not permit an easy navigation of the virtual camera with standard geometric transforms. Furthermore, this interaction option does not allow for an easy collision avoidance.
- *Guided interaction* limits the movement of the camera by a number of constraints. If the right constraints are specified, it provides a good mixture of flexibility and guidance to a target structure [53]. For example, the interaction system mimics a submarine that is immersed into a drift to the target, while the kinematic control of the camera mimics the propulsion of the submarine.

15.5.3 User Interface

The Graphical User Interface (GUI) implements the actual connection between the user and the virtual camera. Hence, it must be embedded into the clinical workflow. In particular, the user interface must be clearly structured and address the requirements of the user (e.g., radiologist, surgeon) [54, 55].

In particular, the GUI should contain facilities to switch between different graphical representations and interaction models as well as to parameterize them e.g. with respect to speed and colors so that user can adapt it to their viewing preferences. Often, it is desired to incorporate annotations, such as free-hand drawings to mark information for surgery [56] or measurement facilities, e.g. to show the current diameter of a structure. However, care is necessary not to overwhelm user with too many options. Simplicity and guidance is often more important than maximum flexibility.

15.5.4 Case Study: Virtual Colonoscopy

Cancer of the colon and rectum is the second leading cause of cancer deaths in the USA. Approximately 150,000 new cases of colorectal cancer are diagnosed every year [57], and similar numbers are assumed for western Europe and parts of Asia. Consequently, it is imperative that an effective diagnostic procedure is found to detect colonic polyps or tumors at an early stage (smaller than 5 mm in diameter), usually using video or optical endoscopy. Here, a fiber optical probe is introduced into the colon through the rectum. By manipulating the tiny camera attached to the tip of the probe, the physician examines the inner surface of the colon to identify abnormalities. This invasive procedure takes about one hour and requires intravenous sedation, resulting in high costs.

However, this endoscopic method is typically rather expensive or to circumstantial for prophylactic screening, somewhat unpleasant, and results in a low patient acceptance. Consequently, virtual colonoscopy was proposed to limit optical colonoscopy to cases in which either a suspicious polyp was found (which induced a biopsy or removal of the polyp) or which were inconclusive in virtual colonoscopy. An inconclusive result typically happens if (shape) defects of the graphical representation of the inner colon surface cannot be identified as either polyps or residual stool.

After cleansing and inflating of the colon (both actions are also required for video colonoscopy), a CT or MRI scan is performed. The resulting image stack is pre-processed and examined using a the virtual endoscopy system.

Hong et al. compared the results of video/optical and virtual endoscopy based on polyps found in both procedures (Fig. 15.22). A similar study has been presented by Fenlon et al. [58]. The authors found that the performance of virtual colonoscopy is comparable to video colonoscopy, as long as the data resolution is sufficient to detect polyps of the respective size. Problems arose from residual stool, which often was the cause of a false positive finding. More recently, Pickhardt et al. [47] found several positively identified polyps in virtual colonoscopy, which have not been seen in the initial video colonoscopy. Overall, virtual colonoscopy achieved a sensitivity of more than 88% for polyps larger than 6 mm, and a specificity of close to 80%.

15.6 Conclusions

In this chapter, we have described the various options available to render complex anatomical image data. Segmentations are used to differentiate different organs, in particular if the intensity contrast of the image data is insufficient to depict the boundaries. Next to the technical aspects, we have

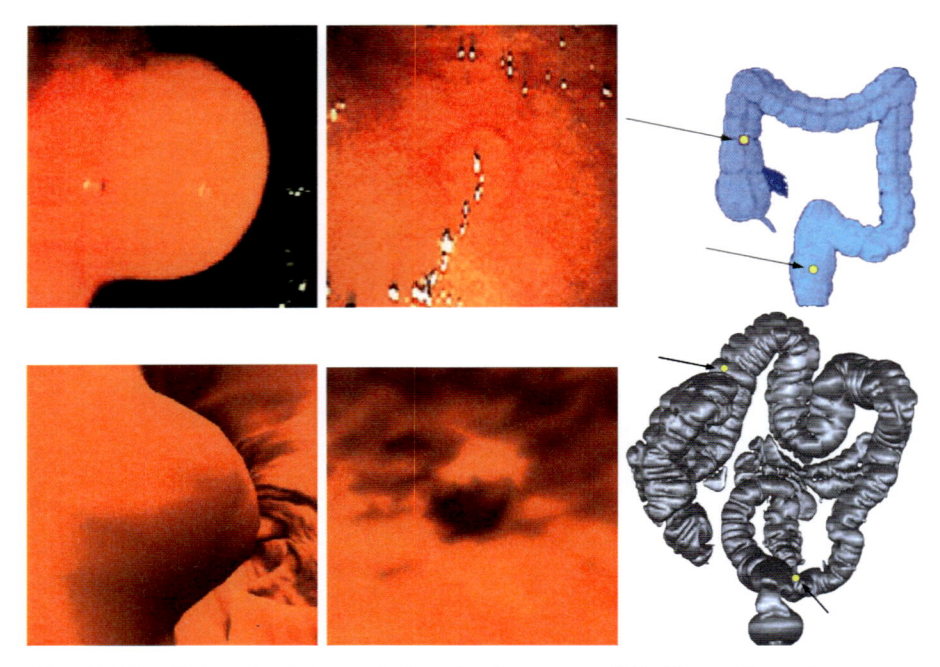

Fig. 15.22. *Polyps in video and virtual colonoscopy [53]. Top*: video colonoscopy; *Bottom*: virtual colonoscopy; *Left*: 8 mm polyp; *Middle*: 4 mm polyp; *Right*: colon overview. The polyp positions (yellow marker) are indicated with arrows

discussed advantages and disadvantages of the different ways of visualizing the image data. Special visualization techniques have been discussed for vascular structures. In particular, model-based techniques for both, direct and indirect volume rendering techniques have been presented.

Finally, we have introduced virtual endoscopy. As a principle limitation, all visualized structures are identified either directly by intensity contrast of the voxels in the image dataset, or through a segmentation. An incompletely segmentation of important anatomical structures (e.g., blood vessels) may lead to critical complications if the physician puts too much trust in this data. Therefore, a segmentation must be performed with utmost diligence, while the use of virtual endoscopy must be accompanied with a healthy level of skepticism.

References

1. Preim B, Bartz D. Visualization in Medicine. Theory, Algorithms, and Applications. Burlington: Morgan Kaufmann Publishers; 2007.
2. Engel K, Hadwiger M, Kniss J, et al. Real-Time Volume Graphics. Natick, MA: A.K. Peters Ltd.; 2006.

3. Lorensen W, Cline H. Marching cubes: a high resolution 3d surface construction algorithm. In: Proc ACM SIGGRAPH; 1987. p. 163–169.
4. Hege H, Seebaß M, Stalling D, et al. A Generalized Marching Cubes Algorithm Based on Non-Binary Classifications. Zuse Institute Berlin (ZIB); 1997. ZIB SC 97-05.
5. Banks D, Linton S. Counting cases in marching cubes: toward a generic algorithm for producing substitopes. In: Proc IEEE Vis; 2003. p. 51–58.
6. Kay D, Greenberg D. Transparency for computer synthesized images. In: Proc ACM SIGGRAPH; 1979. p. 158–164.
7. Meißner M, Huang J, Bartz D, et al. A practical evaluation of four popular volume rendering algorithms. In: Proc IEEE/ACM Symp Vol Vis Graph; 2000. p. 81–90.
8. Krüger W. The applicaton of transport theory to visualization of 3-d scalar data fields. In: Proc IEEE Vis; 1990. p. 273–280.
9. Sabella P. A rendering algorithm for visualizing 3d scalar fields. In: Proc ACM SIGGRAPH; 1988. p. 51–58.
10. Max N. Optical models for direct volume rendering. IEEE Trans Vis Comput Graph. 1995;1(2):99–108.
11. Levoy M. Display of surfaces from volume data. IEEE Comput Graph Appl. 1988;8(3):29–37.
12. Westover L. Footprint evaluation for volume rendering. In: Proc ACM SIGGRAPH; 1990. p. 367–376.
13. Mueller K, Möller T, Crawfis R. Splatting without the blur. In: Proc IEEE Vis; 1999. p. 363–371.
14. Hadwigger M, Sigg C, Scharsach K, et al. Real-time ray-casting and advanced shading of discrete isosurfaces. Comput Graph Forum. 2005;24(3):303–12.
15. Pfister H, Lorensen W, Bajaj C, et al. The transfer function bake-off. IEEE Comput Graph Appl. 2001;21(3):16–22.
16. Kindlmann G, Durkin J. Semi-automatic generation of transfer functions for direct volume rendering. In: Proc IEEE/ACM Symp Vol Vis; 1998. p. 79–86.
17. Chuang J, Weiskopf D, Möller T. Hue-preserving color blending. IEEE Trans Vis Comput Graph. 2009;15(6):1275–82.
18. Chan M, Wu Y, Mak W, et al. Perception-based transparency optimization for direct volume rendering. IEEE Trans Vis Comput Graph. 2009;15(6):1283–90.
19. Lacroute P, Levoy M. Fast volume rendering using a shear-warp factorization of the viewing transformation. In: Proc ACM SIGGRAPH; 1994. p. 451–458.
20. Wittenbrink C, Malzbender T, Goss M. Opacity-weighted color interpolation for volume sampling. In: Proc IEEE/ACM Symp Vol Vis; 1998. p. 135–142.
21. Hauser H, Mroz L, Bischi G, et al. Two-level volume rendering. IEEE Trans Vis Comput Graph. 2001;7(3):242–52.
22. Hastreiter P, Naraghi R, Tomandl B, et al. 3D-visualization and registration for neurovascular compression syndrome analysis. Lect Notes Comput Sci. 2002;2488:396–403.
23. Tiede U, Schiemann T, Höhne K. High quality rendering of attributed volume data. In: Proc IEEE Vis; 1998. p. 255–261.
24. Beyer J, Hadwiger M, Wolfsberger S, et al. High-quality multimodal volume rendering for preoperative planning of neurosurgical interventions. IEEE Trans Vis Comput Graph. 2007;13(6):1696–1703.

25. Allamandri F, Cignoni P, Montani C, et al. Adaptively adjusting marching cubes output to fit a trilinear reconstruction filter. In: Proc Eurographics Workshop Vis Sci Computing; 1998. p. 25–34.
26. Schumann C, Oeltze S, Bade R, et al. Model-free surface visualization of vascular trees. In: IEEE/Eurographics Symp Vis Eurographics; 2007. p. 283–290.
27. Bade R, Haase J, Preim B. Comparison of fundamental mesh smoothing algorithms for medical surface models. In: Proc Simul Vis; 2006. p. 289–304.
28. Vollmer J, Mencel R, Müller H. Improved laplacian smoothing of noisy surface meshes. In: Proc Eurographics; 1999. p. 131–138.
29. Taubin G. A signal processing approach to fair surface design. In: Proc ACM SIGGRAPH; 1995. p. 351–358.
30. Nielson G. Dual marching cubes. In: Proc IEEE Vis; 2004. p. 489–496.
31. Cebral JR, Castro MA, Appanaboyina S, et al. Efficient pipeline for image-based patient-specific analysis of cerebral aneurysm hemodynamics: technique and sensitivity. IEEE Trans Med Imaging. 2004;24(4):457–67.
32. Ohtake Y, Belyaev A, Alexa M, et al. Multilevel partition of unity implicits. ACM Trans Graph. 2003;22(3):463–70.
33. Masutani Y, Masamune K, Dohi T. Region-growing-based feature extraction algorithm for tree-like objects. Lect Notes Comput Sci. 1996;1131:161–71.
34. Hahn HK, Preim B, Selle D, et al. Visualization and interaction techniques for the exploration of vascular structures. In: Proc IEEE Vis; 2001. p. 395–402.
35. Kirbas C, Quek F. A review of vessel extraction techniques and algorithms. ACM Comput Surv. 2004;36(2):81–121.
36. Gerig G, Koller T, Székely G, et al. Symbolic description of 3-d structures applied to cerebral vessel tree obtained from mr angiography volume data. Lect Notes Comput Sci. 1993;687:94–111.
37. Höhne KH, Pflesser B, Pommert A, et al. A realistic model of the inner organs from the visible human data. Lect Notes Comput Sci. 2000;1935:776–85.
38. Felkl P, Wegenkittl R, Bühler K. Surface models of tube trees. In: Proc Comput Graph Int; 2004. p. 70–77.
39. Bloomenthal J, Shoemake K. Convolution surfaces. In: Proc ACM SIGGRAPH; 1991. p. 251–256.
40. Oeltze S, Preim B. Visualization of vascular structures with convolution surfaces. In: Proc IEEE/Eurographics Symp Vis; 2004. p. 311–20.
41. Oeltze S, Preim B. Visualization of vascular structures with convolution surfaces: method, validation and evaluation. IEEE Trans Med Imaging. 2005;25(4):540–49.
42. Fiebich M, Straus CM, Sehgal V, et al. Automatic bone segmentation technique for CT angiographic studies. J Comput Assist Tomogr. 1999;23(1):155–61.
43. Frangi AF, Niessen WJ, Vincken KL, et al. Multiscale vessel enhancement filtering. Lect Notes Comput Sci. 1998;1496:130–37.
44. Vega F, Hastreiter P, Fahlbusch R, et al. High performance volume splatting for visualization of neurovascular data. In: Proc IEEE Vis; 2005. p. 271–278.
45. Vega F, Sauber N, Tomandl B, et al. Enhanced 3d-visualization of intracranial aneurysms involving the skull base. Lect Notes Comput Sci. 2003;2879:256–63.
46. Bartz D. Virtual endoscopy in research and clinical practice. Comput Graph Forum. 2005;24(1):111–26.
47. Pickhardt P, Choi J, Hwang I, et al. Computed tomographic virtual colonoscopy to screen for colorectal neoplasia in asymptomatic adults. N Engl J Med. 2003;349(23):2191–2200.

48. Auer D, Auer L. Virtual endoscopy – a new tool for teaching and training in neuroimaging. Int J Neuroradiol. 1998;4:3–14.
49. Bartz D, Straßer W, Gürvit O, et al. Interactive and multi-modal visualization for neuroendoscopic interventions. In: Proc Eurographics/IEEE Symp Vis; 2001. p. 157–164.
50. Freudenstein D, Wagner A, Gürvit O, et al. Virtual representation of the basal cistern: technical note. Med Sci Monit. 2002;8(9):153–58.
51. Bartz D, Mayer D, Fischer J, et al. Hybrid segmentation and exploration of the human lungs. In: Proc IEEE Vis; 2003. p. 177–184.
52. Dachille F, Kreeger K, Wax M, et al. Interactive navigation for PC-based virtual colonoscopy. Proc SPIE 2001; 4321: 500–4.
53. Hong L, Muraki S, Kaufman A, et al. Virtual voyage: interactive navigation in the human colon. In: Proc ACM SIGGRAPH; 1997. p. 27–34.
54. Preim B. Model-based visualization for intervention planning. In: Proc Sci Visualization: Adv Concepts. Dagstuhl, Germany 2010. p. 163–78.
55. Pickhardt P. Three-dimensional endoluminal CT colonoscopy (virtual colonoscopy): comparison of three commercially available systems. Am J Roentgenol. 2003;181(6):1599–1606.
56. Krüger A, Kubisch C, Strauß G, et al. Sinus endoscopy: application of advanced GPU volume rendering for virtual endoscopy. IEEE Trans Vis Comput Graph. 2008;14(6):1491–98.
57. Cohen L, Basuk P, Waye J. Practical Flexible Sigmoidoscopy. New York: Igaku-Shoin; 1995.
58. Fenlon H, Nunes D, Schroy P, et al. A comparison of virtual and conventional colonoscopy for the detection of colorectal polyps. N Engl J Med. 1999;341(20):1496–1503.

16
Processing and Visualization of Diffusion MRI

James G. Malcolm, Yogesh Rathi, and Carl-Fredrik Westin

Summary. This chapter provides a survey of techniques for processing and visualization of diffusion magnetic resonance imagery. We describe various approaches to modeling the local diffusion structure from scanner measurements. In particular, we differentiate between parametric and non-parametric models, and describe regularization approaches. We then look at techniques to trace out neural pathways and infer global tissue structure. Deterministic, probabilistic, and global trajectories are analyzed, and techniques of validation are introduced. Last, we draw upon these as building blocks for the visualization and analysis of the neural architecture of individuals and groups. Special attention is drawn to volume segmentation, fiber clustering, and tissue analysis.

16.1 Introduction

The advent of diffusion magnetic resonance imaging (dMRI) has provided the opportunity for noninvasive investigation of neural architecture. While structural MRI has long been used to image soft tissue and bone, dMRI provides additional insight into tissue microstructure by measuring its microscopic diffusion characteristics. To accomplish this, the magnetic field induces the movement of water while the presence of cell membranes, fibers, or other macro-molecules hinder this movement. By varying the direction and strength of the magnetic fields, we essentially use the water molecules as a probe to get a sense of the local tissue structure.

At the lowest level, this diffusion pattern provides several insights. For example, in fibrous tissue the dominant direction of allowed diffusion corresponds the underlying direction of fibers. In addition, quantifying the anisotropy of the diffusion pattern can also provide useful biomarkers. Several models have been proposed to interpret scanner measurements, ranging from geometric abstractions to those with biological motivation. In Sect. 16.2, various models and methods for interpreting the diffusion measurements are introduced.

By connecting these local orientation models, tractography attempts to reconstruct the neural pathways. Tracing out these pathways, we begin to see how neurons originating from one region connect to other regions and how well-defined those connections may be. Not only can we examine properties of the local tissue but we begin to see the global functional architecture of the brain, but for such studies, the quality of the results relies heavily on the chosen fiber representation and the method of reconstructing pathways. In Sect. 16.3, several techniques for tracing out pathways are described.

At the highest level, neuroscientists can use the results of local modeling and tractography to examine individuals or groups of individuals. In Sect. 16.4, approaches to segment tissue with boundaries indistinguishable with structural MRI are surveyed, applying network analysis to characterize the macroscopic neural architecture, reconstruct fiber bundles from individual fiber traces, and analyzing groups of individuals.

16.2 Modeling

16.2.1 Imaging the Tissue

The overall signal observed in an dMRI image voxel (millimetric) is the superposition of signals from many underlying molecules probing the tissue (micrometric). Thus, the image contrast is related to the strength of water diffusion. At each image voxel, diffusion is measured along a set of distinct gradients, $u_1, \ldots, u_n \in \mathbb{R}^3$, producing the corresponding signal, $s = [s_1, \ldots, s_n]^T \in \mathbb{R}^n$. A general weighted formulation that relates the measured diffusion signal to the underlying fiber architecture may be written as:

$$s_i = s_0 \sum_j w_j e^{-b_j u_i^T D_j u_i}, \tag{16.1}$$

where s_0 is a baseline signal intensity, b_j is the b-value, an acquisition-specific constant, w_j are convex weights, and D_j is a tensor describing a diffusion pattern. One of the first acquisition schemes developed, diffusion tensor imaging (DTI) uses these measurements to compute a Gaussian estimate of the diffusion orientation and strength at each voxel [1].

Going beyond this macroscopic description of diffusion, various higher resolution acquisition techniques have been developed to capture more information about the diffusion pattern. One of the first techniques, diffusion spectrum imaging (DSI), measures the diffusion process at various scales (multiscale) by sampling densely throughout the voxel [2]. From this, the Fourier transform is used to convert the signal to a diffusion probability distribution. Due to a large number of samples acquired (usually more than 256 gradient directions), this scheme provides a much more accurate description of the diffusion process. However, on account of the large acquisition time (of the order of 1–2 h per subject), this technique is not typically used in clinical scans, and its use is restricted to few research applications.

Instead of spatially sampling the diffusion in a lattice throughout the voxel, a spherical shell sampling could be used. Using this sampling technique, it has been demonstrated that the orientation distribution function (ODF) could be recovered from the images acquired on a single spherical shell [3]. This significantly reduced the acquisition time, while providing most of the information about the underlying diffusion in the tissue. Naturally, this led to the application of techniques for estimating functions on a spherical domain. For example, Q-Ball Imaging (QBI) demonstrated a spherical version of the Fourier transform to reconstruct the probability diffusion as an iso-surface [3].

To begin studying the microstructure of fibers with these imaging techniques, we need models to interpret these diffusion measurements. Such models fall broadly into two categories: parametric and non-parametric.

16.2.2 Parametric Models

One of the simplest models of diffusion is a Gaussian distribution: an elliptic (anisotropic) shape indicates a strong diffusion direction, while a more rounded surface (isotropic) indicates less certainty in any particular direction (Fig. 16.1c). While robust, assuming this Gaussian model is inadequate in cases of mixed fiber presence or more complex orientations where the signal may indicate a non-Gaussian pattern. To handle these complex patterns, higher resolution imaging and more flexible parametric models have been proposed including mixtures of tensors [4, 6–9] and directional functions [10, 11]. While these typically require the number of components to be fixed or estimated separately, more continuous mixtures have also been proposed [12]. Furthermore, biologically inspired models and tailored acquisition schemes have been proposed to estimate physical tissue microstructure [13, 14].

16.2.3 Non-parametric Models

Non-parametric models can often provide more information about the diffusion pattern. Instead of modeling a discrete number of fibers as in parametric models, non-parametric techniques estimate a spherical orientation distribution function indicating potential fiber directions and the relative certainty thereof. For this estimation, an assortment of surface reconstruction methods have been introduced:

- QBI to directly transform the signal into a probability surface [3]
- Spherical harmonic representations [5, 15, 16]
- Higher-order tensors [17, 18]
- Diffusion profile transforms [19, 20]
- Deconvolution with an assumed single-fiber signal response [21, 22]
- And more

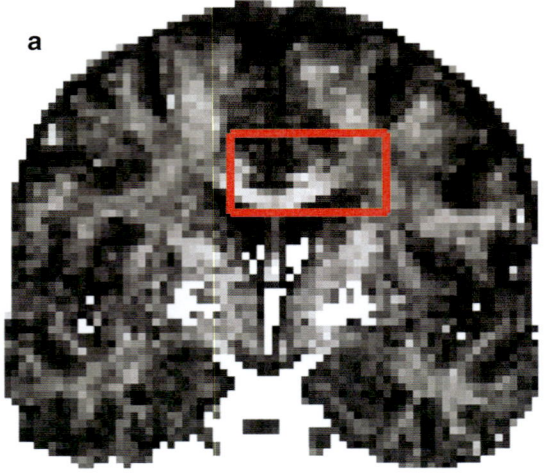

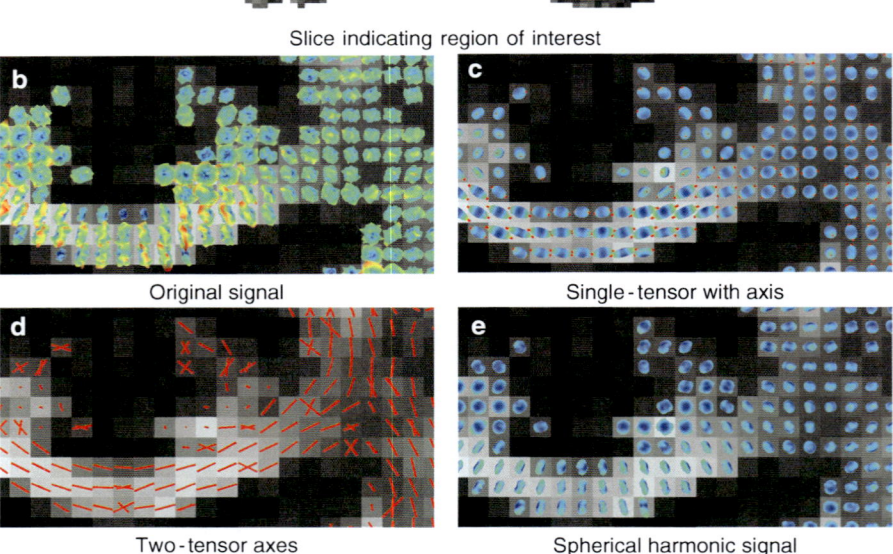

Fig. 16.1. *Model comparison.* Comparison of various models within a coronal slice (**a**) passing through the corpus callosum. In (**b**) the original signal appears noisy. In (**c**) a single tensor fit provides a robust estimate of the principal diffusion direction. In (**d**) a two-tensor model is fit to planar voxels and the two axes are reported [4]. In (**e**) spherical harmonics provide a smoothed non-parametric estimate of the signal surface eliminating much of the noise seen in (**b**) [5]

A comparison of techniques is given in Fig. 16.1. For example, Fig. 16.1e shows a spherical harmonic reconstruction of the signal. Compare this to the original signal in Fig. 16.1b.

It is important to keep in mind that there is a distinction made often between the reconstructed diffusion ODF and the putative fiber ODF; while most techniques estimate the diffusion function, its relation to the underlying fiber function is still an open problem. Spherical convolution is designed to directly transform the signal into a fiber distribution [15, 19, 21], yet diffusion sharpening strategies have been developed to deal with Q-ball and diffusion functions [23].

While parametric methods directly describe the principal diffusion directions, interpreting the diffusion pattern from model-independent representations typically involves determining the number and orientation of principal diffusion directions present. A common technique is to find them as surface maxima of the diffusion function [16, 21, 23], while another approach is to decompose a high-order tensor representation of the diffusion function into a mixture of rank-1 tensors [24].

16.2.4 Regularization

As in all physical systems, the measurement noise plays a nontrivial role, and so several techniques have been proposed to regularize the estimation. One could start by directly regularizing the MRI signal by designing filters based on the various signal noise models [25, 26]. Alternatively, one could estimate the diffusion tensor field and then correct these estimated quantities. For spherical harmonic modeling, a regularization term can be been directly included in the least-squares formulation [5, 16].

Attempts such as these to manipulate diffusion-weighted images or tensor fields have received considerable attention regarding appropriate algebraic and numeric treatments [27, 28].

Instead of regularizing signal or model parameters directly, an alternative approach is to infer the underlying geometry of the vector field. Another interesting approach treats each newly acquired diffusion image as a new system measurement. Since diffusion tensors and spherical harmonics can be estimated within a least-squares framework, one can use a Kalman filter to update the estimate and optionally stop the scan when the model parameters converge [29]. Furthermore, this online technique can be used to alter the gradient set so that, where the scan to be stopped early, the gradients up to that point are optimally spread (active imaging) [30].

16.2.5 Characterizing Tissue

The goal of diffusion imaging is to draw inferences from the diffusion measurements. As a starting point, one often converts the diffusion weighted image volumes to a scalar volume much like structural MRI or CT images. Starting with the standard Gaussian diffusion tensor model, an assortment of scalar measures have been proposed to quantify the size, orientation, and shape of the diffusion pattern [31]. For example, fractional anisotropy (FA) quantifies the deviation from an isotropic tensor, an appealing quantity, because it corresponds to the strength of diffusion while remaining invariant to orientation.

Derivatives of these scalar measures have also been proposed to capture more information about the local neighborhood [32, 33], and these measures have been extended to high-order tensors [34]. Furthermore, a definition of generalized anisotropy has been proposed to directly characterize anisotropy in terms of variance in the signal, hence avoiding an assumed model. While geometric in nature, studies have shown these to be reasonable proxy measures for neural myelination [35, 36]. Some studies have also examined the sensitivity of such measures against image acquisition schemes [37, 38].

Meaningful visualization of diffusion images is difficult because of their multivariate nature, and much is lost when reducing the spectral signal down to scalar intensity volumes. Several geometric abstractions have been proposed to convey more information. Since the most common voxel model is still the Gaussian diffusion tensor, most of the effort has focused on visualizing this basic element. The most common glyph is an ellipsoid simultaneously representing the size, shape, and orientation; however, since tensors have six free parameters, more elaborate representations have been proposed to visualize these additional dimensions using color, shading, or subtle variations in shape [31, 39]. Apart from tensors, visualization strategies for other models have received comparatively little attention, the typical approach being to simply to visualize the diffusion isosurface at each voxel.

A vast literature exists on methods of acquisition, modeling, reconstruction, and visualization of diffusion images. For a comprehensive view, we suggest [31, 40].

16.3 Tractography

To compliment the wide assortment of techniques for signal modeling and reconstruction, there is an equally wide range of techniques to infer neural pathways.

At the local level, one may categorize them either as tracing individual connections between regions or as diffusing out to estimate the probability of connection between regions.

In addition, more global approaches have been developed to consider not only the local orientations but also the suitability of entire paths when inferring connections.

16.3.1 Deterministic Tractography

Deterministic tractography involves directly following the diffusion pathways. Typically, one places several starting points (seed points) in one region of interest (ROI) and iteratively traces from one voxel to the next, essentially path integration in a vector field. One terminates these fiber bundles when the local diffusion appears week or upon reaching a target region.

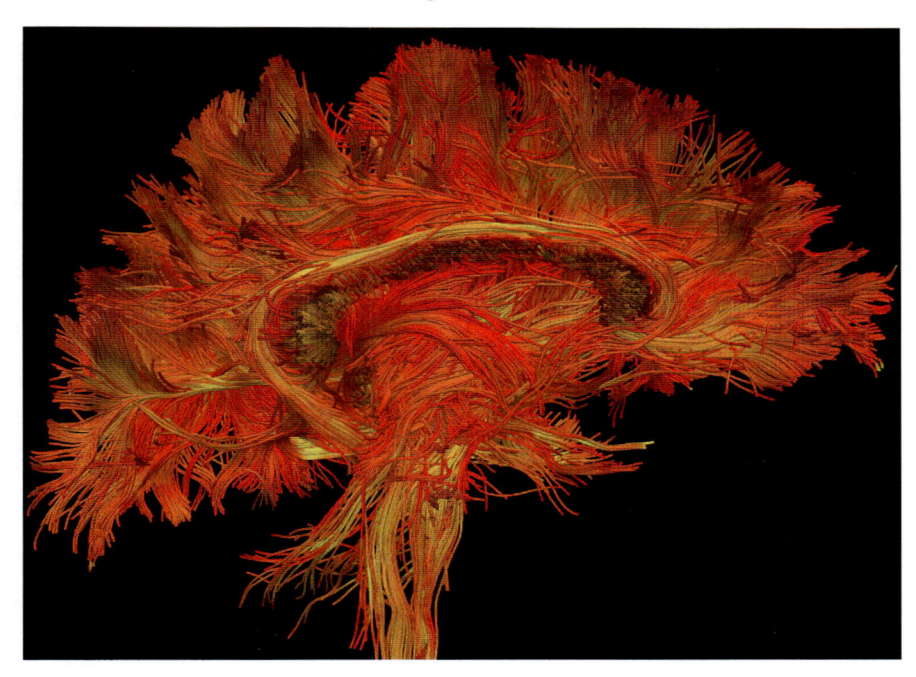

Fig. 16.2. *Tractography.* Cutaway showing tractography throughout the left hemisphere colored by FA to indicate diffusion strength [41]. From this view, the fornix and cingulum bundle are visible near the center

Figure 16.2 offers a glimpse from inside the brain using this basic approach. Often additional regions are used as masks to post-process results, e.g., pathways from Region A but not touching Region B.

In the single tensor model, standard streamline tractography follows the principal diffusion direction of the tensor, while multifiber models often include techniques for determining the number of fibers present or when pathways branch [9, 42]. Since individual voxel measurements may be unreliable, several techniques have been developed for regularization, for example, using the estimate from the previous position [43] as well as filtering formulations for path regularization [44] and model-based estimation [41].

The choice of model and optimization mechanism can drastically effect the final tracts. To illustrate, Fig. 16.3 shows tractography from the center of the corpus callosum using a single-tensor model and a two-tensor model using the filtered technique from [41].

16.3.2 Probabilistic Tractography

While discrete paths intuitively represent the putative fiber pathways of interest, they tend to ignore the inherent uncertainty in estimating the principle

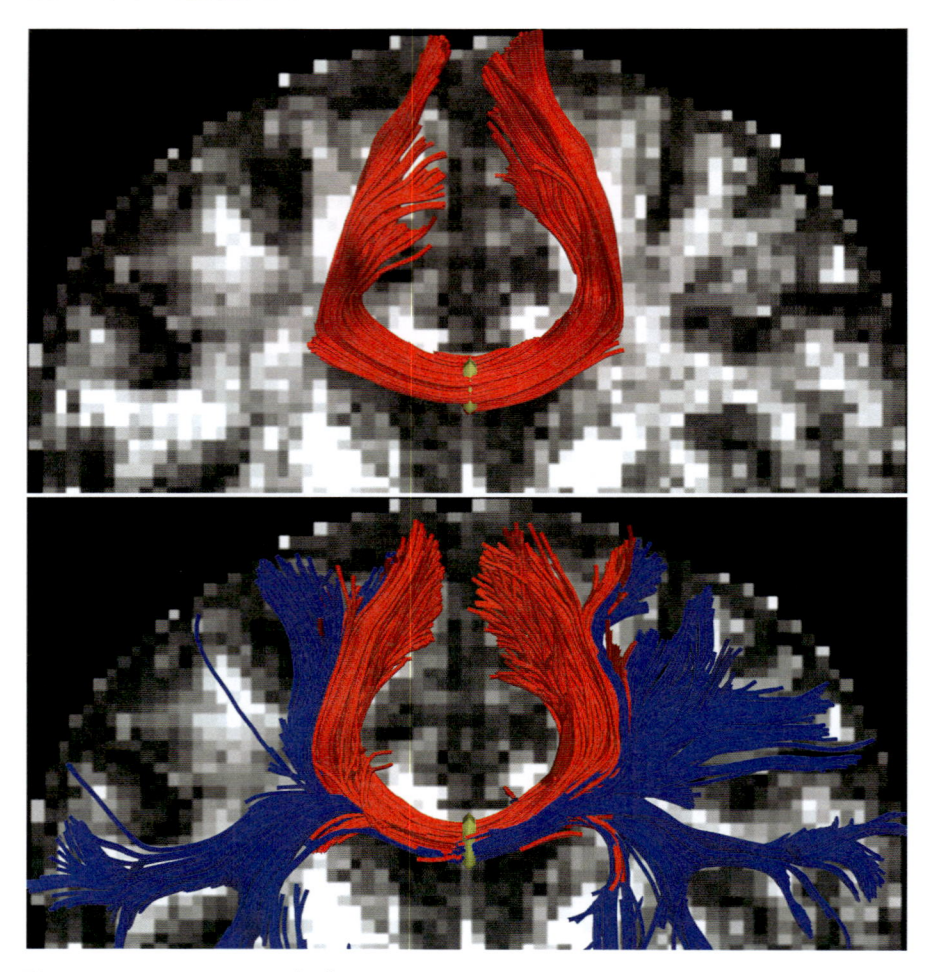

Fig. 16.3. *Tractography [41].* Tractography from the center of the corpus callosum (seed region in yellow). The single-tensor model (*top*) captures only the corona radiata and misses the lateral pathways known to exist. The two-tensor method [41] (*bottom*) reveals many of these missing pathways (*highlighted in blue*)

diffusion directions in each voxel. Instead of tracing discrete paths to connect voxels, one may query the probability of voxel-to-voxel connections, given the diffusion probability distributions reconstructed in each voxel.

Several approaches have been developed based on sampling. For example, one might run streamline tensor tractography treating each as a Monte Carlo sample; the more particles that take a particular path, the more likely that particular fiber pathway [45]. Another approach would be to consider more of the continuous diffusion field from Q-ball or other reconstructions [8, 46–48]. By making high curvature paths unlikely, path regularization can be naturally enforced within the probabilistic framework. Another approach is to propagate an arrival-time isosurface from the seed region out through the diffusion field, the front evolution force being a function of the local diffusivity [49, 50].

Using the full diffusion reconstruction to guide particle diffusion has the advantage of naturally handling uncertainty in diffusion measurements, but for that same reason, it tends toward diffuse tractography and false-positive connections. One option is to constrain diffusivity by fitting a model, thereby ensuring definite diffusion directions yet still taking into account some uncertainty [8, 45, 47]. A direct extension is to introduce a model selection mechanism to allow for additional components where appropriate [6, 51]. However, one could stay with the nonparametric representations and instead sharpen the diffusion profile to draw out the underlying fiber orientations [23, 52].

16.3.3 Global Tractography

Despite advances in voxel modeling, discerning the underlying fiber configuration has proven difficult. For example, looking at a single voxel, the symmetry inherent in the diffusion measurements makes it difficult to tell if the observed pattern represents a fiber curving through the voxel or a fanning pattern. Reliable and accurate fiber resolution requires more information than that of a single voxel. For example, instead of estimating the fiber orientation, one could infer the geometry of the entire neighborhood [53].

Going a step further, one could say that reliable and accurate connectivity resolution requires even more information, beyond simply a voxel neighborhood. In some respects, probabilistic tractography can be seen to take into account more global information. By spawning thousands of particles, each attempting to form an individual connection, probabilistic techniques are able to explore more possibilities before picking those that are likely. However, if these particles still only look at the local signal as they propagate from one voxel to the next, then they remain susceptible to local regions of uncertainty. Even those with resampling schemes are susceptible since the final result is still a product of the method used in local tracing [48].

A natural step to address such problems is to introduce global connectivity information into local optimization procedures of techniques mentioned above. The work of [54] does this by extending the local Bayesian formulation in [6] with an additional prior that draws upon global connectivity information in regions of uncertainty. Similarly, one could use an energetic formulation still with data likelihood and prior terms, but additionally introduce terms governing the number of components present [55].

Another approach is to treat the entire path as the parameter to be optimized and use global optimization schemes. For example, one could model pathways as piecewise linear with a data likelihood term based on signal fit and a prior on spatial coherence of those linear components [56]. One advantage of this path-based approach is that it somewhat obviates the need for a multi-fiber voxel model; however, such a flexible global model dramatically increases the computational burden.

An alternative formulation is to find geodesic paths through the volume. Again using some form of data likelihood term, such methods then use techniques for front propagation to find globally optimal paths of connection [57–61].

Tractography is often used in group studies which typically require a common atlas for inter-subject comparison. Beginning with the end in mind, one could determine a reference bundle as a template and use this to drive tractography. This naturally ensures both the general geometric form of the solution and a direct correspondence between subjects [62, 63]. Alternatively, the tract seeding and other algorithm parameters could be optimized until the tracts (data driven) approach the reference (data prior) [64]. Since this requires pre-specifying such a reference bundle, information that may be unavailable or difficult to obtain, one could even incorporate the formulation of the reference bundle into the optimization procedure itself [65].

16.3.4 Validation

In attempting to reconstruct neural pathways virtually, it is important to keep in mind the inherent uncertainty in such reconstructions. The resolution of dMRI scanners is at the level of 3–$10\,\text{mm}^3$; while physical fiber axons are often an order of magnitude smaller in diameter – a relationship that leaves much room for error. Some noise or a complex fiber configuration could simply look like a diffuse signal and cause probabilistic tractography to stop in its tracks, while a few inaccurate voxel estimations could easily send the deterministic tracing off course to produce a false-positive connection. Even global methods could produce a tract that fits the signal quite well but incidentally jumps over an actual boundary in one or two voxels it thinks are noise. Consequently, a common question is: Are these pathways really present?

With this in mind, an active area of study is validating such results. Since physical dissection often requires weeks of tedious effort, many techniques have been used for validating these virtual dissections. A common starting point is to use synthetic and physical phantoms with known parameters when evaluating new methods [66]. When possible, imaging before and after injecting radio-opaque dyes directly into the tissue can provide some of the best evidence for comparison [67, 68]. Another powerful approach is to apply bootstrap sampling or other non-parametric statistical tests to judge the sensitivity and reproducibility of resulting tractography against algorithm parameters, image acquisition, and even signal noise [37, 38, 64, 69, 70].

16.4 Applications

Having outlined various models and methods of reconstructing pathways, we now briefly cover several methods of further analysis.

16.4.1 Volume Segmentation

Medical image segmentation has a long history, and much of it focused on scalar intensity-based segmentation of anatomy. For neural segmentation, structural MRI easily reveals the boundaries between gray-matter and white-matter, and anatomic priors have helped further segment some internal structures [71]; however, the boundaries between many structures in the brain remain invisible with structural MRI alone. The introduction of dMRI has provided new discriminating evidence in such cases where tissue may appear homogeneous on structural MRI or CT but contain distinct fiber populations.

To begin, most work has focused segmentation of the estimated tensor fields. Using suitable metrics to compare tensors, these techniques often borrow directly from active contour or graph cut segmentation with the approach of separating distributions. For example, one could define a Gaussian distribution of tensors to approximate a structure of interest [72]. For tissues with more heterogeneous fiber populations, e.g., the corpus callosum as it bends, such global parametric representations are unsuitable. For this, non-parametric approaches are more appropriate at capturing the variation throughout such structures [73, 74]. Another approach to capture such variation is to limit the parametric distributions to local regions of support, essentially robust edge detection [75].

In Fig. 16.4, a graph cut segmentation of the corpus callosum is visualized [74]. The color-coded FA image is shown for visualization, while segmentation was performed on the underlying tensor data.

When computing regional statistics for segmentation, one needs to calculate the distance between any two tensors in the field. To do so, one must take into account that tensor parameters do not lie in a Euclidean vector space, i.e., addition of two tensors' coefficients does not necessarily still produce a new valid tensor. Ignoring this and using the standard L_2-norm produces a poor segmentation (Fig. 16.4, middle), while properly accounting for the nonlinearity via a Riemannian mapping produces a more accurate segmentation (Fig. 16.4, bottom).

An altogether different approach to segmenting a structure is to divide it up according to where portions connect elsewhere. For example, the thalamus contains several nuclei indistinguishable in standard MR or even with contrast. After tracing connections from the thalamus to the cortex, one study demonstrated that grouping these connections revealed the underlying nuclei [76].

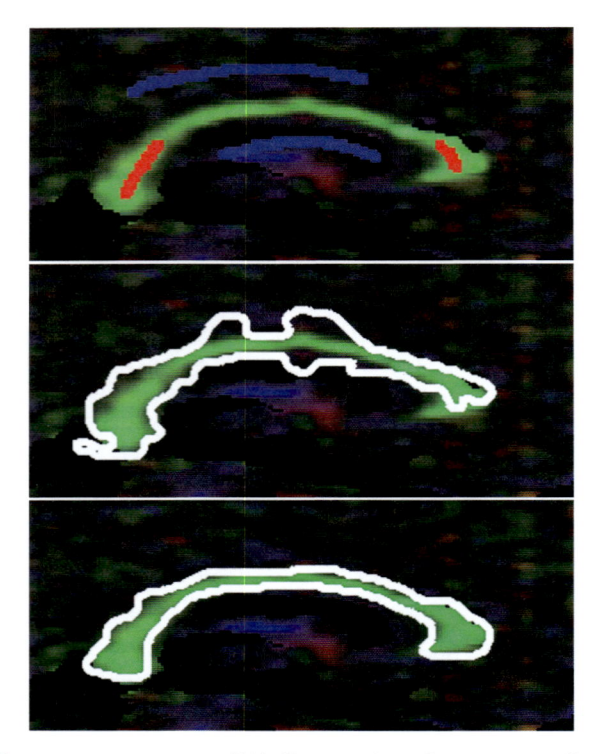

Fig. 16.4. *Tensor segmentation [74].* Segmenting the corpus callosum using the graph cut technique from [74] (side view) is visualized as color-coded anisotropy, where the color intensity is based on FA (between 0 and 1) and the color *red*, *blue*, or *green* indicates the orientation of that voxel along the x-, y-, or z-axis, respectively. *Top*: initial seed regions; *Middle*: Euclidean mapping not taking into account the structure of the underlying tensor manifold; *Bottom*: Riemannian mapping taking this structure into account when computing statistics and so produces a correct segmentation

16.4.2 Fiber Clustering

The raw output of full-brain tractography can produce hundreds of thousands of such tracings, an overwhelming amount of information. One approach to understanding and visualizing such results is to group individual tracings into fiber bundles. Such techniques are typically based around two important design choices: the method of comparing fibers and the method of clustering those fibers.

In comparing two fibers, one often starts by defining a distance measure, these typically being based on some point-to-point correspondence between the fibers [77–79]. With this correspondence in hand, one of the most common distances is then to take the mean closest point distance between the two fibers (Hausdorff distance). An alternative is to transform each fiber to a new vector space with a natural norm, e.g., a fiber of any length can be encoded with only the mean and covariance of points along its path and then use the L_2 distance [80]. An altogether different approach is to consider the spatial overlap between fibers [81]. Since full-brain tractography often contains many small broken fragments as it tries to trace out bundles, such fragments are often separated from their actual cluster. Measures of spatial overlap may be more robust in such cases. In each of these methods, fibers were only considered as sequences of points, i.e., connections and orientations were ignored. Recent work demonstrates that incorporating such considerations provides robust descriptors of fiber bundles [82].

Based on these distances, several methods have been developed to cluster the fibers. Spectral methods typically begin by computing the pairwise distance (affinity) between any two fibers and encode this as an $n \times n$ Gram matrix, after which normalized cuts can be applied to partition the Gram matrix and hence the fibers [80]. Affinity has recently been demonstrated as an efficient and robust alternative which automatically determines the number of clusters to support a specified cluster size preference [83]. In Fig. 16.5, shows how clustering can automatically reveal known structures and provide a more coherent view of the brain. In addition, clustering can be used to judge outliers. For example, Fig. 16.6 reveals several streamlines that appear to have gone off track relative to the cluster centers.

Another clustering approach is to use the inner product space itself. For example, one can efficiently group directly on the induced manifold by iteratively joining fibers most similar until the desired clustering emerges. To avoid construction of the large Gram matrix, variants of expectation maximization (EM) have been demonstrated to iteratively cluster fibers, an approach that naturally lends itself to incorporating anatomic priors [79, 81, 84]. Alternatively, one can begin with the end in mind by registering a reference

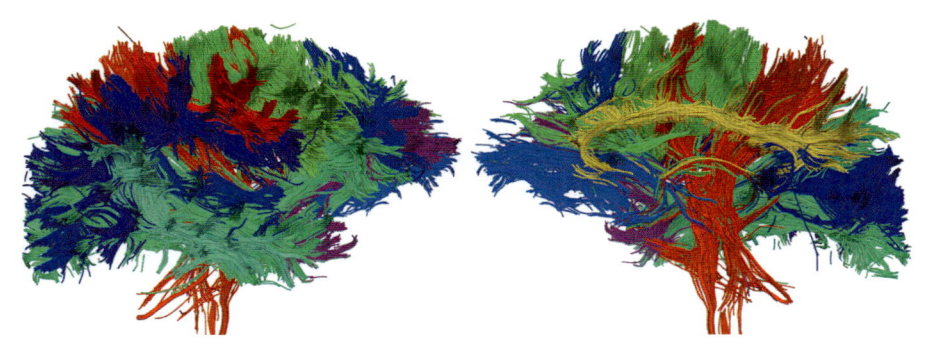

Fig. 16.5. *Clustered tractography.* Full-brain streamline tractography clustered using affinity propagation. Viewed from the outside (*left*) and inside cutting away the left hemisphere (*right*). Among the visible structures, we see the cingulum bundle (*yellow*), internal capsule (*red*), and arcuate (*purple*).

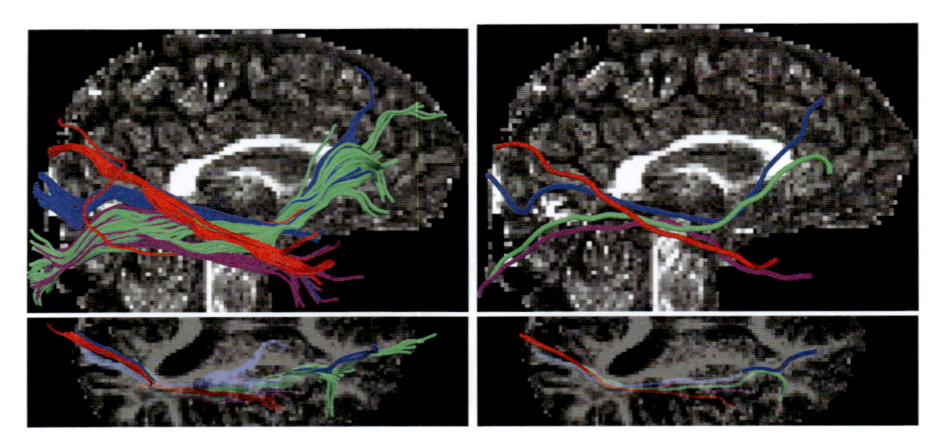

Fig. 16.6. *Clustered fronto-occipital fibers.* Fronto-occipital fibers from the right hemisphere using streamline tractography and clustered into bundles (*left*). Viewing the most representative fiber in each bundle (*right*), we see a fiber from one cluster (*red*) that appears to have wandered off the pathway

fiber bundle template to patients thus obviating any need for later spatial normalization or correspondence [65].

16.4.3 Connectivity

While tissue segmentation can provide global cues of neural organization, it tells little of the contribution of individual elements. Similarly, while clustered tracings are easily visualized, deciphering the flood of information from full-brain tractography demands more comprehensive quantitative analysis. For this, much has been borrowed from network analysis to characterize the neural topology. To start, instead of segmenting fibers into bundles, one can begin by classifying voxels into hubs or subregions into subnetworks [85, 86].

Dividing the brain up into major functional hubs, one can then view it as a graphical network as in Fig. 16.7. Each of these edges is then often weighted as a function of connection strength, but may also incorporate functional correlation to give further evidence of connectivity.

One of the first results of such analysis was the discovery of dense hubs linked by short pathways, a characteristic observed in many complex physical systems (small-world phenomena). Another interesting finding came from combining anatomic connections from dMRI with neuronal activity provided by fMRI [87]. They found that areas which are functionally connected are often not structurally connected; hence, tractography alone does not provide the entire picture.

For a recent review of this emerging field of structural and functional network analysis, we recommend [88].

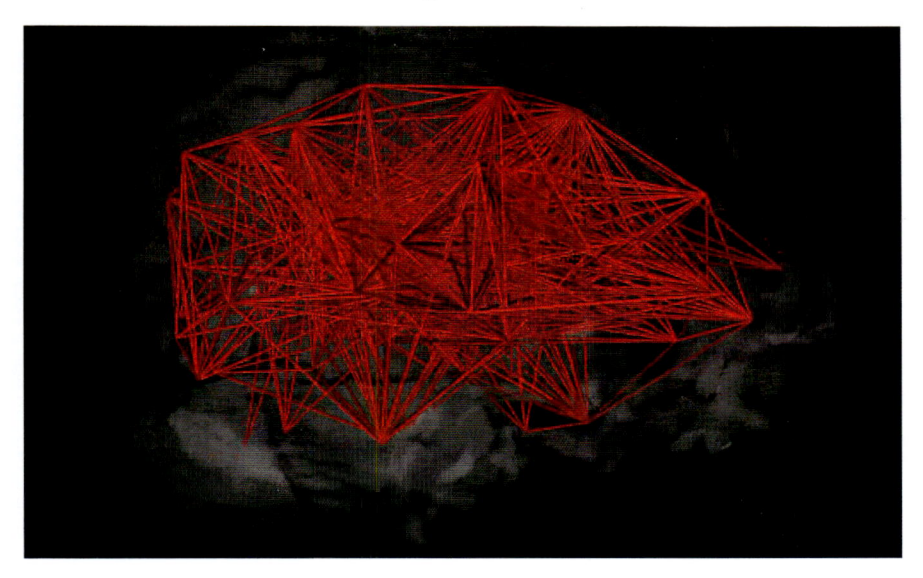

Fig. 16.7. *Neural network.* The brain viewed as a network of weighted connections. Each edge represents a possible connection and is weighted by the strength of that path. Many techniques from network analysis are applied to reveal hubs and subnetworks within this macroscopic view

16.4.4 Tissue Analysis

Several reviews exist documenting the application and findings of using various methods [89–91].

In forming population studies, there are several approaches for framing the analysis among patients. For example, voxel-based studies examine tissue characteristics in regions of interest [92]. Discriminant analysis has been applied to determine such regions [93]. Alternatively, one could also perform regression on the full image volume taking into account variation not only in diffusion but also in the full anatomy [94]. In contrast, tract-based studies incorporate the results of tractography to use fiber pathways as the frame of reference [77, 95], and several studies have demonstrated the importance of taking into account local fluctuations in estimated diffusion [63, 78, 84, 96, 97].

A common approach in many of these studies is to focus on characterizing individual pathways or bundles. To illustrate this analysis, Fig. 16.8 shows fibers connecting a small region in each hemisphere. We then average FA plotted along the bundle as a function of arc-length. Furthermore, we plot the FA from both single- and two-tensor models to show how different models often produce very different tissue properties.

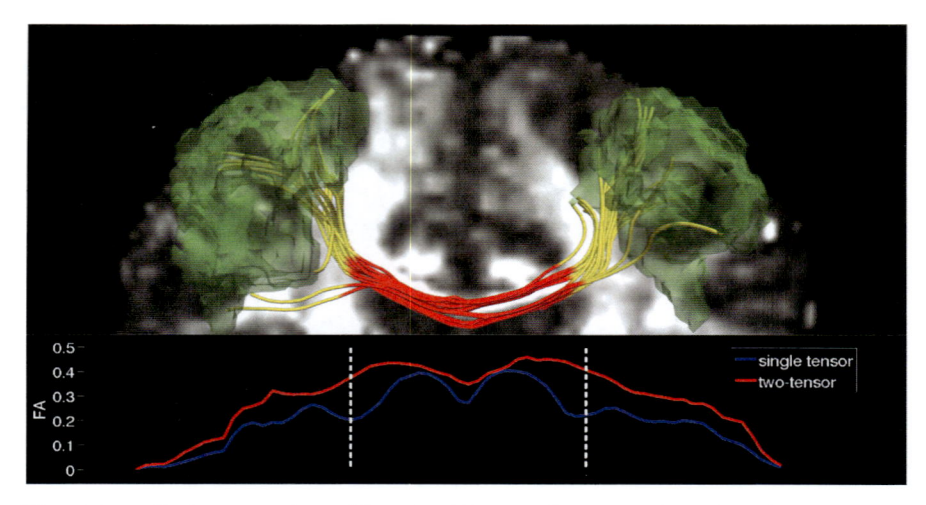

Fig. 16.8. *Pathway analysis.* Plotting FA as a function of arc-length to examine local fluctuations. Fibers are selected that connect the left and right seed regions (*green*). Note how the FA from single-tensor (*blue*) is lower in regions of crossing compared to two-tensor FA (*red*)

16.5 Summary

Diffusion MRI has provided an unprecedented view of neural architecture. With each year, we develop better image acquisition schemes, more appropriate diffusion models, more accurate pathway reconstruction, and more sensitive analysis.

In this survey, we began with an overview of the various imaging techniques and diffusion models. While many acquisition sequences have become widely distributed for high angular resolution imaging, work continues in developing sequences and models capable of accurate resolution of biological properties such as axon diameter and degree of myelination [14]. We then reviewed various parametric models starting with the diffusion tensor on up to various mixture models as well as high-order tensors. Work continues to develop more accurate and reliable model estimation by incorporating information from neighboring voxels [41, 53]. Furthermore, scalar measures derived from these models similarly benefit from incorporating neighborhood information [33].

Next we outlined various methods of tractography to infer connectivity. Broadly, these techniques took either a deterministic or probabilistic approach. We also documented the recent trend toward global approaches, those that combine local voxel-to-voxel tracing with a sense of the full path [55]. Even with such considerations, tractography has proven quite sensitive to image acquisition and initial conditions; so much work has gone into validation. Common techniques are the use of physical phantoms [66] or statistical tests like bootstrap analysis [64, 69, 70].

Finally, we briefly introduced several machine-learning approaches to make sense of the information found in diffusion imagery. Starting with segmentation, several techniques for scalar intensity segmentation have been extended to dMRI. With the advent of full-brain tractography providing hundreds of thousands of fiber paths, the need to cluster connections into bundles has become increasingly important. The application of network analysis to connectivity appears to be an emerging area of research, especially in combination with alternate imaging modalities [88]. Finally, we noted several approaches to the analysis of neural tissue itself in ROIs or along pathways.

References

1. Bihan DL. Looking into the functional architecture of the brain with diffusion MRI. Nat Rev Neurosci. 2003;4:469–80.
2. Wedeen VJ, Hagmann P, Tseng WY, et al. Mapping complex tissue architecture with diffusion spectrum magnetic resonance imaging. Magn Reson Med. 2005;54:1377–86.
3. Tuch D. Q-ball imaging. Magn Reson Med. 2004;52:1358–72.
4. Peled S, Friman O, Jolesz F, et al. Geometrically constrained two-tensor model for crossing tracts in DWI. Magn Reson Med. 2006;24(9):1263–70.
5. Descoteaux M, Angelino E, Fitzgibbons S, et al. Regularized, fast, and robust analytical Q-ball imaging. Magn Reson Med. 2007;58:497–510.
6. Behrens T, Johansen-Berg H, Jbabdi S, et al. Probabilistic diffusion tractography with multiple fibre orientations: what can we gain? NeuroImage 2007;34:144–55.
7. Hosey T, Williams G, Ansorge R. Inference of multiple fiber orientations in high angular resolution diffusion imaging. Magn Reson Med. 2005;54:1480–89.
8. Parker G, Alexander DC. Probabilistic anatomical connectivity derived from the microscopic persistent angular structure of cerebral tissue. Phil Trans R Soc B. 2005;360:893–902.
9. Kreher B, Schneider J, Mader I, et al. Multitensor approach for analysis and tracking of complex fiber configurations. Magn Reson Med. 2005;54:1216–25.
10. Kaden E, Knösche T, Anwander A. Parametric spherical deconvolution: Inferring anatomical connectivity using diffusion MR imaging. NeuroImage 2007;37:474–88.
11. Rathi Y, Michailovich O, Shenton ME, et al. Directional functions for orientation distribution estimation. Med Image Anal. 2009;13:432–44.
12. Jian B, Vemuri B, Özarslan E, et al. A novel tensor distribution model for the diffusion-weighted MR signal. NeuroImage 2007;37(1):164–76.
13. Assaf Y, Basser P. Composite hindered and restricted model of diffusion (CHARMED) MR imaging of the human brain. NeuroImage 2005;27:48–58.
14. Assaf Y, Blumenfeld-Katzir T, Yovel Y, et al. AxCaliber: a method for measuring axon diameter distribution from diffusion mri. Magn Reson Med. 2008;59:1347–54.
15. Anderson A. Measurement of fiber orientation distributions using high angular resolution diffusion imaging. Magn Reson Med. 2005;54(5):1194–1206.

16. Hess C, Mukherjee P, Han E, et al. Q-ball reconstruction of multimodal fiber orientations using the spherical harmonic basis. Magn Reson Med. 2006;56:104–17.
17. Özarslan E, Mareci T. Generalized diffusion tensor imaging and analytical relationships between diffusion tensor imaging and high angular resolution diffusion imaging. Magn Reson Med. 2003;50:955–65.
18. Basser PJ, Pajevic S. Spectral decomposition of a 4th-order covariance tensor: applications to diffusion tensor MRI. Signal Process. 2007;87:220–36.
19. Jansons K, Alexander DC. Persistent angular structure: new insights from diffusion mri data. Inverse Probl. 2003;19:1031–46.
20. Özarslan E, Shepherd T, Vemuri B, et al. Resolution of complex tissue microarchitecture using the diffusion orientation transform. NeuroImage. 2006;31(3):1086–1103.
21. Tournier JD, Calamante F, Gadian D, et al. Direct estimation of the fiber orientation density function from diffusion-weighted MRI data using spherical deconvolution. NeuroImage 2004;23:1176–85.
22. Jian B, Vemuri B. A unified computational framework for deconvolution to reconstruct multiple fibers from diffusion weighted MRI. IEEE Trans Med Imag. 2007;26(11):1464–71.
23. Descoteaux M, Deriche R, Knoesche T, et al. Deterministic and probabilistic tractography based on complex fiber orientation distributions. IEEE Trans Med Imag. 2009;28(2):269–86.
24. Schultz T, Seidel H. Estimating crossing fibers: a tensor decomposition approach. IEEE Trans Vis Comput Graph. 2008;14(6):1635–42.
25. Koay CG, Basser PJ. Analytically exact correction scheme for signal extraction from noisy magnitude MR signals. J Magn Reson. 2006;179:317–22.
26. Aja-Fernandez S, Niethammer M, Kubicki M, et al. Restoration of DWI data using a Rician LMMSE estimator. IEEE Trans Med Imag. 2008;27:1389–1403.
27. Batchelor PG, Moakher M, Atkinson D, et al. A rigorous framework for diffusion tensor calculus. Magn Reson Med. 2005;53(1):221–25.
28. Fletcher PT, Joshi S. Riemannian geometry for the statistical analysis of diffusion tensor data. Signal Process. 2007;87(2):250–62.
29. Poupon C, Roche A, Dubois J, et al. Real-time MR diffusion tensor and Q-ball imaging using Kalman filtering. Med Image Anal. 2008;12(5):527–34.
30. Deriche R, Calder J, Descoteaux M. Optimal real-time Q-ball imaging using regularized Kalman filtering with incremental orientation sets. Med Image Anal. 2009;13(4):564–79.
31. Westin CF, Maier S, Mamata H, et al. Processing and visualization for diffusion tensor MRI. Med Image Anal. 2002;6:93–108.
32. Kindlmann G, Ennis DB, Whitaker RT, et al. Diffusion tensor analysis with invariant gradients and rotation tangents. IEEE Trans Med Imag. 2007;26(11):1483–99.
33. Savadjiev P, Kindlmann G, Bouix S, et al. Local white matter geometry indices from diffusion tensor gradients. In: Proc MICCAI; 2009. p. 345–352.
34. Özarslan E, Vemuri BC, Mareci TH. Generalized scalar measures for diffusion MRI using trace, variance, and entropy. Magn Reson Med. 2005;53:866–76.
35. Beaulieu C. The basis of anisotropic water diffusion in the nervous system: a technical review. NMR Biomed. 2002;15:438–55.
36. Jones DK. Determining and visualizing uncertainty in estimates of fiber orientation from diffusion tensor MRI. Magn Reson Med. 2003;49:7–12.

37. Whitcher B, Tuch D, Wisco J, et al. Using the wild bootstrap to quantify uncertainty in diffusion tensor imaging. Hum Brain Mapp. 2008;29(3):346–62.
38. Chung S, Lu Y, Henry R. Comparison of bootstrap approaches for estimation of uncertainties of DTI parameters. NeuroImage 2006;33(2):531–41.
39. Ennis DB, Kindlmann G, Rodriguez I, et al. Visualization of tensor fields using superquadric glyphs. Magn Reson Med. 2005;53:169–76.
40. Minati L, Węogonglarz WP. Physical foundations, models, and methods of diffusion magnetic resonance imaging of the brain: a review. Concepts Magn Reson A. 2007;30(5):278–307.
41. Malcolm JG, Shenton ME, Rathi Y. Neural tractography using an unscented Kalman filter. In: Inf Process Med Imaging; 2009. p. 126–138.
42. Guo W, Zeng Q, Chen Y, et al. Using multiple tensor deflection to reconstruct white matter fiber traces with branching. In: Int Symp Biomed Imaging; 2006. p. 69–72.
43. Lazar M, Weinstein DM, Tsuruda JS, et al. White matter tractography using diffusion tensor deflection. Hum Brain Mapp. 2003;18:306–21.
44. Gössl C, Fahrmeir L, P utz B, et al. Fiber tracking from DTI using linear state space models: detectability of the pyramidal tract. NeuroImage 2002;16:378–88.
45. Behrens T, Woolrich M, Jenkinson M, et al. Characterization and propagation of uncertainty in diffusion-weighted MR imaging. Magn Reson Med. 2003;50:1077–88.
46. Perrin M, Poupon C, Cointepas Y, et al. Fiber tracking in Q-ball fields using regularized particle trajectories. In: Inf Process Med Imaging; 2005. p. 52–63.
47. Friman O, Farnebäck G, Westin CF. A Bayesian approach for stochastic white matter tractography. IEEE Trans Med Imag. 2006;25(8):965–78.
48. Zhang F, Hancock E, Goodlett C, et al. Probabilistic white matter fiber tracking using particle filtering and von Mises-Fisher sampling. Med Image Anal. 2009;13:5–18.
49. Campbell JSW, Siddiqi K, Rymar VV, et al. Flow-based fiber tracking with diffusion tensor and Q-ball data: Validation and comparison to principal diffusion direction techniques. NeuroImage 2005;27(4):725–36.
50. Tournier JD, Calamante F, Gadian D, et al. Diffusion-weighted magnetic resonance imaging fibre tracking using a front evolution algorithm. NeuroImage. 2003;20:276–88.
51. Freidlin RZ, Özarslan E, Komlosh ME, et al. Parsimonious model selection for tissue segmentation and classification applications: a study using simulated and experimental DTI data. IEEE Trans Med Imag. 2007;26(11):1576–84.
52. Tournier JD, Calamante F, Connelly A. Robust determination of the fibre orientation distribution in diffusion MRI: non-negativity constrained super-resolved spherical deconvolution. NeuroImage 2007;35:1459–72.
53. Savadjiev P, Campbell JS, Descoteaux M, et al. Labeling of ambiguous subvoxel fibre bundle configurations in high angular resolution diffusion MRI. NeuroImage 2008;41:58–68.
54. Jbabdi S, Woolrich M, Andersson J, et al. A bayesian framework for global tractography. NeuroImage 2007;37:116–29.
55. Fillard P, Poupon C, Mangin JF. A novel global tractography algorithm based on an adaptive spin glass model. In: Proc MICCAI; 2009. p. 927–934.
56. Kreher B, Madeer I, Kiselev V. Gibbs tracking: a novel approach for the reconstruction of neuronal pathways. Magn Reson Med. 2008;60:953–63.

57. O'Donnell L, Haker S, Westin CF. New approaches to estimation of white matter connectivity in diffusion tensor MRI: elliptic PDE's and geodesics in tensor-warped space. In: Proc MICCAI; 2002. p. 459–466.
58. Pichon E, Westin CF, Tannenbaum A. A hamilton-jacobi-bellman approach to high angular resolution diffusion tractography. In: Proc MICCAI; 2005. p. 180–187.
59. Prados E, Lenglet C, Pons J, et al. Control theory and fast marching techniques for brain connectivity mapping. In: Comput Vis Pattern Recognit; 2006. p. 1076–1083.
60. Lenglet C, Deriche R, Faugeras O. Inferring white matter geometry from diffusion tensor mri: application to connectivity mapping. In: Eur Conf Comput Vis; 2004. p. 127–140.
61. Fletcher PT, Tao R, Joeng WK, et al. A volumetric approach to quantifying region-to-region white matter connectivity in diffusion Tensor MRI. In: Inf Process Med Imaging; 2007. p. 346–358.
62. Eckstein I, Shattuck DW, Stein JL, et al. Active fibers: matching deformable tract templates to diffusion tensor images. NeuroImage 2009;47:T82–T89.
63. Goodlett CB, Fletcher PT, Gilmore JH, et al. Group analysis of DTI fiber tract statistics with application to neurodevelopment. NeuroImage 2009;45:S133–S142.
64. Clayden J, Storkey A, Bastin M. A probabilistic model-based approach to consistent white matter tract segmentation. IEEE Trans Med Imag. 2007;26(11):1555–61.
65. Clayden JD, Storkey AJ, Maniega SM, et al. Reproducibility of tract segmentation between sessions using an unsupervised modelling-based approach. NeuroImage 2009;45:377–85.
66. Poupon C, Rieul B, Kezele I, et al. New diffusion phantoms dedicated to the study and validation of high-angular-resolution diffusion imaging (HARDI) models. Magn Reson Med. 2008;60(6):1276–83.
67. Lin CP, Wedeen VJ, Chen JH, et al. Validation of diffusion spectrum magnetic resonance imaging with manganese-enhanced rat optic tracts and ex vivo phantoms. NeuroImage 2003;19:482–95.
68. Dauguet J, Peled S, Berezovskii V, et al. Comparison of fiber tracts derived from in-vivo DTI tractography with 3D histological neural tract tracer reconstruction on a macaque brain. NeuroImage 2007;37:530–38.
69. Lazar M, Alexander A. Bootstrap white matter tractography (Boot-Tract). NeuroImage 2005;24:524–32.
70. Jones DK, Pierpaoli C. Confidence mapping in diffusion tensor magnetic resonance imaging tractography using a bootstrap approach. Magn Reson Med. 2005;53(5):1143–49.
71. Pohl KM, Kikinis R, Wells WM. Active mean fields: solving the mean field approximation in the level set framework. In: Inf Process Med Imaging; 2007. p. 26–37.
72. de Luis-Garcia R, Alberola-Lopez C. Mixtures of Gaussians on tensor fields for the segmentation of DT-MRI. In: Proc MICCAI; 2007. p. 319–326.
73. Rathi Y, Michailovich O, Tannenbaum A. Segmenting images on the tensor manifold. In: Comput Vis Pattern Recognit; 2007.
74. Malcolm J, Rathi Y, Tannenbaum A. A graph cut approach to image segmentation in tensor space. In: Component Analysis Methods (in CVPR); 2007. p. 1–8.

75. Lankton S, Melonakos J, Malcolm J, et al. Localized statistics for DW-MRI fiber bundle segmentation. In: Math Methods Biomed Image Anal; 2008. p. 1–8.
76. Behrens TEJ, Johansen-Berg H, Woolrich MW, et al. Non-invasive mapping of connections between human thalamus and cortex using diffusion imaging. Nat Neurosci. 2003;6(7):750–57.
77. Ding Z, Gore J, Anderson A. Classification and quantification of neuronal fiber pathways using diffusion tensor MRI. Magn Reson Med. 2003;49:716–21.
78. Corouge I, Fletcher PT, Joshi S, et al. Fiber tract-oriented statistics for quantitative diffusion tensor mri analysis. Med Image Anal. 2006;10(5):786–98.
79. O'Donnell LJ, Westin CF. Automatic tractography segmentation using a high-dimensional white matter atlas. IEEE Trans Med Imag. 2007;26:1562–75.
80. Brun A, Knutsson H, Park HJ, et al. Clustering fiber traces using normalized cuts. In: Proc MICCAI; 2004. p. 368–375.
81. Wang X, Grimson WEL, Westin CF. Tractography segmentation using a hierarchical Dirichlet processes mixture model. In: Inf Process Med Imaging; 2009. p. 101–113.
82. Durrleman S, Fillard P, Pennec X, et al. A statistical model of white matter fiber bundles based on currents. In: Inf Process Med Imaging; 2009. p. 114–125.
83. Leemans A, Jones DK. A new approach to fully automated fiber tract clustering using affinity propagation. In: Int Symp Magn Reson Med. vol. 17; 2009. p. 856.
84. Maddah M, Grimson WEL, Warfield SK, et al. A unified framework for clustering and quantitative analysis of white matter fiber tracts. Med Image Anal. 2008;12(2):191–202.
85. Gong G, He Y, Concha L, et al. Mapping anatomical connectivity patterns of human cerebral cortex using in vivo diffusion tensor imaging tractography. Cereb Cortex. 2009;19(3):524–36.
86. Hagmann P, Cammoun L, Gigandet X, et al. Mapping the structural core of human cerebral cortex. PLoS Biol. 2008;6(7):e159.
87. Honey CJ, Sporns O, Cammoun L, et al. Predicting human resting-state functional connectivity from structural connectivity. Proc Nat Acad Sci. 2009;106(6):2035–40.
88. Bullmore E, Sporns O. Complex brain networks: graph theoretical analysis of structural and functional systems. Nat Rev Neurosci. 2009;10:186–98.
89. Lim K, Helpern J. Applications of diffusion-weighted and diffusion tensor MRI to white matter diseases: a review. NMR Biomed. 2002;15:570–77.
90. Horsfield MA, Jones DK. Neuropsychiatric applications of DTI: a review. NMR Biomed. 2002;15:587–93.
91. Kubicki M, McCarley R, Westin CF, et al. A review of diffusion tensor imaging studies in schizophrenia. J Psychiatr Res. 2007;41:15–30.
92. Ashburner J, Friston K. Voxel-based morphometry: the methods. NeuroImage 2000;11(6):805–21.
93. Caan M, Vermeer K, van Vliet L, et al. Shaving diffusion tensor images in discriminant analysis: A study into schizophrenia. Med Image Anal. 2006;10(6):841–49.
94. Rohlfing T, Sullivan EV, Pfefferbaum A. Regression models of atlas appearance. In: Inf Process Med Imaging; 2009. p. 151–162.
95. Smith SM, Jenkinson M, Johansen-Berg H, et al. Tract-based spatial statistics: voxelwise analysis of multi-subject diffusion data. NeuroImage 2006;31:1487–1505.

96. O'Donnell LJ, Westin CF, Golby AJ. Tract-based morphometry for white matter group analysis. NeuroImage 2009;45(3):832–44.
97. Yushkevich PA, Zhang H, Simon TJ, et al. Structure-specific statistical mapping of white matter tracts. NeuroImage 2008;41:448–61.

Part VII

Image Management and Integration

17
Digital Imaging and Communications in Medicine

Michael Onken, Marco Eichelberg, Jörg Riesmeier, and Peter Jensch[†]

Summary. Over the past 15 years Digital Imaging and Communications in Medicine (DICOM) has established itself as the international standard for medical image communication. Most medical imaging equipment uses DICOM network and media services to export image data, thus making this standard highly relevant for medical image processing. The first section of this chapter provides a basic introduction into DICOM with its more than 3,600 pages of technical documentation, followed by a section covering selected advanced topics of special interest for medical image processing. The introductory text familiarizes the reader with the standard's main concepts such as information objects and DICOM media and network services. The rendering pipeline for image display and the concept of DICOM conformance are also discussed. Specialized DICOM services such as advanced image display services that provide means for storing how an image was viewed ("Softcopy Presentation States") and how multiple images should be aligned on an output device ("Structured Display" and "Hanging Protocols") are described. We further describe DICOM's sophisticated approach ("Structured Reporting") for storing structured documents such as CAD information, which is then covered in more detail. Finally, the last section provides an insight into a newly developed DICOM service called "Application Hosting", which introduces a standardized plug-in architecture for image processing, thus permitting users to utilize cross-vendor image processing plug-ins in DICOM applications.

17.1 DICOM Basics

This section provides an overview of the Digital Imaging and Communications in Medicine (DICOM) standard and, therefore, lays a foundation for discussion of advanced DICOM services later in this chapter.

[†] On April 15, 2010, our colleague and friend Peter Jensch passed away unexpectedly. We will remember him for his evident contributions to this field and for his outstanding personal character.

T.M. Deserno (ed.), *Biomedical Image Processing*, Biological and Medical Physics, Biomedical Engineering, DOI: 10.1007/978-3-642-15816-2_17,
© Springer-Verlag Berlin Heidelberg 2011

17.1.1 Introduction and Overview

DICOM is a standard in the area of medical informatics and has successfully penetrated wide areas of medical image communication [1]. Today, most imaging modalities are digital and offer a DICOM interface. Originating from radiology, the standard has established in other medical fields such as oncology or cardiology and has been extended beyond the communication of images, e.g., by providing workflow and security services.

DICOM emerged in 1993 from the American College of Radiology (ACR) – National Electrical Manufacturers Association (NEMA) standard [2], which was not very successful due to some conceptual weaknesses. Grandfathering some building blocks of ACR/NEMA but eliminating most of its flaws, DICOM was able to widely supersede early proprietary protocols and to prevent "private" communication solutions from being established. The DICOM standard is being developed together by industry and users. Since the standard's initial release in 1993, it has been continuously evolving over the years. Thus, the standard has grown from originally around 750 pages in 1993 to currently more than 3,600 pages. Starting as an industry standard, DICOM has also become an international standard over the last years. In 2004, DICOM was published under the title Medical Image Communication (MEDICOM) as European Standard (EN) [3] and in 2006 as international ISO standard [4].

The DICOM standard text currently consists of 18 documents each focusing on different aspects. The standard is being extended regularly by the so-called supplements and correction proposals. Supplements are used for introducing new services and objects into the standard as necessary. In rare cases objects or services are also retired, meaning that they should not be implemented by modern systems any more but may be supported for backward compatibility. Correction proposals are applying only small changes to the standard, e.g., they correct ambivalent wordings in the text that may lead to non-interoperable implementations. After being balloted, supplements and correction proposals immediately become part of the standard (status "Final Text"). Frequently, every 1 or 2 years all final text documents are applied to the full standard text which is then published. Thus when talking about the "current standard", the last edition of the standard must be taken into account (at this time "DICOM 2008"), plus all documents that received status "Final Text" since then.

17.1.2 Information Objects

Part 3 describes information objects representing data structures for medical images or other documents. There are object definitions for CT, ultrasound, MRI and so on. Also some objects are dedicated to non-images such as electrocardiograms (ECGs) and raw data. DICOM image objects – being the most common kind of DICOM objects and being most relevant in the context of this book – do contain lots of information besides the actual pixel data.

Thus, a DICOM image contains information regarding the patient examined (e.g., name, sex, date of birth), device parameters, or rendering information. Because of intermixing information from different fields, these kind of objects are called "Composite Objects". They can be thought of persistent documents that can be stored to an archive, in comparison to so-called Normalized Objects (not further discussed here) which in most cases are transient and may be more associated with a message than a document.

In part 3, DICOM defines the structure of objects by Information Object Definitions (IODs). An IOD consists of a short description followed by a so-called module table. A module consists of a list of attributes from a specific field, e.g., the patient module contains attributes such as patient's name or patient's sex. Since many modules like the patient module are needed in different object types (i.e., different IODs), they are only printed once in the standard and then referenced from the module table. Modules as well as their contained attributes can be mandatory, optional or conditional. Attributes sometimes have further restrictions on the values to be stored.

An attribute is uniquely defined by its so-called Tag, a number composed of two 16-bit numbers written in hexadecimal form. The first and second number are often referred to as group and element number, respectively. An attribute also has a name, like patient's name being the name of the attribute with tag (0010,0010). A value for a specific attribute is not only constrained by its description in the corresponding module but also due to the fact that every attribute in DICOM is tied to one of 27 data types. These are called value representation (VR). Each VR is abbreviated with two capital letters. There are VRs for numbers, texts, codes, person names, binary data, etc. The full ruleset for every VR can be found in part 5 of the standard. Tag, Name and VR can be looked up in part 6 of the standard which provides a list of all defined DICOM attributes sorted by tag (Table 17.1).

All attributes defined in the standard have even group numbers. However, it is permitted to include vendor-specific attributes into an IOD. Those always have an odd group number and are called "Private"(attributes).

Table 17.1. *Excerpt from DICOM Data Dictionary.* The last column Value Multiplicity (VM) describes how many values can be stored into that attribute. Most frequently, VM is exactly one, but attributes such as "Other Patient Names" can store one to many items

Tag	Name	VR	VM
(0010,0010)	Patient's Name	PN	1
(0010,0020)	Patient ID	LO	1
(0010,0021)	Issuer of Patient ID	LO	1
(0010,0030)	Patient's Birth Date	DA	1
(0010,0040)	Patient's Sex	PN	1
(0010,1000)	Other Patient IDs	LO	1-n
(0010,1001)	Other Patient Names	PN	1-n

As described, a DICOM image object is composed of a list of DICOM attributes collected from different modules. Sometimes it can be useful to structure attributes not only in a flat list but also in a more tree-like fashion. This is achieved by attributes having a special VR called "Sequence of Items"(SQ). Like other attributes, sequence attributes have a name and a VR. However, the value of a sequence attribute is made of a list of so-called items. An item is a block structure containing a list of (generally) arbitrary attributes. Often the number of items and also their content is restricted in terms that the permitted attributes inside are constrained in the corresponding module definition.

In DICOM, all composite objects are sorted into a four-level hierarchy. Every object belongs to exactly one so-called Series. Each Series is contained in exactly one "Study" and a Study relates to exactly one patient. The other way round, a patient can have multiple Studies which may consist of one or more Series including one or more objects (e.g., images) each. DICOM does not clearly define whether to start an additional Study or Series when new images are acquired. Therefore, a Study is never "completed" from the technical point of view. However, it is not possible to place images from different modalities or devices (e.g., CT and MR) into the same Series.

17.1.3 Display Pipeline

DICOM offers a generic image model that allows for storing a wide range of image types, including movies, color and grayscale, high and low resolution images. The Image Pixel Module is included in any DICOM Image IOD. It contains attributes for the actual pixel data and information about how to interpret it for visualization. Besides other information, this module includes attributes for:

- *Image Resolution:* The attributes Rows and Columns are used to denote width and height of the image. If pixels are not square, the proportion of pixels can be defined in the Pixel Aspect Ratio attribute.
- *Photometric Interpretation:* This attribute stores the color model to be used for interpretation of the pixel data. Possible values range from RGB, YCbCr, monochrome to palette-based color models and others.
- *Pixel Representation:* The values in the Pixel Data attribute can be stored to be interpreted signed or unsigned as denoted by the Pixel Representation attribute. Unsigned pixel values are typically needed for CT images, where the pixel values represent (the signed range of) Hounsfield Units.
- *Pixel Data (uncompressed images):* Each pixel sample value, representing either a grayscale or a color component value, is stored here in a "pixel cell". However, the sample value must not necessarily fill a whole cell but some bits may remain unused. Three attributes, Bits Allocated, Bits Stored, and High Bit, determine the alignment of sample values in cells.

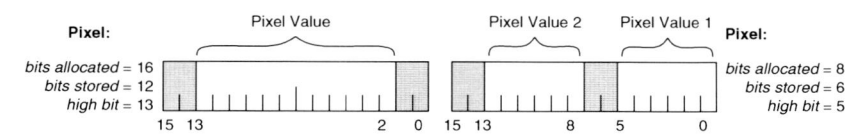

Fig. 17.1. *Bits Allocated, Bits Stored and High Bit*

Bits Allocated denotes the number of bits for one cell, Bits Stored the number of those bits actually used, and High Bit the highest bit position used within the cell. Figure 17.1 shows two examples. All pixels of the image are stored one after another, from left to right and in top-down direction. If multiple frames are stored, the next frame just starts immediately after the last pixel of the previous frame.

- *Pixel Data (compressed images):* In case the pixel data is compressed, the content of the attribute contains a special sequence called Pixel Sequence which usually contains one so-called pixel item per frame[1]. The very first pixel item is dedicated to an (optionally empty) offset table holding byte offsets to each frame. Each pixel item containing a frame consists of a "blob" of encoded data. The type of compression used can be derived from the Transfer Syntax (Sect. 17.1.4) of the image object. Attributes such as Image Resolution, Photometric Interpretation and so on then describe the characteristics of the *uncompressed* image. Pixel Data being compressed are often referred to as "encapsulated".

Rendering an image to an output device is based on the attributes from the Image Pixel Module, but they are not sufficient for resulting in similar viewing impressions of a single image on different output devices. That is why over the years, DICOM also standardized the rendering process resulting in the display pipeline shown in Fig. 17.2. The overall idea is to transform the implementation-specific input values from the Pixel Data attribute into a standardized color space: For grayscale those output values are called Presentation Values (P-Values), and for color they are named Profile Connection Space Values (PCS-Values), which then can be interpreted by a rendering system supporting appropriate calibration means. However, to produce P- or PCS-Values the stored pixels pass a chain of transformations. For the majority of images, most of them are optional.

The following transformation steps apply to grayscale (monochrome) images:

- *Modality LUT Transformation:* This transformation can be used for transforming originally stored manufacturer-dependent pixel values into manufacturer-independent values that are appropriate for the corresponding modality, e.g., Hounsfield Units for CT images or optical density for film

---

[1] It is also permitted to have frames spanning more than one item but that is not very common

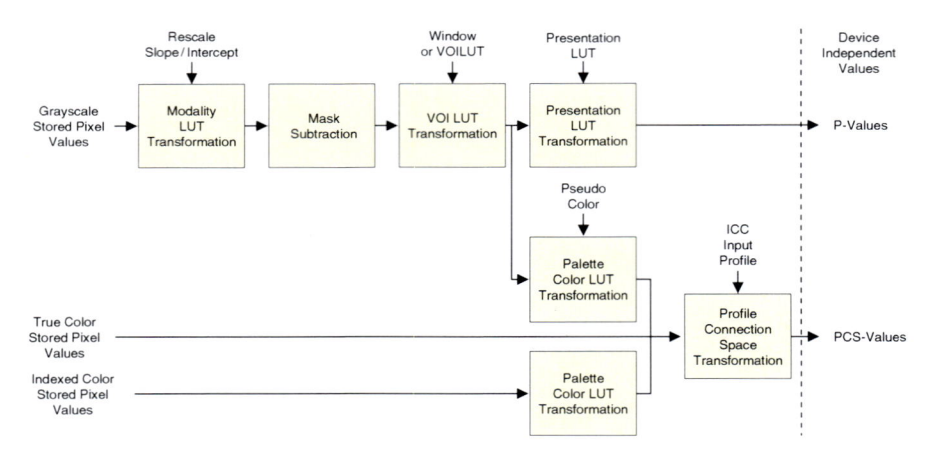

Fig. 17.2. *DICOM Grayscale and Color Image Transformation Models*

digitizers. Either a linear transformation (attributes Rescale Slope and Rescale Intercept) can be defined or a Lookup Table (LUT) for a non-linear transformation.
- *Mask Subtraction:* The Mask Subtraction Module can be used to define how frames of one or more images can be subtracted from each other to construct a meaningful result image. This transformation makes sense especially for X-ray angiography (XA) or X-ray radio-fluoroscopy (XRF) images to compare frames taken from a body part acquired one time with and another time without contrast media. When subtracted, the resulting image nicely highlights the vessel system.
- *VOI LUT Transformation:* Offering the same operations as the modality LUT transformation (using different attributes, e.g., Window Width and Window Center instead of Rescale Slope and Intercept), the Value of Interest (VOI) LUT Module transforms the incoming grayscale values in such manner that relevant details of the image are selected and highlighted. This is due to the fact that usually a medical grayscale image contains more shades of gray than the human eye (and the monitor) is able to distinguish simultaneously. Hence, it is necessary to either focus on a specific range of grayscale values and/or map the input values to appropriate output values. VOI LUT transformations are often used as presets highlighting either bones, soft tissue or other image content.
- *Presentation LUT Transformation:* For use in calibrated environments, the Presentation LUT may be used as a final step after the VOI LUT to produce P-Values. P-Values then can be interpreted by systems calibrated according to the grayscale standard display function defined in DICOM part 14. For a human observer, similar viewing impressions on different output devices are obtained.

The module for the Presentation LUT Transformation is available for the DICOM print service and can also be defined in Presentation State objects (Sect. 17.2.1). For some grayscale images, it may be interesting to introduce pseudo colors that replace the original shades of gray. This can be done using the Palette Color LUT module.

For color images, the rendering pipeline is comparably short. True color images can be converted to PCS-Values by the Profile Connection Space Transformation which makes use of an International Color Consortium (ICC) profile [5] stored within the attributes of the image object. For indexed color models, first the Palette Color LUT Transformation has to be evaluated before applying the Connection Space Transformation.

17.1.4 Network and Media Services

Section 17.1.2 has outlined how DICOM image objects are composed out of attributes. In the following, the network and media services are discussed, which operate on the defined IODs, i.e., by sending an image object to an archive or burning it to CD. Therefore, it is not sufficient to define which attributes (like Patient's Name or the image resolution) must be included, but also how the data are encoded for network transmission or media storage. This chapter selects and summarizes the most (practically) relevant DICOM services and describes how encoding of objects is characterized by means of so-called Transfer Syntaxes.

A service in DICOM is called Service Object Pair (SOP) Class. This notion already implies that a specific service is always paired with a specific object (i.e., IOD). For example, there is one service for transmitting CT Images that are based on the CT Image IOD (CT Image Storage SOP Class) and another for MRI objects (MRI Storage SOP Class). In the standard, related SOP Classes are grouped together in a so-called Service Class. For example, the above CT and MR storage services belong to the group named "Storage Service Class".

Transfer Syntaxes

When two parties are about to exchange DICOM information (e.g., images), they must agree which kind of information to exchange (SOP Class) but also how these data are encoded. DICOM offers 35 Transfer Syntaxes with 14 of them being already retired from the standard. All Transfer Syntaxes share a common approach (Fig. 17.3) and differ only in three aspects:

- *Implicit vs. Explicit VR:* One Transfer Syntax does not send the VR field but leaves it up to the receiver to look up the VR from a data dictionary (electronic counterpart of part 6 of the standard). This behavior is called "Implicit VR" while sending with VR information is named "Explicit VR".
- *Little vs. Big Endian:* When transferring binary values that need more than one byte for transmission, it must be decided which byte should be

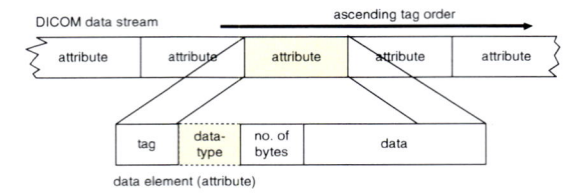

Fig. 17.3. *Attribute encoding for DICOM data streams.* All attributes of an object are sorted by tag in ascending order and then transmitted with their VR (exception: Little Endian Implicit VR), the attribute value and its length

sent or stored first. When choosing Little Endian, the least significant byte is stored first, for Big Endian vice versa.
- *Compression of pixel data:* In practice, transmission and storage in DICOM often happens uncompressed. However, DICOM also offers compression methods, that are only common for some medical disciplines (e.g., ultrasound movies) but generally can be used for any kind of DICOM image object. All those methods only compress the contents of the Pixel Data attribute inside the DICOM object but leave attributes like Patient's Name, etc. untouched. Compression schemes offered include different JPEG variants (e.g., JPEG Baseline, JPEG-LS, JPEG2000), RLE and others.

Not all combinations of these three aspects are valid (see part 5 of the standard). Basically, all DICOM systems communicating over a network must support at least one Transfer Syntax, which is Little Endian Implicit VR (uncompressed). As depicted in Fig. 17.3, only attributes are written and not any module information, because these are only means for organizing IODs in part 3.

Network Services

While part 3 of the standard describes data structures, part 4 specifies high-level network services (thus, SOP Classes grouped in Service Classes) working on these objects that can be used for communication between parties. Generally, DICOM defines its own low-level network protocol which is based on the OSI paradigm [6]. Basically, it is designed for being used together with different low-level protocols. However, the only protocol binding currently defined is based on TCP/IP (called "DICOM Upper Layer Protocol for TCP/IP").

A DICOM node on the network is called Application Entity (AE). Each AE is either a client, Service Class User (SCU) that is using or a server, Service Class Provider (SCP) that is providing services. Every AE uses two kind of messages for communication: Association Control Service Element (ACSE) messages are used for connection management and DICOM Message Service Element (DIMSE) messages are used for transmitting the payload data,

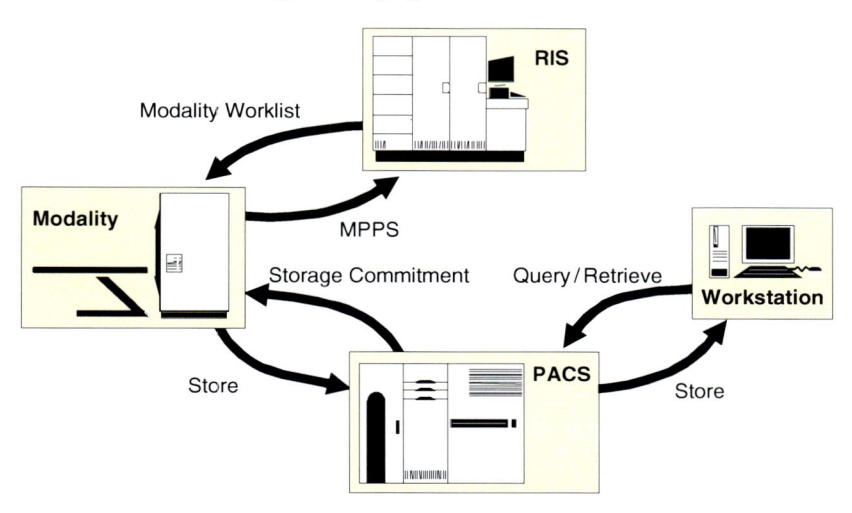

17 Digital Imaging and Communications in Medicine

Fig. 17.4. *Example scenario with DICOM services*

e.g., a DICOM image object. An SCU starts communication by sending an ACSE message offering the SCP the desired DICOM services by referring to selected SOP Classes. For each SOP Class, a list of Transfer Syntaxes is proposed. The SCP receives this proposal and assembles a response message in which every proposed SOP Class is either denied or accepted together with a single selected Transfer Syntax. When receiving this response package, the SCU knows which services can be used on that connection. This phase of DICOM communication is called Association Negotiation. If one of either party wants to terminate the communication, it can do so by sending an appropriate ACSE message.

An example scenario including the described services is shown in Fig. 17.4. The most common DICOM services used in practice are:

- *Storage and Storage Commitment Services*: The DICOM Storage SOP Classes (e.g., CT Image Storage SOP Class) are used for transmitting a composite DICOM object over the network. For each kind of object, there is a dedicated Storage SOP Class. An MR modality, for example, would probably negotiate the MR Image Storage SOP Class with an SCP and then send acquired MR images on that connection. The receiver could be a Picture Archiving and Communication System (PACS), a system that is used for archiving composite DICOM objects. Also, the modality could be able to create a more generic kind of DICOM image based on the Secondary Capture Image SOP Class. It then could additionally offer to transfer the acquired MR images as Secondary Capture images.

  The Storage SOP Classes transmit objects from one AE to another, but there are no rules what the receiver is about to do with the received objects, e.g., a workstation may delete it after image review. With Storage Commitment, a further message is exchanged between SCU and SCP for assuring that the SCP (probably a PACS) has received the images for archival, i.e. that the SCU (e.g., imaging modality) itself must not keep a copy of the objects but may delete them.

- *Query/Retrieve Service*: Like the Storage Service Class, the DICOM Query/Retrieve (Q/R) service and its corresponding SOP Classes have been in DICOM since 1993. Q/R services allow for searching an archive for DICOM images and to initiate the transmission of selected objects. Therefore, Q/R is typically provided by PAC systems. Q/R offers different SOP Classes for searching and retrieving varying in what information can be searched and how transmission of objects is performed. In the query phase, an archive can be searched for images based on filter mechanisms, e.g., it is possible to ask for a list of studies based on a defined patient name. If the SCU decides to download any DICOM objects (retrieve phase), the Q/R protocol is used to initiate the transfer. However, the transmission itself is done via the Storage SOP Classes. It is also possible for an SCU to tell the SCP not to transfer the DICOM objects to the SCU itself but to a third system.
- *Worklist Service*: Over the years, DICOM was extended around its core services, among others in the area of work flow support. Using the Modality Worklist SOP Classes (belonging to the Basic Worklist Management Service Class), a modality can query an information system for a so-called worklist. For example, in a radiology department a CT modality may ask the Radiology Information System (RIS), which CT exams are planned for a specific time of day or which patients are about to be examined, thus querying a timetable of pending examinations that reside on the information system. The patient's name and other information can be taken over by the modality into the acquired images which then may be stored to the archive. As a result, Modality Worklist Management brings patient and other information available in the RIS to the modality and into the image objects without the need of re-entering or even re-typing information, thus leading to consistent data in RIS and PACS.
- *Modality Performed Procedure Step (MPPS) Service*: However, the IS does not receive any responses about what the connected modalities are actually doing – which might substantially differ from what was originally planned. Hence, the MPPS Service was introduced consisting of three SOP Classes. The most common one is the Modality Performed Procedure Step SOP Class that permits modalities to send status, patient, billing and other information about an ongoing examination to an IS.

Media Services

Unlike the exchange of DICOM objects over the network, for media exchange the communicating parties are not able to negotiate SOP Classes, Transfer Syntaxes or the kind of media and file system to be used. Also, in the same kind as for networking, it should be possible to describe the kind of objects and encoding exchanged in the Conformance Statement (Sect. 17.1.5). For those reasons, DICOM defines so-called Media Storage Application Profiles. Each profile exactly defines a set of SOP Classes, Transfer Syntaxes and media that can be used for exchange. For example, the STD-GEN-CD Profile permits all kind of SOP Classes, encoded using Transfer Syntax Little Endian Explicit VR on a CD medium with ISO 9600 Level 1 file sytem. Every medium exchanged must contain a special DICOM object called DICOMDIR in the root directory that serves as an index of DICOM objects on the medium. There are currently more than 100 Standard Application Profiles, supporting DVD, MODs, USB flash drives and so on.

17.1.5 Conformance

As already noted, DICOM offers lots of different objects, services and other specifications such as display calibration requirements and so on. Of course, only a subset of these services is usually supported by a specific system. Every device only implements those services that make sense in its application domain; e.g., there is no reason for a CT modality to implement the MR Image Storage SOP Class. On the other hand, there are no guidelines in DICOM which require specific devices to implement a defined set of services, e.g., whether the above-mentioned CT also supports worklist and/or MPPS. Users and possible buyers of those systems must know in advance which services are supported to fit into the planned workflow and integrate into the existing infrastructure. That is why DICOM *requires* every system claiming DICOM conformance to publish a document called DICOM Conformance Statement (CS), which describes all services and options implemented by that device.

Form and content of a CS are quite exactly defined in part 2 of the DICOM Standard; e.g., it dictates the order and naming of chapters, table layouts and some required drawings. After a title page, overview and table of content, two important chapters have to follow: network-related and media-related services. Both chapters must contain detailed information about which SOP Classes and Transfer Syntaxes in which roles (SCU/SCP) are supported, system-specific status codes, options supported (e.g., supported query keys in Q/R) and configuration parameters. For media services, it must also be documented whether the system works as a File Set Creator (FSC), Updater FSU or Reader FSR. The following sections of the document summarizes character sets, security profiles (e.g., signing and encrypting of objects), private extensions like private attributes, coding schemes (e.g., SNOMED) and calibration means supported by the system. Overall, a Conformance Statement offers indispensable information about a system's DICOM capabilities.

17.2 Advanced DICOM Services

17.2.1 Advanced Image Display Services

When the DICOM standard was conceived in the early 1990s, it was primarily intended as a technology enabling an image transfer between imaging modalities, image archives and display workstations of multiple vendors. However, the undisputed success of DICOM at this task highlighted a weakness of the standard in a related area. Even though DICOM enabled to display the same image on different workstations, in many cases the resulting display looked quite different. A *consistent* image display across multiple devices was not guaranteed [7].

There are multiple reasons for this problem, the most important one being the fact that the color models used in the DICOM standard are rather vaguely defined. For example, "MONOCHROME2" refers to a monochrome image where the lowest possible image pixel value corresponds to black, and the highest possible image pixel value corresponds to white. Unfortunately, DICOM does not define the meaning of pixel values between black and white. If for example pixel values would range from 0 to 255, the mean value 127 could be interpreted as 50% of the maximum luminance of the display device, as 50% of the optical density of an X-ray film to which the image is to be printed, as 50% of the maximum tissue density encountered in the image, or as 50% of the maximum "brightness" of a monitor as perceived by a human observer. Different interpretations have been implemented by different vendors, causing inconsistent image display and, in some cases, a very poor quality of displayed images.

Another problem is related to the clinical workflow. It is quite likely that the physician interpreting the images will adjust the image display such that the medically relevant image features are optimally visible. However, when these adjustments are made, the original image is most probably already stored in an irreversible manner in the image archive; so the adjustments cannot be stored as part of the image unless every image is duplicated, which is usually not acceptable.

A number of DICOM extensions have been developed to address these issues.

Softcopy Presentation States

DICOM Softcopy Presentation States (in short: "Presentation States") are DICOM documents that store precise instructions on how to display a specific image or a related set of images. A Presentation State contains the parameters for all grayscale or color transformations that apply to a specific image type according to the DICOM image transformation model (Fig. 17.2), overriding values encoded in the image itself. In addition, a Presentation State can contain graphical annotations and a number of spatial transformations to be applied to the image (Fig. 17.5):

Device Independent Values → Shutter Transformation → Image Relative Annotation → Spatial Transformation → Displayed Area Relative Annotation → Display

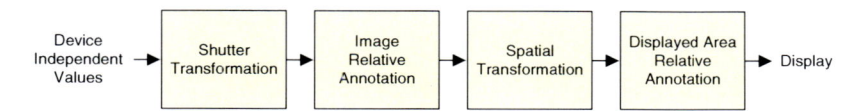

Fig. 17.5. *DICOM Common Spatial and Annotation Transformation Model*

- *Shutter Transformation:* Display shutters allow for masking unwanted parts of the image, e.g., unexposed parts of an X-ray that would otherwise be displayed very brightly and negatively affect the visibility of relevant parts of the image.
- *Image Relative Annotation:* To highlight regions of interest or to show measurements, images can be annotated with graphical and textual comments, which are displayed as overlays. Since they are not "burned" into the image, annotations in a Presentation State can be switched on and off interactively.
- *Spatial Transformation:* Images can be rotated and flipped (often required for Computed Radiography images). Images can be zoomed to a defined "displayed area" or by a specified factor. Also, an image can be displayed at its true physical size, if the size is known.
- *Displayed Area Relative Annotation:* This type of graphical annotation is not "attached" to the image but to the display (view port). This permits, for example, certain textual comments to be displayed in the corners of the screen independent from the zoom factor or rotation applied to the image.

Over time, several different types of Presentation States (i.e., different SOP Classes in DICOM terminology) have been defined by the DICOM committee to support different use cases:

- *Grayscale Softcopy Presentation States (GSPS)* [8] were the first type of Presentation State to be standardized, in 1999. GSPS apply to monochrome images only and contain all parameters of the grayscale display pipeline depicted in the upper row of Fig. 17.2, plus the transformations described above. The output of a GSPS is defined in P-Values, i.e., grayscale values in a perceptionally linearized space where equal differences in pixel value correspond to equal perceived changes in brightness by an average human observer. Such grayscale values can be displayed on a monitor calibrated according to the DICOM Grayscale Standard Display Function or the CIELAB curve [9].
- *Color Softcopy Presentation States (CSPS)* [10] are the counterpart to GSPS for color images, both true color and palette color images. Instead of grayscale transformations, a CSPS contains an ICC Input Profile [5] that describes the conversion of the color values of the image (or, more precisely, the device that created the image) to a device-independent Profile Connection Space (PCS) defined by the ICC, which is either the CIE

1931 XYZ curve [11] or CIELAB [9]. Such PCS color values are used as input to an ICC color management system, which describes the capabilities of a monitor through an ICC Output Profile and provides a mapping to the color space of the display device that retains the original colors as good as possible, given the physical limitations of the display device.

- *Pseudo-Color Softcopy Presentation States (PCSP)* [10] are used to display monochrome images such as nuclear medicine or Positron Emission Tomography (PET) images in pseudo-color using a color LUT that is applied after the VOI LUT transform, i.e., after adjustment of the grayscale window center and width (Fig. 17.2). Like CSPS, pseudo-color presentation states contain an ICC profile enabling a consistent color display on calibrated displays.
- *Blending Softcopy Presentation States (BSPS)* [10] are used to encode the blending of two spatially co-registered sets of grayscale images (e.g., CT, PET) such that one set of images is displayed as grayscale, and the other one superimposed as color of varying opacity. BSPS do not describe the actual spatial registration of the two image sets. This is done by "'Spatial Registration' objects". The transformation model for BSPS is shown in Fig. 17.6.
- *XA/XRF Grayscale Softcopy Presentation States* [12] are an extension of GSPS specifically for use with X-ray angiography and X-ray radiofluoroscopy images. Unlike GSPS, different shutters can be defined for different frames of a multi-frame image, and for mask operations (subtraction of a mask generated by averaging non-contrast image frames), multiple image regions within the mask with different pixel shift offsets per region can be defined.

A Presentation State only contains references to the images it applies to and, therefore, does not duplicate the image data. Presentation States are relatively small (typically only few Kbytes) and can be stored and transmitted with a minimal resource increase. Presentation States fit well into the established DICOM information model (they are just a separate DICOM series within the study containing the images) and can be transmitted, stored and retrieved

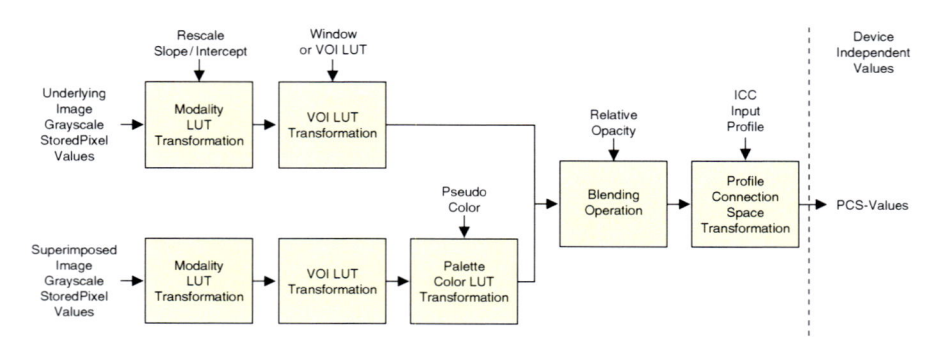

Fig. 17.6. *DICOM Grayscale to Color Blending Transformation Model*

with the existing DICOM Storage and Q/R services, requiring few changes in existing systems. It is possible to have a single Presentation State for a complete series of images or to have different Presentation States ("views") of the same image.

Structured Display

DICOM Structured Display objects [13] are similar in concept to DICOM Presentation States. While Presentation States describe how exactly a specific image should be displayed on screen, a Structured Display object describes the layout of a display that may be composed of multiple images or Presentation States, including cine or stack mode displays.

A Structured Display object mainly consists of a list of "image boxes", i.e., rectangular areas on the screen each of which is filled with one image or another DICOM document. For each of these boxes, the following parameters are defined:

- Position and size of the image box on screen (relative to the screen size)
- A list of images or a single Presentation State, stereometric image or non-image object (e.g. structured report) to be displayed in the image box
- Horizontal and vertical alignment of images displayed in the image box
- Layout type (stack mode display, cine display or single image)
- Priority in case of overlapping image boxes
- Cine related parameters: playback type (loop, sweep or stop), playback speed and initial state (playing or stopped)
- Synchronization information for a synchronized display of images in multiple image boxes

Furthermore, a Structured Display may contain a list of text annotations to be displayed on screen (possibly as overlays over the images). Finally, it contains the resolution and minimum bit depth of the monitor for which the Structured Display was generated.

Hanging Protocols

While a DICOM Structured Display object specifies the layout and display of a specific set of DICOM images, which are referred to by their unique identifiers, a Hanging Protocol [14] can be seen as an abstract template that describes how a certain *type of study* should be arranged on screen. The result of this layout ("hanging") process could then be stored as a Structured Display.

This means that Hanging Protocols are not related to a specific patient, but to an imaging modality, study type or body part. DICOM defines specific network services enabling the storage and retrieval of Hanging Protocols based on these properties; a search by user name is also supported. Once a user selects an image study at a diagnostic workstation, the workstation can extract sufficient information from the images to query for the most appropriate Hanging Protocol, retrieve and apply the best match, while still providing the user with the capability of changing the layout to his or her preferences, and possibly saving these preferences as a new Hanging Protocol.

Once a Hanging Protocol has been selected, a three-phase process takes place: selection, processing and layout. In the first phase, so-called image sets are selected. Each image set is either a part of the study for which the Hanging Protocol was selected, or part of a related prior study (e.g. the last prior or a prior study that is at least 6 but not more than 24 months old). It is up to the diagnostic workstations to locate and retrieve priors based on these abstract definitions. In the second phase, so-called Display Sets are defined. Each Display Set is a rectangular area on one of the possible multiple screens of the display workstation into which either a single image or a rectangular grid of images ("tile mode") taken from a single image set is to be rendered. Filter criteria describe which images from the image set should be displayed in each Display Set (for example, in Mammography one Display Set could display only the left-hand side images, and another one could select the right-hand side images from the same image set). In addition to the layout parameters available for Structured Display, in a Hanging Protocol certain image processing techniques can be requested, e.g., a multi-planar reconstruction (MPR), 3D volume/surface rendering or Maximum Intensity Projection (MIP) of a volume defined by a set of CT, MRI or PET images. Furthermore, the order in which multiple images should be displayed (e.g., in "tile" or "stack" mode) in one Display Set can be defined. Finally, the so-called Presentation Intent describes how the images should be rendered within the Display Set – this information is essentially a subset of a Presentation State and can be used to generate Presentation States for the selected images. In the third and final phase, the screen layout is generated, based on the display set definitions, and images are rendered into each image box.

In summary, as [14] states, Hanging Protocols enable users to conveniently define their preferred methods of presentation and interaction for different types of viewing circumstances once, and then to automatically layout image sets according to the users' preferences on workstations of similar capability.

17.2.2 DICOM Structured Reporting

The idea of exchanging structured medical reports electronically is at least as old as computers are used in medicine. However, it took more than a decade from the initial release of DICOM's predecessor ACR/NEMA before the DICOM committee started to deal with this topic. Before, the main focus was to push the "imaging world" where these standards also have their seeds.

In the mid of the 1990s when the work started, another extension of the DICOM standard had major influence on the introduction of DICOM Structured Reporting (SR): "Visible Light Image for Endoscopy, Microscopy, and Photography" [15]. Besides the actual pixel data and describing attributes, the new image objects for pathology, dermatology and ophthalmology also contain information on the "image acquisition context". In order to avoid the introduction of dedicated DICOM attributes for each acquisition technique, a generic approach was developed: medical concepts and context-dependent terms are specified indirectly by means of codes.

After an unpromising draft that was regarded as too complex and hard to implement, the extension for Structured Reporting (SR) was completely revised and published for Final Text in April 2000 [16]. The coding part was sourced out and released as a separate extension about 1 year later [17]. The following sections give an overview of the basic concepts of DICOM SR, the codes and document templates as well as selected SR applications.

SR Concepts

DICOM Structured Reporting allows for storing almost arbitrary information in a structured way in DICOM format. Not only medical reports and clinical records but also measurements, procedure logs and the like can be archived and exchanged in a standardized manner. Therefore, the name "Structured Reporting" is somewhat misleading since it suggests a rather limited scope. "Structured Data" would be a more appropriate name, applicable to many fields of application. For all applications, the same basic principles apply:

- Pieces of structured information are related to each other
- In addition to simple text, codes and/or numerical values are used
- External DICOM objects like images are referenced

In this context, it is already sufficient if a single principle applies. For example, if a couple of DICOM objects are to be flagged in some way, DICOM Structured Reporting is suited in the same way as for the standardized storage of less structured reports. SR documents also do not have to be complex or qualified for being transformed into a human readable form: the exchange of measurements from an ultrasound machine is an example for this. However, the DICOM standard only specifies the semantical level of such a structured document, aspects of visualization are more or less excluded [18].

Basically, a Structured Reporting document is constructed in the same way as other composite objects, i.e., they consist of a list of attributes. However, instead of pixel data like in DICOM images, SR documents contain a so-called document tree that represents the actual structured content. Compared to other DICOM objects, sequence of items elements are more frequently used, e.g., to map the hierarchical structure of the document tree adequately to the DICOM data structures. And, for Structured Reporting there is no limitation for the nesting of sequences; the document can be structured with arbitrary complexity. However, to keep the requirements for a certain SR application low, there are three general SR IODs: "Basic Text" is the simplest IOD and mainly used for simple, text-oriented documents; "Enhanced IOD" also allows for coding numerical values; and "Comprehensive IOD" comprises all aspects of DICOM Structured Reporting. Recently, more specialized SR IODs have been introduced (see Sect. "SR Applications").

In addition to the general modules like Patient and Study which are also used for other composite objects, the SR IODs contain a special Series and two SR Document modules. The "SR Document General" module includes information that refers to the whole document, e.g., for document management and workflow support. Among others, there is a flag indicating whether the content of the document is completed or partial and a flag indicating whether or not the document is attested by a verifying observer, who is accountable for its content. There is also a mechanism for storing references to previous versions of the document like a preliminary draft.

The "SR Document Content" module contains the hierarchically structured document tree that consists of a number of content items which are related to each other. Each content item carries a piece of the whole information of the document. The type of the information is described explicitly by the value type. The DICOM standard currently defines 15 different value types, e.g., for texts, numerical values, codes, dates, times, person names, spatial and temporal coordinates, references and containers. In addition to the actual value, each content item has a concept name that identifies the purpose of the item, e.g., whether a numerical value describes a diameter or an area. The root node of the document tree always has to be a container and its concept name describes the title of the document.

The type of the relationship between two content items is also explicitly specified. The standard currently defines seven different relationship types (e.g., contains has observation context, has concept modifier, has properties and inferred from). The relationship is directed and points from the higher-level content item (source) to the lower-level content item (destination). As an exception, for some SR IODs it is also possible to refer from one content item to a content item in another sub-tree of the document tree. This kind of relationship is called "by-reference" in contrast to the usual "by-value" relationship. In doing so, the document tree becomes a Directed Acyclic Graph (DAG) because references to ancestor content items are forbidden to prevent loops.

Figure 17.7 shows an extract from a DICOM SR document tree in which the malignancy of a mass is inferred from the observation that the mass has an irregular margin. The rectangles in this figure represent the content items where the first line denotes the value type, the second line the concept name and the third line the actual value. The diamonds represent the relationships between the content items. The dashed line illustrates a by-reference relationship.

Codes and Templates

The use of controlled terminology is an integral part of any approach for structured documentation. Coded entries have a distinct advantage over clear

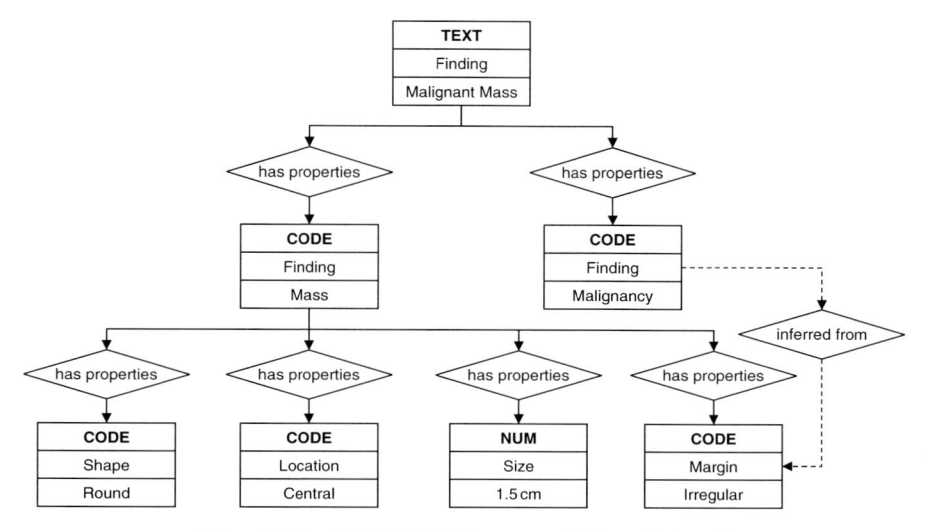

Fig. 17.7. *DICOM SR Document Tree (Extract)*

text: they are unique. In DICOM Structured Reporting, codes are, therefore, used in many places. In some places, they are even required.

A code in DICOM consists of at least three components: the coding scheme designator, the code value and the code meaning. The coding scheme designator identifies the system of concepts or catalog in which the code is defined (e.g., ICD-10, SNOMED). Within such a scheme, the code value has to be unique; outside of the scheme, it has no value because different coding schemes can use the same code value. Finally, for each code the associated meaning has to be specified. The benefit of this requirement is that applications even if they do not support a certain scheme still can use the textual description instead of the typically cryptic code value.

In DICOM Structured Reporting, codes are mainly used for specifying concept names, values in code content items and measurement units in numerical value content items. In addition, they are used to uniquely identify persons, medical procedures, anatomic regions, etc. In order to avoid that, e.g., ICD-10 instead of SNOMED codes are used to specify an anatomic region, codes can be grouped by semantic aspects. These groups of codes that are to be used in a particular context are called "context groups". For example, DICOM describes in the context group with ID 4 which codes are to be used for anatomic regions. When referring to a certain context group in the standard, the list of codes could either be mandatory ("Defined Context Group") or a recommendation only ("Baseline Context Group").

A context group may contain codes from different coding schemes and may even include other context groups. Furthermore, a context group can be designated as extensible or non-extensible, i.e., whether it is allowed to extend this group for a certain application or not. There is also a version number for each context group which can be used to reference a particular context group in an unambiguous manner. This is because even a non-extensible context group can be extended with new codes in a future edition of the DICOM standard.

Usually, the standard only defines its own codes when there is no appropriate code in a common coding scheme like SNOMED. All these DICOM-specific codes are listed and defined in part 16 of the standard, the "DICOM Content Mapping Resource (DCMR)". The numerous context groups used throughout the standard are also defined in this part. Furthermore, for certain local applications, it might be useful to define private codes. DICOM explicitly supports this by providing a mechanism that avoids conflicts between different locally defined coding schemes.

Another important aspect of DICOM SR is the flexibility for structuring the content of a document. On the one hand, this can be considered as an advantage since it allows for storing almost any kind of document. On the other hand, it has the disadvantage that the same information can be encoded very differently by arranging the various content items and connecting them with relationships in different ways. In particular, applications that need to read and process SR documents would profit from a more constrained tree structure.

The DICOM standard, therefore, defines so-called templates, i.e., "patterns that specify the concept names, requirements, conditions, value types, value multiplicity, value set restrictions, relationship types and other attributes of content items for a particular application" [1]. A template can either specify a substructure of the document tree (e.g., the properties of a lesion) or a complete document (e.g., radiology report). The latter is called "root template" because it starts at the root of the document tree.

The template definitions are also part of the DCMR in part 16 of the DICOM standard. Similar to the context groups, all templates are identified by a number, e.g., 2000 for the "Basic Diagnostic Imaging Report", and they can also be extensible or non-extensible. The content of a template is defined in a table where each line represents a content item or specifies which other template is to be included at this position. A few templates can also be parametrized which is in particular useful for generic templates like measurements.

In practice, the use of templates really helps to unify the structure and content of SR documents. Unfortunately, the mechanism for identifying the templates used in a document is very limited. Therefore, DICOM SR documents are usually defined completely based on templates, i.e., using a root template. However, the decision to specify the structure and content of an SR document in the DCMR almost independently from the underlying DICOM data structures has significantly supported integrating special knowledge from different medical fields.

SR Applications

The flexible format of DICOM SR allows for many different applications. The extensive use of codes enhances the machine-readability and facilitates the automatic analysis of the stored data. Furthermore, the use of templates brings a consistent structure of the document content and, therefore, improves the automatic processing by a reading application. For these reasons, SR documents are best-suited for the standardized device communication, especially for the exchange of measurements and related data.

For example, the various measurement values determined at an ultrasound machine can be transmitted to another system in a standardized manner using an appropriate SR document and the DICOM storage service. Another example of use is the standardized storage of the results of a CADx or CADe system. Currently, the standard defines templates and SOP Classes for mammography, chest and colon CAD. DICOM SR is also used to document and exchange radiation dose information of X-ray and CT exams, supporting more details than the MPPS service.

Of course, DICOM SR is also used for medical reports and other clinical documents. Since the level of structuring and coding can be chosen individually, almost any document can be converted into the DICOM SR format. SR documents are, therefore, suitable for both radiology where less-structured documents are still predominant and other medical fields where more fine-grained information is collected (e.g., cardiology). This flexibility also allows for a smooth migration of today's free-text reports to more structured and standardized documents.

In addition, there are a couple of SOP Classes that use the concepts of DICOM Structured Reporting for other purposes. A famous example is the Key Object Selection Document which allows for flagging particular DICOM images and other composite objects for various reasons, e.g., to select the relevant images of a CT series or to sort out images for quality reasons. And finally, the Procedure Log IOD is used "for the representation of reports or logs of time-stamped events occurring during an extended diagnostic or interventional procedure, typical of the cardiac catheterization lab" [1].

17.2.3 Application Hosting

Over the past years, the possibilities in medical image processing have evolved enormously. This development is due to different reasons: New modalities like the modern generation of CT and MR devices provide more data for post-processing of images. Accompanied by the massively increased computational power of computer systems, this has led to new, complex analysis methods, e.g., in the area of 3D visualization, multi-modality fusion or quantitative biomarker software. Against this background, there is a rising demand for the consistent, comparable and vendor-neutral exchange of image processing results but also of the post-processing applications itself. Hence, the latter offers the chance to facilitate the exchange of these applications between different commercial systems to ease the comparability of processing results from different sites in clinical trials and to close the gap between application of such tools in the research and commercial area [19]. A new approach aiming at these targets is currently addressed by DICOM Working Group 27 (Web Technology for DICOM) with the development of Supplement 118 "Application Hosting" [20]. The extension is currently in public comment phase, i.e., it soon could be balloted and thus be incorporated (with minor changes if applicable) into the standard as the new document part 19.

Basic Concepts

Supplement 118 defines a plug-in concept permitting the development of post-processing applications which plug into an imaging application by utilizing specified launching and communication services. This approach tries to overcome the problem that a third-party image processing plug-in (e.g., fusion of CT and PET images) must be adapted for each vendor separately due to the lack of a standardized communication between plug-in and hosting software. By defining a generic API, a single plug-in should be able to work "out of the box" on top of different vendor's software products (Fig. 17.8).

A system making use of such a standardized plug-in is called "Hosting System" while the plug-in itself is named a "Hosted Application". For the said CT and MR image fusion, a Hosting System would launch a Hosted Application which then retrieves the corresponding data (e.g., CT/PET image objects) from the Hosted System. It then starts the fusion computations and notifies the Hosting System about any result images as soon as they are available.

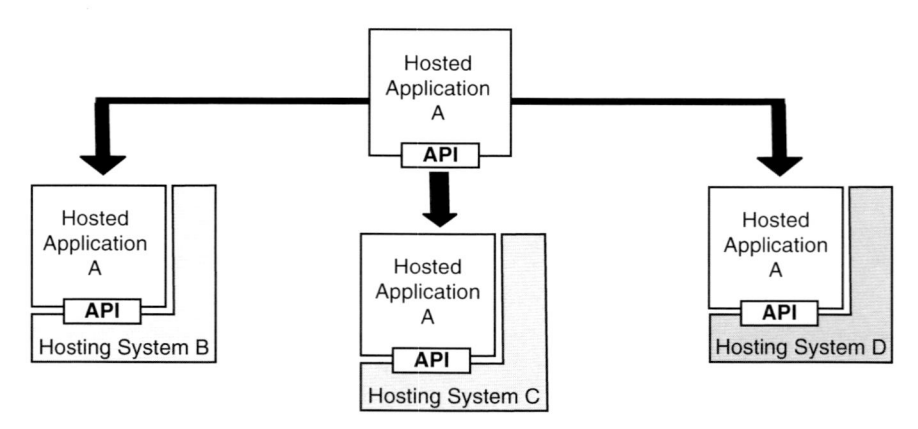

Fig. 17.8. *Single plug-in implementation (A) utilized by different vendor's products (B-D)*

Web Service API

It was tried for the API to find a sufficient trade-off between ease of use (and implementation) and mechanisms being sufficient and generic enough to support a broad range of possible applications. Besides other aspects, the API is designed to be implementable for any common programming language and operating system and to leverage existing technology. Also, the API should be extensible in future versions of the Application Hosting concept while staying backward compatible. The first version of the API concentrates on the exchange of input and result data (DICOM and non-DICOM) between Hosting System and Hosted Application. Implementations could run on Windows, Unix or Mac OS X systems and their corresponding GUIs (if applicable for the Hosted Application). It is also foreseen that the Hosted Application may use part of the screen assigned by the Hosting System as a playground for manual data input or notification messages. Later versions of the API may enrich this functionality; thus the API is said to be developed and rolled out in "stages" over the next years.

The API between Hosted Application and Hosted System can be divided into two categories:

- *Life Cycle Management*: The Hosting System must be able to launch and terminate the Hosted Application.
- *Interaction*: While running, Hosting System and Hosted Application must at least be able to exchange input, processing and output data and to communicate status information.

In this first version of the Application Hosting API, a Hosted Application is launched in a pre-configured, vendor-specific way, e.g. from the command line or by selecting a menu entry in the Hosting System's GUI. When running, a Hosted Application may switch between the states (Table. 17.2). All possible state transitions are also defined in the supplement.

There are three interfaces defined which must be implemented by a Hosting System and/or a Hosted Application:

Table 17.2. *States of Hosted Application*

State	Description
IDLE	In IDLE state the Hosted Application is waiting for a new task assignment from the Hosting System. This is the initial state when the Hosted Application starts.
INPROGRESS	The Hosted Application is performing the assigned task.
SUSPENDED	The Hosted Application is stopping processing and is releasing as many resources as it can, while still preserving enough state to be able to continue processing.
COMPLETED	The Hosted Application has completed processing, and is waiting for the Hosting System to access and release any output data that the Hosted Application has informed it of.
CANCELED	The Hosted Application is stopping processing, and is releasing all resources with no chance to resume processing.
EXIT	The terminal state of the Hosted Application

- *Application Interface*: It is implemented by a Hosted Application and permits the Hosting System to set and get the Hosted Application's state. Also the Hosted Application can be asked to bring its GUI windows (if any) to the front.
- *Host Interface*: It is implemented by the Hosting System and is used by the Hosted Application during plug-in execution to request services (like a screen area for drawing) from the Hosting System or to notify it about any events and state transitions. Furthermore, the plug-in can ask the Hosting System to provide a newly created Unique Identifier (UID). Other API calls are devoted to the Hosted Application's file system interactions. The Hosting System may be asked to return a directory for temporary files as well as one to store the final output data.
- *Data Exchange Interface*: It must be supported by both Hosting System and Hosted Application. It offers functionality for exchanging data using either a file-based or a model-based mechanism.

The model-based mechanism supports information exchange over XML Infosets [21]. At the moment two models are defined for use in DICOM, the Native DICOM Model and the Abstract Multi-Dimensional Image Model. A recipient of such data must not know how to parse the original native format (which may be binary as for DICOM), but instead works on the abstraction of such data in the form of an XML Infoset. Therefore, existing standard tools can be used for querying the XML Infoset model using XPath [22] expressions without the need of accessing the original format. For both models, it is also possible to link to binary bulk data instead of a text string as usually needed for the XML Infoset approach. This is useful if a text string encoding of a specific data value (pixel data, lookup tables etc.) is not appropriate because of performance issues, etc.

While the DICOM Native Model uses an XML structure being organized to directly represent the structure of an encoded DICOM dataset (thus, arranging DICOM elements with its sequences and items in a tree), the Abstract Multi- Dimensional Image Model offers a way of exchanging multidimensional image data. The dimensions are not necessarily spatial coordinates but may also represent measure points in time or measurements of other physical units. For example, the latter approach could be used for storing 3D MR reconstruction volumes or, by adding time as a fourth dimension, 3D MR volumes over time.

All API calls are realized as Web Services specified using the Web Service Description Language (WSDL) [23].

Use Cases

There are a lot of use cases for processing DICOM objects in a standardized environment. Here, the two examples chosen are anonymization of DICOM images and screening applications with the latter taken from the supplement itself.

A feature often desired by users of PACS workstations or other systems is the anonymization of DICOM images. Usually, such a functionality is not even offered as an add-on by the original vendor. Supposed that the system provides an Application Hosting interface, a user could easily buy an anonymization plug-in from a third party, configure it and start it each time DICOM objects should be anonymized. In this setting, the add-on is the Hosted Application running on top of the original software, thus, the Hosting System. Before, the Hosted Application must be installed and announced to the Hosting System manually (may change in later API stages). The same applies for the startup which may be done by clicking the Hosting System's GUI controls. After startup, further processing steps can be done using the described API. The Hosted Application is activated by setting its state to "INPROGRESS" and then is notified about DICOM images to be anonymized using (probably) the file-based method. The Hosted Application accesses the announced objects by loading them from the reported files. After (or while processing) the plug-in asks for an output directory where it should store the resulting, anonymized objects, stores the data in that directory and notifies the Hosting System about the availability of anonymized result images. Accordingly, the plug-in notifies the Hosting System about its new status "COMPLETED". The Hosting System then asks the Hosted Application for a descriptive list of files where the announced result objects can be found and receives this information from the plug-in. After reading the anonymized images (and, e.g., sending them to a case study archive), the Hosting System sets the state of the anonymization plug-in to its default state, "IDLE".

The other example deals with the CAD. Applications in that area are often designed for supporting the physician reading a set of images for diagnosis. Therefore, the software tries to identify abnormalities or issues of interest the physician may further check. One problem preventing CAD software for becoming much more popular is that one vendor's workstation software often only permits CAD implementations from the same vendor. Also, some companies are only offering very specialized CAD solutions, which only work for very specific body parts and tasks, e.g. for detection of cancer in mammography images. Thus, a user must buy different workstations or even servers from different companies to run different CAD applications. With the API defined for Application Hosting, the different third-party CAD plug-ins may run on top of a single platform.

Synopsis

The concept defined with Application Hosting is mostly valuable for users who will have a greater palette of applications and vendors to choose from. Furthermore, also vendors can avail the opportunity to have their applications (sold and) running at sites that till then were fully tight to products of a competing company. Especially for smaller specialized vendors, Application Hosting offers a good chance by easing market access.

The current API is quite simple and, therefore, should be relatively easy to implement. However, the future will show whether the proposed interface is specific enough to handle all workflows, e.g., whether the Hosting System can be designed in a way that resulting image objects or other results can be handled in a reasonable fashion. Nevertheless, by staging the development it becomes possible to refine the Application Hosting concept based on the experiences collected with earlier versions.

Overall, Application Hosting definitely offers great chances for users and vendors and might prove itself first as a very valuable extension to DICOM and second as a further step of bringing Web Technology to the standard.

17.3 Conclusions and Outlook

The DICOM standard has enormously grown since its initial release in 1993 and will probably keep evolving into new medical and technological areas. As the previous development shows, demands for new objects and services often come up with new modality and technology trends appearing on the market.

Built around core image management services (transmission, query, printing), the standard introduced concepts for procedure planning (e.g., Modality Worklist and MPPS) and also was extended to take more control over the complete image life cycle. The latter was and still is accomplished by specifying a full rendering pipeline for image visualization. The accompanied objects and services such as Hanging Protocols, Structured Display and different kinds of Presentation States are very valuable concepts for achieving this goal and hopefully will be widely found in products in the near future. The same applies to Structured Reporting which provides powerful means for storing and transmitting structured and coded information.

With the proposed Application Hosting concept, DICOM seriously opens up for Web Service communication which is already playing an important role in areas outside medical image communications. The DICOM standard is freely available; nevertheless, sometimes it is criticized for defining dedicated binary data encodings and network protocols instead of relying on existing technology, especially XML. However, when DICOM established more than 15 years ago and originating from ACR/NEMA standard from the 1980s, recent developments such as XML and Web Services have not been around at all or at least to that extent. The aforesaid DICOM-specific concepts have been exposed to be a powerful and reliable base for establishing sustainable interoperability between systems and many (partly open source) solutions are available greatly supporting DICOM-based implementations. Besides the expected benefits of Application Hosting already discussed in this chapter, the underlying Web Service communication may also facilitate access to the world of DICOM for people being more familiar with Web Service technology.

It can be expected that DICOM will keep *the* standard for medical image communication and related issues over the next years. The standard's permanently growing complexity may seem challenging to follow – however, understanding the basic concept should greatly help approaching specialized topics within the standard.

References

1. NEMA. Digital imaging and communications in medicine (DICOM). Rosslyn, VA, USA: National Electrical Manufacturers Association; 2008. PS 3.x–2008.
2. NEMA. Digital imaging and communications standard (ACR-NEMA 1.0). Washington DC, USA: National Electrical Manufacturers Association; 1985. 300-1985.
3. CEN/TC 251. Health informatics – Digital imaging – Communication, workflow and data management. Brussels, Belgium: European Committee for Standardization; 2004. EN 12052.
4. ISO/IEC. Health informatics – Digital imaging and communication in medicine (DICOM) including workflow and data management. Geneva, Switzerland: International Organization for Standardization; 2006. 12052:2006.
5. ICC. Specification ICC.1:2003-09, File Format for Color Profiles (Version 4.1.0). International Color Consortium; 2003.
6. ISO/IEC. Information technology – Open Systems Interconnection – Basic Reference Model – Part 1: The Basic Model. Geneva, Switzerland: International Organization for Standardization; 1994. 7498-1.
7. Eichelberg M, Riesmeier J, Kleber K, et al. Consistency of softcopy and hardcopy: preliminary experiences with the new DICOM extensions for image display. Procs SPIE. 2000;3980:57–67.
8. NEMA. DICOM Supplement 33: Grayscale Softcopy Presentation State Storage. National Electrical Manufacturers Association; 1999.
9. CIE. CIE Publication 15.2-1986: Colorimetry, Second Edition. Commission Internationale de l'Eclairage; 1986.
10. NEMA. Supplement 100: Color Softcopy Presentation State Storage SOP Classes. National Electrical Manufacturers Association; 2005.
11. CIE. Commission internationale de l'Eclairage proceedings 1931. Cambridge: Cambridge University Press; 1932.
12. NEMA. Supplement 140: XA/XRF Grayscale Softcopy Presentation State Storage SOP Class. National Electrical Manufacturers Association; 2009.
13. NEMA. Supplement 123: Structured Display. National Electrical Manufacturers Association; 2008.
14. NEMA. Supplement 60: Hanging Protocols. National Electrical Manufacturers Association; 2005.
15. NEMA. Supplement 15: Visible Light Image for Endoscopy, Microscopy, and Photography. National Electrical Manufacturers Association; 1999.
16. NEMA. Supplement 23: Structured Reporting Storage SOP Classes. National Electrical Manufacturers Association; 2000.
17. NEMA. Supplement 53: DICOM Content Mapping Resource (DCMR). National Electrical Manufacturers Association; 2001.
18. Riesmeier J, Eichelberg M, Wilkens T, et al. A unified approach for the adequate visualization of structured medical reports. Procs SPIE. 2006;6145:148–55.

19. Pearson J, Tarbox L, Paladini G, et al. Emerging radiological software standards and development technologies: impact on clinical translation and trials. In: Mulshine JL, Baer TM, editors. Quantitative Imaging Tools for Lung Cancer Drug Assessment. Hoboken, NJ, USA: Wiley; 2008. p. 67–94.
20. NEMA. DICOM supplement 118: Application Hosting, Public Comment Version – January 23, 2009. National Electrical Manufacturers Association; 2009.
21. W3C. XML information set (Second Edition). World Wide Web Consortium; 2004. http://www.w3.org/TR/xml-infoset/.
22. W3C. XML Path Language (XPath) 2.0. World Wide Web Consortium; 2007. http://www.w3.org/TR/xpath20/.
23. W3C. Web Services Description Language (WSDL) 1.1. World Wide Web Consortium; 2001. http://www.w3.org/TR/wsdl/.

18

PACS-Based Computer-Aided Detection and Diagnosis

H.K. (Bernie) Huang, Brent J. Liu, Anh HongTu Le, and Jorge Documet

Summary. The ultimate goal of Picture Archiving and Communication System (PACS)-based Computer-Aided Detection and Diagnosis (CAD) is to integrate CAD results into daily clinical practice so that it becomes a second reader to aid the radiologist's diagnosis. Integration of CAD and Hospital Information System (HIS), Radiology Information System (RIS) or PACS requires certain basic ingredients from Health Level 7 (HL7) standard for textual data, Digital Imaging and Communications in Medicine (DICOM) standard for images, and Integrating the Healthcare Enterprise (IHE) workflow profiles in order to comply with the Health Insurance Portability and Accountability Act (HIPAA) requirements to be a healthcare information system. Among the DICOM standards and IHE workflow profiles, DICOM Structured Reporting (DICOM-SR); and IHE Key Image Note (KIN), Simple Image and Numeric Report (SINR) and Post-processing Work Flow (PWF) are utilized in CAD-HIS/RIS/PACS integration. These topics with examples are presented in this chapter.

18.1 Introduction

Picture Archiving and Communication System (PACS) technology for healthcare enterprise delivery has become a part of the daily clinical imaging service and data management operations for most health care institutions. Alongside PACS, new technologies have emerged including Computer-Aided Diagnosis (CAD), which utilizes computer methods to obtain quantitative measurements from medical images and clinical information to assist clinicians to assess a patient's clinical state more objectively. However, CAD needs image input and related information from PACS to improve its accuracy; and PACS benefits from CAD results online and available at the PACS workstation as a second reader to assist physicians in the decision making process. Currently, these two technologies remain as two separate independent systems with only minimal system integration. This chapter addresses the challenges and solutions encountered by both technologies.

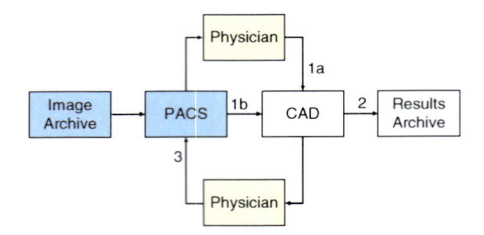

Fig. 18.1. *PACS and CAD not integrated.* The physician must manually transfer the image to the CAD workstation (1a), initiate the CAD processing, which is archived in the CAD system only (2), and transfer the results back to the PACS by means of natural language writing the report (3)

Figure 18.1 depicts the PACS environment (shaded boxes) and a CAD workstation or server location that is outside the realm of PACS. These two systems are usually disjoint. When an image is needed for CAD processing, the workflow is as follows:

1a. A technologist or radiologist transmits the original images from the PACS server or PACS workstation to CAD workstation for processing
1b. CAD processing of the exam is ordered through RIS, or directly from its creating modality
2. The results are stored within the CAD domain, since the CAD workstation or server is a closed system
3. A clinician needs to physically go to the CAD workstation to view results and transfer into the clinical report with natural language worded by the investigator writing the report

18.2 The Need for CAD-PACS Integration

In most CAD systems, the analyzed images need to reside on the local storage of the workstation running the applications. In the current best practice clinical workflow, medical images are stored in PACS. Therefore, the images must be queried for and retrieved by the workstation for a CAD system to process. The DICOM Query and Retrieve (Q/R) begins by sending a DICOM query command that contains query keys, such as patient name, medical record, modality, etc. to PACS and then waits for a response. Once the workstation receives the response, which contains a patient name or a list of patients satisfying the query keys, it then sends another DICOM command to PACS to retrieve the images back to the workstation. If the CAD application is not implemented with the Q/R functionality from PACS, one must manually load the images to the workstation for CAD process or manually push the images from PACS. After the images are loaded, the CAD performs two tasks:

- Automatic segmentation to detect the location of possible abnormalities in images
- Quantification to classify the detected regions or lesions

The main purpose for integrating CAD with PACS for clinical operations is to utilize CAD as a second reader for diagnosis of medical images [1,2]. In order to utilize CAD results more efficiently for this purpose, the CAD should be integrated within the daily clinical PACS environment. Currently, some PACS and CAD vendors have had some success integrating several CAD applications within a PACS environment, but the solution is either CAD-specific or in a closed PACS environment with proprietary software.

18.2.1 Approaches of CAD-PACS Integration

Computer-aided detection (CADe) is based on images, which must be received from the archive for analysis. This is usually done using a DICOM Query/Retrieve (Q/R) command. Conceptually, integration of CAD with DICOM PACS can have four approaches, which differ in the systems performing the query and retrieve commands. In the first three, the CAD is connected directly to the PACS, while the fourth approach is to use a CAD server to connect with the PACS:

- *PACS Workstation Retrieves and CAD Workstation Performs Detection*: In this approach, the PACS workstation queries and retrieves images from the PACS database while the CAD workstation performs the detection. Figure 18.2a illustrates the steps of the integration. This method involves the PACS server, the PACS workstation, and the CAD workstation. A DICOM C-store function must be installed in the CAD workstation.

    The major disadvantage to this approach is that the particular studies must be queried for by the PACS workstation and manually pushed to the CAD workstation for processing, which is a complex workflow. In addition, once the results are generated, they reside only on the CAD workstation.
- *CAD Workstation Retrieves and Performs Detection*: In this approach, the CAD workstation performs both, querying and retrieving of the image data from the archive, and thereafter the detection within the image data. This method only involves the PACS server and the CAD workstation. The function of the PACS server is almost identical to that of the last method. The only difference is that the last method uses the PACS workstation for querying and retrieving images, whereas in this method the CAD workstation performs this step. For this reason DICOM Q/R must be installed in the CAD workstation (Fig. 18.2b).

    Although for this approach, the CAD workstation can directly query and retrieve from the PACS to obtain the particular image study for processing, the workflow is still manual and a disadvantage. In addition, once the results are generated, they reside only on the CAD workstation.

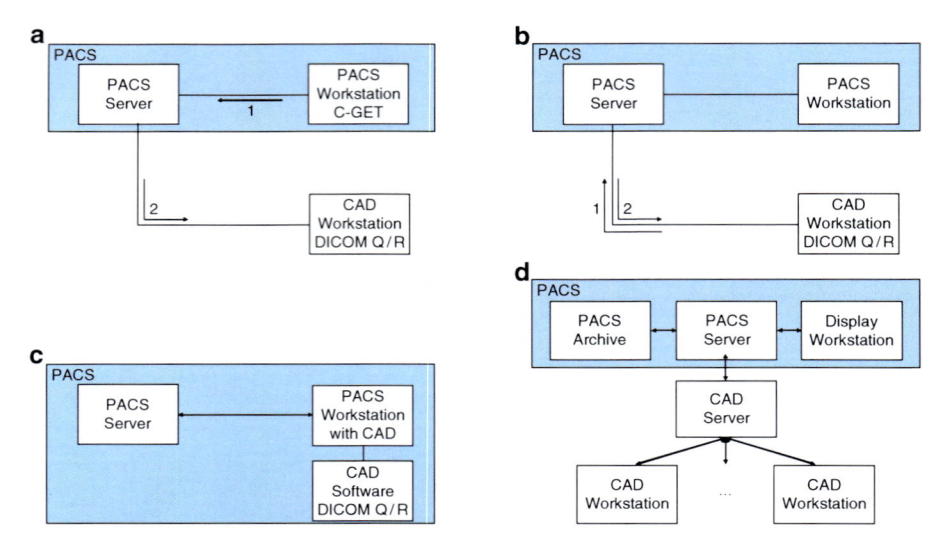

Fig. 18.2. *Conceptual methods of integrating CAD with PACS.* (**a**) PACS workstation queries/retrieves and the CAD workstation performs the detection (C-GET is a DICOM service); (**b**) the CAD workstation queries/retrieves and performs detection; (**c**) the PACS workstation has the CAD software integrated; and (**d**) a CAD server is integrated with the PACS

- *PACS Workstation with Integrated CAD Software*: A more advanced approach is to install the CAD software within the PACS workstation. This method eliminates all components in the CAD system and its connection to the PACS (Fig. 18.2c).

  Most of the CAD components can be eliminated which is an advantage. However, the major disadvantage is that because the CAD must be integrated directly with PACS, the CAD manufacturer must work very closely with the PACS manufacturer, or vice versa, to open up the software which rarely happens due to the competitive market.
- *Integration of CAD Server with PACS*: In this method, the CAD server is connected to the PACS server. The CAD server is used to perform CAD for PACS workstations (Fig. 18.2d).

  This is the most ideal and practical approach to a CAD-PACS integration. The CAD server can automatically manage the clinical workflow of image studies to be processed and can archive CAD results back to PACS for the clinicians to review directly on PACS workstations. This also eliminates the need for both the PACS manufacturer and the CAD manufacturer to open up their respective software platforms for integration.

18.2.2 CAD Software

CAD software [1] can be implemented within a stand-alone CAD workstation, a CAD server, or integrated in PACS as PACS-based CAD. Currently several PACS and CAD companies have successfully integrated their CAD applications within the PACS operation, but these applications are either in a CAD-specific workstation or in a closed PACS operation environment using proprietary software. For example in mammography, cf. Chapter 13, page 329, CAD has become an integral part of a routine clinical assessment of breast cancer in many hospitals and clinics across the United States and abroad. However, the value and effectiveness of CAD usefulness are compromised by the inconvenience of the stand-alone CAD workstation or server, certain DICOM standards and IHE workflow profiles are needed, which will be described in the next section.

18.3 DICOM Standard and IHE Workflow Profiles

In order to integrate CAD and HIS/RIS/PACS efficiently, certain basic ingredients are needed from Health Level Seven (HL7) standard[1] for textual data, DICOM standard for image communication [3], and Integrating the Healthcare Enterprises (IHE) profiles [2] in order to comply with the Health Insurance Portability and Accountability Act (HIPAA) requirements. These requirements include:

- Health Care Access
- Portability
- Renewability
- Preventing Health Care Fraud and Abuse
- Administrative Simplification
- Medical Liability Reform, containing five rules:
  – The Privacy Rule
  – The Transactions and Code Sets Rule
  – The Security Rule
  – The Unique Identifiers Rule
  – The Enforcement Rule
- The HITECH Act addressing privacy and security concerns associated with the electronic transmission of health information

Among the DICOM standard and IHE workflow profiles, DICOM Structured Reporting (DICOM-SR), and IHE Key Image Notes (KINs), IHE Simple Image and Numeric Reports (SINRs), and IHE Post-processing Work Flows (PWFs) are important components in CAD-HIS/RIS/PACS integration [4,5].

---

[1] http://www.hl7.org
[2] http://www.ihe.net

18.3.1 DICOM Structured Reporting

The scope of DICOM Structured Reporting (DICOM-SR) is the standardization of structured clinical reports in the imaging environment [6]. DICOM-SR documents record observations made for an imaging-based diagnostic or interventional procedure, particularly information that describes or references images, waveforms, or a specific Regions of Interest (ROI). DICOM-SR was introduced in 1994 and achieved major recognition when Supplement 23 was adopted into the DICOM standard in 1999 as the first DICOM-SR for clinical reports. The DICOM Committee has initiated more than 12 supplements to define specific DICOM-SR document templates, cf. Sect. 17.2.2, page 442. Among these supplements, two that relate to capturing CAD results have been ratified:

- The Mammography CAD SR (Supplement 50, 2000)
- The Chest CT CAD SR (Supplement 65, 2001)

In practice, the use of structured forms for reporting is known to be beneficial in reducing the ambiguity of natural language format reporting by enhancing the precision, clarity, and value of the clinical document.

DICOM-SR is generalized by using DICOM Information Object Definitions (IODs) and services for the storage and transmission of structured reports. Figure 18.3 provides a simplified version of the DICOM model of the real world showing where DICOM-SR objects reside. The most important part of an DICOM-SR object is the report document content, which is a DICOM-SR template that consists of different design patterns for various applications. Once the CAD results with images, graphs, overlays, annotations, and text have been translated into a DICOM-SR template designed for this application, the data in the specific template can be treated as a DICOM object stored in the worklist of the data model (Fig. 18.3, shaded boxes), and it can be displayed for review by a PACS workstation with the DICOM-SR display function. The viewing requires the original images from which the CAD results were generated so that the results can be overlaid onto the images.

Fig. 18.3. *Real world model of DICOM.* The DICOM-SR document is located in the DICOM data module (*shaded box*), which is at the same level as the DICOM image

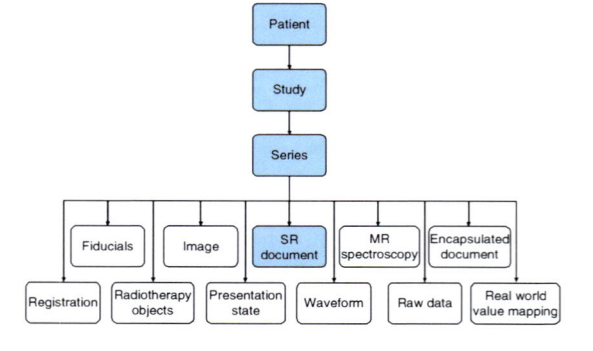

The DICOM-SR display function can link and download these images from the PACS archive and display them as well on the workstation.

18.3.2 IHE Profiles

IHE profiles provide a common language to discuss the integration needs of healthcare sites and integration capabilities of healthcare IT products. They organize and offer clear implementation paths for communication standards, such as DICOM, HL7, and World Wide Web Consortium (W3C), and security standards to meet specific clinical needs. The first large-scale demonstration (IHE connectathon) was held at the Radiological Society of North America (RSNA) annual meeting in 1999, and in subsequent meetings thereafter. In these demonstrations, manufacturers came together to show how their products could be integrated together according to IHE protocols.

There are three IHE profiles useful for CAD-PACS integration:

1. *KIN* allows users to flag images as significant (e.g., as reference, for surgery) and to add a note explaining the content
2. *SINR* specifies how diagnostic radiology reports (including images and numeric data) are created, exchanged, and used
3. *PWF* provides a worklist, its status and result tracking for post-acquisition tasks, such as CADe, Computer-Aided Diagnostics (CADx), or other image processing tasks

18.4 The CAD-PACS™ Toolkit

In the beginning of this chapter, we have discussed the current workflow of CAD in clinical use. To overcome the several bottlenecks, a CAD-PACS toolkit (Fig. 18.4, elliptic box), which can integrate with the PACS server and/or workstation with the CAD server and/or workstation via the DICOM standard and IHE profiles, passes the CAD results to the PACS server for archiving and the PACS workstation for viewing; and query/retrieves original images from PACS server to PACS workstation to be overlaid with the CAD results. In addition, it can automatically pass images directly from the PACS server or PACS workstation to the CAD workstation for processing.

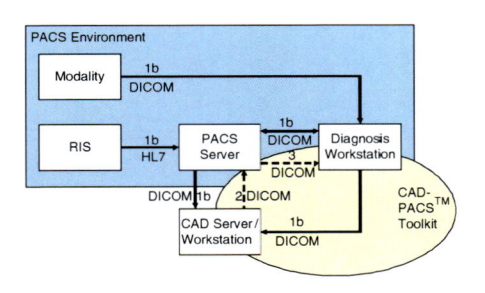

Fig. 18.4. *PACS and CAD integrated.* The CAD workflow (*dotted lines*) is integrated in the PACS environment (*shaded box*) using the CAD-PACS™ Toolkit (*elliptic box*)

18.4.1 Concept

The CAD-PACS^{TM} Toolkit is a software toolkit using the HL7 standard for textual information; the DICOM standard for various types of data formats, including images, waveforms, graphics, overlays, and annotations; and IHE workflow profiles described in the aforementioned section for the integration of CAD results within the PACS workflow [5]. This CAD software toolkit is modularized and its components can be installed in five different configurations:

1. A stand-alone CAD workstation
2. A CAD server
3. A PACS workstation
4. A PACS server or
5. A mix of the previous four configurations

In general, a CAD manufacturer would be more comfortable with the first two approaches because there is very little collaboration needed for the PACS software, which is too complex for most CAD manufacturers. On the other hand, a PACS manufacturer would prefer to use an in-house CAD or acquire the CAD from a third party and integrate it with its own PACS using the latter three approaches.

18.4.2 Structure, Components, and Editions

The CAD-PACS^{TM} Toolkit has five software modules:

1. i-CAD-SC^{TM} creates the screen shot for any CAD application, converts it to a DICOM object and sends it to PACS for storage
2. i-CAD^{TM} resides in the CAD workstation and provides key functions for CAD-PACS integration, including DICOM-SR object creation and archival, query and retrieval of images for CAD processing, and communication with the i-PPM^{TM} module
3. i-PPM^{TM} residing in the PACS server provides functions to schedule and track status of CAD-PACS workflow. This module is also used as a supplement for those PACS manufacturers which do not support post-processing management in order to be DICOM and IHE-compliant for CAD-PACS integration
4. Receive-SR^{TM} resides in the PACS server and performs the functions of archiving, query and retrieving DICOM-SR objects from the PACS server
5. Display-SR^{TM} resides in the PACS workstation. This module is used when PACS does not support DICOM-SR C-Store Service Class User (SCU) and C-Find. It is built as a display Web server with DICOM-SR C-Store and C-Find features

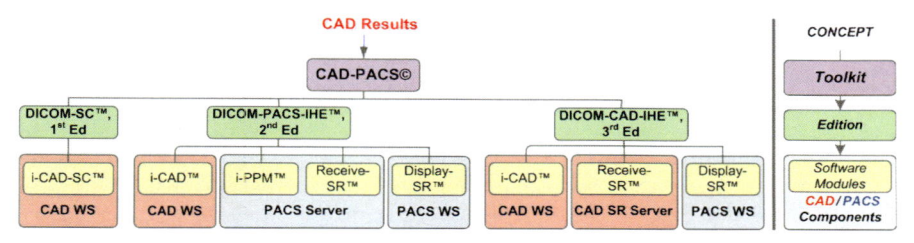

Fig. 18.5. *Architecture of the CAD-PACS^{TM} Toolkit [5].* The five modules are combined to three different editions (*left*). The concept of four levels of integration with the CAD-PACS^{TM} Toolkit is shown on the right

Furthermore, the CAD-PACS^{TM} Toolkit has three editions for the different levels of PACS integration requirements. Each edition contains some or all of the software modules (Fig. 18.5) [5]:

- *DICOM-SC^{TM}*: The first edition converts a simple screen capture output, and the CAD data are not stored for future use.
- *DICOM-PACS-IHE^{TM}*: The second edition is for full CAD-PACS integration requiring elaborate collaboration between the CAD developer and the PACS manufacturer.
- *DICOM-CAD-IHE^{TM}*: The third edition does not require the elaborate integration efforts of the two parties, and proper use of the CAD-PACS toolkit is sufficient, which favors the independent CAD developer.

18.5 Example of CAD-PACS Integration

In this section, we provide a step-by-step procedure to integrate a CAD with PACS using the Bone Age Assessment (BAA) of children on a hand and wrist radiograph as an example.

The classical method of BAA is a clinical procedure in pediatric radiology to evaluate the stage of skeletal maturity based on a left hand and wrist radiograph through bone growth observations. The determination of skeletal maturity ("bone age") plays an important role in diagnostic and therapeutic investigations of endocrinological abnormality and growth disorders of children. In clinical practice, the most commonly used BAA method is atlas matching by a left hand and wrist radiograph against the Greulich and Pyle atlas, which contains a reference set of normal standard hand images collected in 1950s with subjects exclusively from middle and upper class Caucasian populations. The atlas has been used for BAA around the world for more than 50 years [7].

18.5.1 The Digital Hand Atlas

Over the past 30 years, many studies have raised questions regarding the appropriateness of using the Greulich and Pyle atlas for BAA of contemporary children [8]. However, these studies did not provide a large-scale and systematic method for validation. A digital hand atlas with normal children collected in the United States along with a CAD-BAA method has been developed during the past 10 years in our laboratory as a means to verify the accuracy of using the Greulich and Pyle atlas to assess today's children bone age [9].

The digital hand atlas consists of eight categories, where each category contains 19 age groups, one group for subjects younger than 1 year, and 18 groups at 1-year intervals for subjects aged 1–18 years. The case distribution within each of these 18 groups is as even as possible during the case collection of gender and ethnicities (Table 18.1).

The total is 1,390 cases. For each case, at least two pediatric radiologists had verified the normality and chronological age, and assessed the bone age of the child based on the Greulich and Pyle atlas matching method [10].

18.5.2 CAD Evaluation in a Laboratory Setting

After the CAD was completed, the system needed to be integrated with the PACS. The integration is then evaluated first in a laboratory setting, followed by the clinical environment. After image acquisition using Computed Radiography (CR), Digital Radiography (DR), or film scanner, the image is archived in the PACS server. The laboratory set up then mimics the clinical workflow as shown with four steps (Fig. 18.6):

0. The PACS workstation query/retrieves the hand image from the PACS archive and displays it on the monitor
1b. The modality/PACS server also sends a second copy of the hand image to the CAD server which generates CAD results

Table 18.1. *The digital hand atlas.* A breakdown of cases according to gender and ethnics

Ethnics/Gender	Female	Male	Total
Asian	167	167	334
African-American	174	184	358
Caucasian	166	167	333
Hispanic	183	182	365
Sum	690	700	1,390

Fig. 18.6. *Laboratory setting for BAA evaluation [11].* The BAA-CAD system in the laboratory environment using a PACS simulator is composed of four steps

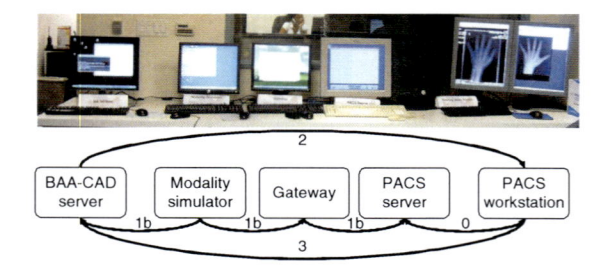

2. The CAD server sends CAD results to the PACS workstation. The radiologist reviews both the image and CAD result on the PACS workstation
3. The diagnosis from the radiologist assisted by CAD results is sent back to the CAD server for storage

18.5.3 CAD Evaluation in a Clinical Environment

After laboratory validation, the BAA-CAD system and the PACS workstation were installed in a clinical environment for further evaluation. In this example, the clinical environment is located at the Radiology Department of Los Angeles County Hospital (LAC) and University of Southern California (USC), where the CAD workstation can access the PACS and CR images. The clinical workflow is similar to the laboratory workflow (Fig. 18.7):

1. The CR modality sends a copy of the hand image to the Web-based CAD server located in the radiology reading room. The PACS workstation also receives a copy of the image from the PACS server
2. The CAD program at the CAD server receives the image, performs BAA and records the results in the CAD server database
3. The CAD server searches the PACS workstation to locate the original image and links up with the CAD result, as well as the best-matched image from the digital hand atlas in the CAD database
4. The Graphical User Interface (GUI) in the PACS workstation displays the original image and the best-matched image (Fig. 18.6, the right most image set on the duel monitors), and assists the radiologist to take advantage of the CAD results to make the final diagnosis

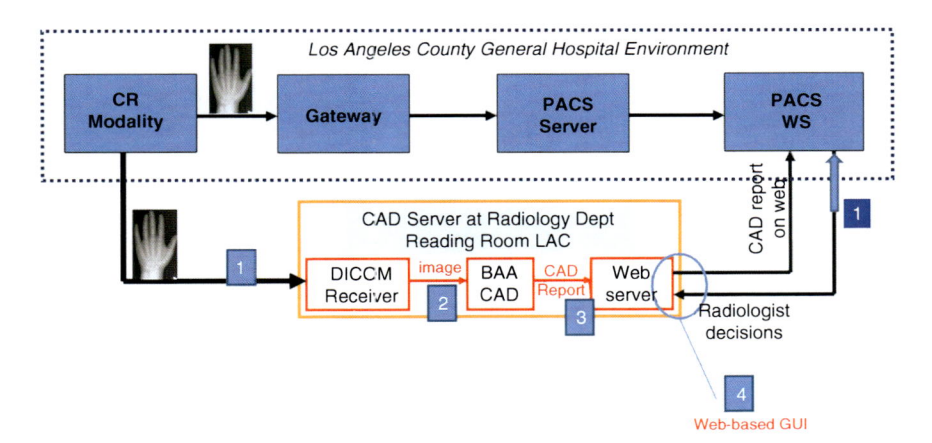

Fig. 18.7. *Clinical BAA evaluation setup.* The diagram depicts the BAA-CAD system in clinical environment and the according workflow implemented in the LAC Hospital with the clinical PACS and the CAD server

18.5.4 CAD-PACS Integration Using DICOM-SR

In Sect. 18.3.1, we presented the concept of DICOM-SR and the need for converting a text file CAD report to DICOM-SR format in order to overlay the contents within the DICOM-SR onto the original image and to display it on the PACS workstation. Referencing to two ratified CAD DICOM-SR templates for mammography and chest CT in the DICOM standard Supplement 50 and 65, a DICOM-SR object for the BAA-CAD based on a tree structure was designed and implemented.

Figure 18.8 illustrates the DICOM-SR template for the BAA-CAD. This design, which utilizes the DICOM standard Supplement 23, has a Document Root BAA-CAD which branches into four parent nodes: Detection Performed (DP), Analysis Performed (AP), Findings summary and Image library. Each DP and AP parent nodes can have one or more children nodes. In this case, DP describes one imaging processing algorithm, which is the BAA algorithm. The AP parent node has two children nodes; each describes methods of quantitative analysis that were performed on the hand image. Each analysis performed can be further branched out to one or multiple grandchild nodes called Single Image Findings (SIF). As shown in Fig. 18.8, each AP children node only has one SIF. The findings summary parent node is the most important part of an SR which includes the BAA-CAD results. The Image Library parent node is optional; however, in the BAA-CAD SR, it is used to reference the images from the digital hand atlas. The data structure format for each child can be obtained directly from the DICOM standard, Supplement 23.

Figure 18.9 depicts the first page of the BAA-CAD report in DICOM-SR format of Patient 1. To the right is the plot of the bone age (vertical axis) against the chronological age (horizontal axis) of Patient 1 (red dot) within the ± two standard deviations of the normal cases in the digital hand atlas.

Figure 18.10 shows an image page of the DICOM-SR report of Patient 2 including the original image (left) from which the BAA-CAD result was obtained, the Greulich and Pyle atlas best-matched image (middle), and the digital hand atlas best-matched image (right). The chronological age and the CAD-assessed age of Patient 2, and the chronological age of the best-matched

Fig. 18.8. *Nested DICOM-SR templates for BAA-CAD.* The template is designed based on the types of output radiologists are required to review

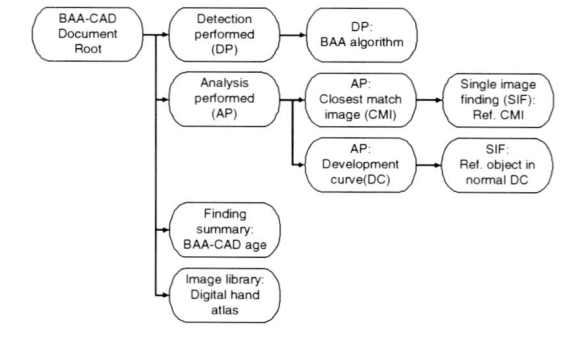

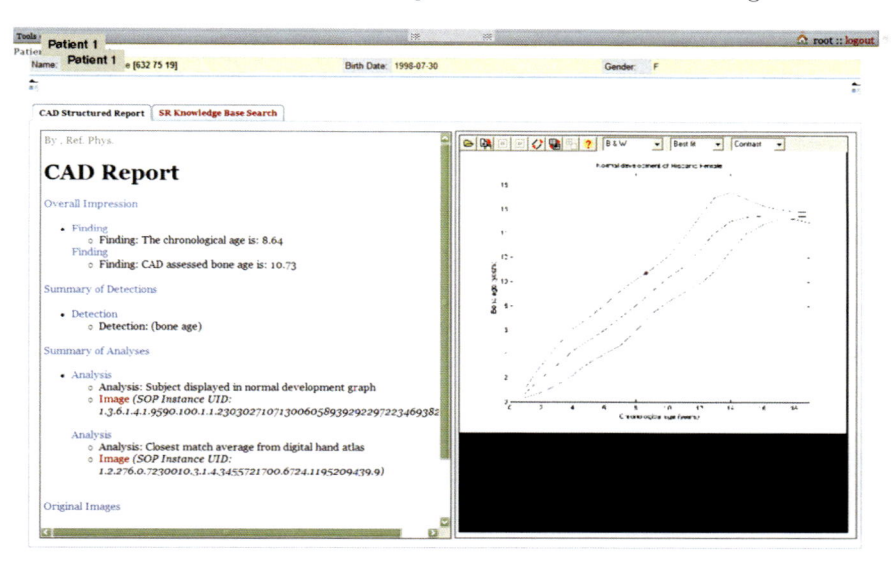

Fig. 18.9. *Integrating BAA-CAD with DICOM-SR. Left*: The CAD report in DICOM-SR format is based on the design of the DICOM-SR template as shown in Fig. 18.8; *Right*: The plot of the CAD-BAA results of a patient (*red dot*) compared with the normals and ± two standard deviations in the digital hand atlas is a component in the DICOM-SR template

image in the digital hand atlas are enclosed inside the green ellipse in the upper right corner. The plot of the CAD-assessed bone age of the patient within the ± two standard deviations of the normal cases in the digital hand atlas is shown in the upper right corner. The best-matched digital hand atlas image is obtained by using the CAD age of Patient 2 to search the digital hand atlas in the order of race, sex and age. The image with the closest chronological age (the best matched age) is the matched image in the digital hand atlas. The chronological age, BAA bone age, and the matched digital hand atlas age are shown at the upper right of the screen within the green ellipse.

18.6 Conclusion

In order for CAD to be useful to aid diagnosis and/or detection, it has to be integrated into the existing clinical workflow. In the case of image-based CAD, the integration is with the PACS daily workflow. We have presented the rationale and methods of CAD-PACS integration with emphasis in PACS workflow profiles using the DICOM standard and IHE workflow profiles.

In the PACS-based workflow approach, the CAD results do not reside in the PACS server and storage; instead they are in the CAD server. PACS

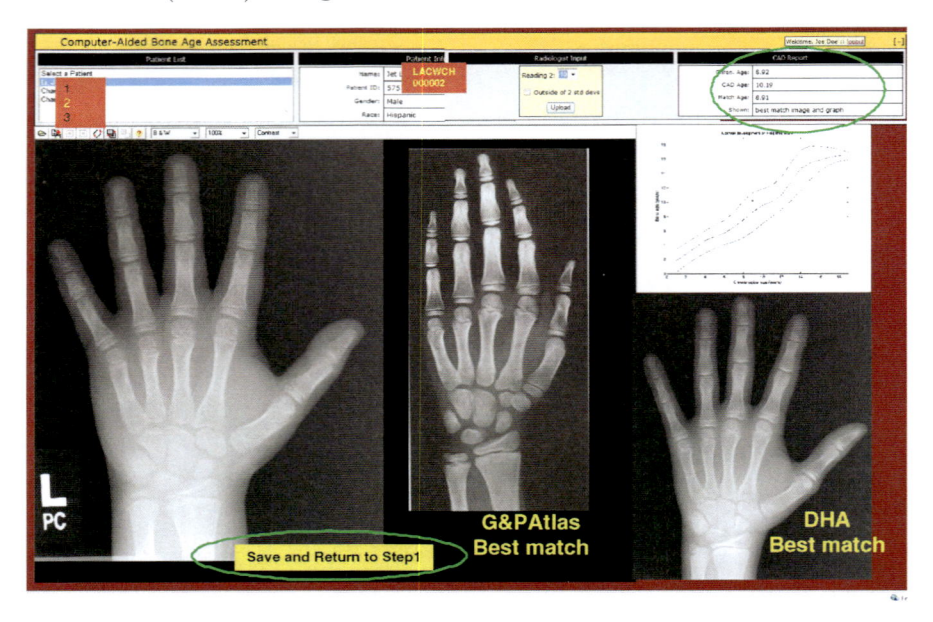

Fig. 18.10. *BAA-CAD GUI on the PACS workstation. Left*: Original image; *Center*: Best matched Greulich and Pyle atlas image; *Right*: Best matched digital hand atlas image. The CAD-assessed bone age of the patient compared to the children in the normal range in the DHA is shown in the plot

images used by the CAD are linked with the CAD results so that both images and CAD results in DICOM format can be displayed on the PACS workstation. We use an example in BAA on hand and wrist joint radiographs as an introduction to the advantage of CAD and PACS integration for daily clinical practice. In general, physicians can assess the bone age of a child using the Greulich and Pyle method, but the question is whether the classic method is still valid for assessing the bone age of children of today. With the integration of BAA-CAD directly into the PACS workflow, the radiologist has the CAD results as the second opinion to assist his/her BAA and to confirm the diagnosis.

In conclusion, the integration of CAD to PACS clinical workflow has many distinct advantages:

- PACS technology is mature. Integrating CAD with the PACS can take advantage of the powerful computers and high-speed networks utilized in PACS to enhance the computational and communication power of the CAD
- The DICOM-SR and IHE workflow profiles can be readily applied to facilitate the integration of CAD results to PACS workstations

- PACS-based query/retrieve tools can facilitate the CAD user to obtain images and related patient data more directly from PACS for CAD algorithm enhancement and execution
- CAD-PACS integration results can be directly viewed and utilized at the PACS workstation together with relevant PACS data
- The very large, dynamic, and up-to-date PACS databases can be utilized by CAD to improve its diagnostic accuracy

To utilize the DICOM-SR content more efficiently, the current trends for CAD and PACS integration is to promote the development of DICOM-compliant databases and services which combine CAD findings and DICOM key image references [5]. This incorporation allows content-based query/retrieval of DICOM imaging studies based on DICOM-SR with its quantitative findings rather than header information of DICOM objects and/or disease category. The benefits of querying/retrieving content-based imaging data could have a large impact on medical imaging research and clinical practice. However, there are many challenges in the development of data mining methodology for CAD including the following:

- Collaboration with PACS vendors at multiple medical centers to open access to both PACS and CAD data
- Acknowledgment of Institutional Review Board (IRB) and personal health information requirements for using human subjects for research with information within the PACS
- Adoption and utilization of DICOM-SR templates in all PACS vendors

References

1. Doi K. Computer-aided diagnosis in medical imaging: historical review, current status and future potential. Comput Med Imag Graph. 2007;31(4–5):198–211.
2. Huang HK, Doi K. CAD and image-guided decision support. Comput Med Imag Graph. 2007;31(4):195–7.
3. NEMA. Supplement 23: Structured Reporting Storage SOP Classes. Rosslyn, VA, USA: National Electrical Manufactures Association; 2000.
4. Zhou Z, Liu BJ, Le A. CAD-PACS integration tool kit-based on DICOM screen capture (SC) and structured reporting (SR) and IHE workflow profiles. J Comput Med Imag Graph. 2007;31(4):346–52.
5. Le AHT, Huang HK, Liu B. Integration of computer-aided diagnosis/detection (CAD) results in a PACS environment using CAD-PACS toolkit and DICOM SR. Int J Comp Asst Rad Surg. 2009;4(4):317–29.
6. DICOM Standards Commitee. http://medicalnemaorg/medical/dicom/2008. 2008.
7. Greulich WW, Pyle SI. Radiographic Atlas of Skeletal Development of Hand Wrist. 2nd ed. Stanford, CA: Stanford University Press; 1959.
8. Huang HK. PACS and Imaging Informatics: Basic Principles and Applications. 2nd ed. Hoboken, NJ: Wiley and Blackwell; 2010.

9. Gertych A, Zhang A, Sayre J, et al. Bone age assessment of children using a digital hand atlas. Comput Med Imag Graph. 2007;31(4):322–31.
10. Zhang A, Sayre JW, Vachon L, et al. Cross-racial differences in growth patterns of children based on bone age assessment. J Radiology. 2009;250(1):228–35.
11. Zhou Z, Law M, Huang HK, et al. An educational RIS/PACS simulator. InfoRAD exhibit. Radiol Soc N. 2002;753.

19
Content-Based Medical Image Retrieval

Henning Müller and Thomas M. Deserno

Summary. This chapter details the necessity for alternative access concepts to the currently mainly text-based methods in medical information retrieval. This need is partly due to the large amount of visual data produced, the increasing variety of medical imaging data and changing user patterns. The stored visual data contain large amounts of unused information that, if well exploited, can help diagnosis, teaching and research. The chapter briefly reviews the history of image retrieval and its general methods before technologies that have been developed in the medical domain are focussed. We also discuss evaluation of medical content-based image retrieval (CBIR) systems and conclude with pointing out their strengths, gaps, and further developments. As examples, the MedGIFT project and the Image Retrieval in Medical Applications (IRMA) framework are presented.

19.1 Introduction

Content-Based Visual Information Retrieval (CBVIR) or Content-Based Image Retrieval (CBIR) has been one on the most vivid research areas in the field of computer vision over the past almost 20 years. The availability of large and steadily growing amounts of visual and multimedia data, and the development of the Internet underline the need to create access methods that offer more than simple text-based queries or requests based on matching exact database fields. Many programs and tools have been developed to formulate and execute queries based on the visual or audio content and to help browsing large multimedia repositories. Still, no general breakthrough has been achieved with respect to large varied databases with documents of differing sorts and with varying characteristics. Answers to many questions with respect to speed, semantic descriptors or objective image interpretations are still open and wait for future systems to fill the void [1].

In the medical field, images, and especially digital images, are produced in ever-increasing quantities and used for diagnosis and therapy. The Radiology Department of the University Hospitals of Geneva alone produced more than 114,000 images a day in 2009, risen form 12,000 in 2002. Large hospital

T.M. Deserno (ed.), *Biomedical Image Processing*, Biological and Medical Physics,
Biomedical Engineering, DOI: 10.1007/978-3-642-15816-2_19,
© Springer-Verlag Berlin Heidelberg 2011

groups such as Kaiser Permanente that manage several hospitals had by early 2009 even 700 TB of data stored in the institutional archives and very large hospitals such as the University Hospital of Vienna currently produces over 100 GB of image data per day.

With Digital Imaging and Communications in Medicine (DICOM), a standard for image communication has been set and patient information can be stored with the actual image(s), although still a few problems prevail with respect to the standardization. In several articles, content-based access to medical images for supporting clinical decision-making has been proposed [1, 2]. Still, only very few systems are usable and used in clinical practice as most often development takes place in computer science departments, which are totally disconnected from clinical practice.

19.1.1 Motivation and History

Image retrieval has been an extremely active research with first review articles on access methods in image databases appearing already in the early 1980s [3]. The following review articles explain the state-of-the-art and contain references to a large number of systems and descriptions of the technologies implemented [4–7]. The most complete overview of technologies to date is given by Smeulders et al. in [8]. This article faces common problems such as the *semantic gap* or the *sensory gap* and gives links to a large number of articles describing the various techniques used in the domain. In a more recent review, the developments over the past 5–10 years are described [9].

Although early systems existed already in the beginning of the 1980s [10], the majority would recall systems such as IBM's Query by Image Content (QBIC)[1] as the start of CBIR [11].

Most of the available systems are, however from academia. It would be hard to name or compare them all but some well-known examples include Photobook [12] and Netra [13] that all use simple color and texture characteristics to describe the image content. Using higher level information, such as segmented parts of the image for queries, was introduced by the Blobworld[2] system [14, 15]. PicHunter [16] on the other hand is an image browser that helps the user to find a certain image in the database by showing to the user images on screen that maximize the information gain in each feedback step. A system that is available free of charge is the GNU Image Finding Tool (GIFT)[3] [17].

19.1.2 Query-by-Example(s) Paradigm

One of the biggest problems in CBIR is the formulation of queries without text. Everyone is used to formulate queries with text (as 90 % of Internet

---

[1] http://wwwqbic.almaden.ibm.com/
[2] http://elib.cs.berkeley.edu/photos/blobworld/
[3] http://www.gnu.org/software/gift/

users are using Google) and explain one's information needs but with visual elements this is far from trivial. Drawing small designs is one possibility requiring artistic skills and being unsuitable for the majority of users. Formulating a Query by Image Example (QBE) is currently the most common way to search for similar images. This is used by most image retrieval systems. Thus, a system can search for visually similar images with one or several example image(s). The problem remaining is to find a suitable example image, which is not always obvious ("page zero problem") [18].

In the medical domain images are usually one of the first examinations performed on patients, and thus query examples are available. Once the user has received a results set of images or cases similar to a given example image or case, systems most often offer the possibility to mark images/cases as relevant and irrelevant and thus refine the search through what is called "relevance feedback"[19].

19.2 General Image Retrieval

General image retrieval started with the main concepts already in 1980 [3]. Still, the real research did not start before the late 1980s, when several systems using simple visual features became available [11].

19.2.1 Classification vs. Retrieval

One of the first and basic questions in image retrieval is whether it is rather an information retrieval task or a classification task. While there are many similarities between them, there are two principle differences [20]:

- *Classification* tasks have a limited number of classes of topics/items and training data for each of the classes that allow training of class-specific parameters.
- *Retrieval* tasks have no fixed classes of items/objects in the database and usually no training data available; documents can be relevant for a particular retrieval task or information need, with relevance being potentially user-dependent.

In general, the techniques according to the classification paradigm follow the general machine learning literature and its approaches, whereas the (information) retrieval approaches follow techniques from general information retrieval.

In the end, when used for CBIR, both represent images by visual features and then find similar images using a distance measure, showing the most similar images to the user ordered by their visual similarity.

Fig. 19.1. *Retrieval system architecture.* Overview of the main components that most image retrieval systems are constituted of

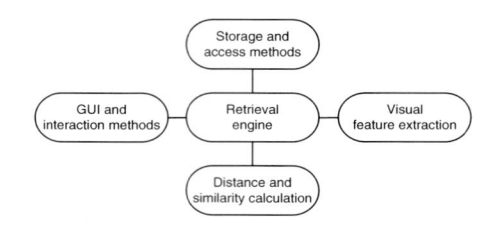

19.2.2 System Components and Computation

Most of these systems have a very similar architecture for browsing and archiving/indexing images comprising tools for the extraction of visual features, for storage and efficient retrieval of these features, for distance measures or similarity calculation, and a type of Graphical User Interface (GUI). This general system setup is shown in Fig. 19.1.

Computational efficiency is another often regarded question. Particularly the visual analysis can take an enormous time for large databases and as the challenge is to scale to millions of images, tools such as grid networks and parallel processing have been used for feature processing. This is mainly done for the off-line step of representing images by features, whereas for the query processing efficient indexing structures are used for quick response times $t_r < 1\,\text{s}$.

19.2.3 Features and Signatures

Visual features were classified into *primitive* features such as color or shape, *logical* features such as identity of objects shown and *abstract* features such as significance of depicted scenes [6]. However, basically all currently available systems only use primitive features such as:

- *Color*: In stock photography (large, varied databases for being used by artists, advertisers and journalists), color has been the most effective feature. The Red, Green, Glue (RGB) color space is only rarely used as it does not correspond well to the human color perception. Other spaces such as Hue, Saturation, Value (HSV) or the Commission Internationale de L'Eclairage (CIE) Lab and Luv spaces perform better because distances in the color space are similar to the differences between colors that humans perceive. Much effort has also been spent on creating color spaces that are optimal with respect to lighting conditions or that are invariant to shades and other influences such as viewing position [21].
- *Texture*: Texture measures try to capture the characteristics of the image with respect to changes in certain directions and the scale of the changes. This is most useful for images with homogeneous texture. Some of the most common measures for capturing the texture of images are wavelets and Gabor filters. Invariances with respect to rotation, shift or scale can

be included into the feature space but information on the texture may get lost in this process [22]. Other popular texture descriptors contain features derived from co-occurrence matrices [23, 24], the Fourier transform [22], and the so-called *Wold* features [25].
- *Local color and texture*: Both, color and texture features can be used also on a local or regional level, i.e. on parts of the image. To use blocks of fixed size, so-called *partitioning*, is the easiest way employing regional features [26]. These blocks do not take into account any semantics of the image itself. When allowing the user to choose a Region of Interest (ROI) [27], or when segmenting the image into areas with similar properties [28], local features capture more information about relevant image structures.
- *Shape*: Fully automated segmentation of images into objects itself is an unsolved problem. Even in fairly specialized domains, automated segmentation causes many problems. In image retrieval, several systems attempt to perform an automatic segmentation for feature extraction [29]. The segmentation process should be based on color *and* texture properties of the image regions [28]. The segments can then be described by shape features, usually being invariant to shift, rotation and scaling [30]. Medical image segmentation with respect to browsing image repositories is frequently addressed in the literature as well [31].
- *Salient points*: In recent years' salient point-based features have had best performances in most of the image retrieval and object classification tasks [32]. The idea is to find representative points (or points that attract the attention) in the images and then analyze the relationships of the points. This permits to extract features that possess several invariants such as invariance to shifts, rotations, scale and even view-point. A large number of such techniques exist for detecting the points and then for extracting features from the salient points.
- *Patches and visual words*: Patches and visual words are closely linked to salient point-based features. As the patches and/or visual words are most often extracted from regions in the images that were identified to contain changes or high gradients and then local features are extracted in these regions. It is also possible to put a regular grid on the image and then extract patches around the points of the grid to well represent the entire image. The term visual words stems from the fact that the features extracted around the selected points are often clustered into a limited number of homogeneous characteristics that can have distributions similar to the distribution of words in text allowing to use techniques well known from text retrieval [33].

All of these features have their benefits and domains where they operate superiorly, but all these features are low-level visual features and might not correspond to semantic categories. For this reason, text, whenever available should be used for the retrieval of images as well, as semantic information is conveyed very easily. All benchmarks show that text has superior performance compared to visual characteristics, but can be complemented efficiently by visual retrieval.

19.2.4 Distance and Similarity Measures

Basically all systems use the assumption of equivalence of an image and its representation in feature space. These systems often use measurements systems such as the easily understandable Euclidean vector space model [11] for measuring distances between a query image (represented by its features) and possible results representing all images as feature vectors in an n-dimensional vector space. This is done although metrics have been shown to not correspond well to human visual perception [34]. Several other distance measures do exist for the vector space model such as the city-block distance, the Mahalanobis distance [11] or a simple histogram intersection [35]. Still, the use of high-dimensional feature spaces has shown to cause problems and great care needs to be taken with the choice of distance measurement to retrieve meaningful results [36,37]. These problems with a similarity definition in high-dimensional feature spaces is also known as the "curse of dimensionality" and has also been discussed in the domain of medical imaging [38].

Another approach is a *probabilistic framework* to measure the probability that an image is relevant [39]. Another probabilistic approach is the use of a Support Vector Machine (SVM) [40,41] for grouping of images into classes of relevant and non-relevant items. In most visual classification tasks, SVMs reach best performance in general.

Various systems use methods that are well known from the field of text retrieval and apply them to visual features, where the visual features have to correspond roughly to words in text [26,42]. This is based on the two principles:

- A feature frequent in an image describes this image well.
- A feature frequent in the collection is a weak indicator to distinguish images from each other.

Several weighting schemes for text retrieval that have also been used in image retrieval are described in [43]. A general overview of pattern recognition methods and various comparison techniques are given in [44].

19.3 Medical Image Retrieval

The number of digitally produced medical images has rising strongly, mainly due to large tomographic series. Videos and images produced in cardiology are equally multiplying and endoscopic videos promise to be another very large data source that are planned to be integrated into many Picture Archiving and Communication Systems (PACS). The management and the access to these large image repositories become increasingly complex. Most accesses to this

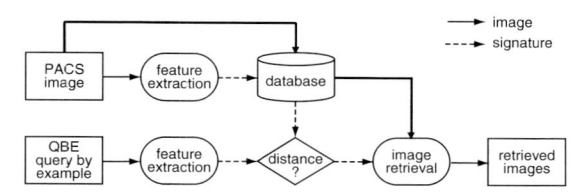

Fig. 19.2. *Medical CBIR system architecture.* All images in the PACS archive and the QBE image are described by a signature. Comparing signatures instead of images allows fast CBIR response

data is based on the patient identification or study characteristics (modality, study description) [45].

Imaging systems and image archives have often been described as an important economic and clinical factor in the hospital environment [46, 47]. More than ten years ago, several methods from the fields of computer vision and image processing have already been proposed for the use in medicine more than 10 years ago [48]. Several radiological teaching files exist [49] and radiology reports have also been proposed in a multimedia form [50].

19.3.1 Application Fields

Content-based retrieval has also been proposed several times from the medical community for the inclusion into various applications [2,51], often without any implementation. Figure 19.2 shows the general system architecture.

Almost all sorts of images have already been used for image retrieval at one point or another. The first separation is on whether systems use a large and varied set of images [52] or work on a very focused domain as diagnosis aid [53].

Typical application domains for CBIR-based image management is case-based reasoning and evidence-based medicine, in particular in fields where diagnosis is regarded as hard and where purely visual properties play an important role, such as mammography [54] or the diagnosis of interstitial lung diseases [55, 56]. CBIR-based eLearning has also been discussed [57].

19.3.2 Types of Images

The medical domain yields an extremely large amount of varying images, and only very few have so far been exploited fully for visual similarity retrieval. When thinking of medical images, clearly radiographs and maybe Computed Tomography (CT) come instantly to mind but there is much more than this usually gray scale set of images.

Here is a list of some of the types of visual data that is available in hospitals and often stored in the PACS:

- 1D signals: EEG, ECG
- 2D gray scale images: X-ray radiography

- 2D color images: microscopy, photography, dermatology
- Gray scale video: ultra-sonography
- Color video: monitoring in a sleep laboratory
- Pseudo-3D (slices): CT, MRI, PET, SPECT
- 3D models: reconstructions of tomographic images
- 4D data: temporal series of tomographic images such as CT images of a beating heart
- nD data: Multi-modal images obtained from combined PET/CT or PET/MRI scanners

It is obvious that medical imaging is much more varied then the images of the general CBIR domains, such as photographs in the Internet.

19.3.3 Image Preprocessing

Image pretreatment is most often used to harmonize the content in a database and thus make feature extraction from the images based on the same grounds. Such preprocessing can be the normalization of gray levels or colors in images.

Another application of pretreatment in the medical domain is the background removal from images and automatic detection of the field of view [58] to concentrate the search on the important objects. Although medical images are taken under relatively controlled conditions, there is a fairly large variety remaining particularly in collections of scanned images.

Some typical images from our database are shown in Fig. 19.3 (top row). The removal is mainly done through a removal of specific structures followed by a low pass filter (median) and then by thresholding and a removal of small unconnected objects. After the object extraction phase, most of the background is removed but only a few images had part of the main object removed (Fig. 19.3, bottom row).

19.3.4 Visual and Non-Visual Image Features

Most of the visual features used in medical images are based on those existing for non-medical imaging as well [59]. For radiographs, there is clearly a need to highlight gray level instead of the color values in non medical image, which can make the search harder. On the other hand, most of the medical images are taken under fairly standardized conditions, requiring fewer invariances and allowing direct comparisons of downscaled versions of the images.

In contrast to non-medical image archives, all medical images do have meta data attached to them as the images are part of a clinical record, that consists of large amounts of structured data and of free text such as laboratory results, anamnesis and release letter. Without this meta information, interpretation of medical cases is impossible. No radiologist would read an image without a minimum of meta data on the patient (e.g., age, sex) and a basic anamnesis as many of the clinical parameters do have a strong influence on the visual

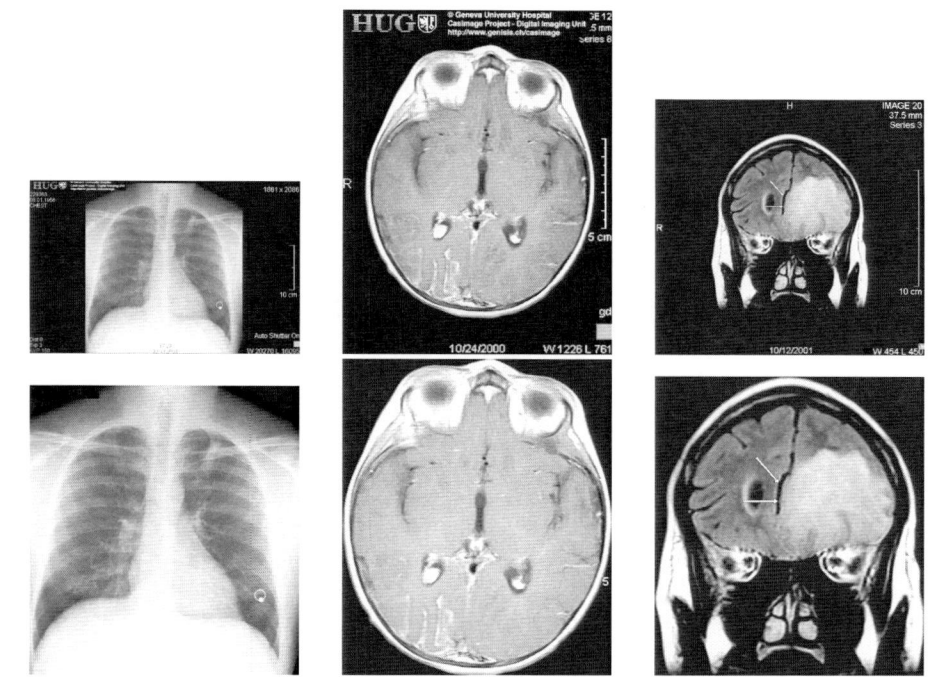

Fig. 19.3. *Image pretreatment.* Images before (*top row*) and after (*bottom row*) the removal of logos and text

characteristics of the images. For instance, old patients have less dense bones, and the lung of a smoker differs significantly from that of a non-smoker.

One of the largest problems is how to combine structured/free text data with visual features. Several fusion approaches have been proposed in [56]. Most often, late fusion is considered best as there are potentially many features and there can be negative interactions between certain of the clinical data and certain visual features. It is also clear that the data quality in patient records is often far from optimal. In an anamnesis for instance, parameters are asked unsystematically, leaving often incompleteness, e.g., if the patient was a smoker or not. Incompleteness of data must be handled appropriately for classification or retrieval [59].

19.3.5 Database Architectures

Many tools and techniques have been used for quick access to large collections of images, similar to access models in general database architectures. Frequently, the goal is to accept a rather long off-line phase of data pretreatment followed by a rather short time of query response. Techniques from text retrieval have proven fast response in sparsely populated spaces and are frequently applied.

Parallel access to databases and grid networks are also used for the off-line phase, i.e., the computationally most heavy phase. For on-line processing this is often too slow, though, as often there is an overhead in grid networks, for example, for the job submission and load balancing part.

19.3.6 User Interfaces and Interaction

Most of the current user interfaces follow the QBE paradigm and allow to upload images to start with, or have a random function to browse images in the database to find a starting point. Most interfaces show a ranked list of image results ordered by similarity. A clear distinction is required for how visual and how textual queries can be formulated. Both together form the most powerful framework [60].

Another important aspect of the user interface is the possibility to obtain more information about the users information need by marking images as positive and/or negative feedback. Many techniques exist for calculating similarity between several positive and negative input images, from combining all features for a joint pseudo-image to performing separate queries with each image and then combining the results.

19.3.7 Interfacing with Clinical Information Systems

The use of content-based techniques in a PACS environment has been proposed several times [61]. PACS are the main software components to store and access the large amount of visual data used in medical departments [62]. Often, several layer architectures exist for quick short-term access and slow long-term storage, but these are steadily replaced by fully hard disk-oriented solutions. The general scheme of a PACS within the hospital is shown in Fig. 19.4. The Integrating the Healthcare Enterprise (IHE) initiative is aiming at data integration in healthcare including all system components.

An indexing of the entire PACS causes problems with respect to the sheer amount of data that needs to be processed allowing efficient access by content to all the images. This issue of the amount of data that needs

Fig. 19.4. *System interconnection.* The PACS is connected with imaging modalities such as CT or MRI, the Radiology (RIS) and the Hospital Information System (HIS)

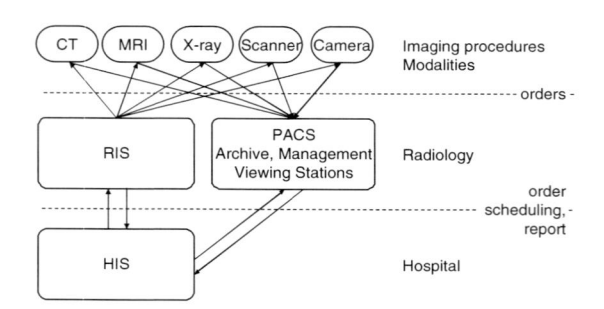

to be indexed is not discussed in any of the articles. Qi and Snyder have proposed to use CBIR techniques in a PACS as a search method but no implementation details are given [63]. Bueno et al. extend a database management system for integrating content-based queries based on simple visual features into PACS [64]. A coupling of image classification with PACS is given in [45]. Here, it is possible to search for certain anatomic regions, modalities or views of an image. A simple interface for coupling PACS with CBIR is also proposed. The identification is based on the DICOM Unique Identifier (UID) of the images. An IHE compliant procedure calling external CBIR application as well as returning the CBIR results into the PACS is described [65].

19.4 Evaluation

Whereas early evaluation in image retrieval was only base on small databases showing a few example images, evaluation in text retrieval has always been a very experimental domain. In CBIR, a first real standardization was achieved with the ImageCLEF[4] medical image retrieval task that started in 2004 and has been organized every year since, including a classification task and a retrieval task based on a data set of the Image Retrieval in Medical Applications (IRMA)[5] group.

19.4.1 Available Databases

Medical image databases have increasingly become available for researcher in the past 5 years. Some of the prominent examples is the Lung Image Database Consortium (LIDC) data, the IRMA database with many different classes and an increasing number of images and the images of the ImageCLEF competition taken first from medical teaching files and then from the scientific medical literature.

Nowadays, the National Institutes of Health (NIH) and the National Cancer Institute (NCI) require funded research to make their data available, and several databases indeed have become available for the public.

19.4.2 Tasks and User Models

When evaluating image retrieval, a certain usage model and information must be defined. A few research groups have actually conducted surveys on the use of images for journalists [66] and in other domains such as libraries or cultural heritage institutions [4].

---

[4] http://www.imageclef.org/
[5] http://irma-project.org/

For ImageCLEF 2005, the topic development was based on two surveys performed in Portland, OR and in Geneva [67, 68]. In total, about 40 medical professionals were surveyed on their image use and search behavior to learn more on how they use images and how they would like to search for them. It became clear that depending on the role of the person (clinician, lecturer, researcher) the information needs are significantly different, so each person who had more than one role had to respond to the questions for all roles. Librarians and students were also included into the survey. Most frequently, people said that they would like to be able to search for pathology and then for modality and the anatomic region. People prefer web engines to search for interesting images for lectures, but were concerned, while on the contrary being about image reliability. Based on these surveys, topics for ImageCLEFmed were developed along the following axes:

- anatomic region shown in the image
- imaging modality (e.g., X-ray, CT, MRI, microscopy)
- pathology or disease shown in the image
- abnormal visual observation (e.g., enlarged heart)

It was tried that topics covered at least two of these axes. A visual query topic is shown in Fig. 19.5, and a query topic requiring more than purely visual features is shown in Fig. 19.6. As ImageCLEF is on multilingual information retrieval and as the collection is in three languages, the topics were also developed in these three languages.

19.4.3 Ground Truth and Gold Standards

One of the most important aspects of evaluation is that there is a clear idea of what a good or perfect query result would be like. In the case of the IRMA collection, this ground truth (or gold standard) is given by the IRMA code that is attributed to each image by a clinician [69]. Its mono-hierarchical multi-axial architecture allows unambiguous ground truth labeling. Therefore,

Fig. 19.5. *Visual query.* An example of a query (topic) of ImageCLEF 2005 that is at least partly solvable visually, using the image and the text as query. Still, use of annotation can augment retrieval quality. The query text is presented in three languages, *English*: "Show me chest CT images with emphysema"; *German*: "Zeige mir Lungen CTs mit einem Emphysem"; *French*: "Montre-moi des CTs pulmonaires avec un emphysème"

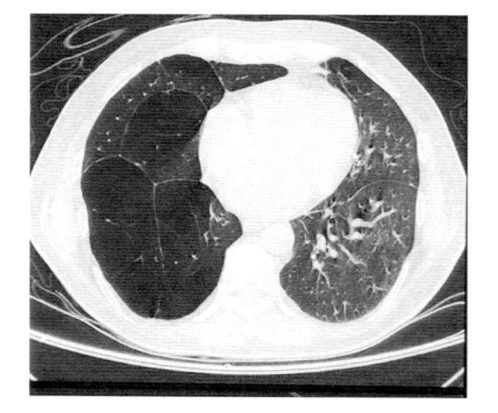

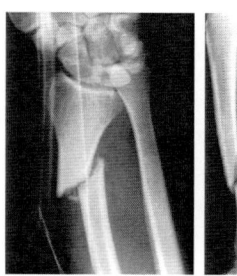

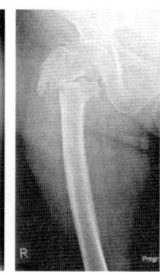

Fig. 19.6. *Semantic query.* A query of ImageCLEF 2005; *English*: "Show me all X-ray images showing fractures"; *German*: "Zeige mir Röntgenbilder mit Brüchen"; *French*: "Montres-moi des radiographies avec des fractures", which requires more than only visual retrieval. Visual features, however, can deliver hints to good results

depending on the data sets, classes can be generated using the entire hierarchy or a partial hierarchy. Image classification systems can then be evaluated by comparing them to the correct class labels.

For image retrieval as in the ImageCLEFmed benchmark, evaluation is slightly different since fixed classes do not exist. Based on well-defined information such as those in Fig. 19.6 experts can judge whether an image is relevant to this query or not. In images three categories were used, relevant, irrelevant, or indeterminable. Based on the judgments of clinicians on such relevance, several retrieval systems can well be compared.

Performance measures for the evaluation of information retrieval in general and image retrieval in particular have initiated intensive discussion for many years. Whereas in image classification the choice is smaller (correctly classified, incorrectly classified), there are many measures existing for retrieval tasks.

19.4.4 Benchmarks and Events

Information retrieval benchmarks have been established in the 1960s with the Cranfield tests. Since 1991, the Text Retrieval Conference (TREC) has created a strong testbed for information retrieval evaluation. For several years, TREC contained a biomedical retrieval called TRECgenomics.

The Cross Language Evaluation Forum (CLEF) started within TREC in 1997 and has been independent since 2000. With ImageCLEF that started in 2003, a new medical task was introduced as well, promoting the search for medical images with textual and visual means combined. From a small database of 8,000 images in 2004 the data sets and tasks have grown larger and more complicated every year. Also regarding the IRMA database and the image classification task, the complexity over 4 years was increased annually.

19.5 Examples for Medical CBIR Systems

In this section, we describes two example projects for medical CBIR.

19.5.1 Medical Gnu Image Finding Tool

Initially, the Medical GIFT (MedGIFT)[6] project was based on the GIFT, which resulted from the Viper[7] project at the University of Geneva [26]. The visual features used are meant for color photography and include a simple color histogram as well as color blocks in various areas of the images and at several scales. To separate the actual query engine from the GUI, the Multimedia Retrieval Markup Language (MRML)[8] was developed. This query language is based on direct communication of search engine and interface via sockets and eases a variety of applications such as meta-search engines and also the integration of a retrieval tool into a variety of environments and applications.

After a while, however, it became clear that new techniques were necessary in the medical domain, and the build components were grouped around five axes:

- Data access, ontologies, data annotation
- Techniques for retrieval and efficient structures to use them on large data sets
- Applications in the medical field such as lung image retrieval, fracture retrieval
- Inclusion of higher dimensional data sources into the retrieval process such as the use of 3D and 4D data
- Evaluation, mainly with the ImageCLEF benchmark describes in Sect. 19.4

Figure 19.7 shows a typical web interface after a query was executed. The query results are displayed ordered by their visual similarity to the query, with a similarity score shown underneath the images as well as the diagnosis. A click on the image links with the case database system and allows to access the full-size images.

In the context of heading towards indexing of higher-dimensional images an interface for browsing 3D repositories was developed [70] (Fig. 19.8).

19.5.2 Image Retrieval in Medical Applications

In Sect. 19.4, we have already introduced the IRMA framework. This research-driven project has been activated for almost 10 years, combining inter-disciplinary expertise from diagnostic radiology, computer science, and medical informatics.

IRMA aims at developing and implementing high-level methods for CBIR including prototypical application (e.g., [41, 71, 72]) to medico-diagnostic tasks on a radiological image archive. They want to perform semantic and formalized queries on the medical image database, which includes intra- and inter-individual variance and diseases.

---
[6] http://www.sim.hcuge.ch/medgift/
[7] http://viper.unige.ch/
[8] http://www.mrml.net/

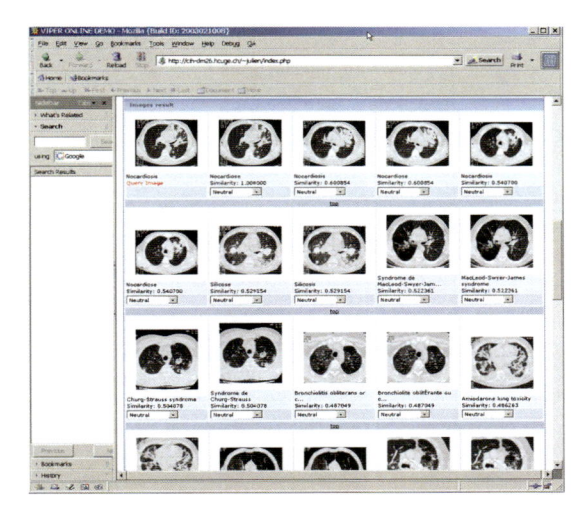

Fig. 19.7. *MedGIFT user interface.* A screen shot of a typical web interface for medical image retrieval system allowing QBE with the diagnosis underneath the image is shown

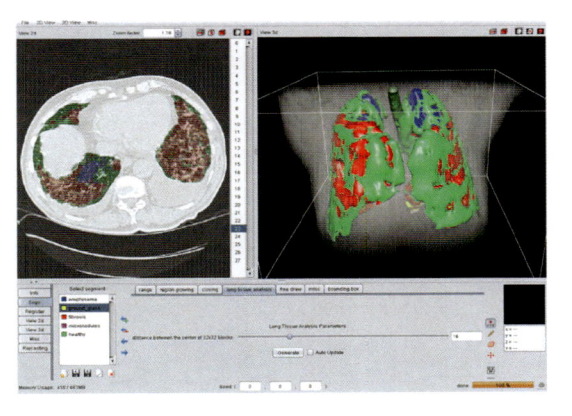

Fig. 19.8. *CBIR user interface supporting 3D data.* An interface that allows searching in 3D databases by visual content and then visualises the images with abnormal regions marked in various colors is shown

IRMA is based on a (i) central database that hold images, features, and the processing methods, (ii) a scheduler that provides distributed processing, (iii) a communicator that is used to interconnect CBIR with RIS and PACS, and (iv) web-based GUIs are provided for applications [45]. Three levels of image content similarity are modeled:

- *Global features* are linked to the entire images and used to automatically classify an image according to the anatomy, biosystem, creation, and direction (registered data layer) [69].
- *Local features* are linked to prominent image regions and used for object recognition (feature layer).
- *Structural features* are linked to spatial or temporal relations between the objects and used for high-level image interpretation (object layer).

A pipeline of image processing is suggested. Iterative region merging is used to build up a Hierarchical Attributed Region Adjacency Graph (HARAG), the data structure that is used to represent images, Objects of Interest (OOIs), and object constellations (scene analysis). Hence, image retrieval is transformed to graph matching. Object comparison operates on the HARAG nodes, while scenes are modeled by graph to sub-graph comparison.

Extended query refinement is provided to the user and allows for undo and redo commands and logical combinations of individual query responses [73]. Figure 19.9 visualized the interaction loops that are all encapsulated within one browser window. Parameter modules are used to transfer the input and parameters from the user to the system (e.g., QBE), and the output modules are used to display the query result (Fig. 19.9, green). Query refinement is supported by the orange loop, and yellow indicates undo, and redo options. The outer loop (Fig. 19.9, blue) allows combining individual queries by AND and OR. Here, the user can seek images having a certain characteristic in one local area and another elsewhere.

A typical IRMA user interface is shown in Fig. 19.10. Here, a spine X-ray databased is searched by shape and shape similarity [74]. The slider bars below the images allow the user to evaluate the retrieval result (query refinement).

Currently, the IRMA group is working on integration of CBIR into the clinical workflow. Figure 19.11 shows the dataflow for CBIR-assisted pre-fetching of previous radiographs supporting the radiologist in reading the scheduled exam. Both, Health Level Seven (HL7) and DICOM interfaces are provided by the IRMA communicator module. The communication steps are performed (**i**) at time of examination scheduling (steps 1–4); (**ii**) in the night before the exam (step 4); and (**iii**) on the day of the examination (steps 5–12). The additional communication steps that have been added to the communication because of CBIR integration are: 2c, 3, 6c, 7, 8, 9, 10a. To support CBIR-based hanging protocols, steps 10b and 11b are required additionally.

Fig. 19.9. *IRMA extended query refinement [73].* Four nested loops are integrated within one browser interface. *Green*: simple QBE; *Orange*: query refinement; *Yellow*: undo and redo; *Blue*: AND and OR

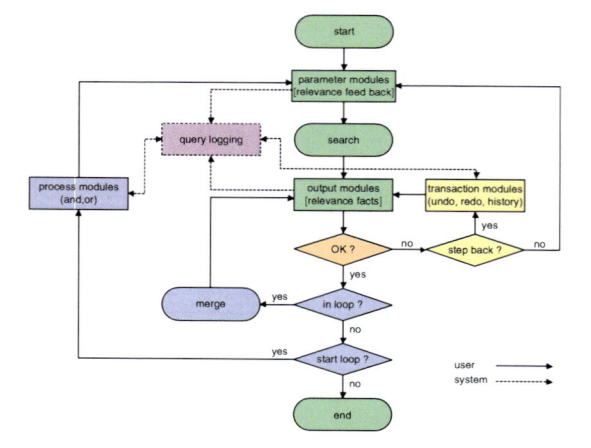

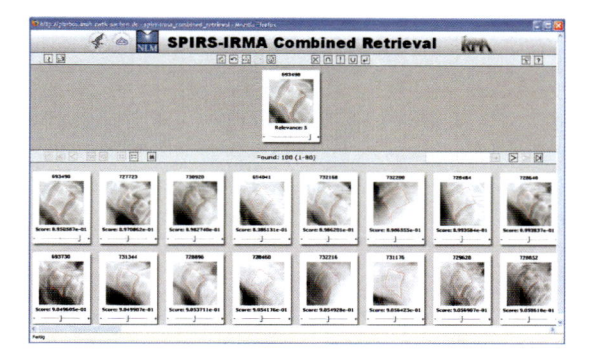

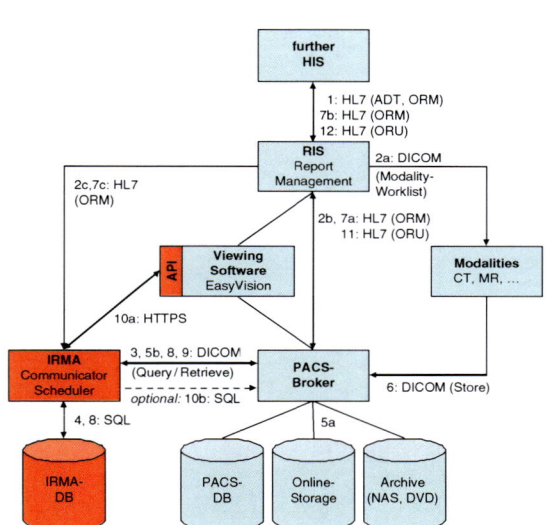

Fig. 19.10. *IRMA user interface.* A typical IRMA web interface supporting QBE, relevance feedback, and extended query refinement. Here, a shape retrieval interface in collaboration with the National Library of Medicine (NLM), National Institutes of Health (NIH), USA is shown

Fig. 19.11. *IRMA integration with HIS and PACS [75].* The regular workflow and its components are visualized in blue; the IRMA system addons in red. Interfacing is based on the Application Programming Interface (API)

19.6 Discussion and Conclusions

Medical images have often been used for retrieval systems and the medical domain is frequently cited as one of the principal application domains for content-based access technologies [76, 77] in terms of potential impact. Still, there has rarely been an evaluation of the performance and the description of the clinical use of systems is even rarer. Two exceptions are the Assert system on the classification of high resolution CTs of the lung [53] and the IRMA system for the classification of images into anatomical areas, modalities and view points [52].

Still, for a real medical application of content-based retrieval methods and the integration of these tools into medical practice, a very close cooperation between the two fields is necessary for a longer period of time. This cannot simply be substituted by an exchange of data or a list of the necessary functionality.

19.6.1 Strengths and Weaknesses of Current Systems

It was pointed out in this chapter that image retrieval has gone a long way from purely theoretical laboratory style developments, where single images were classified into a small number of classes without any clinical application, towards tools that combine visual and clinical data to really aid diagnosis and deliver valuable information to the clinicians. Tools have shown to improve diagnosis in real settings when properly applied [78]. With ImageCLEF there is also a benchmark to compare techniques for visual classification as well as for multi-modal medical information retrieval combining text and image data. Such benchmarks are necessary to proof the performance of techniques and entire systems.

Still, there is currently a total lack of system that are used in clinical practice and in close collaboration with clinicians.

19.6.2 Gaps of Medical CBIR Systems

In [59, 79], several technical gaps in medical image retrieval have been identified (Fig. 19.12). However, there are several other levels of gaps that need to be mentioned in this context. Legal constraints currently limit the application domain as the secondary use of medical data is ruled by nationally different laws that are not always easy to follow. In general, informed consent is required even if data is anonymized. This limits the amount of data potentially accessible and thus also the usefulness of the approach. Tools as the one described in [80] to access research data in patient records with an on-the-fly anonymization should limit these effects, but at the moment, it is still far from being usable in many institutions.

All these gaps finally lead to a usage gap. Clinicians rather use Google to search for images on the web than to search in patient records, where the access is limited via the patient Indentifier (ID). User interface, speed and retrieval quality seem to provide advantages with simple tools such as Google and this needs to be taken into account for new medical CBIR interfaces.

19.6.3 Future Developments

Image retrieval does have a bright future as does information retrieval in general. Information is produced in ever-increasing quantities and it also becomes increasing available, whether through patient record or via the Internet in teaching files or the scientific literature. One of the future challenges is to navigate in a meaningful way in databases of billions of images, allowing for effective and efficient retrieval, and at the same time a diversity in the results displayed and not simply duplicate images. Modern hospitals produced in the order of 100 GB or 120,000 images per day, and only few image retrieval systems could index this data providing a high response speed.

19 Content-Based Medical Image Retrieval   489

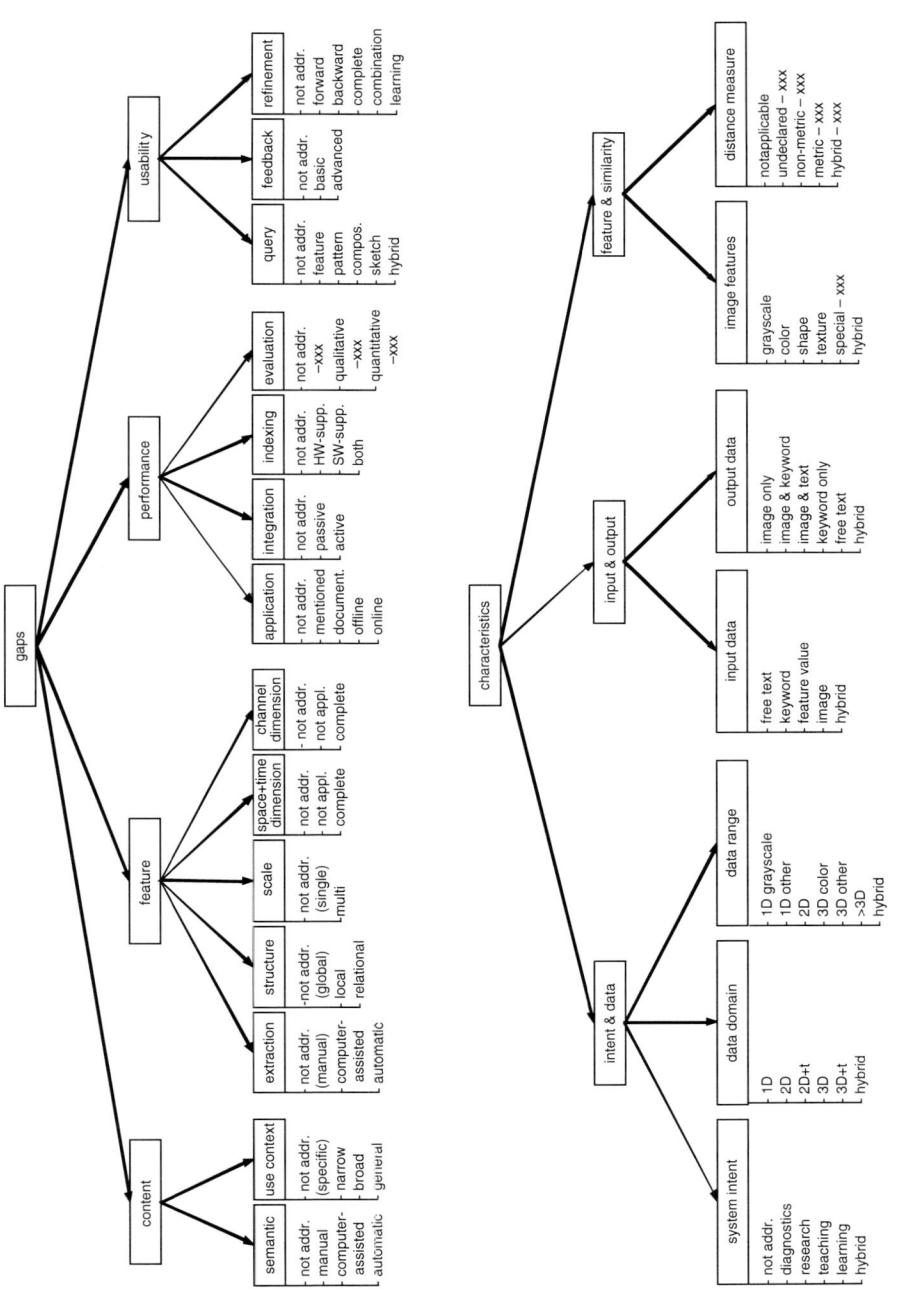

Fig. 19.12. *Gaps in medical CBIR [79]*

By far the largest data volume is produced with 3D and 4D tomographic devices, and there is still little research in this domain although a few approaches mainly for surface models do exist. To better integrate the entire amount of available information, it also seems necessary to merge visual, textual and structured data retrieval into unique systems. Currently the research domains are totally separated, and a closer collaboration is necessary for working systems. The goal in the end should be to deliver the right information, to the right people at the right time, and this information needs to include the visual data.

Another important future task is comprehensive evaluation of retrieval systems in clinical practice and thus in collaboration with clinicians to show their practical benefit. This is required to quantity the impact of CBIR and to determine its limits. Component-level evaluation is necessary to better understand what is currently working and what is not. Having all components accessible via standard interfaces could also help to optimize the overall system performance, which itself will impact CBIR system acceptance by the physicians.

References

1. Müller H, Michoux N, Bandon D, et al. A review of content-based image retrieval systems in medicine: clinical benefits and future directions. Int J Med Inform. 2004;73(1):1–23.
2. Tagare HD, Jaffe C, Duncan J. Medical image databases: a content-based retrieval approach. J Am Med Inform Assoc. 1997;4(3):184–98.
3. Chang SK, Kunii T. Pictorial data-base applications. IEEE Comput. 1981;14(11):13–21.
4. Enser PGB. Pictorial information retrieval. J Doc. 1995;51(2):126–70.
5. Gupta A, Jain R. Visual information retrieval. Commun ACM. 1997;40(5):70–79.
6. Eakins JP, Graham ME. Content-based image retrieval. Tech Report JTAP-039, JISC Technol Appl Program, Newcastle upon Tyne; 2000.
7. Venters CC, Cooper M. Content-based image retrieval. JTAP-054. J Inf Sci Res. 2000.
8. Smeulders AWM, Worring M, Santini S, et al. Content-based image retrieval at the end of the early years. IEEE Trans Pattern Anal Mach Intell. 2000;22(12):1349–80.
9. Datta R, Joshi D, Li J, et al. Image retrieval: ideas, influences, and trends of the new age. ACM Comput Surv. 2008;40(2):1–60.
10. Chang NS, Fu KS. Query-by-pictorial-example. IEEE Trans Softw Eng. 1980;6(6):519–24.
11. Niblack W, Barber R, Equitz W, et al. Storage and retrieval for image and video databases: QBIC project: querying images by content, using color, texture, and shape. Proc SPIE. 1993;1908:173–87.
12. Pentland A, Picard RW, Sclaroff S. Photobook: tools for content-based manipulation of image databases. Int J Comput Vis. 1996;18(3):233–54.

13. Ma WY, Deng Y, Manjunath BS. Human vision and electronic imaging II: tools for texture- and color-based search of images. Proc SPIE. 1997;3016:496–507.
14. Carson C, Thomas M, Belongie S, et al. VISUAL: blobworld: a system for region-based image indexing and retrieval. Lect Notes Comput Sci. 1999;1614:509–16.
15. Belongie S, Carson C, Greenspan H, et al. Color- and texture-based image segmentation using EM and its application to content-based image retrieval. Proc Int Conf Comput Vis. 1998; p. 675–82.
16. Cox IJ, Miller ML, Omohundro SM, et al. Target testing and the PicHunter bayesian multimedia retrieval system. Adv Digit Libr. 1996; p. 66–75.
17. Squire DM, Müller H, Müller W, et al. Design and evaluation of a content-based image retrieval system. In: Rahman SM, editors. Interactive Multimedia Systems. Hershey, PA, USA: IGI Global Press; 2001. p. 125–51.
18. Sclaroff S, La Cascia M, Sethi S, et al. Unifying textual and visual cues for content-based image retrieval on the World Wide Web. Comput Vis Image Underst. 1999;75(1/2):86–98.
19. Rocchio JJ. Relevance feedback in information retrieval. In: Salton G, editors. The SMART Retrieval System: Experiments in Automatic Document Processing. Englewood cliffs: Prentice-Hall; 1971. p. 313–23.
20. Müller H, Zhou X, Depeursinge A, et al. Medical visual information retrieval: state of the art and challenges ahead. Proc ICME. 2007; p. 683–86.
21. Geusebroek JM, van den Boogaard R, Smeulders AWM, et al. Color invariance. IEEE Trans Pattern Anal Mach Intell. 2001;23(12):1338–50.
22. Milanese R, Cherbuliez M. A Rotation, translation and scale-invariant approach to content-based image retrieval. J Vis Commun Image Represent. 1999;10:186–96.
23. Weszka JS, Dyer CR, Rosenfeld A. A comparative study of texture measures for terrain classification. IEEE Trans Syst Man Cybern C Appl Rev. 1976;6(4):269–85.
24. Kuo WJ, Chang RF, Lee CC, et al. Retrieval technique for the diagnosis or solid breast tumors on sonogram. Ultrasound Med Biol. 2002;28(7):903–09.
25. Lu CS, Chung PC. Wold features for unsupervised texture segmentation. Proc Int Conf Pattern Recognit Lett. 1998; p. 1689–93.
26. Squire DM, Müller W, Müller H, et al. Content-based query of image databases: inspirations from text retrieval. Pattern Recognit Lett. 2000;21(13–14):1193–98.
27. Comaniciu D, Meer P, Foran D, et al. Bimodal system for interactive indexing and retrieval of pathology images. Proc IEEE Workshop Appl Comput. 1998; p. 76–81.
28. Winter A, Nastar C. Differential feature distribution maps for image segmentation and region queries in image databases. Proc CBAIVL. 1999; p. 9–17.
29. Lucchese L, Mitra SK. Unsupervised segmentation of color images based on k-means clustering in the chromaticity plane. Proc CBAIVL. 1999; p. 74–78.
30. Loncaric S. A survey of shape analysis techniques. Pattern Recognit. 1998;31(8):983–1001.
31. Lapeer RJ, Tan AC, Aldridge R. A combined approach to 3D medical image segmentation using marker-based watersheds and active contours: the active watershed method. Lect Notes Comput Sci. 2002;2488:596–603.
32. Mikolajczyk K, Schmid C. A performance evaluation of local descriptors. IEEE Trans Pattern Anal Mach Intell. 2005;27(10):1615–30.

33. Sivic J, Zisserman A. Video google: efficient visual search of videos. Lect Notes Comput Sci 2006;4170:127–44.
34. Tversky A. Features of similarity. Psychol Rev. 1977;84(4):327–52.
35. Swain MJ, Ballard DH. Color indexing. Int J Comput Vis. 1991;7(1):11–32.
36. Aggarwal CC, Hinneburg A, Keim DA. On the surprising behavior of distance metrics in high dimensional space. Lect Notes Comput Sci. 2001;1973:420–34.
37. Hinneburg A, Aggarwal CC, Keim DA. What is the nearest neighbor in high-dimensional spaces? Proc VLDB. 2000; p. 506–16.
38. Hanka R, Harte TP. Curse of dimensionality: classifying large multi-dimensional images with neural networks. CIMCSP. 1996.
39. Vasconcelos N, Lippman A. A probabilistic architecture for content-based image retrieval. IEEE Comput Soc. 2000;216–221.
40. Goh KS, Chang E, Cheng KT. Support vector machine pairwise classifiers with error reduction for image classification. J Assoc Comput Machinery. 2001;32–37.
41. de Oliveira JEE, Machado AMC, Chavez GC, et al. MammoSys: a content-based image retrieval system using breast density patterns. Comput Methods Programs Biomed. 2010;99(3):289–97.
42. Westerveld T. Image Retrieval: Content Versus Context. vol. 1. Paris, France: CID; 2000.
43. Salton G, Buckley C. Term weighting approaches in automatic text retrieval. Inf Process Manage. 1988;24(5):513–23.
44. Jain AK, Duin RPW, Mao J. Statistical pattern recognition: a review. IEEE Trans Pattern Anal Mach Intell. 2000;22(1):4–37.
45. Lehmann TM, Güld MO, Thies C, et al. Content-based image retrieval in medical applications. Methods Inf Med. 2004;43(4):354–61.
46. Greenes RA, Brinkley JF. Imaging Systems. New York: Springer; 2000. p. 485–538.
47. Kulikowski C, Ammenwerth E, Bohne A, et al. Medical imaging informatics and medical informatics: opportunities and constraints. Methods Inf Med. 2002;41:183–9.
48. Sarvazyan AP, Lizzi FL, Wells PNT. A new philosophy of medical imaging. Med Hypotheses. 1991;36:327–35.
49. Rosset A, Ratib O, Geissbuhler A, et al. Integration of a multimedia teaching and reference database in a PACS environment. Radiographics 2002;22(6):1567–77.
50. Maloney K, Hamlet CT. The clinical display of radiologic information as an interactive multimedia report. J Digit Imaging. 1999;12(2):119–21.
51. Lowe HJ, Antipov I, Hersh W, et al. Towards knowledge-based retrieval of medical images. The role of semantic indexing, image content representation and knowledge-based retrieval. Proc Annu Symp Comput Appl Med Inform. 1998;882–86.
52. Keysers D, Dahmen J, Ney H, et al. A statistical framework for model-based image retrieval in medical applications. J Electron Imaging. 2003;12(1):59–68.
53. Shyu CR, Brodley CE, Kak AC, et al. ASSERT: a physician-in-the-loop content-based retrieval system for HRCT image databases. Comput Vis Image Underst. 1999;75(1/2):111–32.
54. El-Naqa I, Yang Y, Galatsanos NP, et al. A similarity learnign approach to content–based image retrieval: application to digital mammography. IEEE Trans Med Imaging. 2004;23(10):1233–44.

55. Wong JSJ, Zrimec T. Classification of lung disease pattern using seeded region growing. Proc AI. 2006;233–42.
56. Depeursinge A, Iavindrasana J, Cohen G, et al. Lung tissue classification in HRCT data integrating the clinical context. In: Proc CBMS; 2008. p. 542–547.
57. Müller H, Rosset A, Vallée JP, et al. Integrating content-based visual access methods into a medical case database. Proc MIE. 2003;95:480–85.
58. Lehmann TM, Goudarzi S, Linnenbrügger N, et al. Automatic localization and delineation of collimation fields in digital and film-based radiographs. Proc SPIE. 2002;4684:1215–23.
59. Müller H, Kalpathy-Cramer J. Analyzing the content out of context: features and gaps in medical image retrieval. Int J Healthc Inf Syst Inform. 2009;4(1):88–98.
60. Névéol A, Deserno TM, Darmonic SJ, et al. Natural language processing versus content-based image analysis for medical document retrieval. J Am Soc Inf Sci. 2009;60(1):123–34.
61. Welter P, Deserno TM, Gülpers R, et al. Exemplary design of a DICOM structured report template for CBIR integration into radiological routine. Proc SPIE. 2010;7628:0B1–10.
62. Lemke HU. PACS developments in europe. Comput Med Imaging Graph. 2002;27:111–20.
63. Qi H, Snyder WE. Content-based image retrieval in PACS. J Digit Imaging. 1999;12(2):81–83.
64. Bueno JM, Chino F, Traina AJM, et al. How to add content-based image retrieval capacity into a PACS. Proc CBMS. 2002; p. 321–26.
65. Welter P, Deserno TM, Fischer B, et al. Integration of CBIR in radiological routine in accordance with IHE. Proc SPIE. 2009;7264:041–48.
66. Markkula M, Sormunen E. Searching for photos: journalists' practices in pictorial information retrieval. In: Proc The Challenge of Image Retrieval. Newcastle upon Tyne; 1998. p. 1–13.
67. Müller H, Despont-Gros C, Hersh W, et al. Health Care Professionals Image Use and Search Behaviour. MIE. 2006; p. 24–32.
68. Hersh W, Müller H, Gorman P, et al. Task analysis for evaluating image retrieval systems in the ImageCLEF Biomedical image retrieval task. Proc SOL. 2005.
69. Lehmann TM, Schubert H, Keysers D, et al. The IRMA code for unique classification of medical images. Proc SPIE. 2003;5033:440–51.
70. Depeursinge A, Vargas A, Platon A, et al. 3D case-based retrieval for interstitial lung diseases. Lect Notes Comput Sci. 2009;18–25.
71. Fischer B, Brosig A, Welter P, et al. Content-based image retrieval applied to bone age assessment. Proc SPIE. 2010;7624:121–10.
72. Deserno TM, Antani S, Long RL. Content based image retrieval for scientific literature access. Methods Inf Med. 2009;48(4):371–80.
73. Deserno TM, Güld MO, Plodowski B, et al. Extended query refinement for medical image retrieval. J Digit Imaging. 2008;21(3):280–89.
74. Hsu W, Antani S, Long LR. et al. SPIRS: a Web-based image retrieval system for large biomedical databases. Int J Med Inform. 2009;78 Suppl 1:S13–24.
75. Fischer B, Deserno TM, Ott B, et al. Integration of a research CBIR system with RIS and PACS for radiological routine. Proc SPIE. 2008;6919:I41–10.
76. Beretti S, Del Bimbo A, Pala P. Content-based retrieval of 3D cellular structures. IEEE Comput Soc. 2001;1096–99.

77. Kelly PM, Cannon M, Hush DR. Storage and retrieval for image and video databases III: query by image example: the CANDID approach. Proc SPIE. 1995;2420:238–48.
78. Aisen AM, Broderick LS, Winer-Muram H, et al. Automated storage and retrieval of thin-section CT images to assist diagnosis: system description and preliminary assessment. Radiology 2003;228(1):265–70.
79. Deserno TM, Antani S, Long R. Ontology of gaps in content-based image retrieval. J Digit Imaging. 2008;22(2):202–15.
80. Iavindrasana J, Depeursinge A, Ruch P, et al. Design of a decentralized reusable research database architecture to support data acquisition in large research projects. Proc MedInfo. 2007;12.

Part VIII

Evaluation and Customizing

20
Systematic Evaluations and Ground Truth

Jayashree Kalpathy-Cramer and Henning Müller

Summary. Every year, we see the publication of new algorithms for medical image analysis including segmentation, registration, classification and retrieval in the literature. However, in order to be able to translate these advances into clinical practice, the relative effectiveness of these algorithms needs to be evaluated.

In this chapter, we begin with a motivation for systematic evaluations in science and more specifically in medical image analysis. We review the components of successful evaluation campaigns including realistic data sets and tasks, the gold standards used to compare systems, the choice of performance measures and finally workshops where participants share their experiences with the tasks and explain the various approaches. We also describe some of the popular efforts that have been conducted to evaluate retrieval, classification, segmentation and registration techniques. We describe the challenges in organizing such campaigns including the acquisition of databases of images of sufficient size and quality, establishment of sound metrics and ground truth, management of manpower and resources, motivation of participants, and the maintenance of a friendly level of competitiveness among participants. We conclude with lessons learned over the years of organizing campaigns, including successes and road-blocks.

20.1 Introduction

Medical images are being produced in ever-increasing quantities as a result of the digitization of medical imaging and advances in imaging technology in the last two decades. The assorted types of clinical images are critical in patient care for diagnosis and treatment, monitoring the effect of therapy, education and research. The previous chapters have described a number of techniques used for medical image analysis from 3D image reconstruction to segmentation and registration to image retrieval. The constantly expanding set of algorithms being published in the computer vision, image processing, machine learning and medical image analysis literature underscores the need for sound evaluation methodology to demonstrate progress based on the same data and tasks.

T.M. Deserno (ed.), *Biomedical Image Processing*, Biological and Medical Physics, Biomedical Engineering, DOI: 10.1007/978-3-642-15816-2_20,
© Springer-Verlag Berlin Heidelberg 2011

It has been shown that many of these publications provide limited evaluation of their methods using small or proprietary data sets, making a fair comparison of the performance of the proposed algorithm with previous algorithms difficult [1, 2]. Often, the difficulty in obtaining high quality data sets with ground truth can be an impediment to computer scientists without access to clinical data. We believe that any newly proposed algorithms must be compared to the existing state-of-the-art techniques using common data sets with application-specific, validated metrics before they are likely to be incorporated into clinical applications. By providing all participants with equal access to realistic tasks, validated data sets (including ground truth), and forums for discussing results, evaluation campaigns can enable the translation of superior theoretical techniques to meaningful applications in medicine.

20.2 Components for Successful Evaluation Campaigns

Evaluation is a critical aspect of medical image analyses and retrieval. In the literature, many articles claim superior performance compared to previously-published algorithms. However, in order to be able to truly compare and contrast the performance of these techniques, it is important to have a set of well-defined, agreed-upon tasks performed on common collections using meaningful metrics. Even if the tasks are very different from a technical standpoint (segmentation vs. retrieval, for example), their evaluations share many common aspects. Evaluation campaigns can provide a forum for more robust and equitable comparisons between different techniques.

20.2.1 Applications and Realistic Tasks

First of all, the goal of the algorithms being evaluated must be well understood. A technique such as image segmentation is useful in many clinical areas; however, to perform a thorough evaluation of such algorithms, one must keep the ultimate application in mind. For example, consider the following two segmentation tasks: tumor segmentation to monitor a response to cancer therapy, and anatomical segmentation of the brain from an fMRI study. The nature of each task informs the choice of the optimal evaluation metric. In the first case, missing portions of the tumor (and thereby under-estimating its size) can have serious consequences, and therefore penalties for under-segmentation might be more appropriate. In other applications, however, under-segmentation and over-segmentation may be considered to be equally inconvenient.

An image retrieval system used for performing a systematic review might have different goals than a system used to find suitable images for a lecture or scientific presentation. In the first case, the goal might be to find every relevant article and image, while in the second case a single image that meets the search need might be sufficient. For some applications accuracy might be more important while for those being used in real-time, speed can be critical.

Evaluation campaigns are usually geared toward a specific clinical application. For instance, the Medical Image Computing and Computer Assisted Intervention (MICCAI) grand challenges for segmentation [3] target very specific tasks (e.g., segmentation of prostate, liver etc.). The goal for the image retrieval task in the Cross Language Evaluation Forum (CLEF),[1] medical retrieval campaign (ImageCLEF 2009) is to retrieve images from the medical literature that meet information needs of clinicians [4].

Once the overall goal of the algorithm has been well understood, it is important to identify a set of realistic, meaningful tasks towards that goal. For evaluating an image retrieval system this might consist of a set of reasonable search topics (often derived from user studies or log file analyses [5–7]). For the evaluation of a registration algorithm, an appropriate task might be to register structures in an atlas to equivalent structures in a set of patients. For segmentation challenges the task might be to segment normal anatomical organs (e.g., lung, liver, prostate, vasculature) or abnormalities (e.g., lung nodule, liver tumor, lesion). Classification tasks might include classifying radiographs based on the anatomical location [8], or separating voxels in the brain into white, gray matter and Cereborspinal Fluid (CSF) in Magnetic Resonance Imaging (MRI) data [9]. The number and scale of these tasks (how many topics, how many structures for how many different patient studies, etc.) must be carefully chosen to support the derivation of statistically meaningful metrics.

20.2.2 Collections of Images and Ground Truth

In order to perform a fair comparison of different algorithms, ideally all techniques must be compared on the same database or collection of images. Additionally, these data must be of a sufficient variety, so as to encompass the full range of data found in realistic clinical situations.

Often, computer scientists wishing to evaluate state-of-the-art algorithms do not have access to large amounts of clinical data, thereby limiting the scope of their evaluations. In general, getting access to the large collections necessary for a robust evaluation has been challenging, even for researchers associated with clinical facilities due to issues of cost, privacy and resources.

Recently, there has been a growing trend towards making databases of images available openly towards the goal of promoting reproducible science. Many governmental agencies, including the National Institutes of Health (NIH) in the United States have funded initiatives like the Lung Imaging Database Consortium (LIDC) [10] and the Alzheimer's Disease Neuroimaging Initiative (ADNI)[2] [11] that create well-curated collections of images and clinical data. These collections are typically anonymized to preserve patient privacy, and openly available to researchers. These and other similar initiatives

---

[1] http://www.clef-campaign.org/
[2] http://www.loni.ucla.edu/ADNI

foster collaboration between groups across the world and researchers from different domains including clinicians, imaging scientists, medical physicists and computer scientists.

The task of attaining ground truth or a *gold standard* continues to be challenging. For most applications, the best references are manually generated, and therefore their construction is an extremely time consuming and resource-intensive task. However, often the absolute truth is unknown or unknowable. For instance, it would be quite difficult to absolutely verify the registration of a brain atlas to the MRI of a patient. Similarly, in order to evaluate the performance of segmentation algorithms, experts usually manually delineate the Regions of Interest (ROI). However, the *true* segmentation of a tumor that is not physically resected may never be definitively established.

Additionally, even if there theoretically exists an "objective truth", experts often disagree on what constitutes that truth. In cases with more than one human rater, these questions of inter-observer agreement make the creation of a gold standard difficult. By providing segmentation in the form of annotations of lung nodules by four independent raters, the LIDC database exemplifies this difficulty in obtaining ground truth. Recent research has demonstrated that all four raters agreed on the presence of a nodule at a given location in only approximately 40% of the cases [12].

The problem is not limited to segmentation gold standards. When evaluating the effectiveness of information retrieval systems, relevance judgments are typically performed by domain experts. However, the kappa-measures (used to quantify inter-observer agreement) between experts in relevance judgment tasks often indicate significant levels of disagreement as to which documents count as "relevant". The concept of relevance as applied to images is particularly problematic, as the relevance of a retrieved image can depend on the context in which the search is being performed. An additional source of judgment difficulty is that domain experts tend to be more strict than novices [4], and so the validity of their judgments for a particular task may depend on the nature of the intended users.

20.2.3 Application-Specific Metrics

The choice of metrics should depend on the clinical goal of the algorithm being evaluated. In classification tasks, *error rate* is often the metric of choice. However, if the cost of a miss (e.g., missed detection of a lung nodule) is high, a non-symmetric measure of cost can be used. For registration and segmentation, measures related to volumetric overlap or surface distances can be used. If the goal of an image retrieval system is to find a few good images to satisfy the information need, early precision might be a good measure. On the other hand, if the goal of the task is to find every relevant image in the database, recall-oriented measures might be better suited.

In most evaluation campaigns, the evaluation measures are specified at the outset. Often, a single measure that combines different aspects of the evaluation is preferred, as this makes comparisons between participants straightforward (see Sect. 20.3).

20.2.4 Organizational Resources and Participants

Evaluation campaigns are usually conducted on a voluntary basis as funding for such efforts can be hard to obtain. Organizing such campaigns can be quite resource and time-intensive as the organizers need to acquire databases of images of sufficient size and quality, establish sound performance metrics and ground truth, provide the tabulation of the results, potentially organize the publications of the proceedings and motivate participation by balancing competitiveness with a friendly spirit of collaboration and cooperation.

Having a diverse set of loyal participants is a hallmark of a good evaluation campaign. Often, significantly larger number of groups register for and obtain data to evaluation campaigns than actually submit results and participate in the workshops. It is important to strive to increase the number of actual participants as the collaborative atmosphere, as found in the evaluation campaigns, engenders strides in the field by enabling participants to leverage each other's techniques. One of the challenges of organizing an evaluation campaign is providing tasks that are appropriate for research groups with varying levels of expertise and resources. If the task is too challenging and requires massive computing resources, participation by groups without access to such facilities can be limited. On the other hand, if the task is regarded as being too trivial, the sought-after participation by the leading researchers in the area can be difficult to attract. Options explored by some of the campaigns include providing multiple tasks at different levels, providing baseline runs or systems that can be combined in a modular fashion with the participants' capabilities (ImageCLEF) or providing the option of submitting both fully automatic and semi-automatic runs. Participants can generally be motivated by the opportunity to publish, by providing access to large collections of images that they might otherwise not have access to, as well as the spirit of the competition.

Many evaluation campaigns (see Sect. 20.4) organize workshops at the end of the evaluation cycle where participants are invited to present their methods and participate in discussions. They are often, but not exclusively, held in conjunction with larger conferences.

These workshops are an important part of the evaluation cycle and can be a great opportunity for researchers from across the globe to meet face-to-face in an effort to advance their fields. In addition to the technical aspects, the workshops also provide a chance for participants to provide feedback to the organizers about the collections, the nature of the task as well as the level of difficulty and organizational issues. They also provide a forum where participants can offer suggestions for future tasks, collections, and metrics. Furthermore, an in-person workshop is an excellent opportunity to recruit new organizers, thereby aiding the sustainability of the campaign.

20.3 Evaluation Metrics and Ground Truth

This section describes several of the commonly used performance metrics in medical imaging tasks including registration, segmentation and retrieval.

20.3.1 Registration

One of the first steps in the evaluation of the performance of registration algorithms is simply a visual check. This can be accomplished using image fusion in which one image is overlaid on top of the other with partial transparency and potentially different colors. Alternatively, the images can be evaluated using a checkerboard pattern.

The intensities of the registered images, cf. Chap. 5, page 130, can be used as metric [13]. The rationale behind this approach is that the better the registration performance, the sharper the composited image is expected to be as the registered image will be closer to the target image. With respect to the template image j, the intensity variance is given as

$$\text{IV}_j(x) = \frac{1}{M-1} \sum_{i=1}^{M} (T_i(h_{ij}(x)) - \text{ave}_j(x))^2 \qquad (20.1)$$

where $\text{ave}_j(x) = \frac{1}{M} \sum_{i=1}^{M} T_i(h_{ij}(x))$ denotes the average, T_i is the i-th image of the population, $h_{ij}(x)$ is the transform from image i to j with respect to a Eulerian coordinate system and M is the number of images being evaluated.

Other methods include comparing the forward and reverse transforms resulting from the registration. In a perfect situation, the forward transform would be the inverse of the reverse. The inverse consistency error measures the error between a forward and reverse transform compared to an identity mapping [13]. The voxel-wise Cumulative Inverse Consistency Error (CICE) is computed as

$$\text{CICE}_j(x) = \frac{1}{M} \sum_{i=1}^{M} \|h_{ji}(h_{ij}(x)) - x\|^2 \qquad (20.2)$$

where $\|\ \|$ denotes the standard Euclidean norm. The CICE is a necessary but not sufficient metric for evaluating registration performance [13].

In addition, Christensen et al. [13] note that the transforms resulting from registration algorithms should satisfy the transitivity property. If H_{AB} is the transform from A to B, transitivity implies that $h_{CB}(h_{BA}(x)) = h_{CA}(x)$ or $h_{AC}(h_{CB}(h_{BA}(x))) = x \ \forall A, B, C$ T_i is the ith image of the set and h_{ij} is the registration transform.

The Cumulative Transitive Error (CTE) is defined as

$$\text{CTE}_k(x) = \frac{1}{(M-1)(M-2)} \sum_{\substack{i=1 \\ i \neq k}}^{M} \sum_{\substack{j=1 \\ j \neq i \\ j \neq k}}^{M} \|h_{ki}(h_{ij}(h_{jk}(x))) - x\|^2 \qquad (20.3)$$

Another common approach in registration is to define a structure in the initial image (e.g., in an atlas), register the initial image to the final image (e.g., actual patient image), and deform the structure using the resulting deformation field. If manual segmentation is available on the final image, then many of the metrics defined in the following subsection can be used to compare the manual segmentation to that obtained using registration of the atlas.

20.3.2 Segmentation

Image segmentation, the task of delineating an image into meaningful parts or objects, is critical for many clinical applications. One of the most challenging aspects in evaluating the effectiveness of segmentation algorithms is the establishment of ground truth against which the computer-derived segmentations are to be compared.

Metrics Without Ground Truth

In real-life clinical images, establishing true segmentation often is difficult due to poor image quality, noise, non-distinct edges, occlusion and imaging artifacts. Physical and digital "phantoms" have been used to establish absolute ground truth; however, they do not contain the full range of complexity and variability of clinical images [14].

To avoid the use of phantoms, Warfield et al. [14] proposed the Simultaneous Truth and Performance Level Estimation (STAPLE) procedure, an expectation-maximization algorithm that computes a probabilistic estimate of true segmentation given a set of either automatically generated or manual segmentations. STAPLE has been used for establishing ground truth in the absence of manual segmentations as well as to provide a quality metric for comparing the performance of segmentation algorithms.

However, it should be pointed out that manual segmentations are not reproducible, i.e., they suffer from inter- as well as intra-observer variability, and hence, their usefulness in absolute evaluation of medical image segmentation is limited.

Volume-Based Metrics

Consider the case where the results of a segmentation algorithm are being compared to ground truth using binary labels (i.e., a label of "1" is given to a voxel that belongs to the object being segmented and a label of "0" otherwise). Let A indicate the voxels belonging to the object according to the segmentation under consideration (as determined by either another user or an automatic algorithm) and G refers to the ground truth (Fig. 20.1). A commonly used simple measure is based on the volumes enclosed by the respective

Fig. 20.1. *Venn diagram.* The diagram shows the intersection between the segmented label A and the gold standard G

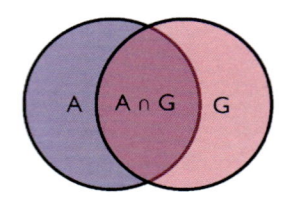

segmentations. The Volumetric Difference (VD) [15] is defined as

$$\text{VD} = \frac{V_a - V_g}{V_g} \times 100 \qquad (20.4)$$

The Absolute Volumetric Difference (AVD) is the absolute value of the above measure. However, these measures do not take into account the spatial locations of the respective volumes, and hence have limited utility when used alone. Additionally, they are not symmetric.

The Dice [16] and Jaccard coefficients [17] are the most commonly used measures of spatial overlap for binary labels. In both cases, the values for the coefficients range from zero (no overlap) to one (perfect agreement).

$$D = \frac{2|A \cap G|}{|A| + |G|} \times 100 \qquad J = \frac{|A \cap G|}{|A \cup G|} \times 100 \qquad (20.5)$$

This is also sometimes known as the *relative overlap measure*. As all these measures are related to each other, typically only one or the other is calculated.

$$J = \frac{D}{2 - D} \qquad (20.6)$$

The Dice coefficient has been shown to be a special case of the kappa coefficient [18], a measure commonly used to evaluate inter-observer agreement. As defined, both of these measures are symmetric, in that over- or under-segmentation errors are weighted equally. To characterize over- and under-segmentations in applications where these might be important (e.g., tumor delineation where the cost for missing the tumor is higher), false positive and false negative Dice measures can be used. The False Positive Dice (FPD) is measure of voxels that are labeled positive (i.e., one) by the segmentation algorithm being evaluated but not the ground truth and hence is a measure of over-segmentation. The False Negative Dice (FND) is a measure of the voxels that were considered positive according to the ground truth but missed by the segmentation being evaluated. Let $\bar{A}$ and $\bar{G}$ be the complements of the segmentation and the ground truth (i.e., they are the voxels labeled 0).

$$\text{FPD} = \frac{2|A \cap \bar{G}|}{|A| + |G|} \times 100 \qquad \text{FND} = \frac{2|\bar{A} \cap G|}{|A| + |G|} \times 100 \qquad (20.7)$$

The above-mentioned spatial overlap measures depend on the size and shape of the object as well as the voxel size relative to the object size. Small differences in the boundary of the segmentation can result in relatively large errors in small objects compared to large objects.

Additionally, the measures discussed above assume that we are comparing the results of one algorithm with one set of ground truth data. However, often there is either no ground truth available, or alternatively, manual segmentations from multiple human raters are available. In these cases, many approaches have been considered, ranging from fairly simplistic majority votes for the class membership of each voxel to the STAPLE algorithm mentioned above [14] or the Williams index [19].

The Williams index [19, 20] considers a set of r raters labeling a set of n voxels with one of l labels. D is the label map of all raters where D_j is the label map for rater j and D_{ij} represents the label of rater j for voxel i. Let $a(Dj, D_{ij})$ be the agreement between rater j and j_i over all n voxels. Several agreement measures can be used. The Williams index I_j as defined below, can be used to assess if observer j agrees at least as much with other raters as they agree with each other.

$$I_j = \frac{(r-2)\sum_{j'\neq j}^{r} a(D_j, D_{j'})}{2\sum_{j'\neq j}^{r}\sum_{j''\neq j}^{j'} a(D_{j'}, D_{j''})} \quad (20.8)$$

All of the metrics discussed thus far have assumed that the class labels were binary, i.e. each voxel belonged to either the structure or the background. Although this has been the case historically and continues to be the predominant mode for classification, more recently, methods as well as probabilistic methods have required the use of partial labels for class membership. Crum et al. [21] discussed the lack of sufficient metrics to evaluate the validity of the algorithms in these cases. They proposed extensions of the Jaccard similarity measure, referred to as Generalized Tanimoto Coefficient (GTC) using results from fuzzy set theory. These overlap measures can be used for comparison of multiple fuzzy labels defined on multiple subjects.

Surface-Based Metrics

Unlike the region-based approaches, surface distance metrics are derived from the contours or the points that define the boundaries of the objects. The Hausdorff Distance (HD) is commonly used to measure the distance between point sets defining the objects. The HD (a directed measure as it is not symmetric) between A and G, $h(A, G)$ is the maximum distance from any point in A to a point in G and is defined as

$$h(A, G) = \max_{a \in A}(d(a, G)) \quad (20.9)$$

where $d(a, G) = \min_{G \in G} \|a - g\|$. The symmetric HD, $H(A, G)$ is the larger of the two directed distances, defined more formally as

$$H(A, G) = \max(h(A, G), h(G, A)) \quad (20.10)$$

The Hausdorff distance, although commonly used, has a few limitations. It is highly susceptible to outliers resulting from noisy data. However, many variations of a more robust version of this measure have been used for applications in segmentation as well as registration.

Software Tools

The Valmet software tool, although no longer actively supported, incorporated many of these measures and has been used for evaluation and visualization of 2D and 3D segmentation algorithms [22]. It includes the measures: volumetric overlap (true and false positives, true and false negatives), probabilistic distances between segmentations, Hausdorff distance, mean absolute surface distance, and interclass correlation coefficients for assessing intra-, inter-observer and observer-machine variability. The software also enabled the user to visualize the results.

20.3.3 Retrieval

Information Retrieval (IR) has a rich history of evaluation campaigns, beginning with the Cranfield methodology in the early 1960s [23] and the System for the Mechanical Analysis and Retrieval of Text (SMART) [24], to more recent Text Retrieval Conference (TREC)[3] campaigns [25].

Precision, Recall, and F-Measure

Precision and recall are two of the most commonly used measures for evaluation retrieval systems, both for text and images. Precision is defined the fraction of the documents retrieved that are relevant to the user's information need. For binary relevance judgments, precision is analogous to positive predictive value. Consider a 2 × 2 table for relevant and retrieved objects (Table 20.1) where A is the set of relevant objects and B is the set of retrieved objects

$$\text{precision} = \frac{\text{relevant documents retrieved}}{\text{retrieved documents}} \qquad P = \frac{|A \cap B|}{|B|} \qquad (20.11)$$

Precision is often calculated for a given number of retrieved objects. For instance P_{10} (precision at 10) is the number of relevant objects in the first ten objects retrieved. Recall, on the other hand, is the ratio of the relevant objects retrieved to the total number of relevant objects in the collection

$$\text{recall} = \frac{\text{relevant documents retrieved}}{\text{relevant documents}} \qquad R = \frac{|A \cap B|}{|A|} \qquad (20.12)$$

---

[3] http://trec.nist.gov/

	Relevant	Not relevant	
Retrieved	$A \cap B$	$\bar{A} \cap B$	B
Not retrieved	$A \cap \bar{B}$	$\bar{A} \cap \bar{B}$	$\bar{B}$
	A	$\bar{A}$	

Table 20.1. *Fourfold table.* A 2 × 2 table for relevant and retrieved objects

Recall is equivalent to sensitivity. It is important to note that recall does not consider the order in which the relevant objects are retrieved or the total number of objects retrieved.

A single effectiveness measure E, based on both precision and recall was proposed by van Rijsbergen [26]

$$E = 1 - \frac{1}{\alpha/P + (1-\alpha)/R} \qquad (20.13)$$

where α denoting a fraction between zero and one can be used to weigh the importance of recall relative to precision in this measure.

The weighted F-score (F-measure) is related to the effectiveness measure as $1 - E = F$

$$F = \frac{1}{\alpha/P + (1-\alpha)/R} = \frac{(\beta^2+1)PR}{\beta^2 P + R} \qquad (20.14)$$

where $\beta^2 = \frac{1-\alpha}{\alpha}$ and $\alpha \in [0,1]$, $\beta^2 \in [0,\infty]$.

In the balanced case where both precision and recall are weighted equally, $\alpha = 1/2$ and $\beta = 1$. It is commonly written as F_1, or $F_{\beta=1}$. In this case, the above equation simplifies to the harmonic mean

$$F_{\beta=1} = \frac{2PR}{P+R}$$

However, α or β can be used to provide more emphasis to precision or recall as values of $\beta < 1$ emphasize precision, while values of $\beta > 1$ emphasize recall.

Average Precision

Overall, precision and recall are metrics based on the set of objects retrieved but not necessarily the position of the relevant objects. Ideal retrieval systems should retrieve the relevant objects ahead of the non-relevant ones. Thus, measures that consider the order of the returned items are also important. Average precision, defined as the average of the precisions computed for each relevant item, is higher for a system where the relevant documents are retrieved earlier.

$$\text{AP} = \frac{\sum_{r=1}^{N}(P(r) \times \text{rel}(r))}{\text{number of relevant documents}} \qquad (20.15)$$

where r is the rank, N the number retrieved, rel() a binary function on the relevance of a given rank, and $P()$ precision at a given cut-off rank.

In evaluation campaigns with many search topics, the Mean Average Precision (MAP) is a commonly used measure. The MAP is the mean of the average precisions for all the search topics and is meant to favor systems that return more relevant documents at the top of the list. However, the maximum MAP that a system can achieve is limited by its recall, and systems can have very high early precision despite having low MAP.

Software

Trec_eval, a software package created by Chris Buckley[4] is commonly used for retrieval campaigns. This package computes a large array of measures including the ones specified above [27]. The ideal measure depends on the overall objective, but many information retrieval campaigns, both text-based (TREC) and image-based (ImageCLEF) use MAP as the lead metric but also consider the performance of early precision.

20.4 Examples of Successful Evaluation Campaigns

20.4.1 Registration

Image registration is another critical aspect of medical image analysis. It is used to register atlases to patients, as a step in the assessment of response to therapy in longitudinal studies (serial registration), and to superimpose images from different modalities (multi-modal registration). Traditionally, rigid and affine techniques were used for registration. More recently, deformable or non-rigid registration techniques have been used successfully for a variety of application including atlas-based segmentation, and motion tracking based on 4D CT. The evaluation of non-rigid registration can however be quite challenging as there is rarely ground truth available.

The original Retrospective Registration Evaluation Project (RREP) and the more recent Retrospective Image Registration Evaluation (RIRE)[5] are resources for researchers wishing to evaluate and compare techniques for CT-MR and PET-MR registration. The "Vanderbilt Database" is made freely available for participants. Although the "truth" transforms remain sequestered, participants can choose to submit their results on-line, enabling them to compare the performance of their algorithms to those from other groups and techniques.

The Non-rigid Image Registration Evaluation Project (NIREP)[6] is an effort to "develop, establish, maintain and endorse a standardized set of relevant benchmarks and metrics for performance evaluation of nonrigid image

---

[4] http://trec.nist.gov/trec_eval/
[5] http://www.insight-journal.org/RIRE/index.php
[6] http://www.nirep.org/index.php?id=22

registration algorithms". The organizers are planning to create a framework to evaluate registration that does not require ground truth by utilizing a diverse set of metrics instead. The database consists of 16 annotated MR images from eight normal adult males and eight females acquired at the University of Iowa. The metrics that are currently implemented include: squared intensity error, relative overlap, inverse consistency error and transitivity error.

20.4.2 Segmentation

MICCAI Grand Challenges are the most prominent of the evaluation events for segmentation. In 2007, a Grand Challenge workshop was held in conjunction with MICCAI to provide a forum for researchers to evaluate their segmentation algorithms on two anatomical sites, liver and caudate, using a common data sets and metrics. This popular workshop has continued to grow with three and four different sub-tasks in 2008 and 2009, respectively.

MICCAI Segmentation in the Clinic: A Grand Challenge

The liver is a challenging organ for CT-based segmentation as it lies near other organs that are of similar density. Additionally, in the case of diseases there can be significant non-homogeneity within the liver itself, adding to the challenge. The MICCAI Grand Challenge Workshop was one of the most prominent efforts to provide an opportunity for participants to compare the performance of different approaches to the task of liver segmentation. Twenty studies were provided as training data, while ten studies were used for the testing and an additional ten were used for the on-site portion of the evaluation. Participants were allowed to submit results from both completely automated techniques as well as interactive methods.

The training data in the caudate part (33 data sets) were acquired from two different sites using different protocols: 18 healthy controls from the Internet Brain Segmentations Repository (IBSR)[7] from Massachusetts General Hospital and 15 studies consisting of healthy and pathological subjects from Psychiatry Neuroimaging Laboratory at the Brigham and Women's Hospital, Boston. The test data consisted of 34 studies from a challenging mix of ages (adult, pediatric, elderly), sites (Brigham and Women's Hospital, Boston, UNC's Parkinson research group, University of North Carolina at Chapel Hill, Duke Image Analysis Laboratory) and acquired along different axes (axial, coronal) The gold standard was established by manual segmentation of experts.

The organizers were interested in establishing a single score that combined many of the commonly used metrics for segmentation described above. They included volumetric overlap error (or Jaccard coefficient), the relative volume

---

[7] http://www.cma.mgh.harvard.edu/ibsr/

Table 20.2. *MICCAI Grand Challenges.* For all experiments, the ground truth was provided manually

Year	Topic	Data Type	Training	Test
2007	Liver	MRI	20	10
	Caudate	MRI	33	34
2008	Lumen line	CTA	8	24
	MS lesion	MRI	20	25
	Liver tumor	CT	10 tumors from 4 patients	10 tumors from 5 patients
2009	Lumen segmentation and stenosis grading	CTA	15	31 + 10 on site
	Prostate	MRI	15	5
	Head and neck	CT	10	8 off-site, 7 online
	Left ventricle	MRI	15	15 online, 15 testing

difference, average surface symmetric distance, root mean square surface distance and the maximum symmetric surface distance. This common score was provided for both the liver and the caudate cases. In addition, the caudate evaluation consisted of a test of reproducibility by providing a set of scans for the same subject on different scanners. The variability of the score across these scans was evaluated. The Pearson correlation coefficient between the reference and the segmentation volumes was another metric provided for the caudate set.

The organizers have continued to make available all the test and training data, enabling new algorithms to be evaluated against the benchmarks established in 2007. Furthermore, the success of the Grand Challenge in 2007 lead to the continuation of this endeavor in 2008 and 2009 with more clinically-relevant segmentation tasks [28, 29], including coronary artery central lumen line extraction in CT angiography (CTA), Multiple Sclerosis (MS) lesions, and others (Table 20.2).

Extraction of Airways from CT

The Extraction of Airways from CT (EXACT)[8] challenge was held as part of the Second International Workshop on Pulmonary Image Analysis in junction with MICCAI 2009. It provides participants with a set of 20 training CTs that had been acquired at different sites using a variety of equipment, protocols, and reconstruction parameters. Participants were to provide results of algorithms for airway extraction on the 20 test sets. The results were evaluated using the branch count, branch detection, tree length, tree length detected, leakage count, leakage volume and false positive rate. Fifteen teams participated in this task. The organizers noted that "there appears to be a trade off between sensitivity and specificity in the airway tree extraction" as "more

---

[8] http://image.diku.dk/exact/information.php

complete trees are usually accompanied by a larger percentage of false positives." They also noted that the semi-automatic methods did not significantly outperform the automatic methods.

Volume Change Analysis of Nodules

Again performed in junction with MICCAI as part of the Second International Workshop on Pulmonary Image Analysis, the goal for the Volume Change Analysis of Nodules (VOLCANO)[9] challenge was to measure volumetric changes in lung lesions longitudinally using two time-separated image series. This was motivated by the notion that measuring volumetric changes in lung lesions can be useful as they can be good indicators of malignancy and good predictors of response to therapy.

The images were part of the Public Lung Database provided by the Weill Cornell Medical College. 53 nodules were available such that the nodule was visible on at least three slices on both scans. These nodules were classified into three categories: 27 nodules ranging in diameter from 4 to 24 mm visible on two 1.25 mm slice scans with little observed size change, 13 nodules ranging in size from approximately 8–30 mm, imaged using different scan slice thicknesses to evaluate the effect of slice thickness and 9 nodules ranging from 5 to 14 mm on two 1.25 mm scans exhibiting a large size change. The participants were provided with information to locate the nodule pairs. The participants were to submit the volumetric change in nodule size for each volume pair, defined as $\frac{(V_2-V_1)}{V_1}$ where V_1 and V_2 are the volumes of the nodule on the initial and subsequent scan.

20.4.3 Annotation, Classification and Detection

ImageCLEF IRMA

The automatic annotation task at ImageCLEFmed ran from 2005 until 2009 [30]. The goal in this task was to automatically classify radiographs using the Image Retrieval in Medical Applications (IRMA) code along for dimensions: acquisition modality, body orientation, body region, and biological system. The IRMA code is a hierarchical code that can classify radiographs to varying levels of specificity. In 2005, the goal was flat classification in to 57 classes while in 2006 the goal was again a flat classification into 116 unique classes. Error rates based on the number of misclassified images was used as the evaluation metric. In 2007 and 2008, the hierarchical IRMA code was used where errors were penalized depending on the level of the hierarchy at which they occurred. Typically, participants were provided 10,000–12,000 training images and were to submit classification for 1,000 test images. In 2009[10], the

---
[9] http://www.via.cornell.edu/challenge/details/index.html
[10] http://www.imageclef.org/2009/medanno/

goal was to classify 2000 test images using the different classification schemes used in 2005–2008, given a set of about 12,000 training images.

Automatic Nodule Detection

Lung cancer is a deadly cancer, often diagnosed based on lung CT's. Algorithms for the automated Computer Aided Diagnosis (CAD) for lung nodules are a popular area of research. The goal for the Automatic Nodule Detection (ANODE)[11] challenge in 2009 was the automated detection of lung nodules based on CT scans. The database consisted of 55 studies. Of these, five were annotated by expert radiologists and were used for training. Two raters (one expert and one trainee) reviewed all the scans, and a third rater was used to resolve disputes. The evaluation was based on a hit rate metric using the 2000 most suspicious hits. The results were obtained using Free-Response Receiver Operating Characteristic (FROC) curves.

Another effort towards the detection of lung nodules in the Lung Imaging Database Consortium (LIDC). The LIDC initiative provides a database of annotated lung CT images, where each image is annotated by four clinicians. This publicly available database enables researchers to compare the output of various Computer Aided Diagnosis (CAD) algorithms with the manual annotations.

20.4.4 Information Retrieval

In information retrieval, evaluation campaigns began nearly fifty ago with the Cranfield tests [23]. These experiments defined the necessity for a document collection, query tasks and ground truth for evaluation, and set the stage for much of what was to follow. The SMART experiments [24] then further systematized evaluation in the domain. The role model for most current evaluation campaign is clearly TREC [25], a series of conferences that started in 1992 and has ever since organized a variety of evaluation campaigns in diverse areas of information retrieval. A benchmark for multilingual information retrieval is CLEF [31], which started within TREC and has been an independent workshop since 2000, attracting over 200 participants in 2009. In addition to its other components, CLEF includes an image retrieval track (called ImageCLEF) which features a medical image retrieval task [32].

20.4.5 Image Retrieval

Image retrieval is a burgeoning area of research in medical informatics [33]. Effective image annotation and retrieval can be useful in the clinical care of patients, education and research. Many areas of medicine, such as radiology, dermatology, and pathology are visually-oriented, yet surprisingly little research has been done investigating how clinicians use and find images [6]. In

---

[11] http://anode09.isi.uu.nl/

particular, medical image retrieval techniques and systems are underdeveloped in medicine when compared with their textual cousins [34].

ImageCLEF[12], first began in 2003 as a response to the need for standardized test collections and evaluation forums and has grown to become today a pre-eminent venue for image retrieval evaluation. ImageCLEF itself also includes several sub-tracks concerned with various aspects of image retrieval; one of these tracks is the medical retrieval task. This medical retrieval task was first run in 2004, and has been repeated each year since.

Image Databases

The medical image retrieval track's test collection began with a teaching database of 8,000 images. For the first several years, the ImageCLEF medical retrieval test collection was an amalgamation of several teaching case files in English, French, and German. By 2007, it had grown to a collection of over 66,000 images from several teaching collections, as well as a set of topics that were known to be well-suited for textual, visual or mixed retrieval methods.

In 2008, images from the medical literature were used for the first time, moving the task one step closer towards applications that could be of interest in clinical scenarios. Both in 2008 and 2009, the Radiological Society of North America (RSNA) made a subset of its journals' image collections available for use by participants in the ImageCLEF campaign. The 2009 database contained a total of 74,902 images, the largest collection yet. All images were taken from the journals Radiology and Radiographics, both published by the RSNA. The ImageCLEF collection is similar in composition to that powering the Goldminer[13] search system. This collection constitutes an important body of medical knowledge from the peer-reviewed scientific literature, and includes high quality images with textual annotations.

Images are associated with specific published journal articles, and as such may represent either an entire figure or a component of a larger figure. In either event, the image annotations in the collection contain the appropriate caption text. These high-quality annotations enable textual searching in addition to content-based retrieval using the image's visual features. Furthermore, as the PubMed IDs of each image's article are also part of the collection, participants may access bibliographic metadata such as the Medical Subject Headings (MeSH) terms created by the National Library of Medicine for PubMed.

Goals of ImageCLEF

A major goal of ImageCLEF has been to foster development and growth of multi-modal retrieval techniques: i.e., retrieval techniques that combine

---

[12] http://www.imageclef.org/
[13] http://goldminer.arrs.org/

visual, textual, and other methods to improve retrieval performance. Traditionally, image retrieval systems have been primarily text-based, relying on the textual annotations or captions associated with images [35]. Several commercial systems, such as Google Images[14] and Yahoo! images,[15] employ this approach. Although text-based information retrieval methods are mature and well-researched, they are limited by the quality of the annotations applied to the images. There are other important limitations facing traditional text retrieval techniques when applied to image annotations:

- Image annotations are subjective and context sensitive, and can be quite limited in scope or even completely absent.
- Manually annotating images is labor- and time-intensive, and can be very error prone.
- Image annotations are very noisy if they are automatically extracted from the surrounding text.
- There is far more information in an image than can be abstracted using a limited number of words.

Advances in techniques in computer vision have led to a second family of methods for image retrieval: Content-Based Image Retrieval (CBIR). In a CBIR system, the visual contents of the image itself are mathematically abstracted and compared to similar abstractions of all images in the database. These visual features often include the color, shape or texture of images. Typically, such systems present the user with an ordered list of images that are visually most similar to the sample (or query) image.

However, purely visual methods have been shown to have limitations and typically suffer from poor performance for many clinical tasks [36]. On the other hand, combining text- and image-based methods has shown promising results [37].

Several user studies have been performed to study the image searching behavior of clinicians [6, 38]. These studies have been used to inform the development of the tasks over the years, particularly to help ImageCLEF's organizers identify realistic search topics.

User Studies

The goal in creating search topics for the ImageCLEF medical retrieval task has been to identify typical information needs for a variety of users. In the past, we have used search logs from different medical websites to identify topics [39, 40]. The starting point for the 2009 topics was a user study conducted at Oregon Health & Science University (OHSU) during early 2009. This study was conducted with 37 medical practitioners in order to understand their needs, both met and unmet, regarding medical image retrieval.

---

[14] http://images.google.com/
[15] http://images.yahoo.com/

During the study, participants were given the opportunity to use a variety of medical and general-purpose image retrieval systems, and were asked to report their search queries. In total, the 37 participants used the demonstrated systems to perform a total of 95 searches using textual queries in English. We randomly selected 25 candidate queries from the 95 searches to create the topics for ImageCLEFmed 2009. We added to each candidate query 2–4 sample images from the previous collections of ImageCLEFmed, which represented visual queries for content-based retrieval. Additionally, we provided French and German translations of the original textual description for each topic to allow for an evaluation of multilingual retrieval.

Finally, the resulting set of topics was categorized into three groups: 10 visual topics, 10 mixed topics, and 5 semantic topics. This classification was performed by the organizers based on their knowledge of the capabilities of visual and textual search techniques, prior experience with the performance of textual and visual systems at ImageCLEF medical retrieval task, and their familiarity with the test collection. The entire set of topics was finally approved by a physician. An example of a visual topic can be seen in Fig. 20.2 while that of a textual topic is shown in Fig. 20.3.

In 2009, we also introduced case-based topics [4] as part of an exploratory task whose goal was to generate search topics that are potentially more aligned with the information needs of an actual clinician in practice. These topics were meant to simulate the use case of a clinician who is diagnosing a difficult case, and has information about the patient's demographics, list of present symptoms, and imaging studies, but not the patient's final diagnosis. Providing this clinician with articles from the literature that deal with cases similar to the case (s)he is working on (similar based on images and other clinical data on the patient) could be a valuable aide to creating differential diagnosis or identifying treatment options, for example, with case-based reasoning [41]. These case-based search topics were created based on cases from the teaching file Casimage, which contains cases (including images) from radiological practice.

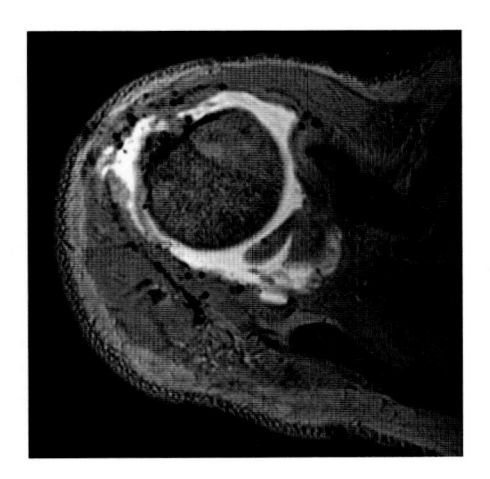

Fig. 20.2. *Example image of a visual query task.* The tasks consists of two or three images and a textual description in three languages, in this case representing the information need "MRI of a rotator cuff"

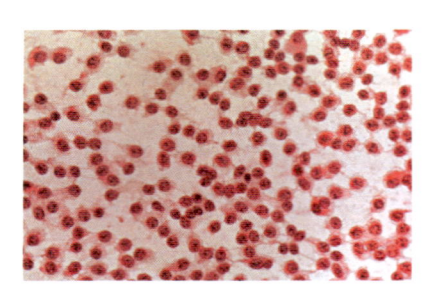

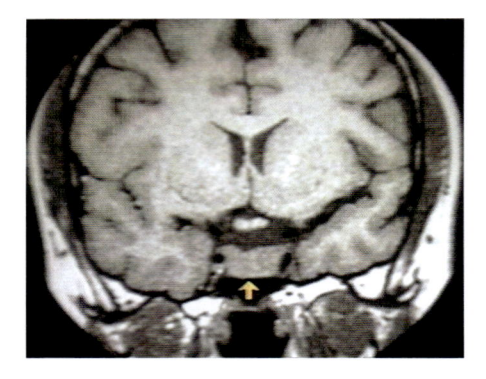

Fig. 20.3. *Sample images of a semantic retrieval task.* The need of information "images of pituitary adenoma" yields a semantic query topic with potentially a large variety of visually quite different images of various modalities and, thus, it is better suited for techniques of textual information retrieval

Ten cases were pre-selected, and a search with the final diagnosis was performed against the 2009 ImageCLEF data set to make sure that there were at least a few matching articles. Five topics were finally chosen. The diagnoses and all information about the chosen treatment were removed from the cases to simulate the aforementioned situation of a clinician dealing with a difficult diagnosis. However, in order to make the judging more consistent, the relevance judges were provided with the original diagnosis for each case.

Relevance Judgments

During 2008 and 2009, relevance judgments were made by a panel of clinicians using a web-based interface. Due to the infeasibility of manually reviewing 74,900 images for 30 topics, the organizers used a TREC-style pooling system to reduce the number of candidate images for each topic to approximately 1,000 by combining the top 40 images from each of the participants' runs. Each judge was responsible for between three and five topics, and sixteen of the thirty topics were judged multiple times (in order to allow evaluation of inter-rater agreement). For the image-based topics, each judge was presented with the topic as well as several sample images.

For the case-based topics, the judge was shown the original case description and several images appearing in the original article's text. Besides a short description for the judgments, a full document was prepared to describe the judging process, including what should be regarded as relevant versus non-relevant. A ternary judgment scheme was used, wherein each image in each pool was judged to be "relevant", "partly relevant", or "non-relevant". Images clearly corresponding to all criteria were judged as "relevant", images whose relevance could not be safely confirmed but could still be possible were marked as "partly relevant", and images for which one or more criteria of the topic were not met were marked as "non-relevant". Judges were instructed in these criteria and results were manually verified during the judgment process.

As mentioned, we had a sufficient number of judges to perform multiple judgements on many topics, both image-based and case-based. Inter-rater agreement was assessed using the kappa metric, given as:

$$\kappa = \frac{Pr(a) - Pr(e)}{1 - Pr(e)} \tag{20.16}$$

where $Pr(e)$ is the observed agreement between judges, and $Pr(a)$ the expected (random) agreement. It is generally accepted that a $\kappa > 0.7$ is good and sufficient for an evaluation. The score is calculated using a 2×2 table for the relevances of images or articles. These were calculated using both lenient and strict judgment rules. Under the lenient rules, a partly relevant judgment was counted as relevant; under strict rules, partly relevant judgments were considered to be non-relevant. In general, the agreement between the judges was fairly high (with a few exceptions), and our 2009 overall average κ is similar to that found during other evaluation campaigns.

20.5 Lessons Learned

Conducting the ImageCLEF campaigns has been a great learning opportunity for the organizers. Most evaluation campaigns are run by volunteers with meager resources. However, a surprising number of researchers willingly donate their data, time and expertise towards these efforts as they truly believe that progress in the field can only come as a result of these endeavors.

Participants have been quite loyal for the ImageCLEFmed challenge, an annual challenge that has been running since 2004. Many groups have participated for four or more years although each year sees newcomers, a welcome addition. A large proportion of participants are actually PhD students who obtain valuable data to validate their approaches. The participants have been quite cooperative, both at the workshops and during the year. They have provided baseline runs or allowed their runs to be used by others in collaborative efforts. Many of the new organizers were participants, thus ensuring a steady stream of new volunteers willing to carry on the mantle of those that have moved away. By comparing the relative performance of a baseline run through the years, we have seen the significant advances being made in the field.

20.6 Conclusions

Evaluation is an important facet of the process of developing algorithms for medical image analysis including for segmentation, registration and retrieval. In order to be able to measure improvements resulting from new research in computer vision, image processing and machine learning when applied to medical imaging tasks, it is important to have established benchmarks against which their performance can be compared. Computer scientists are making huge strides in computer vision, image processing and machine learning, and clinicians and hospitals are creating vast quantities of images each day. However, it can still be quite difficult for the researchers developing the algorithms to have access to high quality, well curated data and ground truth. Similarly, it can also be quite difficult for clinicians to get access to state-of-the-art algorithms that might be helpful in improving their efficiency, easing their workflow and reducing variability.

Evaluation campaigns have provided a forum to bridge this gap by providing large, realistic and well annotated datasets, ground truth, meaningful metrics geared specifically for the clinical task, organizational resources including informational websites and software for evaluation and often workshops for researchers to present their results and have discussions. Examples of successful evaluation campaigns include ImageCLEFmed for medical image retrieval and annotation, the VOLCANO challenge to assess volumetric changes in lung nodules, the EXACT airway extraction challenge and the popular set of MICCAI segmentation grand challenges. Other efforts to provide publicly accessible data and ground truth include the LIDC set of images for the detection of chest nodules based on CTs, the CT and PET images from the ADNI initiative, and the RIRE and NIREP efforts to evaluate registration. Many of these efforts are continuing beyond the workshops by still enabling participants to download data, submit results, evaluating and posting the results, thereby providing venues for the progress in the field to be documented.

References

1. Müller H, Müller W, Squire DM. Performance evaluation in content–based image retrieval: overview and proposals. Pattern Recognit Lett. 2001;22(5):593–601.
2. Price K. Anything you can do, I can do better (no you can't)... Comput Vis Graph Image Process. 1986;36(2–3):387–91.
3. Heimann T, Styner M, Ginneken B. 3D segmentation in the clinic: a grand challenge. MIDAS J. 2007. http://mbi.dkfz-heidelberg.de/grand-challenge2007/.
4. Müller H, Kalpathy-Cramer J, Eggel I, et al. Overview of the CLEF 2009 medical image retrieval track. In: Working Notes of CLEF 2009. Corfu, Greece; 2009.
5. Markkula M, Sormunen E. Searching for photos: journalists' practices in pictorial information retrieval. In: Proc The Challenge of Image Retreival. Newcastle upon Tyne. vol. 56;1998. p. 1–13.
6. Müller H, Despont-Gros C, Hersh W, et al. Health care professionals's image use and search behaviour. Proceedings of Medical Informatics Europe (MIE 2006), Maastricht, Netherlands. 2006; p. 24–32.

7. Hersh W, Müller H, Gorman P, et al. Task analysis for evaluating image retrieval systems in the ImageCLEF biomedical image retrieval task. In: Proc SOL 2005. Portland, OR, USA; 2005.
8. Deselaers T, Deserno TM, Müller H. Automatic medical image annotation in ImageCLEF 2007: overview, results, and discussion. Pattern Recognit Lett. 2008;29(15):1988–95.
9. Davatzikos C, Xu F, An Y. Longitudinal progression of Alzheimer's-like patterns of atrophy in normal older adults: the SPAREAD index. Brain 2009;132(8):2026–35.
10. Armato SG, McNitt-Gray MF, Reeves AP. The lung image database consortium (LIDC): an evaluation of radiologist variability in the identification of lung nodules on CT scans. Acad Radiol. 2007;14(11):1409–21.
11. Müller SG, Weiner MW, Thal LJ. Ways toward an early diagnosis in alzheimer's disease: the alzheimer's disease neuroimaging initiative (ADNI). Alzheimers Dement. 2005;1(1):55–66.
12. Rubin GD, Lyo JK, Paik DS. Pulmonary nodules on multi–detector row CT scans: performance comparison of radiologists and computer–aided detection. Radiology 2005;234(1):274–83.
13. Christensen G, Geng X, Kuhl J, et al. Introduction to the non-rigid image registration evaluation project (NIREP). Lect Notes Comput Sci. 2006;4057:128–35
14. Warfield S, Zou K, Wells W. Simultaneous truth and performance level estimation (STAPLE): an algorithm for the validation of image segmentation. IEEE Trans Med Imaging. 2004;23(7):903–21.
15. Babalola KO, Patenaude B, Aljabar P, et al. Comparison and evaluation of segmentation techniques for subcortical structures in brain MRI. Lect Notes Comput Sci. 2008;5241:409–16.
16. Dice LR. Measures of the amount of ecologic association between species ecology. J Ecology. 1945;26:297–302.
17. Jaccard P. The distribution of the flora in the alpine zone. New Phytol. 1912;11(2):37–50.
18. Zijdenbos AP, Dawant BM, Margolin RA. Morphometric analysis of white matter lesions in MR images: method and validation. Med Imaging. 1994;13(4):716–24.
19. Williams GW. Comparing the joint agreement of several raters with another rater. Biometrics 1976;32(4):619–27.
20. Martin-Fernandez M, Bouix, Ungar L, et al. Two methods for validating brain tissue classifiers. Lect Notes Comput Sci. 2005; p. 515–22.
21. Crum W, Camara O, Hill D. Generalized overlap measures for evaluation and validation in medical image analysis. IEEE Trans Med Imaging. 2006;25(11):1451–61.
22. Gerig G, Jomier M, Chakos A. VALMET: a new validation tool for assessing and improving 3D object segmentation. Lect Notes Comput Sci. 2001;2208:516–23.
23. Cleverdon CW. Report on the Testing and Analysis of an Investigation into the Comparative Efficiency of Indexing Systems. Cranfield, USA: Aslib Cranfield Research Project; 1962.
24. Salton G. The SMART Retrieval System, Experiments in Automatic Document Processing. Englewood Cliffs, New Jersey, USA: Prentice Hall; 1971.
25. Voorhees EM, Harmann D. Overview of the seventh text retrieval conference (TREC–7). In: The Seventh Text Retrieval Conference Gaithersburg, MD, USA. 1999; p. 1–23.

26. van Rijsbergen CJ. Information Retrieval. Englewood Cliffs, New Jersey, USA: Prentice Hall; 1979.
27. Voorhees EM. Variations in relevance judgements and the measurement of retrieval effectiveness. Inf Process Manage. 2000;36(5):697–716.
28. Styner M, Lee J, Chin B. 3D segmentation in the clinic: a grand challenge II: MS lesion. MIDAS J. 2008.
29. Hameeteman. Carotid lumen segmentation and stenosis grading challenge. MIDAS J. 2009;1–15.
30. Tommasi T, Caputo B, Welter P, Guld M, Deserno TM. Overview of the CLEF 2009 medical image annotation track. Lect Notes Comput Sci. 2010;6242:85–93.
31. Savoy J. Report on CLEF-2001 experiments. Lect Notes Comput Sci. 2002;2406:27–43.
32. Müller H, Deselaers T, Kim E. Overview of the ImageCLEFmed 2007 medical retrieval and annotation tasks. Lect Notes Comput Sci. 2008;5152:473–91.
33. Müller H, Michoux N, Bandon D. A review of content-based image retrieval systems in medicine–clinical benefits and future directions. Int J Med Inform. 2004;73(1):1–23.
34. Hersh W. Information Retrieval: A Health and Biomedical Perspective. 2n ed. New York: Springer; 2003.
35. Enser PGB. Pictorial information retrieval. J Doc. 1995;51(2):126–70.
36. Müller H, Kalpathy-Cramer J, Kahn CE Jr. Overview of the ImageCLEFmed 2008 medical image retrieval task. Lect Notes Comput Sci. 2009;5706:500–10.
37. Kalpathy-Cramer J, Bedrick S, Hatt W, et al. Multimodal medical image retrieval: OHSU at ImageCLEF 2008. In: Working Notes of the 2008 CLEF Workshop. Aarhus, Denmark; 2008.
38. Hersh W, Jensen J, Müller H. A qualitative task analysis for developing an image retrieval test collection. In: ImageCLEF/MUSCLE workshop on image retrieval evaluation. 2005; p. 11–16.
39. Müller H, Boyer C, Gaudinat A. Analyzing web log files of the health on the net honmedia search engine to define typical image search tasks for image retrieval evaluation. Stud Health Techn Inform. 2007;12:1319–23.
40. Müller H, Kalpathy-Cramer J, Hersh W, et al. Using Medline queries to generate image retrieval tasks for benchmarking. In: Medical Informatics Europe (MIE2008). Gothenburg, Sweden: IOS press; 2008. p. 523–528.
41. Aamodt A, Plaza E. Case–based reasoning: foundational issues, methodological variations, and systems approaches. Artiff Intell Commun. 1994;7(1):39–59.

21

Toolkits and Software for Developing Biomedical Image Processing and Analysis Applications

Ivo Wolf

Summary. Solutions in biomedical image processing and analysis usually consist of much more than a single method. Typically, a whole pipeline of algorithms is necessary, combined with visualization components to display and verify the results as well as possibilities to interact with the data. Therefore, successful research in biomedical image processing and analysis requires a solid base to start from. This is the case regardless whether the goal is the development of a new method (e.g., for segmentation) or to solve a specific task (e.g., computer-assisted planning of surgery).

This chapter gives an overview of toolkits and software that support the development of biomedical image processing and analysis applications. After the initial introduction, Sect. 21.2 outlines toolkits and libraries that provide sets of already implemented methods for image processing, visualization, and data management. Section 21.3 covers development environments that offer a specialized programming language or visual programming interface. Section 21.4 describes ready-to-use software applications allowing extensions by self-written code. All sections begin with open-source developments.

21.1 Introduction

The development of biomedical image processing and analysis applications can be supported on different levels. The available support and tools can be categorized as follows:

- *Toolkits*, which are basically (class) libraries that provide support for specific tasks and offer the highest level of adaptability
- *Development environments*, which provide a comprehensive application for use during development
- *Extensible software*, which are ready-to-use end-user applications that can be extended by custom code

Which category to use depends on the problem at hand. Solutions to real world problems require all levels: starting with the exploration of algorithms, ending at a tailored solution running within a graphical front-end.

T.M. Deserno (ed.), *Biomedical Image Processing*, Biological and Medical Physics,
Biomedical Engineering, DOI: 10.1007/978-3-642-15816-2_21,
© Springer-Verlag Berlin Heidelberg 2011

The boundaries between the categories are not always clear. Development environments and extensible software applications are usually based on pre-existing toolkits. Some provide additions to the underlying respective toolkits. Conversely, some toolkits offer powerful showcase applications, partly with the possibility to extend them with custom code.

21.2 Toolkits

Toolkits provide sets of already implemented methods for image processing, visualization, and data management. Typically, toolkits are object-oriented class libraries implemented in standard programming languages. Most frequently, C++ is the base language, but quite often wrappings are provided for scripting languages such as Python, the Tool Command Language (TCL) or Java.

The use of toolkits typically requires a rather high level of programming experience and knowledge, but they provide the highest level of flexibility and adaptability to the problem at hand, especially when the development is open-source. Figure 21.1 tries to visualize the usage of toolkits, which can range from building small scale tools to large multi-purpose applications.

21.2.1 The NA-MIC Kit

The multi-institutional National Alliance for Medical Image Computing (NA-MIC) is one of the seven National Centers for Biomedical Computing (NCBC) funded under the National Institutes of Health (NIH) *Roadmap for Bioinformatics and Computational Biology*. To enable research in medical image computing, NA-MIC develops and maintains the NA-MIC Kit [1] and organizes training events. The NA-MIC Kit is a set of tools and toolkits (Table 21.1),

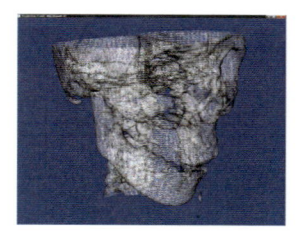

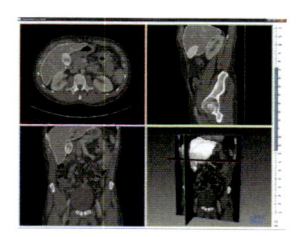

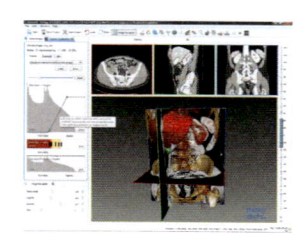

Fig. 21.1. *Solving small to large problems using toolkits. Left*: The VTK-based tool for surface extraction and visualization requires about 50 lines of code. *Middle*: The viewer for multi-planar reformation of images and surface data is based on MITK (which in turn uses VTK and ITK) and requires about 80 lines of code. *Right*: The extensible, multi-purpose end-user application (MITK ExtApp) is additionally based on the BlueBerry toolkit. The version shown has 17 modules consisting of about 16,000 lines of code

Table 21.1. *NA-MIC Kit*

Language:	Mainly C++, parts with wrappings for TCL, Python, Java
Platform:	Windows, Linux, Unix, Mac
License:	BSD-style
Developer:	Multi-institutional
Availability:	Open-source
URL:	www.na-mic.org/Wiki/index.php/NA-MIC-Kit

Table 21.2. *ITK*. The developer consortium consists of six principal organizations, three commercial (including Kitware) and three academic

Language:	C++, wrappings for Tcl, Python, Java
Platform:	Linux, Windows, Mac, Unix
License:	BSD-style
Developer:	Insight Software Consortium
Availability:	open-source
URL:	www.itk.org

which are all free open source software distributed under a Berkeley Software Distribution (BSD)-style license without restrictions (including possible commercial use). Thus, the NA-MIC Kit is itself not a single toolkit or software, but a collection of toolkits and software. It includes:

- 3D Slicer, an extensible software package for visualization and medical image computing (cf. Sect. 21.4.1)
- The Insight Segmentation and Registration Toolkit (ITK) and the Visualization Toolkit (VTK) (cf. Sects. 21.2.2 and 21.2.3)
- KWWidgets, which is a Graphical User Interface (GUI) class library based on Tcl/Tk with a C++ Application Programming Interface (API)

Additionally, it defines a software engineering methodology and provides a number of tools to support this methodology, including support for automatic testing for quality assurance and multi-platform implementations.

21.2.2 Insight Segmentation and Registration Toolkit

The Insight Segmentation and Registration Toolkit (ITK) is probably the most widely used toolkit in medical image processing. It provides an extensive set of (multi-dimensional) algorithms for almost all kinds of image processing tasks with a special focus on segmentation and registration (Table 21.2).

The design of the registration framework separates the registration process into four pluggable components, which can easily be interchanged: similarity metric, transform, optimizer, and interpolation (e.g. nearest-neighbor, linear, B-spline, windowed-sinc). Available metrics include simple measures such as mean squares and normalized cross-correlation as well as different types of mutual information and a Kullback–Leibler distance measure. Transforms include rigid, similarity, affine as well as parametric deformable methods such as B-spline and kernel-based transforms (e.g., elastic body splines and thin-plate splines). ITK's optimization algorithms are generic and can be used for applications other than registration. Examples include gradient descent, conjugate gradient, Nelder–Mead downhill simplex, Powell optimization and many more.

The segmentation framework includes standard methods such as Otsu thresholding and different types of region growing as well as advanced methods like level sets, including a variant with shape guidance, watershed segmentation, and fuzzy connectedness.

The execution of ITK algorithms is based on a demand-driven pipeline concept: an algorithm is only executed when its output data objects are requested and are out-of-date with respect to its input data objects.

ITK is based on advanced C++ constructs, especially templates. For example, the ITK class for image data is templated over the dimension of the image and its data type.

ITK does *not* include methods for displaying images, nor a development environment or an end-user application for exploring the implemented algorithms. Many of the features of the toolkit are described with source code examples in the book entitled *The ITK Software Guide* (with more than 780 pages), which is available for free as a PDF document from the ITK website (also available in a printed version). Additionally, several example applications demonstrating sub-sets of the algorithms included in ITK are available.

Most of the development environments described in Sect. 21.3 provide wrappings of ITK algorithms. An end-user application providing several ITK algorithms is VolView (cf. Sect. 21.4.8).

21.2.3 The Visualization Toolkit

The Visualization Toolkit (VTK) is for visualization what ITK is for image processing algorithms: one of the most popular toolkits in its area. It offers methods for scalar, vector, tensor, texture, and volume visualization (Table 21.3).

Additionally, a large number of algorithms are provided for 3D computer graphics (like modeling methods) as well as for image processing. Examples include implicit modeling, polygon reduction, mesh smoothing, cutting, contouring, marching cubes surface extraction, and Delaunay triangulation. Data processing in VTK is based on a pipeline concept, which was originally similar to that of ITK (or vice versa), but has since been reworked and generalized with VTK version 5. It now allows implementation of updating mechanisms other than the demand-driven concept, which is still principally used.

Table 21.3. *VTK*

Language:	C++, wrappings for Tcl, Python, Java
Platform:	Windows, Linux, Unix, Mac
License:	BSD-style
Developer:	Kitware Inc.
Availability:	open-source
URL:	www.vtk.org

Mechanisms for interaction with the visualized data are available including predefined classes for common tasks.

As is the case with ITK, VTK does neither come with a development environment nor an end-user application. There are, however, several frontends available that can be used to interact with VTK's algorithms and build visualizations: ParaView (cf. Sect. 21.4.5) is an example for an end-user type of application for this purpose, development environments with VTK-support are OpenXIP (cf. Sect. 21.3.2), DeVIDE (cf. Sect. 21.3.3), and MeVisLab (cf. Sect. 21.3.6).

21.2.4 Open Inventor

Open Inventor^{TM} is an object-oriented wrapping of OpenGL® written in C++ (Table 21.4). The central paradigm of Open Inventor is the scene graph. A scene graph is an ordered collection of *nodes* [2]. Each node holds some piece of information about the scene, such as material, shape description, transformation, or light. So-called *actions* can be performed on a scene graph, the most important being the rendering of the scene graph. Other actions are picking, searching, computation of bounding boxes, and writing to files. When performing an action (e.g., rendering), the scene graph is traversed, starting from the root node, from top to bottom and left to right. Open Inventor manages a so-called *traversal state*, which is a collection of elements or parameters in the action at a given time. During traversal, the nodes modify the traversal state, depending on their particular behavior for that action.

This type of scene graph paradigm is quite different from the one VTK uses. In VTK, an object in a scene is a shape and "owns" its (relative) transformation and properties (e.g., color, texture). In Open Inventor, a shape object does not have a transformation or properties without the context of a scene graph traversal.

Open Inventor also has a data-flow pipeline concept, whose objects are called *engines*. Notifications of changed values are pushed through the scene graph, but the evaluation of an engine (processing object) is pulled through the scene graph on demand [3].

Scene graphs including engines with their respective parameters can be serialized into files. This also works for custom extensions. When de-serializing and the code for an extension is not available, the respective node/engine is included as an unknown-node/unknown-engine object. In this case, one can at least perform reading, writing and searching on unknown nodes [2].

Open Inventor has originally been developed by SGI Inc. in the 1990s. The SGI version is open source under the Lesser General Public License (LGPL) of the GNU's Not Unix! (GNU) organization. Commercially supported, extended

Table 21.4. *Open Inventor*

Language:	C++
Platform:	Linux, IRIX; ports for Windows, Mac
License:	LGPL (original version, for commercial versions: see text)
Developer:	Silicon Graphics Inc.
Availability:	open-source
URL:	oss.sgi.com/projects/inventor

Table 21.5. *MITK*. The package is developed by the Division of Medical and Biological Informatics, Deutsches Krebsforschungszentrum (DKFZ)

Language:	C++
Platform:	Windows, Linux, Mac
License:	BSD-style
Developer:	German Cancer Research Center (DKFZ)
Availability:	open-source
URL:	www.mitk.org

versions of the API are available from the Visualization Sciences Group (VSG)[1] and Kongsberg SIM (Coin3D)[2].

As in the case of VTK, Open Inventor does not come with an application and was not specifically designed for biomedical imaging, but it is successfully used as the base for OpenXIP and MeVisLab (Sects. 21.3.2 and 21.3.6).

21.2.5 Medical Imaging Interaction Toolkit

The Medical Imaging Interaction Toolkit (MITK) supports the development of interactive medical imaging software [4] (Table 21.5). Based on ITK and VTK, MITK adds features required in interactive systems that are out of the scope of ITK and VTK.

MITK uses a data-centric scene concept. Data objects (e.g., image, surface) are added to a data repository called "data storage", which is then passed to one or (typically) several views for visualization and interaction. This allows multiple, consistent 2D and/or 3D views on the same data (e.g., three orthogonal 2D views and a 3D rendering) without additional custom code for coordinating the contents of the views. The data storage also allows the definition of semantic relationships between data object, for example, from which original image a segmentation has been derived or that the left atrium is part of the heart. MITK provides support for time-resolved (3D+t) data, an undo/redo concept for interactions, data object properties of arbitrary type, serialization of the data storage, and a module for Image-Guided Therapy (IGT) with support for several tracking systems.

MITK itself is *not* an application framework and can be used within existing applications. Additionally, MITK has an optional layer on the application level, which is based on BlueBerry. BlueBerry, which can also be used independently of MITK, is a modular, cross-platform, C++ application framework

---

[1] www.vsg3d.com/vsg_prod_openinventor.php
[2] www.coin3d.org

based on the well-established ideas from the Open Services Gateway initiative (OSGi)[3] and the Eclipse Rich Client Platform (RPC). Except for a small kernel, all the framework functionality is located in plug-ins. Each plug-in has a *manifest* file declaring its interconnections to other plug-ins: Any number of named *extension points*, and any number of extensions to one or more extension points in other plug-ins can be defined. Thus, a plug-in's extension points can be extended by other plug-ins. BlueBerry itself and the MITK-based application layer provide several useful *services*, also implemented as plug-ins, for typical tasks like logging.

On top of the BlueBerry-based MITK application layer, an open-source, extensible end-user application using Qt (version 4) for the GUI is provided (Sect. 21.4.2).

Medical Imaging Toolkit

There is another development with the same abbreviation: the Medical Imaging Toolkit[4], which comes with an extensible application system called 3DMed [5] (Table 21.6). Contrary to the aforementioned MITK, the Medical Imaging Toolkit is *not* based on VTK or ITK, but inspired by them, and independently implements similar concepts such as a data flow pipeline. It consists of a computational framework providing processing algorithms and a visualization and interaction framework for displaying data. 3DMed is a Qt-based application with a plug-in interface. The basic functions include data I/O, 2D manipulation, image segmentation and registration, 3D visualization and measurement, and virtual cutting [5]. Both, toolkit and application, are implemented in C++ and free for use in research and education. 3DMed is open-source whereas the underlying toolkit is open-interface only [5].

21.2.6 The Image-Guided Surgery Toolkit

The Image-Guided Surgery Toolkit (IGSTK) is dedicated to providing common functionality for Image-Guided Surgery (IGS) applications (Table 21.7).

Language:	C++
Platform:	Windows, Linux, Mac
License:	proprietary
Developer:	Chinese Academy of Sciences (CAS)
Availability:	open-interface, 3D-Med: open-source
URL:	www.mitk.net

Table 21.6. *Medical Imaging Toolkit.* The toolkit is developed by the Medical Image Processing Group, Institute of Automation, Chinese Academy of Sciences (CAS), and is free for research and education

---

[3] www.osgi.org
[4] www.mitk.net

Table 21.7. *IGSTK*. The developer consortium consists of six participants including Georgetown University and Kitware Inc.

Language:	C++
Platform:	Windows, Linux, Mac
License:	BSD-style
Developer:	Insight Software Consortium
Availability:	open-source
URL:	www.igstk.org

Table 21.8. *OpenMAF*. The developers include the BioComputing Competence Centre, the Rizzoli Institute, the CINECA Supercomputing Center (all Italy), and the University of Bedfordshire, United Kingdom

Language:	C++
Platform:	Windows, Linux
License:	BSD-style
Developer:	group of institutions
Availability:	open-source
URL:	www.openmaf.org

With a strong focus on robustness, which is supported by a state machine concept to prevent unintended use of classes, IGSTK provides a set of high-level components integrated with low-level open-source software libraries and APIs from hardware vendors, especially for tracking systems. IGSTK is based on ITK and VTK. Additionally, optional components for graphical user interfaces are available (for Qt, FLTK).

21.2.7 The Multimod Application Framework

OpenMAF or Multimod Application Framework (MAF) provides high level components that can be combined to develop vertical, multi-modal visualization applications (Table 21.8). Based on VTK, the rendering mechanism of OpenMAF allows different, synchronized views of the same data.

Components in OpenMAF control *data entities* and *application services*. Data entities are classes and objects that represent the data. They are called Virtual Medical Entities (VME). Application services are:

- *View*, allowing display and examination of data
- *Operation*, allowing data alteration
- *Interaction*
- *Interface element*, allowing user-interface communication

Data entities and services are managed by associated *manager* components. An OpenMAF application is an instance of the *logic* component, which essentially controls the communication between the underlying components.

Currently, a new version of OpenMAF is in development (MAF3)[5], which will provide a higher level of modularity, a plug-in extension mechanism, scripting integration, and multi-threading support.

---

[5] biomedtown.org/biomed_town/maf3

Language:	C++
Platform:	Windows, Linux
License:	CeCILL2
Developer:	INRIA Sophia Antipolis, France
Availability:	open-source
URL:	www-sop.inria.fr/asclepios/software/ /vtkINRIA3D

Table 21.9. *vtkINRIA3D*

Language:	mixture of ANSI C and C++
Platform:	Windows, Linux, Unix
License:	BSD-style
Developer:	OFFIS e.V., Oldenburg, Germany
Availability:	open-source
URL:	dicom.offis.de/dcmtk.php.en

Table 21.10. *DCMTK*. The developing center is called Oldenburger Forschungs- und Entwicklungsinstitut für Informatik (OFFIS)

21.2.8 vtkINRIA3D

The vtkINRIA3D toolkit is an extension of VTK providing synchronization of views and interaction, and management of spatio-temporal data [6] (Table 21.9). It consists of three libraries:

1. The *vtkRenderingAddOn* library implements the strategy for synchronization of views and interactions, which is based on a cyclic directed tree (each view has a unique parent) structure. When a synchronizing method of any view is called, the synchronization request is transmitted to the children and hence to all nodes of the cyclic structure (avoiding infinite loops).
2. The *vtkVisuManagement* library provides support for specific complex data types, like tensor fields and neural fibers from Diffusion Tensor Imaging (DTI). It is a combination of VTK classes into new classes to simplify the incorporation of complex data manipulation and visualization into custom software.
3. The *vtkDataManagement* library adds support for time sequences by providing a container for several instances of VTK data objects (one per point in time) and simplifying their creation and manipulation.

21.2.9 OFFIS DICOM ToolKit

The OFFIS DICOM ToolKit (DCMTK) is a widely used collection of libraries and applications implementing large parts of the Digital Imaging and Communications in Medicine (DICOM) standard (Table 21.10). This includes DICOM image file handling, sending and receiving images over a network connection, image storage, and a worklist server. DCMTK provides methods for data encoding/decoding, compression/decompression, presentation states, digital signatures, and DICOM Structured Reporting (DICOM-SR). However, DCMTK does *not* include functions for image display.

Table 21.11. *GDCM*. Java and PHP support are currently in testing and experimental state, respectively	Language:	C++; wrappings: Python, C, Java, PHP
	Platform:	Windows, Linux, Unix
	License:	BSD-style
	Developer:	Mathieu Malaterre (project leader)
	Availability:	open-source
	URL:	gdcm.sourceforge.net

Table 21.12. *CTK*. The Common Toolkit is a multi-institutional effort, which is still in its initial phase	Language:	C++
	Platform:	Windows, Linux, Mac
	License:	BSD-style
	Availability:	open-source
	URL:	www.commontk.org

21.2.10 Grassroots DICOM Library

Grassroots DICOM Library (GDCM) supports reading/parsing and writing of DICOM files as specified in Part 10 of the DICOM standard (Table 21.11). GDCM attempts to support all possible DICOM image encodings including lossy and lossless compression. Contrary to DCMTK (cf. previous section), GDCM does *not* include methods for networking. GDCM can be used as part of ITK and provides classes for reading and writing VTK data objects.

21.2.11 The Common Toolkit

Since June 2009, the Common Toolkit (CTK) is organized by a pro-tempore group of like-minded technically oriented software tool builders (Table 21.12). The goals are to:

- Provide a unified set of basic features for use in medical imaging using BSD-style licenses
- Facilitate the exchange and combination of code and data
- Document, integrate, and adapt successful solutions
- Avoid duplication of code and data
- Continuously extend to new tasks within the scope of the toolkit (medical imaging) without burdening existing tasks

21.2.12 Simulation Open Framework Architecture

The Simulation Open Framework Architecture (SOFA) is targeted at real-time simulation, with an emphasis on medical simulation (Table 21.13). SOFA supports the development of simulation algorithms and prototyping simulation applications [7].

In SOFA, simulation components, i.e., data objects (e.g., deformable model, instrument) can have several representations (multi-model representation), which are connected through a mechanism called *mapping*. This allows each representation to be optimized for a particular task (e.g., collision detection, visualization) as well as interaction of different model types like rigid

Language:	C++
Platform:	Windows, Linux, Mac
License:	LGPL
Developer:	INRIA, France
Availability:	open-source
URL:	www.sofa-framework.org

Table 21.13. *SOFA*. The toolkit is developed mainly by three INRIA teams: Alcove, Evasion, and Asclepios, all in France

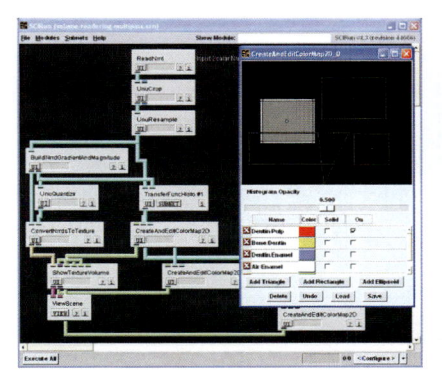

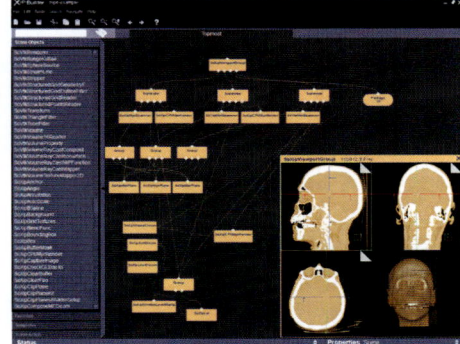

Fig. 21.2. *Visual programming environments (VPEs).* VPEs allow defining dataflows and/or rendering tasks by visually connecting building blocks. *Left*: SCIRun; *Right*: XIP-Builder

bodies, deformable objects, and fluids. SOFA uses a scene-graph concept to organize and process the elements of a simulation while clearly separating the computation tasks from their possibly parallel scheduling.

21.3 Development Environments

Development environments provide a comprehensive application for use during development. Often, a specialized programming language is defined (e.g., MATLAB). There are Visual Programming Environments (VPEs), where the visual programming itself can be regarded as a programming language. The building blocks, which are visually connected in the development environment, are generally implemented in a standard programming language. Again, C++ is the most commonly used language.

VPEs have a long tradition in image processing. The main advantage is that results of different combinations of image processing techniques can be explored without the need for writing code. Thus, no programming experience with a standard language is required – as long as the goal can be achieved with the means available (Fig. 21.2). One of the first systems was Khoros[6] with its visual programming environment called *Cantata*. More recent examples are presented in this section.

---

[6] www.khoral.com

Table 21.14. *SCIRun*. This visual environment is developed at the Scientific Computing and Imaging Institute, U of U, Salt Lake City, UT

Language:	C++, GUI with Tk
Platform:	Windows, Linux, Mac
License:	MIT
Developer:	University of Utah (U of U), UT, USA
Availability:	open-source
URL:	www.scirun.org

Table 21.15. *OpenXIP*. The VPE of OpenXIP is developed by Siemens Corporate Research Inc., Princeton, NJ

Language:	C++
Platform:	Windows (including XIP-Builder); libraries only: Linux, Mac
License:	caBIG (BSD-style)
Developer:	Washington University, St. Luis, MO, USA and Siemens Corporate Research Inc.
Availability:	open-source, XIP-Builder: closed-source
URL:	www.openxip.org

21.3.1 SCIRun

SCIRun is an extensible visual programming environment with a focus on modeling, simulation and visualization [8] (Table 21.14). The VPE allows interactive construction and debugging of dataflow networks composed of *modules*. Visualization is included in the dataflow semantics. Additionally, customized user interfaces (PowerApps) are available that are built on top of a dataflow network, which controls the execution and synchronization of the modules.

Sets of modules are called *packages* in SCIRun and can be added to SCIRun through a plug-in mechanism. Extension packages to the core application include *BioPSE* for bio-electric field calculations and interfaces to ITK, MATLAB and some of the *Teem Libraries*[7] with Nearly Raw Raster Data (NRRD) file handling and DTI routines.

Available applications (PowerApps) include:

- *BioFEM* for finite element problems,
- *BioTensor* for post-processing and visualization of diffusion-weighted MRI
- *BioImage* for the processing and visualization of image volumes
- *Seg3D*, an extensible application for automatic and interactive volume segmentation based on ITK filters with volume rendering capabilities

21.3.2 OpenXIP

The eXtensible Imaging Platform™ (XIP™) project aims at providing an environment for rapidly developing medical imaging applications from an extensible set of modular elements (Table 21.15). The XIP Platform consists of a set of tools and libraries, including the *XIP Reference Host*, *DICOM Application Hosting APIs*, sample applications, and development tools. XIP is based

---

[7] teem.sourceforge.net

on the Open Inventor library (Sect. 21.2.4), which is extended with classes for DICOM data handling and navigation, overlay and region of interest (ROI) definition, as well as ITK and VTK support.

The XIP-Builder development tool is a visual programming environment for constructing and testing scene graphs including field-to-field connections and Open Inventor engines. The Open Inventor concepts of engines and field-to-field connections are similar to pipelines in VTK/ITK. Together with the extensions of Open Inventor provided by the XIP library, the visual programming interface XIP-Builder makes it possible to explore image processing, segmentation, and visualization techniques without the need for writing code.

21.3.3 DeVIDE

The Delft Visualisation and Image processing Development Environment (DeVIDE) is a VPE for medical visualization and image processing algorithms [9] (Table 21.16). DeVIDE, written in Python, allows scripting with Python and access to any object or variable in the system at runtime.

DeVIDE supports a combination of event- and demand-driven scheduling (hybrid scheduling). Modules can either support event- or demand-driven scheduling (more complex to implement). The hybrid scheduling functionality applies demand-driven scheduling where possible and otherwise event-driven scheduling.

DeVIDE comes with support for VTK, ITK, GDCM, and DCMTK, as well as packages for scientific computing and 2D plotting. Extension of DeVIDE is possible using Python code or libraries with a Python wrapping. Each module (block in the visual programming interface) represents a visualization or image processing algorithm implemented as a Python class (or wrapped into a Python class). A special "blank" module called *CodeRunner* is provided that allows the user to add Python code directly. The setup section of a CodeRunner module is executed one time after each modification whereas the execute section is always executed.

Besides the visual programming interface, DeVIDE allows running dataflow networks from the command-line (so-called black-box interface) and applying them to arbitrary datasets.

Language:	Python
Platform:	Windows, Linux
License:	BSD-style
Developer:	Delft University of Technology, Delft, The Netherlands
Availability:	open-source
URL:	graphics.tudelft.nl/Projects/DeVIDE

Table 21.16. *DeVIDE*. At Delft University, the Computer Graphics Group and the CAD/CAM Group are developing this VPE

21.3.4 VisTrails

VisTrails is a scientific workflow and provenance management system [10] (Table 21.17). The goal is to support exploration of data by keeping track of all steps the user performs during the process and hence the provenance of (intermediate and final) results. A workflow in VisTrails is a dataflow and may include functional loops and conditional statements. When creating a workflow, detailed history information is maintained about the steps followed and data derived in the course of an exploratory task. The history information can be stored in XML files or in a relational database. Users can navigate workflow versions, undo changes without loosing results, visually compare different workflows and their results, and examine the actions that led to a result. Workflows can be run interactively in the GUI of VisTrails, or in batch using a VisTrails server.

VisTrails allows the combination of loosely-coupled resources, specialized libraries, grid and web services. It comes with support for VTK and can be extended by other libraries. Additionally, the VisTrails provenance infrastructure can be used within other interactive tools. For example, ParaView (cf. Sect. 21.4.5) includes a VisTrails plug-in.

21.3.5 LONI Pipeline

The pipeline-based VPE from the UCLA's Laboratory Of Neuro Imaging (LONI) [11] allows defining and executing workflows on grid computing architectures, with "workflow" used in the sense of data being processed by a sequence of algorithmic modules (Table 21.18). Workflows can be defined in a visual programming-like GUI. Execution of modules can be performed on the server side of the LONI Pipeline (but also locally, depending on the module definition). Each module is a separate command-line executable with an associated XML file describing its input and output parameters and other

Table 21.17. *VisTrails.* Like SCIRun, this VPE is developed at the Scientific Computing and Imaging Institute, U of U, Salt Lake City, UT, USA

Language:	Python, GUI with Qt
Platform:	Windows, Linux, Mac
License:	GPL
Developer:	U of U, UT, USA
Availability:	open-source
URL:	www.vistrails.org

Table 21.18. *LONI.* This VPE is developed at the Laboratory of Neuro Imaging, UCLA

Language:	Java, modules are executables (any language)
Platform:	Platform with JRE 1.5, some server features on Linux/Unix only
License:	LONI software license
Developer:	University of California at Los Angeles (UCLA), CA, USA
Availability:	closed-source
URL:	pipeline.loni.ucla.edu

meta-data (e.g., description, citation of the implemented algorithm, author, tag). Another GUI is available for creating these XML files. The executable itself does neither need conformance to any specific rules (except for being a command-line tool) nor including any LONI pipeline libraries. It is, therefore, completely independent. The LONI Pipeline handles the passing of data between the executables. In case a grid is available, it can exploit parallelism in workflows, and make processing of large data sets take about as much time as a single one.

The GUI also allows searching the server-side library of modules against the provided meta-data, graphically constructing workflows, as well as execution control and debugging. Optionally, provenance files can be generated, which store the history of data, workflow and execution of all processes. A command-line variant for executing workflows is also available.

The LONI Pipeline primarily targets neuro-imaging researchers, but it is completely independent of a specific processing library. It can run as a light-weight middle-ware and does not include a set of filters or image processing algorithms. The developing institution, the *Laboratory Of Neuro Imaging*, grants access to their LONI Pipeline server (a 600 CPU computing grid) upon request.

21.3.6 MeVisLab

MeVisLab is closed-source, extensible VPE for image processing, and interaction, with a special focus on medical imaging (Table 21.19). MeVisLab allows the definition of hierarchical data flow and visualization networks as well as GUIs. Based on Open Inventor, it provides image processing algorithms, 2D/3D visualization and interaction tools, integration of ITK and VTK, DICOM support, movie and screenshot generation, scripting support (Python, including a debugger, and JavaScript), and wizards for image processing, visualization and macro modules.

MeVisLab comes with the MeVis Image Processing Library (ML), which is a generic framework for image processing consisting of self-descriptive modules. The library implements a request-driven, page-based update paradigm with a priority-controlled page cache. A special memory management for large images allows global, random access to paged images.

Modules in MeVisLab can either be ML, Open Inventor, or ITK modules. MeVisLab can be extended by custom modules in either of these module types using one of the supported C++ development environments and by macro modules. The commercial version allows signing and encrypting modules for redistribution. Free licenses for non-commercial research and evaluation purposes with a few restrictions are available.

21.3.7 MATLAB®

The MATrix LABoratory (MATLAB)® is a specialized vectorized high-level language for numerical applications including an integrated development

Table 21.19. *MeVisLAB*

	Language:	C++, scripting with Python and JavaScript
	Platform:	Windows, Linux, Mac
	License:	proprietary (commercial, non-commercial, unregistered)
	Developer:	MeVis Medical Solutions and Fraunhofer MEVIS, Bremen, Germany
	Availability:	closed-source
	URL:	www.mevislab.de

Table 21.20. *MATLAB®*

	Language:	MATLAB, bindings with other languages
	Platform:	Windows, Linux, Unix, Mac
	License:	proprietary
	Developer:	The MathWorks Inc., USA
	Availability:	closed-source
	URL:	www.mathworks.com

environment. MATLAB provides mathematical functions for linear algebra, statistics, Fourier analysis, filtering, optimization, numerical integration, as well as plotting of functions and tools for building GUIs. Interfacing with programs in other languages is possible (Table 21.20).

The MATLAB language has a particular strength in matrix manipulation, and as images can be regarded as matrices, MATLAB is well-suited for image processing. A large number of MATLAB extensions (toolboxes) are available from the software provider/vendor. Examples are toolboxes for image processing, signal processing, statistics and data analysis, partial differential equations, and computer algebra (*Symbolic Math Toolbox*).

Simulink® is an additional product that is based on MATLAB. It is a tool for modeling, simulating and analyzing multi-domain dynamic systems. Open source alternatives to MATLAB, which are intended to be mostly compatible with the MATLAB language (not with the development environment), are:

- GNU Octave, www.gnu.org/software/octave
- Scilab, www.scilab.org
- FreeMat, freemat.sourceforge.net

21.3.8 Interactive Data Language

As MATLAB, the Interactive Data Language (IDL) is a vectorized high-level language including an integrated development environment, the IDL Workbench. Compared to MATLAB, IDL focuses more on data analysis, data visualization and application development. Image and signal processing features are already included in the base language, whereas some advanced mathematical and statistical routines have to be purchased separately (Table 21.21).

IDL programs are compiled to an interpreted, stack-based intermediate pseudo code. It is run by the IDL Virtual Machine (VL)^{TM}, which is available as a free runtime utility allowing redistribution of compiled code without

Language:	IDL, bindings with other languages
Platform:	Windows, Linux, Unix, Mac
License:	proprietary
Developer:	IT Corp., Boulder, CO, USA
Availability:	closed-source
URL:	www.ittvis.com/IDL

Table 21.21. *IDL*. The vendor of IDL is ITT Visual Information Solutions, USA

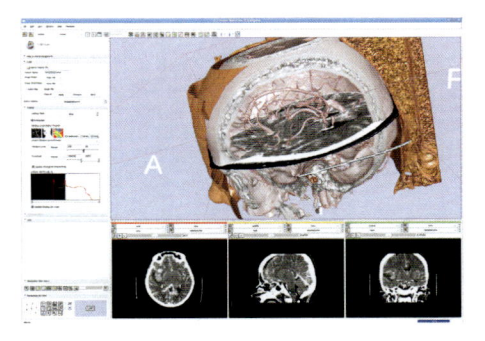

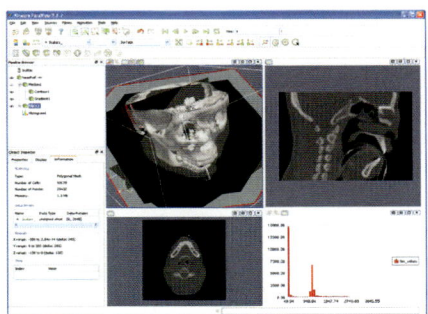

Fig. 21.3. *Extensible end-user applications. Left*: 3D Slicer (image courtesy 3D Slicer, www.slicer.org); *Right*: ParaView

additional licensing fees. The IDL development environment is based on the open-source Java-framework Eclipse[8].

21.4 Extensible Software

Extensible software applications are ready-to-use end-user applications that can be extended by custom code. The applications come with features most commonly required by users. Usually, a certain look and feel is defined and fixed. Developers do not need to care about implementing a front-end for standard tasks. Thus, extensible software applications provide high-level support on the GUI-/front-end-level, but with the least degree of adaptability (Fig. 21.3). The language for implementing fully-integrated extensions is, again, most commonly C++. Additionally, scripting support is sometimes available.

21.4.1 3D Slicer

Slicer or 3D Slicer is an extensible end-user application for visualization and medical image computing [12] (Table 21.22). It comes with functionality for segmentation, registration and 3D visualization of multi-modal image data, as well as advanced image analysis algorithms for DTI, FMRI and IGT. Additional, pre-compiled modules can be downloaded from a web resource with the built-in "Extension Manager Wizard".

---

[8] www.eclipse.org

Table 21.22. *3D Slicer.* The consortium is lead by the Surgical Planning Laboratory, Brigham and Women's Hospital, Boston, USA

Language:	C++, GUI with Qt (formerly Tk and KWWidgets)
Platform:	Windows, Linux, Mac
License:	BSD-style
Developer:	multi-institutional
Availability:	open-source
URL:	www.slicer.org

Table 21.23. *MITK ExtApp and MITK 3M3.* The application is developed at the Division of Medical and Biological Informatics, German Cancer Research Center (DKFZ)

Language:	C++, GUI with Qt
Platform:	Windows, Linux, Mac
License:	BSD-style
Developer:	DKFZ, Heidelberg, Germany
Availability:	open-source; extended 3M3: closed-source (free download)
URL:	www.mitk.org, 3m3.mitk.org

Slicer is based on a data model called *Medical Reality Markup Language* (MRML). A MRML scene is a collection of datasets and their current state, viewing parameters, semantic descriptions, as well as parameterizations of algorithms. MRML scenes can be written to files.

Extensions to Slicer can be implemented on three different levels:

- *Loadable Modules* can access the MRML scene instance of the Slicer application and perform function calls to the slicer core and related APIs. They are dynamically discovered and loaded.
- *Scripted Modules* are like Loadable Modules, but written in Tcl or Python.
- *Command Line Modules* are separate stand alone executables that understand a specific command line syntax; access to the MRML scene is not available. C++ code for parsing the command line syntax can be generated by the "GenerateCLP" tool provided with Slicer. The GUI is automatically generated by the Slicer application based on information requested from the module. Command Line Modules can also be compiled as shared libraries for dynamic linking with Slicer, which omits the need of writing/reading input/output data to/from disk.

21.4.2 MITK ExtApp and MITK 3M3

As mentioned in Sect. 21.2.5, MITK comes with an extensible application (MITK ExtApp), which allows performing image processing tasks such as visualization of 3D and 3D+t images, rigid and deformable fusion (registration) of multiple image volumes, and interactive organ segmentation (Table 21.23). An extended version of this application is called MITK 3M3 and contains additional, closed-source plug-ins, e.g., for vessel tree segmentation, shape-based segmentation of liver and heart, and DTI. It is available for free download.

Language:	C++, GUI with wxWidgets
Platform:	Windows, Linux
License:	BSD-style
Developer:	Universitat Pompeu Fabra (UPF), Barcelona, Spain
Availability:	open-source
URL:	www.gimias.org

Table 21.24. *GIMIAS.* The toolkit is developed at the Center for Computational Image and Simulation Technologies in Biomedicine (CISTIB) at UPF, Spain

21.4.3 Graphical Interface for Medical Image Analysis and Simulation

The Graphical Interface for Medical Image Analysis and Simulation (GIMIAS) is intended as an integrative end-user platform for medical imaging, computational modeling, numerical simulation, and computer graphics, to support multi-disciplinary research and medical studies (Table 21.24). GIMIAS is based on several toolkits described earlier in this chapter: VTK (Sect. 21.2.3), ITK (Sect. 21.2.2), DCMTK (Sect. 21.2.9), and MITK (Sect. 21.2.5). GIMIAS comes with the *Medical Image Modeling and Simulation Toolkit*, which is a set of specialized libraries of medical image modeling and simulation algorithms.

The user is provided with a general visualization tool with DICOM functionality that can be extended through plug-ins for specific applications. GIMIAS provides an API for the plug-ins to interact with the application core and with other plug-ins. Plug-ins can have different complexity, from simple execution of a new algorithm and visualizing the results with the standard viewers, to complex plug-ins that change the entire user interface or allow new types of interaction with the data. For the GUI, WxWidgets is used.

21.4.4 OsiriX

OsiriX is an extensible application primarily dedicated to 2D/3D visualization of multi-modal and multi-dimensional images, but also offering image processing functionality, especially for registration and image fusion. It is available for Mac only (Table 21.25). Besides reading DICOM image files (and many other formats), it is able to receive images transferred by the DICOM communication protocol from any Picture Archiving and Communication System (PACS) or imaging modality.

OsiriX has a plug-in architecture that gives full access to the Cocoa framework and allows accessing image data, DICOM header data, etc.

21.4.5 ParaView

ParaView is an extensible data analysis and visualization application providing qualitative and quantitative techniques (Table 21.26). ParaView offers some functionality for image data, though this is not the primary focus.

Table 21.25. *OsiriX.*
Osirix has been developed at UCLA, and is now owned by Pixmeo, a Switzerland-based company

Language:	Objective-C, GUI with Cocoa
Platform:	Mac
License:	BSD-style
Developer:	Pixmeo Sarl., Bernex, Switzerland
Availability:	open-source
URL:	www.osirix-viewer.com

Table 21.26. *ParaView*

Language:	C++, GUI with Qt (formerly KWWidgets); scripting with Python
Platform:	Windows, Linux, Unix, Mac
License:	BSD-style
Developer:	Kitware Inc.
Availability:	open-source
URL:	www.paraview.org

Although originally developed to analyze extremely large datasets using distributed memory computing resources on supercomputers, ParaView can be run on standard PCs or notebooks.

ParaView is based on VTK and provides access to many VTK filters, which can be combined and parameterized. Additionally, data can be explored and edited interactively in 3D and programmatically using batch processing and scripting.

Besides scripting, ParaView can be extended using a plug-in mechanism. Processing filters (client-side or server-side) as well as extensions of the applications base functionality is possible. For examples, the provenance support of VisTrails (Sect. 21.3.4) has been added as a plug-in and is provided with ParaView since version 3.6.2 (but is not loaded by default).

21.4.6 ImageJ and Fiji

ImageJ is an extensible, Java-based image processing application that is mainly dedicated to (and widely used for) 2D images, but also supports image stacks (Table 21.27). ImageJ supports essentially all standard image processing functions such as convolution, smoothing, median filtering, edge detection, sharpening, logical and arithmetical operations, contrast manipulation, Fourier analysis, scaling, rotating, etc.

ImageJ can be extended by recordable macros and Java plug-ins. More than 500 plug-ins are available from the ImageJ website[9], which differ significantly in perfomance, quality and usability.

Fiji is Just ImageJ (Fiji)[10] is a distribution of ImageJ bundled with a selected set of plug-ins, originally intended for neuro-scientists. Distributed via GPL v2, Fiji is particularly oriented towards image registration, segmentation, 3D reconstruction, and 3D visualization (based on Java3D).

---

[9] `rsb.info.nih.gov/ij/plugins`
[10] `pacific.mpi-cbg.de`

Language:	Java
Platform:	any with Java 1.4 or later
License:	public domain
Developer:	NIH, Bethesda, MD, USA
Availability:	open-source
URL:	rsbweb.nih.gov/ij

Table 21.27. *ImageJ.* ImageJ is developed by the National Institutes of Health (NIH), Bethesda, MD, USA

Language:	Java
Platform:	any with Java 1.4 or later
License:	MIPAV license
Developer:	NIH, Bethesda, MD, USA
Availability:	closed-source
URL:	mipav.cit.nih.gov

Table 21.28. *MIPAV.* The toolkit is developed at the Center for Information Technology (CIT), NIH, USA

21.4.7 MIPAV

The Medical Image Processing, Analysis, and Visualization (MIPAV) application is dedicated to quantitative analysis and visualization of n-dimensional medical images (Table 21.28). MIPAV provides a large number of standard image processing algorithms as well as specialized methods for biomedical imaging, for example for microscopy, abdomen and muscle segmentation, shading correction, and DTI. DICOM support includes reading files, DICOM send and receive, as well as displaying and editing DICOM tags and anonymization of patient information. An extensive documentation comprising two books (in PDF format) with more than 700 pages each is available for MIPAV from the website.

MIPAV can be extended by scripting (including scripts produced by a macro recorder) and plug-ins. Additionally, scripts can be called from the command line (or another program).

21.4.8 VolView

VolView is an extensible application primarily dedicated to volume visualization (Table 21.29). VolView allows reading (one or more) images in many image file formats, oblique reformatting, measurement, annotation, and generation of movies. Additionally, many image filters from ITK (Sect. 21.2.2) are available. VolView can be extended by custom data processing methods written in C/C++ using a plug-in API.

21.4.9 Analyze

Analyze is a software package for multi-dimensional display, processing, and measurement of multi-modal biomedical images (Table 21.30). Analyze is based on the *A Visualization Workshop* (AVW) library, a collection of more than 600 image processing functions.

The *Analyze Developers Add-On*, which needs to be purchased separately, allows extending Analyze by add-on modules.

Table 21.29. *VolView*

Language:	C++, GUI with KWWidgets
Platform:	Windows, Linux, Mac
License:	proprietary
Developer:	Kitware Inc.
Availability:	closed-source
URL:	www.volview.org

Table 21.30. *Analyze.* At Mayo Clinic, the toolkit is developed at the Biomedical Imaging Resource (BIR)

Language:	C, GUI with Tcl/Tk
Platform:	Windows, Linux, Mac
License:	proprietary
Developer:	AnalyzeDirect Inc., Overland Park, USA & Mayo Clinic, Rochester, MN, USA
Availability:	closed-source
URL:	www.analyzedirect.com

Table 21.31. *Amira.* The toolkit was originally developed by the Visualization and Data Analysis Group at Zuse Institute Berlin (ZIB), Germany

Language:	C++, GUI with Qt; scripting with Tcl
Platform:	Windows, Linux, Mac
License:	proprietary
Developer:	Visage Imaging, Richmond, Australia
Availability:	closed-source
URL:	www.amira.com

21.4.10 Amira

Amira is a visualizing, interactive data exploration and processing software for a wide range of applications in life science and biomedicine (Table 21.31). Amira supports multi-dimensional images, surface and volume meshes, vector and tensor fields, CAD data, and other specialized data types and file formats, including time series. A free extension for reading/writing of DICOM files and DICOM Send is available. The processing features of the basic package include image filtering, segmentation, cropping, registration, surface generation and editing. Tools are provided for creating presentations (e.g., animation, movie).

Optional packages are available for microscopy, molecular model visualization and analysis, skeletonization, mesh generation and analysis, additional quantification methods for volume analysis (including labeling, morphology, watershed transform), Virtual Reality (VR) (support for tiled displays and immersive VR configurations such as Cave Automatic Virtual Environments, CAVEs) and very large datasets.

Amira supports scripting in Tcl. The *Developer Option*, which needs to be purchased separately, allows developing custom data types, visualization and processing modules, and input/output routines in C++. Rendering is based on Open Inventor (Sect. 21.2.4).

21.5 Conclusion and Discussion

All toolkits, development environments and extensible applications presented in this chapter have their particular strengths and areas of applications. Some are mutually exclusive; others can be used in combination and benefit from each other. Often, it is a matter of taste which to choose.

As a general rule, development environments with specialized languages and visual programming interfaces have their largest benefits in the exploration of different methods to solve a problem. In case an end-user needs to apply a method, e.g., for further evaluation, implementing that method for use within an extensible applications can save a lot of time. Toolkits provide the highest amount of flexibility and adaptability and form the base of many development environments and extensible applications.

All three categories can complement each other. In the future, a hierarchy of toolkits with different foci may be used as base for a development environment as well as an extensible application, with the development environment providing the possibility to create modules for use within the extensible application. In case the flexibility of a module created by the development environment is insufficient, the underlying toolkits can be used directly – or a completely new, specialized application can be implemented on top of the toolkits, integrating modules created by the development environment or originally written for the extensible application.

References

1. Pieper S, Lorensen B, Schroeder W, et al. The NA-MIC kit: ITK, VTK, pipelines, grids and 3D slicer as an open platform for the medical image computing community. In: Proc IEEE Intl Symp Biomed Imaging ISBI; 2006.
2. Wernecke J. The Inventor Mentor: Programming Object-Oriented 3d Graphics with Open Inventor, Release 2. Boston, MA, USA: Addison-Wesley Longman Publishing Co., Inc.; 1993.
3. Wernecke J. The Inventor Toolmaker: Extending Open Inventor. Release 2. Boston, MA, USA: Addison-Wesley Longman Publishing Co., Inc.; 1994.
4. Wolf I, Vetter M, Wegner I, et al. The medical imaging interaction toolkit (MITK). Med Image Anal. 2005;9(6):594–604.
5. Zhao M, Tian J, Zhu X, et al. The design and implementation of a C++ toolkit for integrated medical image processing and analyzing. Proc SPIE. 2004;5367:39–47.
6. Toussaint N, Sermesant M, Fillard P. vtkINRIA3D: a VTK extension for spatiotemporal data synchronization, visualization and management. In: Proc. of Workshop on Open Source and Open Data for MICCAI. Brisbane, Australia; 2007.
7. Allard J, Cotin S, Faure F, et al. SOFA an open source framework for medical simulation. In: MMVR'15. Long Beach, USA; 2007.
8. Parker SG, Johnson CR. SCIRun: a scientific programming environment for computational steering. In: Proc ACM/IEEE Conf Supercomputing. New York, NY, USA: ACM; 1995. p. 52.

9. Botha CP, Post FH. Hybrid scheduling in the DeVIDE dataflow visualisation environment. In: SimVis; 2008. p. 309–322.
10. Silva CT, Freire J, Callahan SP. Provenance for visualizations: reproducibility and beyond. Comput Sci Eng. 2007;9:82–89.
11. Dinov ID, Horn JDV, Lozev KM, et al. Efficient, distributed and interactive neuroimaging data analysis using the LONI pipeline. Front Neuroinformatics. 2009;3.
12. Pieper S, Halle M, Kikinis R. 3D slicer. In: IEEE Int Symp Biomed Imaging ISBI; 2004.

22

Image Processing and the Performance Gap

Steven C. Horii and Murray H. Loew

Summary. Automated image processing and analysis methods have brought new dimensions, literally and figuratively, to medical imaging. A large array of tools for visualization, quantization, classification, and decision-making are available to aid clinicians at all junctures: in real-time diagnosis and therapy, in planning, and in retrospective meta-analyses. Many of those tools, however, are not in regular use by radiologists. This chapter briefly discusses the advances in image acquisition and processing that have been made over the past 30 years and identifies gaps: opportunities offered by new methods, algorithms, and hardware that have not been accepted by (or, in some cases, made available to) radiologists. We associate the gaps with (a) the radiologists (a taxonomy is provided), (b) the methods (sometimes unintuitive or incomplete), and (c) the imaging industry (providing generalized, rather than optimized, solutions).

22.1 Introduction

Extracting as much information as possible from medical images has long been a goal of those who create the images as well as the physicians who interpret them. In many cases, the effort has been directed at enhancing particular information in the image and suppressing what is regarded as noise.

In imaging techniques for which there was a need, image processing was done in an analog fashion. Mask films in angiography were used for optical subtraction. Analog video processing circuitry was used in fluoroscopic imaging chains for contrast enhancement and inversion. Conventional radiography used asymmetric intensifying screens to exploit more fully the dynamic range of film.

With the advent of inexpensive, fast computing, various forms of image processing have become nearly ubiquitous in radiological imaging. The digital imaging first pioneered in nuclear medicine and computed tomography required computer processing to create viewable images from the signals obtained. As the other chapters of this book have amply explained, applications of image processing methods in radiological imaging are widespread. Given this, what is the "gap" that forms the subject matter of this chapter?

To understand the concept of a gap, what it means, and why it exists, it is perhaps best to examine some use cases where image processing is widely used and accepted clinically (Sect. 22.2). We will then motivate the reason for gaps from the user's (the radiologists) point of view (Sect. 22.3). In Sect. 22.4, we discuss the goals of image processing for medical imaging, before some ideas to bridge the gap are presented in Sect. 22.5.

22.2 Examples of Clinically Useful Image Processing

22.2.1 Windowing and Image Display

One readily apparent problem that was exposed as digital imaging replaced analog was the large difference between the dynamic range of signals that could be captured digitally and what could be presented to the observer – first on film and later on cathode ray tube (CRT) or liquid crystal displays (LCDs). Computed tomography (CT), for example, typically had pixels that could represent a range of 4,096 values (12 bits), whereas typical display devices were limited to approximately 256 (8 bits). Photostimulable phosphor (PSP) plate radiography extended the exposure-to-contrast range over film-screen systems by at least an order of magnitude. Fairly simple mapping of the available display range to a subset of bits in the CT pixel allowed for "window" and "level" adjustments. More complex image processing was done at the CT reconstruction step where filter kernels optimized for different types of imaging (e.g., detail bone vs. brain) were, and are, used.

22.2.2 Contrast and Edge Enhancememt

Digital subtraction angiography (DSA) expanded widely in the early 1980s. It replaced the analog mask film and subtraction with the digital equivalent. The image processing was not terribly complex as it involved contrast inversion, pixel shifting, and image addition, but the imaging technique became very widely used, moving digital image processing into routine clinical use.

PSP radiography was the next large-scale application of digital image processing methods for clinical applications. Unprocessed images from these systems looked very "flat" and radiologists wanted images that more closely resembled those on film. The vendors of the PSP radiography systems responded by developing image processing algorithms that could do this. The algorithms were, and are, proprietary. However, the parameters that can be adjusted to alter the appearance of the resulting images indicate the types of processing used. Simple gamma adjustment for contrast is one. Edge enhancement of varying degrees is another. Some systems perform what amounts to histogram equalization to yield images in which soft tissue and bone (for example) are both visible and are diagnostically useful. Still other vendors use segmentation to turn the area outside the collimated portion of the image black.

22.2.3 Noise Reduction and Color Coding

Ultrasound imaging by nature includes much speckle in the images produced. Analog methods of reducing speckle, such as temporal compounding, had been tried, but not widely used. Spatial compounding was possible with the analog articulated-arm scanners and was used clinically, but with the development of real-time ultrasound imaging, spatial compounding was initially eliminated. However, the ultrasound system vendors provided frame averaging methods (effectively temporal compounding) to reduce speckle. When phased-array ultrasound systems became more widely available, the vendors responded with methods that re-introduced spatial compounding. Using the transducer's multiple elements to generate multiple beams, anatomy was imaged from multiple slightly different angles. Through the coding of the Doppler shift of flowing blood as color and then registering the color signal on the gray-scale image, color flow Doppler was rapidly adopted as a way of showing both blood vessels and qualitative and quantitative blood flow on the same image.

22.2.4 Registration and Segmentation

Multi-detector CT has been the second revolution in CT imaging. These systems are capable of dynamic imaging of the beating heart, long an area on the "wish list" of radiologists and cardiologists. Certainly novel and efficient algorithms for reconstructing the helically acquired data were part of the innovation. The availability of spatially isotropic voxels also enabled three-dimensional (3D) and four-dimensional (4D) imaging of a quality unsurpassed by other imaging methods. Fortunately, image processing methods from the computer graphics community were readily available. CT angiography, virtual colonoscopy, and various forms of surgical and radiation therapy planning all derive from 3D reconstructed information. Segmentation plays a major role in these methods as organs of interest (OOI) are separated from the surrounding structures. Combined with positron emission tomography (PET) images, PET-CT can show not only anatomic structure, but also normal, and particularly abnormal, function. PET-CT relies on image registration, a task simplified by having the PET and CT imaging systems around a common axis, so that the patient is imaged by both systems with only minimal time differences between the image acquisitions.

Magnetic resonance imaging (MRI) has benefited from image processing and techniques such as diffusion tensor imaging (DTI) and functional MRI can provide clinical insight into function and dysfunction in the brain. Image fusion of these imaging methods has been done as a way of improving image-guided neurosurgery. In their paper on the subject, Talos et al.

describe a number of image processing methods required [1]. These included: segmentation, motion correction, rigid and non-rigid registration, and 3D tractography [2].

22.2.5 Image Compression and Management

Image compression has a long and well-documented history. The motivation for compression is to reduce the sizes of the files that contain data from images and sequences of images. A current multi-slice CT scan can easily produce 400 MB or more for a given patient study. To transmit or store files of these sizes becomes costly in time and/or money. Two basic approaches exist for compression: lossless and lossy. Lossless methods (e.g., run-length encoding) allow the original image(s) to be recreated perfectly from their compressed versions. A typical compression ratio (i.e., size of the original compared to the size of the compressed) for a lossless method applied to medical images is approximately 3:1.

Lossy methods such as the Joint Photographic Experts Group (JPEG) or JPEG2000 standard and numerous proprietary variants can yield compression ratios of 20:1 or higher and thus are very attractive when one considers time/space/cost constraints as well as the routine use of tele-radiology. Here, however, the re-created image is not a perfect copy of the original. Many measures have been proposed for assessing the extent of the loss of image content, including mean-square error (aggregated across all pixels in the image), peak signal-to-noise ratio (SNR), and preservation of edge content. None of those, however, addresses directly the central question: is the recreated image useful clinically? The answer depends on several important factors: the modality, the anatomy, the compression ratio, and the "user" of the image – is it a human radiologist, or a computer-assisted detection (CADe) algorithm?

Clearly, it would seem that if the user is a human, the judgment as to usefulness must be made by the radiologist. In practice, this requires a human-observer study that asks a set of expert observers to assess images compressed in various ways and to express preferences. Much of the early work made extensive use of observer studies. In recent years, much effort has gone into the search for automated methods that can *objectively* evaluate the utility to a human observer of a given image for a given task. Such methods would enable the rapid comparison of compression methods across a variety of conditions. No methods have yet been accepted for that purpose, though they are becoming increasingly sophisticated (e.g., [3]) and are including models of human visual perception.

A recent review of the question of whether lossy compression can be used in radiology concluded that small images can be compressed up to 10:1 without perceptible information loss, and that large images can be compressed up to 20:1 or 25:1 without perceptible loss when JPEG or JPEG2000 are used [4]. The ratios ranged from 6:1 (JPEG for angiography) to 50:1 (JPEG2000 for chest computed radiography (CR) or digital radiography (DR)).

22.3 Why are there Gaps?

22.3.1 The Conservative Radiologist

Roentgen's original discovery of X-rays was made on film. From that time until the 1980s, film was the standard for acquiring, displaying, and storing radiographic images. The manufacture, processing, and quality assurance of radiographic film were developed to a high art. Radiologists were so particular as to demand a certain "feel" of the film; it had to have a certain "snap" as it was placed on a light box. Experienced radiologists could often tell when there was a problem with a radiographic film processor by the feel of the film. Besides the physical characteristics of film, the whole of radiologist pattern recognition was based on film images viewed by transillumination. Even with the development of digital imaging, early digital images were printed on film for interpretation. The first PSP systems were supplied with laser film printers. A major objection to these printers in the USA was that the films were printed in a smaller size than conventional radiographic film [5].

The spatial resolution of PSP images was also lower than that of conventional film-screen radiographic systems. The modulation transfer function (MTF) of film-screen systems had a limiting spatial resolution of 5–10 line pairs per mm (1 p/mm). For PSP, the typical value was 2.5 1–5 l p/mm [6]. Early acceptance was among technologists (radiographers) who found the PSP systems much more forgiving of exposure errors because of the wide exposure latitude of these systems.

The actual and perceived difficulties of PSP systems tended to reinforce the radiologist's conservative attitudes when it came to the means by which they provided patient care and earned their livelihood. Digital image processing was yet another thing that altered the images the radiologist was trained to read. For PSP systems, much of the image processing was to achieve a "film-like" look. Acceptance of image processing for digital radiographic images was earned through showing the radiologist the advantages that could be achieved as noted previously. Still, the idea that something is altering the pixels in the digital image the radiologist has to interpret is viewed with some skepticism.

22.3.2 The Busy Radiologist: Digital vs. Analog Workflow

Radiologists are essentially piece workers. Income is often directly tied to the number of examinations that a radiologist interprets over time. Even in academic practices, the trend in the USA is to have radiologists interpret enough examinations to cover their salaries and generate some excess revenue to benefit the department and hospital. Film-based operations were highly optimized for radiologist productivity. Film library clerks put films to be read on multiviewers, put associated paperwork in order, supplied any comparison films needed, and removed films for storage once read. Transcriptionists picked up recorded tapes or logged into digital dictation systems often 24 h a day to type or enter radiographic reports. Physicians who wanted to review films with a radiologist came to the radiology department for such consultations. Digital imaging had the potential to change all this radically, and this has largely come to pass. One author [SH] and colleagues studied the change of behavior as his department transitioned from film-based to digital operation. What they discovered was another factor contributing to radiologist resistance to the change to digital Picture Archiving and Communication Systems (PACSs). They found that work often was "shifted" to radiologists. As an example, prior studies were, as noted, retrieved by film library clerks and put up (in the proper order) with the current examination. In PACS operation, it was the radiologist who had to perform this function by looking at the electronic patient directory and finding the examination to compare [7]. Technologists were responsible for printing films from digital sources and, for those with a larger gray-scale range than could be accommodated on film (e.g., CT), would print the images with two different settings so that the anatomy of interest was optimally displayed (for CT, lung and mediastinum). With PACS, changing the display characteristics fell to the radiologist. These various sorts of work shifting tended to add to the radiologist's interpretation time, an undesirable trend. Adding image processing that would require the radiologist to select a function to perform was, for this reason, regarded with great skepticism. How these objections were, and could be, overcome is the subject of Sect. 22.5.

22.3.3 The Wary Radiologist: Malpractice Concerns

The popular as well as medical literature are replete with both anecdotal stories and scholarly works about malpractice ("medical misadventure" in the United Kingdom (UK)) and the resulting effects on healthcare [8]. In 2000, the Institute of Medicine published a sobering report that described how large a number of patients in the USA died each year as a result of mostly preventable errors [9]. Most practicing physicians in the USA would describe at least some impact on their practice as a result of a rise in the cost of malpractice insurance. There is considerable debate about the causes of this increased cost, with insurance carriers claiming the large sums resulting from verdicts against the physician as a major factor and skeptical physicians citing poor investments on the part of the insurance companies and the resulting need for operating revenue. Whatever the cause, the effect has been to increase the wariness of the physician from working outside of the standard of practice.

For radiologists, using equipment in a manner for which it was not intended, not following well-accepted policies and procedures, and being impaired when performing work are among the reasons cited as root causes for negligence claims [10]. This has made radiologists very wary of anything that potentially alters the image in some way that is not widely used in the radiology community. This tends to limit the introduction of novel image processing methods. Image processing, such as used by the PSP vendors, that has been vetted through the Food and Drug Administration (FDA) processes reduces the radiologist's concern since it then places some liability burden on the equipment vendor. The FDA typically does not evaluate image processing software itself unless, as for computer-aided detection (e.g., Device: Analyzer, medical image,[1] Device: Lung computed tomography system, computer-aided detection[2]), it has the potential to result in harm to a patient (FDA Class 3). What the FDA does is to require a Good Manufacturing Practice (GMP) with the idea that software developed using such processes is likely to function as intended and to have few errors (Medical Device Exemptions 510(k) and GMP Requirements[3]). With some limitations and exceptions, the FDA does include Off-the-Shelf (OTS) software in the general equipment premarket approval process [11].

Software developed in an academic research laboratory and used in experimental imaging is exempt from FDA regulation. However, research use of software on patient images is subject to Institutional Review Board (IRB) requirements. These generally involve assessments of risk to the patient, development of an appropriate protocol including informed consent, and ensuring patient safety. For image processing, the typical research methods involve comparing a novel technique against established ones. Patient risk is minimized as the images are typically interpreted both in a conventional manner and using the proposed new technique. Also, studies are usually done on images that have already been interpreted.

22.3.4 The Skeptical Radiologist: Evidence-Based Requirements

When image processing is unobtrusive, does not add to workload, and provides visible improvement in an image, many radiologists would be willing to accept such a technique. The cautious radiologist would also want to know if the method at least did not result in a deterioration of diagnostic performance and, preferably, would improve it. The radiologist familiar with observer performance studies will even expect that statistically robust methods are used for the evaluation of an image processing method and will look for these in published studies. Most radiologists are familiar with receiver operating characteristic (ROC) techniques, as they are often part of the radiology trainee's curriculum. ROC analysis provides a comprehensive description of diagnostic accuracy because it estimates and reports all of the combinations of sensitivity and specificity that a diagnostic test (e.g., radiograph, radiologist's report) is able to provide [12, 13].

---

[1] http://www.accessdata.fda.gov/scripts/cdrh/cfdocs/cfPCD/classification.cfm?ID=4699
[2] http://www.accessdata.fda.gov/scripts/cdrh/cfdocs/cfPCD/classification.cfm?ID=4676
[3] http://www.accessdata.fda.gov/scripts/cdrh/cfdocs/cfpcd/315.cfm?GMPPart=892#start

That is not to say that radiologists are all so compulsive about what is done to the images they interpret. The proprietary processing algorithms used by the PSP vendors were, and are, largely unknown by practicing radiologists, though the physicists and engineers in radiology often could "figure out" at least what class of image processing was being performed on the raw data. With PSP a nearly ubiquitous replacement for film-screen systems, the use of these digital images has become the standard of practice. More recently, direct capture radiography – using area detectors that can provide digital data directly (e.g., DR) – has been replacing PSP for some applications. These images also rely on image processing.

Radiologists in academic centers tend to be more interested in the evidence for applying particular image processing methods to images. The peer-reviewed articles they then write regarding comparisons and observer performance help influence the production-oriented practicing radiologist.

22.3.5 Tails, Dogs, and Gaps

Scientific contributions to image analysis in radiology almost always make their way to the practicing radiologist by way of industry – commercial devices and systems comprise hardware and/or software that often embody techniques or algorithms from the scientific literature. In 2006, Morgan et al. citing a 2004 editorial [14] observed that the challenges of transferring innovative research into industry solutions were related to Reiner's reference to the industry "tail" wagging the healthcare "dog" [15]. Morgan et al. identified a gap between the generalized solutions industry is willing to provide and the optimized solutions healthcare providers (read radiologists) want. This was called the "value innovation gap". Morgan et al. further wrote that those experiencing this gap have a nagging frustration that the end-users are not driving the process [15].

The authors' approach to a solution involved a paradigm shift from a "passive vendor-driven model" to an innovative and "active user-driven model".

This idea was taken up more recently in a radiology context by Flanders [16]. He proposes a data-driven approach to software design in which there is a critical study of user interactions with applications in a variety of clinical settings. The field of usability could be brought to bear as well, to help in objective evaluation of user interaction. All of this would lead to the capability for real and substantive customization by the user. He argues for the use of new technologies, such as Web services, Application Programming Interfaces (APIs), and Asynchronous Javascript XML (AJAX), which are machine-independent and can provide a method for exchanging data between disparate applications and a customized user interface.

In the application of new image processing algorithms, or the extension/ combination of existing ones, this philosophy of meeting customer-driven requirements seems especially valuable. As new users become familiar with a system, they should be able to experiment safely with alternative approaches to visualizing, measuring, and extracting details from their images.

With customer leadership comes responsibility. Flanders notes that customers must take an active role in the assessment of core applications and services. He concludes that most important, knowledgeable customer domain experts should be allowed to take a more active role in the entire product development cycle [16].

22.4 The Goals of Image Processing for Medical Imaging

22.4.1 Automation of Tasks

As described in Sect. 22.3.2, radiologists are highly driven by the need to be productive, so that any image-processing technique that can speed up the reading process tends to be viewed more positively than a method that does not change, or lengthens, interpretation time. Image-processing techniques that can automate routine, but time-consuming, tasks are of a sort that would be beneficial.

Appropriate Image Presentation

Even the replacement of film-screen systems with CR/PSP or DR has still not eliminated, one problem faced by technologists and radiologists. Radiographs taken at the patient bedside often are done because the patients are too ill to travel to the radiology department and often are also done urgently. Under these circumstances, it is not difficult to understand why the PSP cassette or direct capture plate can be misoriented for the radiograph despite marks on the cassettes or plates. The resulting inverted or rotated image must then be corrected by either the technologist or the radiologist. Automation of this process has been accomplished with a high degree of accuracy and with speed sufficient for incorporation into routine workflow [17]. Although the situation has improved with more manufacturer adherence to Digital Imaging and Communications in Medicine (DICOM), which is a standard for handling, storing, printing, and transmitting information in and about medical images [18], with PSP systems, a conventional two-view (Postero-Anterior (PA) and lateral) chest radiograph could be displayed on a workstation with the PA view on the left monitor and the lateral view on the right, or vice versa. The problem is that most radiologists prefer a particular arrangement of the images, and the choice may vary between radiologists. In what has come to be known as the "hanging protocol", images are arranged automatically as the radiologist would like to view them. In film-based operation, this was done by the film library clerks, but automation of this function in PACS required that either the views be properly identified (one element in DICOM) or to be able to identify a view automatically. Boone et al. developed a neural network classifier to determine automatically if a chest radiograph was a PA or lateral view [19]. They were able to demonstrate both the success of the method and documenting time savings as a result of the radiologist not having to rearrange the images.

Reliable Measurements

A long-established evaluation of a person's heart size on a chest radiograph has been the measurement of the ratio of the transverse diameter of the heart to the transverse diameter of the chest; the Cardio-Thoracic Ratio (CTR). The traditional way to do this quantitatively was to measure these dimensions on the film. With PSP and digital imaging, at least the measuring calipers could be made electronic. A very early application of image processing in radiography was the automation of the computation of the CTR. There are applications of this that were attempted before digital imaging. In 1973, Sezaki and Ukena described an apparatus for scanning and digitizing a chest radiographic film and using detected edges of the heart and lungs to compute the CTR [20] and were granted a patent for the process in 1975 [21]. Digital image processing methods have improved the speed of automatic computation of this ratio and do so without specialized equipment. A paper from 2006 by von Ginneken, Stegmann and Loog [22] describes automated CTR computation (and, in addition, describes some of the segmentation methods discussed subsequently). Despite these (and many other) methods for automatic determination of CTR, the resulting algorithms, even though automated, have not found wide utilization. This is largely because radiologists are very adept at recognizing an enlarged heart without computing this ratio. CTR is sometimes used when the heart size is perceived as being borderline, but is quickly computed using the linear measurements available on workstations. In this instance, despite the ability to compute the CTR quickly and in an automated fashion, it is not widely used by radiologists.

An area of great interest for physicians who study vascular diseases such as arteriosclerosis (a leading cause of morbidity and mortality in the USA [23]) is the quantification of narrowing of vessels (stenoses). Conventional catheter angiography, in which contrast material is injected into vessels through catheters placed into them, opacifies the lumen of the vessels and provides some qualitative assessment of stenoses. Quantitative measurements can be done, but are difficult because of the various geometric magnification aspects of the imaging process. Placing reference objects of known size at the same position from the film (or detector) and X-ray source as the vessel is required for actual measurement. The advent of digital subtraction angiography helped reduce the radiation exposure associated with film-based angiography and the resulting images provided high-contrast vessel lumen images amenable to automated analysis. However, the method did not eliminate the difficulty with actual measurement [24] as the image acquisition was still a projection radiographic technique.

3D Measurements

CT was a major advance for quantitative measurement of many anatomic structures. The relationship of the pixel size of the image to the real-world size could be determined accurately. With the development of multi-detector or multi-slice CT, isotropic voxels could be produced. This allowed for accurate dimensional measurement in all three machine axes and the resulting anatomic planes. An outcome of this has been the ability to measure actual blood vessel diameters, both the external dimension and the lumen measurements. These measurements can be used to compute vascular stenoses much more accurately than has been possible. The digital nature of these images led to the development of algorithms for automatic assessment of blood vessels, particularly coronary arteries [25–27]. Software is available from some vendors that can carry out extensive automated evaluation of the coronary arteries. These methods offer fast, automated results, though some questions about accuracy have been raised. Blackmon et al. showed that there is a high degree of correlation between manually done and automated coronary analyses [27]. Because the reconstruction algorithms can directly affect the way vessel edges and the contrast-enhanced lumens appear in the images, the use of particular reconstruction kernels for coronary imaging has proven useful [28]. Development of examination-specific reconstruction kernels is an active area for image processing research.

22.4.2 Improvement of Observer Performance

Radiologists make errors when interpreting images. Overall, the error rate has been estimated at 4% [29]. For chest radiographs, images of patients initially read as "normal" and later diagnosed with lung cancer showed that in approximately 90% of the cases, the cancer was visible on the chest image in retrospect [30]. Similarly, for "normal" mammograms of patients subsequently found to have breast cancer, the lesion was detectable on the mammogram in retrospect approximately 75% of the time [31]. The problem for the patient and radiologist in these situations is not that the imaging system failed to render the abnormality, but that the radiologist did not detect it. How much a lesion stands out from the background, both in gray-scale value (contrast) and difference from surrounding complexity was termed "conspicuity" by Kundel and Revesz [32]. Building on those results, recent work has begun to objectively characterize conspicuity by combining human visual psychophysics and signal processing [33]. Increasing the conspicuity of abnormalities is one way in which it has been theorized that the miss rate could be reduced. Image processing has had a prominent role in methods to increase conspicuity.

Of reasons why abnormalities are difficult to detect, that they may not "stand out" enough from their surroundings is one cause. Various image processing methods for improving conspicuity have been, and are, being used or evaluated. Koenker provides an overview of various post-processing methods in digital radiography [34], and Prokop et al. published an overview of image processing in chest radiography [35].

Contrast Enhancement

If a lesion is only subtly different in gray-scale value from its surroundings, the lack of contrast will make it difficult to see. Contrast enhancement of various types has long been a staple image-processing technique. One author [SH] recalls that this was done even with film. He was taught that contrast could be increased in light areas of a film by tilting the film and in so doing, increasing the thickness of film emulsion through which light from the light box had to pass. For lesions that were only slightly more exposed than their surroundings, this would increase the attenuation of these areas more than it would for the less exposed background. Digital imaging made it possible to manipulate contrast very readily. For some imaging methods, changing contrast is essential. Since CT produces images with a wider range of digital values than can be represented by directly viewable display devices (and than can be appreciated by the eye), remapping a variable number of bits in the display range to the bit range of the image (the "window width" and "window level" operations) allows for display of the whole range of the CT image, though not on a single static image.

An examination of the gray-scale histogram of many medical images will show that pixel values tend to be clustered around certain values. The image-processing technique of histogram equalization attempts to use the available range of display values by equalizing the histogram. Done in the simplest manner, the flattening of the histogram tends to reduce the image contrast, not increase it. For this reason, adaptive histogram equalization and its variations were developed to maximize the utilization of display gray levels without decreasing contrast [36].

Edge Enhancement

Since the human visual system includes various receptors sensitive to edges, enhancing the boundaries, or edges, of a structure or lesion might increase its conspicuity. Edge enhancement techniques from image processing are well known and are another class currently used in CR/PSP and DR systems. There are also CT reconstruction algorithms that enhance edges. The use of, typically through unsharp masking, though direct high-pass filtering, differentiation, and neural network methods [37] have also been used. Edge enhancement has been shown to improve detection of abnormalities and structures with edge-like properties [38, 39]. Edge enhancement can be overdone, since high-frequency noise also tends to be enhanced resulting in images that are too difficult to interpret. This suggests that noise reduction methods could also be of value in improving detectability.

Noise Reduction

Noise is a companion of all medical imaging and is responsible for at least some component of the complexity that forms part of the conspicuity equation. For radiographic images, the typical source of noise is quantum mottle resulting from the statistics of X-ray photons and detectors [40]. CR/DR systems also introduce noise in the digitization process and electronic noise in the various components of the imaging chain. MRI has several sources of noise, the chief of which is actually the patient. Thermal motion of the various molecules in the body perturb the radio-frequency (RF) signal induced by the imaging process [41]. MRI also has noise arising from the magnet system and coils, the RF generators and amplifiers, and the digitizing process. Ultrasound imaging has speckle as the main type of noise. This arises from the superposition of sound waves reflected by randomly distributed scattering foci in tissues. Ultrasound imaging also has electronic noise from the pulser and analog receiver circuitry and the analog-to-digital conversion process.

Image processing for noise reduction is aimed at the source noise. Readers of this book are likely to be familiar with many forms of noise reduction. Some examples are provided in the paragraphs that follow and are not intended to be a comprehensive review of noise reduction techniques. Since quantum mottle in radiographs is random and results from the quantum statistical process of the exposure (whether analogue film or digital device), the only way to reduce quantum mottle is to increase the number of photons, which means increasing exposure (dose). For a given exposure, however, if a detector is more efficient at capturing the photons that it does receive, it will have lower noise than a less-efficient detector. An advantage of some newer types of PSP phosphors and direct capture digital detectors is increased quantum detection efficiency (QDE) compared to earlier PSP systems or film. Radiographic images are also plagued by scatter. Rather than traveling in a straight line from the X-ray source, a photon may be scattered, striking the detector at an angle that does not represent the anatomic structures through which it passed. This results in an overall decrease in image contrast. Scatter is also random, but can be reduced through the use of an air gap between the patient and detector. This allows some of the scattered photons to miss the detector. A more practical method is the use of a collimating grid (collimator). This is a device composed of X-ray attenuating strips alternating with open (or low attenuating) spaces. Off-angle photons will strike the strips and be absorbed rather than going through to the detector. While grids reduce scatter very effectively, they can introduce their own noise in the form of narrow stripes on the image. Image processing has been applied to reduce the noise resulting from grids [42] as well as some quantum mottle in certain frequency bands.

Reduction of noise has been primarily attempted through various filtering methods [43], though averaging techniques have also been used, the latter to great advantage in ultrasound and fluoroscopy. A review of major techniques is provided by Rangayyan [44].

Processing Context-Dependent Noise

In the clinical use of images, however, another class of noise exists. This is the noise of structures not of interest for the clinical question to be answered. In a chest X-ray, for example, the usual interest is the lungs. In that case, the ribs (which are projected over the lungs) are "noise". However, if the patient has sustained a traumatic injury of the chest, then the ribs (which might have been fractured in the injury) become "signal". This sort of noise problem is much more difficult to approach by image processing, though there are techniques that can help. PSP and DR can potentially record images from different portions of the energy spectrum of the exposing X-ray beam. From a single exposure, through the use of energy filtering (not, in this case, filtering in the sense of signal processing) two different images can be made, one from the higher energy portion of the X-ray beam and the other from the lower energy portion. Because of differential attenuation of tissues at different X-ray energies, it is possible through processing of these images to yield images that are weighted to soft tissue or bone [45]. A conventional image, the soft tissue (lung), and bone images can all be displayed for the radiologist offering images that reduce anatomic "noise" though it is dependent on the radiologist to determine whether or not these are useful. The problem of superimposed anatomy as noise on projection radiographs is one reason that CT provides advantages over projection images. Anterior anatomic structures that would be superimposed over posterior ones on projection images are separated by CT imaging.

Change Detection

A major task that the radiologist has is to detect changes. A typical clinical question is whether a particular disease process is getting worse or better in a patient. Such questions are often based on the size of an abnormality, such as an area of pneumonia in a lung decreasing or increasing. Digital imaging allows for mathematical operations to be performed on images simply, as the images are already in a numeric representation. One method to detect change is to subtract the current image from a prior one; changes will be enhanced by such techniques. Temporal subtraction has been developed and tried and readily demonstrates that the resulting images can direct the radiologist to pay more attention to the areas that show change [46,47] potentially improving detection and making comparisons simpler. The processing power available in current workstations and servers allows for image warping to be included in the subtraction process, providing for reduced artifacts in the subtracted images.

Change detection has not been limited to radiographic images. Patriarche and Erickson developed an algorithm to track the change of brain tumors on MRI and display the changes to the radiologist [48, 49]. Sensitivity, specificity, and accuracy for detection of tumor progression were 0.88, 0.98, and 0.95, respectively.

Computer-Aided Diagnosis

Automated change detection and increasing conspicuity are two classes of methods that point to a successful application of image processing in radiology: computer-aided diagnosis (CAD) – sometimes differed between computer-aided detection (CADe) and computer-aided diagnostics (CADx). The interest in applying computer-based analysis to assist diagnosis has a long history. An early paper by Lodwick et al. in 1966 described some of the then-current attempts at applying various computer-based analysis methods to radiographs [50]. In 1984, the National Aeronautics and Space Administration (NASA) with support from the National Institutes of Health (NIH) National Heart Lung and Blood Institute (NHLBI) sponsored work in surveying computer applications in radiograph enhancement and automated extraction of quantitative image information [51]. CAD has now become a commercial software product for mammography (the first product approved by the FDA in 1998) and lung nodule detection. In general (and the reason for some preference for "computer-aided detection" as the meaning of the acronym), these systems are used as an adjunct to a radiologist's interpretation. The systems make "marks" on the images where the software has detected a suspicious area. The radiologist then evaluates these and may dismiss them or raise a concern about them. For mammography, a large study determined that the use of CAD increased the detection of cancers by 19.5% and increased the proportion of detected early stage cancers from 73 to 78% [52]. These results were achieved with no change in the positive predictive value of biopsies done (a desirable result – the concern about CAD is that it would increase the biopsy rate based on the CAD marks and not on radiologist and referring physician judgment). Although the patient recall rate (bringing patients back because of suspicious findings) did increase (from 6.5 to 7.7%), the authors point out that the increased rate of detecting true abnormalities would justify an increase in recall rate.

Lung nodule detection poses another challenge for radiologists. Lung lesions often have low conspicuity because of low contrast and high surrounding complexity. Even with the use of CT for evaluating the lungs, nodules can be inconspicuous because of small size or adjacency to normal blood vessels. White et al. showed that CAD on CT has some success in improving radiologist sensitivity for detecting lung nodules [53]. Recent work has shown that lung CT CAD can significantly increase radiologists' detection of nodules less than 5 mm in size [54]. Nodules of this size have a greater likelihood of being overlooked by the radiologist. Even though small nodules of this size have a low probability of being malignant (in a low-risk population), it is desirable to find these nodules so they can be followed-up in the high-risk population.

Texture discrimination, particularly in ultrasound, is another important facet of diagnostic interpretation. Diseases of most solid organs in the abdomen result not only in anatomic distortion but also in texture alteration. Theories of human perception of texture were first given a computational basis by Julesz who examined texture using a statistical treatment [55]. A well-known conjecture was put forward in this work: that humans cannot distinguish between textures that have the same second-order statistics (homogeneity). Julesz himself later showed that this conjecture was incorrect [56], but the work did establish one computational method of analyzing and synthesizing texture. The examination of co-occurrence matrices as well as modeling of texture based on scattering physics are two (among a number) of the methods used to provide quantitative comparisons of the texture of normal and abnormal tissues. Garra et al. used the former approach in analyzing breast lesions and yielded a 100% sensitivity for classifying breast lesions as malignant [57]. The methods were not able to exclude all of the benign masses (cysts, s, and fibrocystic nodules) included, but the analyses did result in correctly excluding the majority of these from the malignant category. The implication of the study is that use of texture analysis in breast ultrasound could reduce the biopsy rate by excluding non-malignant lesions. Using scattering modeling and decomposition of the echo signal into coherent and diffuse components, Georgiou and Cohen developed algorithms that successfully discriminated malignant from benign lesions in liver and breast ultrasound images [58]. Statistical ultrasonics was studied extensively by Wagner [59] and Insana [60] and a historical summary and tribute is provided by Insana [61]. Despite these successful applications of texture analysis in ultrasound, the equivalent of CAD for ultrasound has not yet emerged. That is not to say that the years of research on texture and speckle in ultrasound did not yield clinically useful results; most of the ultrasound vendors use various signal processing methods based on statistical ultrasonics research to reduce speckle and other noise in the ultrasound machine itself. One of the authors [SH] has been doing clinical ultrasound since the late 1970s and can readily testify to the vast improvement in ultrasound image quality.

Virtual Imaging

In recent years, CAD has been used in CT colonography (CTC). In this imaging technique, high-resolution CT studies of the colon are used to reconstruct 3D volumetric images of the large bowel. These can then be viewed either in a "fly through" mode, which simulates optical colonoscopy, or with the image virtually opened and flattened. The 2D cross-sectional images are also reviewed. CAD has been successfully applied to CTC (sometimes referred to as "virtual colonoscopy") in the search for polyps. Colon polyps, when small and difficult to detect, are often benign, but can be precursors of malignant lesions, so need to be followed. Yoshida and Dachman reviewed the background of CTC and concluded that it showed promise in detecting polyps but that additional refinement was needed [62]. More recently, Taylor et al. studied the use of CAD in CTC in two modes – concurrent with reading (much as CAD for mammography), and as a "second reader" [63]. They found that using CAD for CTC in a concurrent mode was more time efficient, but use as a second reader increased sensitivity for detection of smaller lesions. While CAD for CTC is not yet as widely used as CAD for mammography, use is increasing. Overall, CAD has proven to be a very successful application of image-processing techniques in radiology.

22.5 Closing the Gap

As the authors hope to have shown, though there are gaps between the large body of research on medical image processing and clinical applications, that there are many instances of the use of image processing in daily clinical work. Much of this is "hidden" from the radiologist and included in such things as the automated preprocessing of PSP radiographic images, the reconstruction kernels for CT, segmentation for 3D volumetric image display, and speckle reduction in ultrasound. The growth of CAD is a more readily visible application, though the internal steps used in the process are not typically well understood by the radiologist.

The two major means by which the gaps between image processing research and clinical practice can be narrowed or bridged are neither unexpected nor (the authors believe) too far outside the experience of researchers and clinicians. These means are education and research – the two building blocks of academic radiology.

22.5.1 Education

While years of educating radiologists about imaging physics included extensive descriptions of fundamentals such as the transfer function of film and means for producing subtraction radiographs, much of the traditional study of radiologic physics by the radiologist was heavily weighted to analog imaging. The preponderance of digital imaging has forced a revision of much of the physics curriculum for radiology residents (for whom the Board examination has a whole part devoted to physics). The opportunity was taken by the American Association of Physicists in Medicine (AAPM) which, in 2009, produced an extensive revision of the radiology physics educational program for residents [64]:

- *Basic Imaging Science and Technology*: Module 7 of the curriculum includes items in the "Fundamental Knowledge", "Clinical Application", and "Clinical Problem-Solving" sections directly involving image processing.

- *Describe the different processes used to convert the acquired raw data into a final image used for interpretation*: Item 3 of the Fundamental Knowledge Section.
- *Determine how changes in each image processing procedure impact the final image produced. Evaluate how these changes affect the image of different objects or body parts and their associated views*: Item 2 of the Clinical Application Section.
- *Choose the appropriate image processing to be used for a specific exam*: Item 3 of Clinical Problem-Solving Section.

While "image processing" in these sections applies largely to PSP radiography, an instructor could use these curricular elements to expand on image processing, what is involved, how the particular algorithms were developed, and why they are valuable. The more fundamental aspects of image processing are not disregarded in the AAPM proposal, in fact, Sect. 7.3 of Module 7 is devoted entirely to image processing. The outline includes many of the topics covered in this volume (e.g., image segmentation, image enhancement, volume rendering).

Radiology trainees have, if not an aversion, at least a dislike of concepts explained with extensive mathematics. However, since their primary interest is in learning to interpret images, demonstrations of the effects of image processing can be a very effective educational tool. In some, this will produce enough curiosity that they will want to know the mathematics behind what they have seen. Showing, or maybe better with an interactive demonstration, such processes as: high- and low-pass filtering, edge enhancement (through various means), histogram equalization, frequency domain representation through the Fourier transform, and segmentation could all be very effective educational tools. Astute residents may, for example, come to understand the relationship of these processing elements and the practical steps the technologists use when selecting an examination type on a PSP plate reader; with experience, residents may come to suggest additional processing steps that will complement their own understanding of an image or study.

22.5.2 Research

To help close the research-practice gap will likely also require a parallel education of the graduate and postdoctoral students (from engineering, computer science, or physics) in clinical aspects of radiology. The authors are not advocating sending these students to medical school, but rather that radiologists should be willing to spend their time explaining the problems they face in the practice of their specialty. A graduate student or postdoctoral trainee with an understanding of all that image processing can do should be able to think of possible (novel) solutions (i.e., research) when sitting with a radiologist who shows the student a problem, for example, finding the tip of a vascular catheter on a chest radiograph.

The bi-directional education processes – image processing researchers educating radiologists about techniques and radiologists teaching image processing experts about clinical problems – is certainly not novel or unique. It can, however, provide a model that can also be extended to more collaborative projects. On the research side, this can mean translational research that is directed at particular clinical needs. Clinical radiologists can aim to understand enough about image processing that they can serve as advisors on grants, or to work with senior research faculty as mentors for graduate students.

It will be important, as suggested above, to involve industry in the process as a full partner. Researchers reach most radiologists through commercial equipment that is used in a production rather than a research setting. Industry can make equipment that builds in usability and flexibility, and includes a robust API [16] that permits extensive customization by the user. Not all users may want to be so heavily involved in design and feedback to the suppliers – but those who wish to should find it straightforward.

22.6 Conclusion

In one of the authors' [SH] experience, an approach that might prove useful in having image-processing be more integrated in clinical imaging is to find a "champion" in radiology who both understands image processing and is enthusiastic about it. This has been a successful approach in the information technology area and has been helpful in making a radiology department's transition from analog to digital imaging much smoother than it otherwise would have been. A champion for image-processing applications in radiology could be either at a departmental level, or even multiple such individuals in the various sections of a department. The role of the champion is as an "educator", instructing colleagues about the potential use of image processing to solve particular problems. The role also includes "translator" working to make the principles behind image processing methods understandable to the clinical radiologist and to explain the importance of various clinical problems to image-processing researchers.

That many image-processing techniques have found their way into products or procedures used in the clinical practice of radiology is an illustration that gaps can be bridged if radiologists and image-processing researchers understand each other's work.

References

1. Talos IF, O'Donnell L, Westin CF, et al. Diffusion tensor and functional MRI fusion with anatomical MRI for image guided neurosurgery. Lect Notes Comput Sci. 2003;2878:407–15.

2. Westin CF, Maier SE, et al. Processing and visualization for diffusion tensor MRI. Med Image Anal. 2002;6(2):93–108.
3. Zhang Y, Pham BT, Eckstein MP. Task-based model/human observer evaluation of SPIHT wavelet compression with human visual system-based quantization. Acad Radiol. 2005;12:324–36.
4. Koff DA, Shulman H. An overview of digital compression of medical images: can we use lossy image compression in radiology? Can Assoc Radiol J. 2006;57(4):211–17.
5. Kangarloo H, Boechat MI, Barbaric Z, et al. Two-year clinical experience with a computed radiography system. Am J Roentgenol. 1988;151:605–08.
6. Andriole KP. Image acquisition. In: Dreyer KJ, Hirschorn DS, Thrall JH, et al., editors. PACS: A Guide to the Digital Revolution. Chapter 11, 2nd ed. New York: Springer; 2006. p. 189–227.
7. Horii SC, Kundel HL, Kneeland B, et al. Replacing film with PACS: work shifting and lack of automation. Proc SPIE. 1999;3662:317–22.
8. Berlin L. Radiologic malpractice litigation: a view of the past, a gaze at the present, a glimpse of the future. Am J Roentgenol. 2003;181:1481–86.
9. Kohn LT, Corrigan JM, Donaldson MSE. To Err is Human: Building a Safer Health System. Washington, DC: National Academies Press; 2000.
10. Berlin L. Radiologic errors and malpractice: a blurry distinction. Am J Roentgenol. 2007;189:517–22.
11. Center for Devices and Radiological Health, FDA, editor. Guidance for industry, FDA reviewers and compliance on off-the-shelf software use in medical devices; 1999.
12. Metz CE. Receiver operating characteristic analysis: a tool for the quantitative evaluation of observer performance and imaging systems. J Am Coll Radiol. 2006;3(6):413–22.
13. Krupinski EA, Jiang Y. Anniversary paper: evaluation of medical imaging systems. Med Phys. 2008;35(2):645–59.
14. Reiner B, Siegel E, Siddiqui K. The tail shouldn't wag the dog. J Digit Imaging. 2004;17(3):147–48.
15. Morgan M, Mates J, Chang P. Toward a user-driven approach to radiology software solutions: putting the wag back in the dog. J Digit Imaging. 2006;19(3):197–201.
16. Flanders AE. Increasing user satisfaction with healthcare software. RadioGraphics 2008;28:1259–61.
17. Grevera G, Feingold E, Phan L, et al. Automatic correction of orientation of CR chest PA images; 1997. RSNA Annual Meeting, Oral Presentation.
18. Kahn CEJ, Carrino JA, Flynn MJ, et al. DICOM and radiology: past, present, and future. J Am Coll Radiol. 2007;4(9):652–7.
19. Boone JM, Hurlock GS, Seibert JA, et al. utomated recognition of lateral from PA chest radiographs: saving seconds in a PACS environment. J Digit Imaging. 2003;16(4):345–49.
20. Sezaki N, Ukeno K. Automatic computation of the cardiothoracic ratio with application to mass creening. IEEE Trans Biomed Eng. 1973;20(4):248–53.
21. Sezaki N, Ukeno K. Apparatus for automatic computation of cardiothoracic ratio. March 4, 1975.
22. von Ginneken B, Stegmann MB, Loog M. Segmentation of anatomical structures in chest radiographs using supervised methods: a comparative study on a public database. Med Image Anal. 2006;10(1):19–40.

23. Heron M, Hoyert DL, Murphy SL, et al. Deaths: final data for 2006; p. 80.
24. Bartlett ES, Walters TD, Symons SP, et al. Carotid stenosis index revisited with direct CT angiography measurement of carotid arteries to quantify carotid stenosis. Stroke 2007;38:286–91.
25. Abada HT, Larchez C, Daoud B, et al. MDCT of the coronary arteries: Feasibility of low-dose CT with ECG-pulsed tube current modulation to reduce radiation dose. Am J Roentgenol. 2006;186:S387–S390.
26. Khan MF, Wesarg S, Gurung J, et al. Facilitating coronary artery evaluation in MDCT using a 3D automatic vessel segmentation tool. Eur Radiol. 2006;16(8):1789–95.
27. Blackmon KN, Streck J, Thilo C, et al. Reproducibility of automated noncalcified coronary artery plaque burden assessment at coronary CT angiography. J Thoracic Imaging. 2009;24(2):96–102.
28. Hoffman U, Ferencik M, Cury RC, et al. Coronary CT Angiography. J Nuc Med. 2006;47(5):797–806.
29. Berlin L. Radiologic errors and malpractice; a blurry distinction. Am J Roentgenol. 2007;189:517–22.
30. Muhm JR, Miller WE, Fontana RS, et al. Lung cancer detected during a screening program using four-month chest radiographs. Radiology 1983;148:609–15.
31. Harvey JA, Fajardo LL, Innis CA. Previous mammograms in patients with impalpable breast carcinoma: retrospective vs. blinded interpretation. Am J Roentgenol. 1993;161:1167–72.
32. Kundel HL, Revesz G. Lesion conspicuity, structured noise, and film reader error. Am J Roentgenol. 1976;126:1233–38.
33. Perconti P, Loew MH. Salience measure for assessing scale-based features in mammograms. J Opt Sec Am A. 2007;24:B81–B90.
34. Koenker R. Improved conspicuity of key x-ray findings using advanced postprocessing techniques: clinical examples. Medica Mundi. 2005;49(3):4–11.
35. Prokop M, Neitzel U, Schaefer-Prokop C. Principles of image processing in digital chest radiography. J Thoracic Imaging. 2003;18(3):148–64.
36. Pizer SM, Amburn EP, Austin JD, et al. Adaptive histogram equalization and its variations. Computer Vis Graph Image Process. 1987;39:355–68.
37. Suzuki K, Horiba I, Sugie N. Neural edge enhancer for supervised edge enhancement from noisy images. IEEE Trans Pattern Anal Machine Intell. 2003;25(12):1582–96.
38. Yoon HW, Kim HJ, Song KS, et al. Using edge enhancement to identify subtle findings on soft-copy neonatal chest radiographs. Am J Roentgenol. 2001;177:437–40.
39. Ludwig K, Link TM, Fiebich M, et al. Selenium-based digital radiography in the detection of bone lesions: preliminary experience with experimentally created defects. Radiology 2000;216:220–24.
40. Wolbarst AB. Image quality: contrast, resolution, and noise. Primary determinants of the diagnostic utility of an image. In: Wolbarst AB. Physics of Radiology. Chapter 18, 2nd ed. Medical Physics Publishing; 2005. p. 204–211.
41. Noll DC. A primer on MRI and functional MRI; 2001. PDF on-line: http://www.eecs.umich.edu/ dnoll/primer2.pdf.
42. Okamoto T, Furui S, Ichiji H, et al. Noise reduction in digital radiography using wavelet packet based on noise characteristics. Signal Process. 2004;8(6):485–94.
43. Wong WCK, Chung ACS. In: Lemke HU, Inamura K, Doi K, et al., editors. A Nonlinear and Non-Iterative Noise Reduction Technique for Medical Images:

Concept and Methods Comparison. International Congress Series. Elsevier; 2004. p. 171–176.
44. Rangayyan RM. Removal of Artifacts. CRC: Boca Raton. 2005.
45. Ergun DL, Mistretta CA, Brown DE, et al. Single-exposure dual-energy computed radiography: improved detection and processing. Radiology 1990;174:243–49.
46. Kinsey JH, Vannelli BD. Application of digital image change detection to diagnosis and follow-up of cancer involving the lungs. Proc SPIE. 1975;70:99–112.
47. Kano A, Doi K, MacMahon H, et al. Digital image subtraction of temporally sequential chest images for detection of interval change. Med Phys. 1994;21:453–61.
48. Patriarche JW, Erickson BJ. Automated change detection in serial MR studies of brain tumor patients, Part 1. J Digit Imaging. 2007;20(3):203–22.
49. Patriarche JW, Erickson BJ. Automated change detection in serial MR studies of brain tumor patients, Part 2. J Digit Imaging. 2007;20(4):321–28.
50. Lodwick GS, Turner JAH, Lusted LB, et al. Computer-aided analysis of radiographic images. J Chronic Dis. 1966;19(4):485–96.
51. Selzer RH. Computer processing of radiographic images. Proc SPIE. 1984;516:16–27.
52. Freer TW, Ulissey MJ. Computer-aided detection: prospective study of 12,860 patients in a community breast center. Radiology 2001;220:781–86.
53. White CS, Pugatch R, Koonce T, et al. Lung nodule CAD software as a second reader: a multicenter study. Acad Radiol. 2008;15(3):326–33.
54. Sahiner B, Chan HP, Hadjiiski LM, et al. Effect of CAD on radiologist's detection of lung nodules on thoracic CT scans: analysis of an observer performance study by nodule size. Acad Radiol. 2009;16(12):1518–30.
55. Julesz B. Visual pattern discrimination. IRE Trans Inf Theory. 1962;8:84–92.
56. Julesz B, Gilbert EN, Shepp LA, et al. Inability of humans to discriminate between visual textures that agree in second-order statistics: revisited. Perception. 1973;2:391–405.
57. Garra BS, Krasner BH, Horii SC, et al. Improving the distinction between benign and malignant breast lesions: the value of sonographic texture analysis. Ultrason Imaging. 1993;15(4):267–85.
58. Georgiou G, Cohen FS. Is early detection of liver and breast cancers from ultrasound possible? Pattern Recognit. 2003;24(4–5):729–39.
59. Wagner RF, Smith SW, Sandrik JM, et al. Statistics of speckle in ultrasound B-scans. IEEE Trans Son Ultrason. 1983;SU-30:156–63.
60. Insana MF, Wagner RF, Garra BS, et al. Analysis of ultrasound image texture via generalized Rician statistics. Proc SPIE. 1985;556:153–59.
61. Insana MF. Statistical ultrasonics: the influence of Robert F. Wagner. Proc SPIE. 2009;7263:1–6.
62. Yoshida H, Dachman AH. CAD for CT colonography: current status and future. Int Congr Ser. 2004;1268:973–7.
63. Taylor SA, Charman SC, Lefere P, et al. CT colonography: investigation of the optimum reader paradigm by using computer-aided detection software. Radiology 2008;246:463–71.
64. AAPM Subcommittee of the Medical Physics Education of Physicians Committee. Diagnostic Radiology Residents Physics Curriculum; 2009. http://www.aapm.org/education/documents/Curriculum.pdf.

Index

AAM, 299, 300
AAPM, 561
ABCD, 311
  rule, 313, 314, 316, 323
Abdomen, 541
Absolute calibration, 20
Absorption, 7, 58, 63, 382
  coefficient, 7
  power, 58
Academic radiology, 561
Accuracy, 142, 219
ACE, 217
Acquisition time, 8
ACR, 331, 344, 428
ACR/NEMA, 428, 442, 452
ACSE, 434
Active
  contour, 35, 36, 271, 281, 347, 413
  imaging, 407
  shim, 64
  surface, 281
Adaptability, 521
Adaptive
  diffusion, 187
  measurement, 374
  segmentation, 28
  threshold, 28, 40
Adenoid, 275
Adhesion border, 36
Adjacency relation, 252
ADNI, 373, 499
AE, 434
AFC, 256, 267, 275

segmentation, 274
Affine
  invariant, 184
  registration, 149
  transform, 133
Affinity
  function, 254, 267
  homogeneity-based, 254
  object feature-based, 255
  propagation, 415
Age spot, 308
Agglomerative
  algorithm, 32
  method, 231
  segmentation, 32
AHA, 366
AIF, 369
Air, 28
Air-free coupling, 12
AJAX, 552
AJCC, 311
Algebraic
  sum-min, 276
  union, 276
Algorithm
  mean shift, 166
Aliasing effects, 15
ALM, 308
Alpha particle, 5
Alzheimer's disease, 66, 76, 145, 147, 148, 173, 234
Ambient light, 23
Amino acid, 67

Amira, 542
AMN, 313
A-mode, 12
Amplitude
    analyzer, 58
    mode, 12
    reduction, 99
Analyze, 541
AnalyzeDirect Inc., 542
Anatomical landmark, 135, 140
Anatomic prior, 413, 415
Anatomy, 485
Aneurysm, 88
Angiogenesis, 331
Animal PET, 69
Anisotropy, 162, 405, 408
    filter, 18
    MRI, 387
    template, 18
ANN, 316, 319, 321
Annihilation, 58
    photon, 58, 74
    process, 57
Annotation, 541
ANODE, 512
AOM, 334, 346, 347
Aorta, 87
Aortic valve, 95
AP, 466
APD, 58
APD-based PET, 69
Aperture, 98, 165
API, 448–450, 487, 523, 528, 532, 539, 541, 552
Appearance
    model, 276, 298, 299
    prior, 284, 297
Application
    development, 536
    Hosting, 448, 449, 452
Approximation, 135
Area
    closing, 114
    opening, 113, 119
Arithmetic reconstruction, 8
Arrival-time, 411
Arteriosclerosis, 554
Arthritis, 363
Artificial intelligence, 40

ASCF, 119, 120
ASD, 87
ASF, 111, 118
Asia, 397
ASM, 148, 299
Aspect ratio, 430
Association
    negotiation, 435
    rule mining, 215, 217
Asymmetry, 316
Atlas, 294, 463
    registration, 298
Atlas-based segmentation, 147, 288, 508
Atomic number, 58
Attenuation, 59
    correction, 84
AUC, 319
Australia, 308, 542
Automatic
    classification, 43
    interaction, 396
    landmarking, 149
    segmentation, 36
Auto-regressive model, 231
AVD, 504
Axons, 412

BAA, 463
Back projection, 8
Background
    marker, 125
    model, 28
Backscatter, 83
Back-to-front, 23, 393
Backward pruning, 212
Ball scale, 266
Balloon, 34, 36
    force, 283
    model, 35
    segmentation, 36, 43
Barrel distortion, 20
Barycenter, 381
Basal cell carcinoma, 307
Basis function, 134
Bayes rule, 200
BDWG, 361
Beam
    broadening, 100
    distortion, 100

focusing, 86, 99
forming, 86
hardening, 7
skewing, 100
steering, 85, 86, 93, 99
warping, 100
Benchmark, 488
Bernstein polynomial, 136
β^+-decay, 57
BF, 59
BGO, 56
BI-RADS, 218
Bicubic interpolation, 367
Big Endian, 433
Bilinear interpolation, 385
Binarization, 26, 30
Binary
  morphology, 19
  reconstruction, 119
Binding potential, 69
Bio-electric field, 532
Biological
  noise, 370
  tissue, 30
Biomarker, 403
Bio-mechanical
  model, 150
  motion, 145
BioPSE, 532
Biosystem, 485
Bipolar window, 172
BIR, 542
BI-RADS, 218, 331, 343, 344
Black-box, 314, 533
Blending
  order, 381
  surface, 391
Blob representation, 233
Blobworld, 472
Block prior, 284
Blood
  flow, 13, 388
  vessel, 381, 384, 555
BlueBerry, 526
Blurring, 73
  adaptive, 187
B-mode, 12, 13, 81, 85, 88, 331, 367
Body imaging, 68
Bohr model, 5, 6

BoneXpert, 37
Bootstrap
  analysis, 418
  sampling, 412
Border
  diffusiveness, 316
  length, 27
Bottom-up, 31
Boundary, 204
Boundary-based feature, 298
Bounding box, 286, 525
Bowel, 560
Brain, 21, 120, 273, 384
  imaging, 68
  MRI, 228, 290
Branemark implant, 40, 43
Breast
  cancer, 329, 555
  imaging, 330
  parenchyma, 333
  ultrasound, 560
Bremsstrahlung, 6, 7
Brightness
  mode, 13
  variation, 20
Brovey method, 72
BSD, 523
B-spline, 137, 286, 290, 294, 523
  FFD, 137
  surface, 390
  tensor, 136
BSPS, 440
Bulging surface, 391

CAD, 131, 198, 221, 451, 455, 458, 512, 533, 559
CADe, 309, 325, 329, 330, 332, 333, 335, 447, 457, 548, 559
CADx, 309, 313, 325, 329, 330, 333–335, 337, 338, 341, 343, 348, 351, 352, 447, 461, 559
Calcium, 365
C4.5 algorithm, 212
Calibration, 16, 41, 134, 325
CAM, 533
Camera
  calibration, 134
  control, 396
Canberra distance, 207, 209

Cantata, 531
Cardboard piece, 119
Cardiac
    CT, 95
    cycle, 81
    MRI, 95
    ultrasound, 81
Cardiovascular imaging, 68
CARS, X
CART, 319, 321
    classifier, 320
Cartesian coordinates, 182
CAS, 131, 527
Case
    database, 325
    table, 380
CASH, 311
CAT, 131
Catchment basins, 122
Catheter angiography, 554
CAVE, 542
CBIR, 47, 190, 197, 333–335, 341, 471, 488, 514
CBVIR, 471
CC, 335
CCD, 323
Cell, 281
    culture, 29
    membrane, 35, 36, 43
Centerline, 182
Centroid, 140
Cerebral
    aneurysm, 389
    cortex, 61
    peduncle, 65
Cerebro-spinal fluid, 228–230, 234, 384
CGMM, 235, 236
Characteristics
    curve, 41
    equation, 185
    radiation, 6, 7
Chebychev distance, 207, 210
Checkerboard pattern, 502
Chest, 447, 466
    radiograph, 553, 554
Cholesterol, 365
Chromosome, 113, 117
CI, 38, 40, 325
CICE, 502

CIE Lab, 474
CIMT, 366
Circular transform, 144
CISTIB, 539
CIT, 541
City block distance, 207
Classification, 25, 32, 37, 325, 332, 333, 473, 483
    method, 319
Classifier over-fitting, 212
CLEF, 483, 499
Clinical
    ABCD rule, 317
    application, 149, 561
    endpoint, 361
    evaluation, 333, 374
    imaging, 362
    trial, 149, 362, 368, 369, 373
    validation, 373
    variability, 370
Clique potential, 160
Close-open filter, 111, 120
Closing, 19, 108, 110, 116
Clustering, 28, 230, 241, 414, 415
CM, 307, 309, 310, 321
CMY, 72
CNMD, 310
CNR, 86
CNS, 363, 372
Coarse-to-fine strategy, 285, 287, 301
Cocoa framework, 539
CodeRunner, 533
Coincident
    board, 60
    detection, 58
    time, 58
Collapsing, 389
Collimating grid, 557
Collimator, 557
Collinear landmark, 140
Colon, 396, 447, 560
Colonic polyp, 397
Color, 384, 474, 525
    adjustment, 21
    bleeding, 383
    blending, 383
    coding, 413
    histogram, 199, 202, 484
    map, 101

transform, 316
variety, 317
Colorectal cancer, 397
Colored glasses, 101
Combined boundary, 109
Communication, 2, 45
Complexity, 3
Component filter(s), 112, 119, 120
Compounding, 547
Compression, 83, 145
  ratio, 548
  syndrome, 384
Compton effect, 7
Computation
  cost, 385
  model, 539
  speed, 261
  time, 3
Computer
  algebra, 536
  dermoscopy, 313
  graphics, 22, 539
  mouse, 396
  scientist, 500
Computer science, 484, 562
Computer vision, 280
Conditional
  random field, 231
  statement, 534
Cone-shaped volume, 13
Confocal mode imaging, 312
Conjugate gradient, 524
Connectathon, 461
Connected component(s), 29, 111, 113, 119, 238
Connectivity, 269, 411, 416
  measure, 255
Consistency error, 509
Conspicuity, 555
Context group, 445
Contour, 524
  prior, 284
Contrast, 18, 161, 170, 204, 395, 440, 555, 556
  adjustment, 21
  agent, 7, 363, 392, 393
  enhancement, 41, 333, 555, 556
  filter, 18
  inversion, 546

  manipulation, 540
Contrast-enhanced
  CT, 389
  imaging, 149
  MRI, 145, 149
Control point, 135
Conventional beam forming, 93
Convexity, 27
Convolution, 16, 18, 178, 191, 540
  kernel, 191
  surface, 391, 392
Co-occurrence, 143
  matrix, 160, 203, 560
    contrast, 161
    generalized, 161, 162
    local, 165
Coordinate system, 394
Cornea tissue, 125
Cornerness, 184
Coronary
  artery, 555
  vessel, 393
Corpus callosum, 120, 127, 406, 410, 413
Correction proposal, 428
Correlation
  fractal dimension, 201
  integral, 200
Correspondence estimation, 149
Corresponding point, 132, 135
Cortical column, 191
Corticospinal tract, 66
Cost function, 254
Coulomb field, 5, 6
Covariance matrix, 298
CPU, 383
CR, 464, 548, 553
Cranfield methodology, 506
Cranfield test, 483
Cranial nerve, 384
Cross-validation, 320, 321
CRT, 546
Crus cerebri, 65
Crystal size, 73
CS, 437
CSF, 74, 273, 499
CSI, 64
CSPS, 439

CT, 1, 21, 28, 55, 68, 95, 131, 134, 146, 202, 273, 294, 312, 362, 363, 367, 386, 392, 413, 428, 431, 442, 448, 477, 478, 480, 546, 555
  angiography, 547
  gantry, 8
  reconstruction, 11
CTA, 510
CTC, 560
CTE, 502
CTK, 530
CTR, 554
Cuberille approach, 22
Curse of dimensionality, 476
Curvedness, 187
CVP, 393
Cyclotron, 68
Cyst, 560

1D, 5, 229, 363
2D, 3, 5, 81, 363
3D, 3, 5, 82, 233, 252, 363, 379, 547
  model, 300
  mouse, 396
  Navigator, 24
  reconstruction, 540
  Slicer, 523, 537
  texture, 300
  visualization, 540
4D, 3, 5, 98, 547
3DMed, 527
DAG, 444
Darwinian paradigm, 41
Data
  analysis, 536, 539
  level, 3
  mining, 216
  prior, 412
Data-adaptive metric, 29
Data-based feature, 25
Data-driven, 412, 552
  initialization, 233
Dataflow network, 532
Daubechies, 339
DBM, 147
DBT, 330, 332
DCE, 367
DCE-MRI, 331, 344–346, 368, 369
DCMR, 446

DCMTK, 529, 530, 533, 539
DDSM, 335, 346, 351
Deblurring, 74, 75
Decision
  making, 317, 361, 365
  support, 208, 323
  tree, 212
Deconvolution, 74, 405
Deep structure, 194
Deformable
  fusion, 538
  model, 235
  object, 531
  super-quadric, 291
  transform, 132
Deformation
  field, 242, 503
  model, 138
Delaunay triangulation, 524
Delineated object, 254
Delineation, 3, 27, 31, 35
Demand-driven
  concept, 524
  pipeline, 524
  scheduling, 533
Dementia, 273
Demon's algorithm, 243
Dendrite, 35, 36
Denmark, 37
Denoising, 195, 333
Density, 382
Dental
  chart, 44
  implantology, 21
  status, 3
Depression, 273
Depth
  cue, 393
  encoding, 101
  shading, 22
Dermatology, 443, 512
Dermatoscopy, 312
Dermoscopy, 311, 312
  ABCD rule, 317
  CADx, 310, 313
  device, 322, 323
  image, 315
  system, 322
DES, 382

Detector
  crystal, 61
  system, 61
Deterministic
  clustering, 231
  tractography, 408
Development environment, 521, 543
DeVIDE, 525, 533
DFT, 163
Diagnosis, 102, 395
Diagnostics, 22
  algorithm, 310, 313, 314
  radiology, 484
  skill, 317
  test, 551
Diamond artifact, 385
Dice coefficient, 504
DICOM, 1, 46, 427, 455, 472, 529, 530, 535, 539, 541, 553
  application hosting, 532
DICOM-SR, 443, 444, 446, 447, 455, 459, 460, 466, 529
Diffeomorphic transform, 139, 149
Differential
  equation, 139
  structures, 317
Differentiation, 556
Diffusion, 195
  equation, 180, 188
  field, 410, 411
  geometry-driven, 187, 189
  MRI, 418
  non-linear, 188
  orientation, 404
  pattern, 403, 404
  process, 139
  profile, 405, 411
  strength, 404
  tensor, 407
Diffusion-weighted MRI, 532
Digital
  image, 1, 253
  sensor, 1
  signature, 529
Dilation, 19, 40, 108, 109, 115
Dimensionality curse, 198, 202, 211, 213
DIMSE, 434
Directed graph, 253
Direct rendering, 398

Discrete filtering, 18
Discriminant analysis, 417
Displacement field, 242
Distance function, 198, 205, 208, 209
Distance-related attenuation, 381
Distortion, 20, 204
Divisive
  algorithm, 32
  segmentation, 32
DKFZ, 526, 538
D-mode sonography, 13
dMRI, 403, 412, 413, 416, 419
DNA, 117
DOF, 132, 133, 280, 286
Dome filter, 121
Dopamine, 61
  transporter, 60
Doppler effect, 13
Doppler imaging, 547
Doppler mode, 13
Doppler ultrasound, 331
DP, 466
DPV, 311
DR, 464, 548, 552, 553
Drug
  development, 362
  screening, 361
DSA, 181, 546
DSI, 404
DTI, 126, 404, 529, 532, 537, 538, 541, 547
DTM, 212
Dualing operator, 388
Duplex mode, 13
DVD, 437
DWT, 72
Dynamics, 121, 127
  imaging, 149
  shim, 64
  thresholding, 29
Dyslexia, 126

Early precision, 500
ECG, 90, 241, 428, 477
  gating, 91
ECG-gated TEE, 96
Echo
  production, 63
  time, 10

Echocardiogram, 246, 248
Eclipse, 527, 537
Edge, 281
　collapsing, 389
　completion, 30
　detection, 413, 540
　　statistical, 172
　enhancement, 546, 556, 562
　extraction, 25, 30, 41
　filtering, 18
　flipping, 389
　focusing, 194
　image, 35
　level, 3
　location, 30
　profile, 30
Edge-based feature, 25, 30
Edge-based segmentation, 30, 34
Edge-off operation, 112, 120
Edge-preserving filter, 119
Edge-preserving smoothing, 187, 189
Education, 395, 562
EEG, 477
Effectiveness measure, 507
Egocentric perspective, 395
Eigendecomposition, 296
Eigenfunction, 139
Eigenimage, 339
Eigenvalue, 296
Eigenvector, 296, 299
Einstein, 178
　convention, 182, 183
Ejection fraction, 88
Elastic
　deformation, 144
　registration, 144
Elasticity, 35, 82
Electrical dipole, 63
Electron, 5
Electronic, 58
　patient record, 323
Ellipse, 33
ELM, 312
EM, 59, 199, 227, 231, 415
EM algorithm, 232, 236, 237, 503
EM/MPM, 200
　algorithm, 208, 209
　descriptor, 210
Emission, 382

Empty space skipping, 393
EN, 428
Endoscopy, 1, 6, 20, 35, 442
Endpoint, 361, 364
End-user application, 537
Energy, 204
　filter, 558
Engineering, 562
Engraving machine, 115
ENT, X
Entropy, 142, 179, 204
Envelope detection, 83
EPI, 64
Epidermis, 307
Epilepsy, 234
Equi angle skewness, 388, 389
Equilibrium state, 5, 9
Erosion, 19, 40, 108, 115
E-step, 232
Euclidean, 191
　boundary, 109
　distance, 29, 198, 207, 210, 252, 293
　geometry, 42
　group, 192
　mapping, 414
　norm, 502
　shortening flow, 188
　space, 413
　vector space, 476
Eulerian
　coordinate, 502
　scheme, 291
Europe, 309, 397
Evaluation, 149, 324, 482
　database requirements, 325
　methodology requirements, 325
Event-driven scheduling, 533
Evolutionary algorithm, 38, 40, 41
EXACT, 510
Examination-specific reconstruction, 555
Excitement, 10
Exclusion criteria, 325
Exocentric perspective, 394
Expert system, 39
Explicit VR, 433, 437
Extensible
　application, 543
　software, 521

External
  boundary, 109
  energy, 35
Extrema, 121
Extremum, 187
Ex-vivo, 96, 131

FA, 407, 409, 414, 417
Factorial subgroup, 192
Fairing, 386
False positive, 333
Fan-shaped aperture, 13
Fast-rotating array, 91
Fat, 28
FB, 59
FC, 251
  algorithm, 255
  iterative relative, 263
FDA, 57, 332, 550, 559
FDG, 62
FDI, 44
Feature
  extraction, 25, 41, 198, 216, 316
  selection, 212, 216, 334
  space, 28, 40, 142, 228, 230, 231, 233
  vector, 38, 198, 230
  visualization, 317
Feature-based affinity, 255, 261, 269, 270
Feature-based registration, 141
FEM, 290
Ferromagnetic material, 64
FFD, 136
FFDM, 330, 332
FFDM-DBT, 331
Fiber clustering, 414
Fibroadenoma, 560
Fibrocystic nodule, 560
FID, 9, 63
Field color asymmetry, 316
Field-of-view, 81, 95, 99
Fiji, 540
Filament, 6
Film-screen system, 553
Filter
  kernel, 546
Filter(ing), 16, 179, 191, 536
Filtered back projection, 8
Finite mixture model, 232

First moment, 179
First-order shim, 64
FISH, 117
Fitness function, 41
Fixed shim, 64
Fix point, 31
Flexibility, 563
Flowline, 184
FLT, 67
Fluid registration, 144
Fluoroscopy, 7, 35, 193, 558
F-measure, 507
fMRI, 69, 363, 416, 537
FND, 504
Focusing, 98
Foreshortening, 89
Formalized query, 484
FOS, 338
Fourier analysis, 536, 540
Fourier basis function, 135
Fourier descriptor, 204, 335, 340, 347
Fourier domain, 9
Fourier theory, 388
Fourier transform, 8, 64, 191, 199, 404, 405, 562
Fourth order moment, 171
FOV, 60
FPD, 504
Fractal, 159, 340
  dimension, 200, 202
  model, 159, 337
France, 529
Free interaction, 396
FreeMat, 536
Free text, 478
Frequency
  compounding, 367
  domain, 199
Frequency-based feature, 318
Frequency-encoding gradient, 64
FROC, 512
Front-end, 537
Front evolution, 411
Front-to-back, 23, 393
FSC, 437
F-score, 507
FSR, 437
FSU, 437
Full-brain tractography, 414, 416, 419

Full volume imaging, 95, 97
Fully-automated algorithms, 373
Functional
  data, 373
  loop, 534
  MRI, 547
Fusion, 22, 448
Fuzzy
  connectedness, 524
  logic, 38, 40, 41, 43
  model, 276
  relation, 276
  segmentation, 505
  set theory, 505
FWHM, 56
FZ Jülich, 11

GA, 334
Gabor family, 180
Gabor filterbank, 298
Gabor function, 166, 170, 177
Gamma
  photon, 57, 58
  quantum, 58
Gantry tilt, 133
Gauge coordinates, 181, 182
Gauss, 185
Gaussian, 136, 139, 180, 199
  cluster, 234
  covariance, 238
  curvature, 186
  derivative, 177
  diffusion, 407, 408
  distribution, 171, 298, 393, 405, 413
  estimate, 404
  filter, 166
  function, 267
  kernel, 180, 194
  model, 405
  PSF, 74
  shape, 114
GC, 251
GDCM, 530, 533
GE Health Care, 97
Generative model, 299
Generic feature, 199
Genus zero shape, 294
Geocentric perspective, 394
Geodesic, 412

  path, 190
  track, 190
Geometric
  aberration, 20
  distance, 29
  feature, 43
  landmark, 135
  measurement, 20
  modeling, 386
  reasoning, 178, 191, 195
  registration, 21
  transform, 396
Geometry-adaptive constraint, 193
Geometry-driven diffusion, 189, 194
Germany, 529
Gestalt, 190
GG, 270
Gibbs distribution, 160
Gibbs modeling, 160
Gibbs sampler, 200
GIF, 46
GIFT, 472
GIMIAS, 539
GLCM, 338
Global
  feature, 198, 485
  threshold, 29
  tractography, 411
  transform, 134, 286
GMM, 199, 227, 228, 231
  modeling, 233
  parameter, 237
GMP, 551
GNU, 525
  Octave, 536
Gold standard, 145, 150, 325, 482, 500, 504, 509
Goodness of fit, 298–300
Google, 488
Gouraud shading, 23, 385
GPA, 296
GPU, 383
Gradient
  coil, 62
  descent, 524
  direction, 384
  frequency-encoding, 64
  operator, 138, 180
  phase-encoding, 64

readout, 64
shading, 22
slice-selection, 64
system, 63
Grammar, 158
Gram matrix, 415
Graph
  cut, 241, 254, 413
  matching, 486
Graphic
  clustering, 228
  network, 416
  representation, 394
  tablet, 396
Graph-theoretic clustering, 228, 231, 241, 246
Grating lobe, 99
Gravitation field, 9
Gray level histogram, 209
Gray matter, 31, 200, 228–230, 234, 273, 413
Gray scale
  encoding, 83
  image, 368
  reconstruction, 119
  shade, 115
  thresholding, 121
Green's function, 180
Greulich and Pyle, 463
Grid
  computing, 534
  network, 480
  service, 534
Ground truth, 36, 145, 241, 482, 500, 503, 508
GSPS, 439
GTC, 505
GUI, 396, 449, 465, 474, 523, 527, 534–537, 539
Guided interaction, 396
Gyri, 62
Gyroscope, 9

Hand atlas, 464
Hanging protocol, 441, 452, 553
HARAG, 485
Haralick
  descriptor, 203, 208, 209
  feature, 203, 209

HARDI, 192
Hard plaque, 393
Hardware
  acceleration, 383
  vendor, 528
Hausdorff distance, 414, 506
HD, 505
Health care, 45
Heart, 538, 554
  attack, 366
  cycle, 5
Heaviside function, 200
Hessian, 190
  eigenvector, 185
  matrix, 185, 190
Heuristic method, 30
Hierarchical watershed, 127
High bit, 431
High-field MRI, 65, 69
High-level
  diagnostic, 318
  feature, 314
  language, 536
  processing, 2, 3
Highlight, 227
High-order
  statistics, 171, 175
  tensor, 405, 418
High-pass filter, 18, 556, 562
High-resolution
  CT, 560
  imaging, 56, 69
  MRI, 363
High voltage, 6
Hilbert pair, 169
Hilbert transform, 83, 169
HIPAA, 455, 459
Hippocampus, 66, 68
HIS, 46, 72, 455, 480, 487
Histogram, 16, 27, 199, 208, 229, 556
  equalization, 556, 562
  matching, 298
  orientation, 162
  prior, 284, 298
  stretching, 16
  thresholding, 230
  transform, 16, 18
History of data, 535
HL7, 47, 455, 459, 486

h-maxima, 121
　filter, 121, 122
h-minima, 121
　filter, 127
Hologram, 101
Homogeneity, 204, 560
Homogeneity-based affinity, 254, 261, 269
Homogeneous coordinates, 133
Hosted application, 449
Hounsfield value, 42
Hourgh transform, 25
HRRT, 55, 60
HRRT-PET, 60, 61
HSI color space, 317
HSV, 474
HU, 27, 202, 390, 431
Human
　dermoscopy, 313
　perception, 560
Huntington's disease, 174
Hybrid
　scheduling, 533
　storage, 45
Hydrogen, 68
Hyper-surface, 291

IBSR, 509
ICA, 168
ICC, 433, 439
　profile, 439
ICCAS, X
ICD-10, 445
Iconic
　description, 3
　image, 42
ICP, 141, 293
ID, 46
IDEA, 197, 215, 217
IDL, 536
　Workbench, 536
IFT, 265, 271
IGS, 527
IGSTK, 527
IGT, 526, 537
IHE, 47, 455, 459, 480
　profile, 461
　protocol, 461
IHS, 72

Image
　acquisition, 323, 367, 412, 443
　analysis, 2, 25, 323, 537
　annotation, 512
　compression, 548
　contrast, 100
　enhancement, 2, 562
　filtering, 542
　formation, 2
　frame, 440
　fusion, 22, 149, 502, 539, 547
　interpretation, 2, 101
　management, 2, 323
　matching, 234
　measurement, 43
　post-processing, 2
　pre-processing, 2
　processing, 524, 533, 535, 536
　quality, 384
　registration, 149, 540
　resolution, 100
　restoration, 72
　retrieval, 47, 234
　search, 234
　segmentation, 262, 562
　visualization, 2, 72
ImageCLEF, 483, 488
ImageCLEFmed, 483
Image-guided neurosurgery, 547
Image-guided surgery, 149
ImageJ, 540
Imaging
　biomarker, 360, 368
　depth, 12
　endpoint, 372
　modality, 539
　scientist, 500
Immersive VR, 542
Implementation, 3
Implicit
　modeling, 524
　surface, 385, 391
　VR, 433, 434
Inclusion criteria, 325
Indirect rendering, 398
Induced manifold, 415
Inertia, 82
Infinity distance, 207
Information retrieval, 473, 488

Inf-reconstruction, 119
Inherent smoothness, 235
Initial centroid, 243
Initialization, 289
Inner
    marker, 125
    product, 415
INRIA, 531
Integral shading, 22, 24
Intensity, 204
    function, 253
Intensity-based segmentation, 413
Interaction, 101, 526, 527
    device, 396
    model, 396
    paradigm, 396
    tool, 102
Inter-class variance, 28
Inter-disciplinary expertise, 484
Interference, 70
Interfering anatomy, 374
Inter-frame motion, 241
Inter-individual variance, 484
Internal
    boundary, 109
    energy, 35
    force, 287
Inter-observer
    agreement, 500, 504
    variability, 38, 234, 506
Interpolation, 135, 367, 523
    artifact, 385
Intersection point, 380
Inter-subject
    comparison, 412
    registration, 132
    variability, 4, 288
Inter-tissue contrast, 234
Inter-user variability, 288
Intervention, 395
Intra-class variance, 28
Intra-individual variance, 484
Intra-observer variability, 234, 506
Intra-operative support, 395
Intra-subject
    registration, 132, 149
    variability, 4
Intra-tissue noise, 234
Intricate structure, 191

Intrinsic
    geometry, 181
    landmark, 140
    resolution, 73
Inverse of variance, 204
In vitro, 95, 361, 362
In vivo, 95, 131, 361, 362
IOD, 429, 433, 443, 447, 460
Ionizing radiation, 68
IP, 47
IR, 506
IRB, 469, 551
IRFC, 263, 267
IRMA, 47, 471, 481, 484, 511
    framework, 48
ISO, 47, 428
ISO 9600, 437
Isodata
    algorithm, 29
    clustering, 28
Isoline, 119
Isophote, 181, 184, 188
    curvature, 184
Isosurface, 380, 405, 411
    rendering, 380
Isotropic, 162, 405
    filter, 18
    tensor, 407
Isovalue, 380
Italy, 528
IT Corp., 537
ITK, 523, 527, 528, 532, 533, 535, 539, 541

Jaccard coefficient, 504, 509
Jaccard similarity, 505
Jacobian determinant, 146
Java, 522
Jeffrey divergence, 207, 210
Jitters, 74
Job submission, 480
Joystick, 396
JPEG, 434, 548
JPEG-LS, 434
JPEG2000, 434, 548
JSW, 364

Kalman filter, 407
Kappa coefficient, 504

Kappa metric, 517
Karhunen-Loève transform, 25, 40, 339
Keratocytes, 307
Kernel, 18, 180
Kernel-based estimation, 173
Kernel-based transform, 523
Khoros, 531
KIN, 455, 459
Kinematic control, 396
Kinetic energy, 6
Kitware Inc., 523, 524, 528, 540, 542
KLT, 168
K-means, 231
  algorithm, 233, 236
  clustering, 28, 238, 243
k-NN, 39, 298, 334
  classifier, 40
  query, 206
Knowledge-based system, 39
Kongsberg SIM, 526
K-space, 8, 64
Kullback–Leibler distance, 523
Kurtosis, 171
KWWidgets, 523

LA, 85
Labeling, 112
Label volume, 380
Laboratory test, 374
LAC, 465
Lagrange multiplier, 179
Lamé's elasticity, 138
Landmark, 132, 135, 148, 150, 293, 294, 299
Landscape model, 119, 120
Laplace operator, 138
Laplacian, 144
  fairing, 387
  filter, 387
  function, 387
  matrix, 388
  model, 144
  smoothing, 387
Larmor frequency, 11, 63
Larmor theorem, 9
Laryngoscope, 20
Larynx, 20
Laser film printer, 549
Lateral
  pathway, 410
  shrinking, 138
Laws mask, 167
3-Layer ANN, 321
LCD, 546
LDA, 334
L_2 distance, 415
Learning
  algorithm, 214
  classification, 38
Least-squares, 407
Left ventricle, 87, 246
Length measurement, 20
Lesion
  burden, 238
  segmentation, 333
Level
  component, 119
  set, 119, 280–282, 291, 524
    function, 290
Levenshtein distance, 39
LGPL, 525
LIDC, 481, 499, 500, 512
Lighting, 22
Light photon, 58
Limbic region, 61
Line of Gennari, 66
Line scanner, 9
Linear
  algebra, 536
  classifier, 284
  elasticity model, 144
  interpolation, 385, 390
  layer perceptron, 321
  system, 14, 300
  transform, 17
Little Endian, 433, 434, 437
Liver, 381, 386, 538
Livewire segmentation, 30
LMM, 308
Load balancing, 480
Local appearance, 298
Local fairing, 386
Local feature, 198, 485
Local gap, 30
Local gradient, 30
Locality parameter, 286
Local MIP, 393
Local orientation, 404

Local search, 298
Localization, 27, 31, 35
LoG, 337
Logarithmic compression, 85
Logistic regression, 319, 320
Log-likelihood, 232
Logreg, 319
Longitudinal stretching, 138
LONI, 534
    pipeline, 535
Lossless compression, 548
Lossy compression, 548
Lower jaw, 40
Low-field MRI, 65
Low-level
    feature, 198, 203, 314
    processing, 2
Low-pass filter, 18, 388, 562
Low-risk population, 560
LRA, 334
LS, 252
LSA, 66
LSO, 56, 58
Lung, 550, 554
    cancer, 555
    nodule detection, 559
LUT, 17, 383, 431, 432, 440
Luv, 474
LV, 85, 96
Lymph node, 387

Machine learning, 280, 314
Mac OS X, 449
Macroscopic imaging, 324
MAF, 528
MAF3, 528
Magnetic
    field, 9, 63
    moment, 9
Mahalanobis distance, 29, 37, 240, 284, 298, 476
Mahalanobis transform, 315
Major axis, 27
Malignant melanoma, 308
Mammography, 6, 168, 447, 466, 559, 561
    lesions, 331
Mandible, 275
Manifold surface, 290

Manual interaction, 396
MAP, 200, 234, 367, 508
Mapping, 530
Marching cubes, 22, 23, 380, 385, 388, 390–392, 524
Marginal distribution, 160
Marker extraction, 121
Markov model, 160
Mask, 18
Matching prior, 284, 297
Material interface, 380
Mathematica, 182
Mathematical
    model, 374
    morphology, 18, 30
MathWorks Inc., 536
MATLAB, 531, 532, 535
Max-star, 276
Maxima, 190
Maximum
    likelihood, 231
    projection, 22
Maxwell's law, 5
Mayo clinic, 542
MC(s), 330, 331, 335, 338, 346–348
MDL, 295
Mean
    curvature, 291
    shape, 286
    shift, 166, 231
    squares, 523
Mean square error, 548
MedGIFT, 484
Median filter, 540
Mediastinum, 550
Medical
    application, 25
    history, 374
    image processing, 76
    imaging toolkit, 527
    informatics, 484, 512
    misadventure, 550
    physicist, 500
    simulation, 530
    ultrasound, 13
MEDICOM, 428
Melanin, 307
Melanocytes, 307
Melanoma, 308

Melanosomes, 307
Membership scene, 269
Membrane model, 144
Menzies method, 311
Mesh, 513
　fairing, 386
　generation, 542
　relaxation, 293
　smoothing, 524
Mesh-to-mesh registration, 295
Mesh-to-volume registration, 294
Meta-data, 535
Meta-information, 478
Meta-rule, 290
Metabolic rate, 61
Metabolism(s), 67, 69
Metallic implant, 30
Metric
　histogram, 202, 208
　space, 206
MeVisLab, 525, 526, 535
MHD, 207
　distance, 210
MI, 143
MICCAI, 499
Micrometric, 404
Microscopy, 6, 20, 28, 29, 35, 193, 442, 541
　imaging, 324
Microstructure, 405
Midbrain, 61
Middle-ware, 535
Millimetric, 404
Mimic approach, 314
Minima, 190
Mining
　approach, 216
　process, 215
Minkowski distances, 206
Minkowski family, 206
MIP, 392, 393, 442
MIPAV, 541
Mirroring, 23
MIT, 56
MITK, 526, 538, 539
　3M3, 538
　ExtApp, 538
Mitral valve, 87, 95, 246
Mixture model, 418

Mixture of Gaussians, 232
ML, 232, 535
MLO, 335
M-mode, 13, 81, 88
MOD, 437
Modality worklist, 452
Model-based approach, 299, 374
Model-based clustering, 231
Model-based estimation, 409
Model-based segmentation, 34, 281
Model-based technique, 390, 398
Model-based visualization, 389
Model-free visualization, 389, 391
Modeling, 230, 532
Model selection, 411
Moiré pattern, 15
Mole, 307
Molecular
　diagnosis, 117
　imaging, 57, 60, 363
Mole mapping, 309, 324
Molybdenum, 6
Moment, 204
　invariant, 204
Monogenetic signal, 170
Monogenic signal, 169, 170
Mono-hierarchical, 482
Mono-modal
　application, 142
　registration, 142
Mono-modality registration, 141
Mono-spectral imaging, 323
Monte Carlo sample, 410
Morphology, 19
　closing, 30
　erosion, 40
　filtering, 16, 19, 40, 43
　gradient, 109, 116, 127
　opening, 30
　operator, 18
　post-processing, 29, 31
　reconstruction, 108, 115, 119
　segmentation, 122
Mosaic panel, 61
Motion
　correction, 548
　mode, 13, 81
　simulator, 145
　tracking, 508

vector, 241
Motoneuron, 35, 36
Moving maximum, 115
Moving medical imagery, 241
Moving minimum, 115
MP, 107
MPM, 200
MPPS, 436, 447, 452
MPU, 390
    implicit, 392
MRA, 273, 393
MRF, 158–160, 165, 199, 200, 231, 235
    model, 234
MRI, 1, 5, 9, 11, 21, 55, 68, 95, 120,
    131, 134, 145, 146, 157, 165, 173,
    186, 192, 273, 275, 294, 312, 330,
    331, 335, 344, 352, 362, 363, 389,
    413, 428, 442, 478, 480, 499, 515,
    532, 547
    gantry, 65
    signal, 407
MRI-based modality, 126
MRM, 145
MRML, 484, 538
MRS, 69, 367
MRSI, 331
MS, 228, 238, 510
M-step, 232
MSW, 127, 128
MTF, 549
Multi-atlas fusion, 148
Multi-axial, 482
Multi-detector CT, 555
Multi-dimensional image, 539
Multi-domain system, 536
Multi-fiber model, 409
Multi-frame, 440
Multi-modal
    CADx, 352
    data, 230
    histogram, 232
    image, 537, 539
    imaging, 149
    registration, 21, 22, 132, 141, 145,
        146, 508
Multi-model representation, 530
Multi-orientation
    analysis, 194
    score, 192
    stack, 191
Multi-protocol MRI, 273
Multi-quadrics, 136
Multi-resolution approach, 285, 294,
    299
Multi-scale, 404
    analysis, 178, 194
    approach, 177
    derivative, 298
    segmentation, 247
    shape analysis, 185
    singularity, 190
Multi-sequence data, 230
Multi-slice CT, 393, 555
Multi-spectral, 326
    imaging, 323
Multi-structure segmentation, 289
Multi-threading support, 528
Multi-variate nature, 408
Multiple sclerosis, 126, 273
Muscle, 541
    growth, 246
Musculoskeletal imaging, 68
Mutual information, 135, 146, 523
MV, 85
Myelination, 408, 418
Myocardial
    infarction, 88
    segmentation, 102

Nabla, 180
NaI(Tl), 58
NA-MIC, 522
    Kit, 522, 523
NASA, 559
Navier equation, 138
Navier–Stokes equation, 139
Navigation, 101
NbTi, 63
NCBC, 522
NCC, 142
NCI, 481
Nearest-neighbor, 523
Neighborhood, 16, 27, 42, 111, 411
4-Neighborhood, 42, 108, 111, 119
8-Neighborhood, 42, 108, 111, 119
Nelder–Meade downhill simplex, 524
NEMA, 428
Netra, 472

Network analysis, 416
Neural
    fiber, 529
    network(s), 38, 40, 284, 556
    pathways, 408
    segmentation, 413
    tractography, 408
Neurodegenerative disorders, 61
Neurogenetic, 147
Neuroimaging, 146, 535
Neurological
    disease, 76
    imaging, 68
Neurology, 21
Neuron, 40
Neuroreceptor, 55, 61
Neuroscience, 76
Neurotransmitter, 55
Neurotransporter, 61
Neutron, 5
Nevus, 307
NHLBI, 559
NIH, 361, 481, 487, 499, 522, 541, 559
NIREP, 508
NLM, 487
NM, 308
NMF, 168
NMI, 143
NMR, 56
NN, 39, 206, 211
NNT, 309
Nobel prize, 56
Noise, 386, 407, 412
    reduction, 29, 390, 557
Non-empty set, 254
Non-imaging data, 374
Non-invertible transform, 144
Non-linear
    diffusion, 188, 193, 195
    scale-space, 193
Non-melanocytic lesion, 326
Non-negativity, 206
Non-parametric
    clustering, 228
    model, 134
Non-rigid
    motion, 246
    registration, 139, 141, 144–147, 149, 289, 293, 508, 548

    transform, 134, 146
Non-symmetric measure, 500
Normal vector, 299
Normalized cut, 241, 415
Normalized cross-correlation, 523
Normalized filter, 179
NP-complete, 245
NRRD, 532
Nuclear imaging, 363
Nuclei, 113
Numerical
    classification, 38
    simulation, 539
NURBS, 290
Nyquist rate, 15
Nyquist theorem, 14

OA, 364
OAI, 373
Object
    appearance, 298
    level, 3
    marker, 125
    occlusion, 227
Objective
    evaluation, 552
    imaging, 373
Oblique reformatting, 541
Observer-machine variability, 506
Observer performance, 552
Occam's razor, 295
Occlusion, 190
OCT, 312
ODF, 405, 406
OFFIS, IX, 529
Omega algorithm, 213, 214, 217
Oncological imaging, 68
OOI, 30, 251, 254, 257, 279, 298, 395, 486, 547
Opacity, 383
    adaption, 383
    function, 380
Open-close filter, 111
OpenGL, 525
Opening, 19, 108, 110, 116
Open Inventor, 525, 533, 535, 542
Open source, 452
Opening
    top-hat, 110, 117

OpenXIP, 525, 526
Operator variability, 373
Ophthalmology, 443
Optical
    colonoscopy, 397, 560
    endoscopy, 397
    flow, 241
    magnification, 324
    model, 382
Optimization, 41, 144, 289, 295, 536
    method, 132
Optimizer, 523
Order-statistic filter, 115
Organ segmentation, 538
Orientation, 27, 170
    bundles, 191
    histogram, 162
    score, 191, 195
Orthogonal radial polynom, 205
OSA, 273, 275
OSGi, 527
OSI, 47
OsiriX, 539
OTS, 551
Otsu, 28
    segmentation, 28
    thresholding, 219, 524
Outer marker, 125
Out-of-plane motion, 89
Over-fitting, 212, 321
Overlap measure, 505
Overlapping anatomy, 374
Over-segmentation, 32, 121, 123, 124, 127, 235, 498, 504

PA, 553
PACS, 1, 45, 435, 455, 458, 476, 480, 487, 539, 550, 553
Page zero problem, 473
Palette, 433
Panoramic radiograph, 3
Pantograph, 115
Parahippocampal region, 66
Parallel
    beam, 93, 99, 100
    implementation, 195
    projection, 134
Parameterization, 36, 135, 294, 295
Parameter vector, 300

Parametric
    clustering, 228
    model, 134, 418
    surface, 290
Param-to-param registration, 294
ParaView, 525, 534, 539
Parenchyma, 330
Parkinson's disease, 66, 76
Partial effect, 15
Partial pixel, 15
Partial volume, 15, 42, 61, 234, 392
Partitioning, 475
Parzen window(ing), 143, 284
Passive shim, 64
Patches, 475
Path, 253
    strength, 255
Pathology, 443, 512
Pathway, 412, 416, 419
    reconstruction, 418
Pattern
    analysis, 310
    recognition, 549
PCA, 37, 72, 168, 284, 286, 290, 316
PCS, 439
PCSP, 440
PCS-Value, 431
PD, 368
PDE, 138
PDM, 37, 280
Pearson correlation, 347, 510
Peer-reviewed article, 552
Perceptron, 321
Performance measure, 483
Perfusion, 369
    imaging, 312
Personal Space Technologies, 102
Perspective projection, 134
PET, 21, 55, 60, 131, 146, 173, 312, 362, 363, 440, 442, 448, 478, 547
PET/CT, 312
PET/MRI fusion, 57
Phantom, 412
Pharmacokinetic model, 369
Phased-array
    transducer, 92
    ultrasound, 547
Phase-encoding gradient, 64
Philips Healthcare, 96

586   Index

Phong
  model, 23
  shading, 23, 36
Photobook, 472
Photography, 6, 134, 311, 442
Photon, 5, 57, 58
  amplifier, 58
Physical phantom, 150, 412
Physics, 562
PicHunter, 472
Pico-molar range, 67
Piezoelectric crystal, 12, 82, 83
Pipeline concept, 524
Pixel
  affinity graph, 244
  clustering, 27–29
  color asymmetry, 316
  compounding, 367
  level, 3
  shifting, 546
  transform, 16
Pixel-adaptive thresholding, 28
Pixel-based analysis, 30
Pixel-based feature, 25
Pixel-based segmentation, 19, 27, 28
Pixmeo Sarl., 540
PK, 368
Plain radiography, 7
Planned interaction, 396
Plaque, 365, 393
Plug-in architecture, 539
PMT, 58
POC, 361
Point-based registration, 135
7-point checklist, 311, 313, 323
Point cloud, 140
Point operation, 16, 18
3-point score, 311
Point set, 290
Poisson's ratio, 138
Polar coordinate, 162
Polarized glasses, 101
Polygonal
  graphics, 380
  isosurface, 384
  model, 381
  reduction, 524
  rendering, 381
  representation, 380

Polynomial approximation, 204
Polyp, 560
POM, 361
Pontine area, 66
Population, 41
Population-based knowledge, 374
Population-based optimization, 295
Portal vein, 392
Position, 27
Positive predictive value, 506, 559
Positron, 5, 57, 58
  emission, 57, 58
Posterior
  distribution, 200
  marginal, 200
  probability, 238
Post-processing, 27, 30, 32, 234
Post-shading, 383
Powell optimization, 524
PowerApps, 532
Power spectrum, 163
ppm, 64
pQCT, 363
P&R curve, 208
Pre-beam forming, 94, 95, 97, 99
Pre-processing, 41, 120, 234
Precession, 9
Precision, 506
Predictive measurement, 374
Presentation
  intent, 442
  state, 438, 440, 452, 529
Pre-shading, 383
Principal curvature, 185, 186
Principal mode, 291
Probabilistic
  clustering, 231
  framework, 411, 476
  segmentation, 147, 233
  tractography, 409, 411
Procrustes
  algorithm, 293
  analysis, 296
Profilometry, 312
Prognosis, 102
Programming language, 531
Projection radiography, 367
Projective transform, 134
Prolapse, 36

Prophylactic screening, 397
Prospective gating, 90, 92
Proton, 5
  density, 63
Proton-weighted
  imaging, 230
  MRI, 229
Prototype, 36
Pruning method, 212
Pseudo-color(ing), 17, 235, 367, 368, 440
Pseudo-landmark, 135
PSF, 74
PSL, 307, 309, 310, 321, 325
  classification, 319–321
  segmentation, 315
PSP, 546, 553
  system, 549, 552
Pulmonary
  artery, 87
  disease, 363
  hypertension, 88
Pulse-echo
  measurement, 83
  signal, 12
Pulse timer, 58
Pulsed-wave Doppler, 13
P-Value, 431
PWF, 455, 459
Python, 522

Q-ball, 410
QBE, 47, 473, 477, 480, 485–487
QBI, 405
QBIC, 472
QDE, 557
Q/R, 436, 437, 441, 456, 457
Quadratic
  distance, 207
  form, 210
Quantitative
  imaging, 373
  measurement, 2
Quantization, 13, 14
  noise, 14
Quantum
  mottle, 557
  noise, 557
Quasi-landmark, 135

Query
  center, 206
  refinement, 486
  response, 479
  result, 482, 486

RA, 85
Radial polynomial, 205
Radiography, 1, 68, 134
  imaging, 363
Radioisotope, 57
Radiology, 512, 562
Radionuclide, 57, 58, 60, 67
Radio-opaque dye, 412
Radiopharmaceutical ligand, 67
Raleigh quotient, 245
Random coincidences, 59
Randomized controlled trial, 325
Range
  data, 281
  query, 206
Rank-1 tensor, 407
Ray
  casting, 24, 383, 393
  tracing, 24
rCMRGlc, 61
R&D, 360
Reading, 541
Readout gradient, 64
Real-time, 94, 97
  imaging, 149
  simulation, 530
  visualization, 23
Recall, 506
Recall-oriented measure, 500
Receive focusing, 86
Receiver coil, 63
Receptive field, 191
Receptor binding, 60
Reconstruction, 119, 382, 556
  from markers, 111
  from opening, 112, 120
  from opening top-hat, 113
  kernel, 555, 561
Red nucleus, 65
Reflection, 11, 109
Region
  growing, 32, 42, 243, 252, 338, 524
  level, 3

merging, 33
prior, 284
Regional
    maxima, 121
    minimum, 121
Region-based feature, 37, 40, 298, 300
Region-based segmentation, 31
Registration, 16, 20, 21, 132, 149, 234, 440, 524, 537–539, 542
    accuracy, 145
    basis, 132
    inter-subject, 132
    intra-subject, 132
    mesh-to-volume, 294
    multi-modal, 132
    param-to-param, 294
    serial, 132
    surface-based, 141
    volume-to-volume, 294
    voxel-based, 141
Regression, 417
Regularization, 139, 144, 194, 281, 291, 409
Regularizing constraint, 282
Related graph, 253
Relative
    calibration, 21
    overlap, 509
Relaxation, 9, 64
Relevance feedback, 473
Reliability, 4
Relief algorithm, 211
Relief-F, 211
Rendering, 101, 398, 442, 525
Repeatability, 325, 363
Repetition time, 10
Report creation, 323
Reproducibility, 412
Research-driven project, 484
Residual, 300
Resolution, 73
Retrieval, 2, 45, 47, 473
Retrospective gating, 92
Rewinding, 63
RF, 9, 62, 83, 557
    coil, 63
    excitement, 9
    shield, 70
    signal, 83

RFC, 258, 267, 270
RGB, 17, 72, 315
Rib artefact, 91
Ridge detector, 182
Riemannian mapping, 413, 414
Riez transform, 170
Right information, 490
Right people, 490
Right time, 490
Right ventricle, 87
Rigid
    registration, 21, 134, 145, 149, 293, 538, 548
    transform, 133, 134, 280, 286
Ring detector, 56, 60
RIRE, 508
RIS, 46, 436, 455, 480, 485
RLE, 434
RNA, 117
Robustness, 4, 142, 256, 261, 299, 528
ROC, 319, 335, 551
    analysis, 551
    curve, 320
ROI, 14, 16, 61, 218, 228, 329, 333, 334, 349, 363, 408, 419, 460, 475, 500, 533
Rosenfeld's connectivity, 269
Rotation, 190, 394
Rotator cuff, 515
RPC, 527
RREP, 145, 508
RSNA, 461, 513
Rule-based segmentation, 289
Rule-based system, 240
Run-length encoding, 548
RV, 85, 199
RWTH Aachen University, 46

Saddle point, 187, 190
Salient point, 475
Sampling, 13, 14, 383, 405
    theorem, 14, 15
Scalar visualization, 524
Scale-based affinity, 266
Scale-based FC, 273
Scale-space, 190, 245
Scaling, 190, 394
Scatter/Scattering, 7, 23, 59, 83, 557
Scene, 252

analysis, 47, 486
  graph, 525
  level, 3
Schizophrenia, 126, 173, 234
Scilab, 536
Scintillation, 58, 61
  detector, 58, 73
Scintillator, 56, 58
SCIRun, 532
Sclerosis, 234
Scouting image, 91
SCP, 434, 437
Screening, 309, 395
Scripting integration, 528
SCU, 434, 437, 462
SD, 337
SE, 335
Secondary capture, 435
Second moment, 179
Second-order
  derivative, 189
  statistics, 560
Sector scan, 13
Seed
  point, 32, 252, 408
  region, 410, 414
Segmentation, 23, 25, 27, 35, 37, 41, 102, 148, 199, 254, 262, 332, 333, 337, 338, 347, 383, 390, 413, 533, 537, 540–542, 547, 548, 561, 562
  atlas-based, 288
  framework, 524
  multi-structure, 289
  rule-based, 289
Selection bias, 325
Self-intersection, 294
Self rotation, 9
Semantics, 3
  gap, 4, 198, 472
  meaning, 32
  network, 44
  query, 483, 484
  retrieval, 516
Semi-transparent, 381, 395
Semiconductor, 58
Sensitivity, 99, 219, 309, 412, 507
Sensory gap, 472
SER, 349
Serial registration, 132, 142, 146, 508

Serie, 430
Service class, 433
Set-theoretic complement, 109
SFM, 330
SGI Inc., 525
SGLD, 203
Shading, 22, 24
  correction, 541
Shadow, 227
Shannon, 14
Shannon-Wiener entropy, 142
Shape, 27, 199, 204, 298, 475
  analysis, 392
  guidance, 524
  index, 186
  model, 148, 149, 295
  parameter, 299
  prior, 102, 282, 291
  space, 299
  vector, 186, 187
Shape-based feature, 204
Shape-based segmentation, 538
Shape-from-shading, 43
Shape-memory force, 283
Shared information, 141
Shear, 133
Shift integral, 191
Shim
  correction, 64
  system, 62
Shrinking
  effect, 282
  problem, 261
Shuttle
  bed, 68
  railway, 71
  system, 70
Side lobe, 99
Siemens Inc., 60, 532
Siemens Medical Solutions, 97
SIF, 466
SIFT, 190
Signal
  enhancement, 369
  noise, 412
  processing, 58, 536
  theory, 388
Signature, 38, 190, 198, 206
Silicon-based US, 99

Silicon Graphics Inc., 526
SIM, 316
Similarity, 205
  measure, 141, 142
  metric, 132, 144, 523
  query, 206
Simplex mesh, 281, 290
Simulation, 532
  algorithm, 530
  application, 530
Simulink, 536
Simultaneity, 58
Single-tensor model, 417
Sink, 260
Sinogram, 59, 74, 173
SINR, 455, 459
Sinus rhythm, 91
SiPM, 58
Size, 27
Skeletal
  imaging, 6
  radiograph, 33
  structure, 391, 393
Skeleton, 19, 30, 41
Skeletonization, 390, 542
Skewness, 171, 388
Skin
  lesion, 323
  peeling, 274
  surface microscopy, 312
Skull-implanted marker, 145
Slice-based viewing, 389
Slicer, 537
Slice selection, 63, 64
Slim-tree metric, 208
Small-world phenomena, 416
SMART, 506
Smoothing, 18, 32, 187, 387
  algorithm, 386
  connected filter, 125
  edge-preserving, 187
  effect, 261
  filter, 119
Snake, 34, 35, 271, 281
SNOMED, 437, 445
SNR, 7, 14, 59, 63, 86, 99, 100, 189, 234, 366, 548
Sobel
  filter, 30

mask, 18
operator, 18, 26
SOFA, 530
Softassign Procrustes, 293
Soft plaque, 393
SOP, 371, 433, 447
SOR, 138, 139
Sound wave, 11
Source, 260
Space elements, 252
Spatial
  compounding, 367, 547
  convolution, 23
  domain, 19, 21, 199
  encoding, 63
  frequency, 8, 64, 163
  information, 230, 380
  registration, 440
  resolution, 12, 55, 68, 98, 362, 393, 549
Spatio feature space, 166
Spatio-temporal
  feature, 228
  information, 241
Specificity, 219, 309
Speckle reduction, 561
SPECT, 21, 131, 362, 363, 478
Spectral
  imaging, 312
  method, 415
Spectrum, 323
Spherical
  harmonic modeling, 405, 407
  shell, 405
Sphericalness, 185
Spin, 9
  flip, 63
Spin-echo sequence, 10, 11
Spine X-ray, 486
Spin-lattice relaxation, 10, 64
Spinning proton, 9
Spin-spin relaxation, 10, 64
Spiral CT, 8
Splatting, 393
Spline-based FFD, 137
Spline-based transform, 135
Spline curve, 280
Split and merge, 32
Splitting, 32

Spoiling, 63
Squamous cell carcinoma, 307
SR, 366, 367, 442, 443
SRN, 338
SSD, 142
SSM, 280, 291, 293, 295, 298, 308
Stack reconstruction, 115, 119
Stain debris, 113
Staircasing artifact, 383, 385, 388
Standard
    affinity, 267
    monitor, 101
    programming, 531
STAPLE, 148, 503
Static thresholding, 27
Statistical
    approach, 26
    association rule, 215
    atlas, 234
    classification, 38
    clustering, 228
    model, 148, 298
    shape variation, 286
    test, 412
Statistics, 536
Steering, 82, 98
Stiffness, 35
Stitching artifact, 97
Storage, 2
Streamline, 410, 415
    tractography, 409, 415
Strel, 19
Stretching, 16
Striatum, 61
Strike artifact, 100, 101
Stroke, 126, 366
    volume, 88
Structel, 19
Structure(al)
    approach, 26
    data, 373, 443, 478
    display, 441, 452
    element, 19, 108, 115
    feature, 485
    imaging, 363, 367
    matrix, 185
    MRI, 403, 413
Study, 430
    setting, 325

Suavity, 204
Subdivision surface, 391
Subjectivity, 363
Sub-millimeter resolution, 57, 70
Subnetwork, 416
Subregion, 416
Subsampling, 15
Substantia nigra, 61, 65–67
Subtraction, 21
Sub-voxel decomposition, 385
Sum-min, 276
Sum-product, 276
Sun spot, 308
Super-conduction, 63
    coil, 65
    wire, 63
Super-resolution, 366, 368
Supervised classification, 38
Supplement, 428
Support vector machine, 284
Sup-reconstruction, 114, 119, 120
Surface, 3, 290
    distances, 500
    extraction, 524
    fairing, 386
    generation, 542
    mesh, 542
    microscopy, 312
    model, 242, 389
    reconstruction, 22, 23
    registration, 243
    rendering, 275, 380, 389
    smoothing, 388
Surface-based registration, 141
Surface-based rendering, 22, 36
Surface-based shading, 24
Surface-based visualization, 24
Surgery planning, 273
Surrogate endpoint, 361
SVD, 140
SVM, 334, 341, 476
Symbolic description, 3
Symmetry, 170, 206
Synaptic bouton, 35, 43
Synovial joint, 364
Syntactic classification, 38
Synthetic phantom, 412

Tailored acquisition, 405

Taubin filter, 389
Taxonomy, 280
Taylor series expansion, 242
TBM, 147
TCL, 522
Tcl/Tk, 523
TCP, 47
TCP/IP, 434
Technical variability, 370
TEE, 96
Teem library, 532
Tegmental area, 61
Tele-consultation, 323
Tele-dermatology, 326
Tele-medicine, 2
Tele-radiology, 548
Template, 18
Temporal
 compounding, 547
 information, 230
 resolution, 98, 99, 362
 subtraction, 558
Temporomandibular joint, 16
Tensor, 405
 contraction, 181, 182
 field, 413, 529, 542
 model, 407
 scale, 266
 tractography, 410
 visualization, 524
Terminals, 260
Testing, 216
Texel, 26
Texton, 26, 163, 173
Text retrieval, 480
Texture, 199, 203, 298, 474, 525
 analysis, 338, 339
 anisotropy, 173
 classification, 337
 clutter, 244
 definition, 157
 description, 158
 element, 26
 level, 3
 matching, 298
 measure, 31
 model, 300
 prior, 284
 visualization, 524

Texture-based feature, 26
TF, 393
Thalamus, 413
Therapy, 22
 planning, 389, 390, 395, 547
Thin-plate
 model, 144
 spline, 136, 293, 294, 523
Third moment, 171, 204
Thomson effect, 7
Threshold-based segmentation, 380
Threshold(ing), 27, 229, 230, 315
 decomposition, 115, 119, 120
 sets, 115
TIFF, 46
Time
 compensation, 84
 motion diagram, 13
Tissue
 density, 83
 elasticity, 82, 83
 segmentation, 416
T-junction, 188
TM diagram, 12
TMG, 126
TM-mode sonography, 13
7.0 T MRI, 66
Toolbox, 536
Toolkit, 521, 543
Top-down, 31
Top-hat
 concept, 113
 filter, 110, 117
 transform, 117
Topic development
 axes, 482
Top-level, 321
Topographic surface, 33
Topology, 291
 scene, 253
 view, 253
Top-point, 177, 190
Top-view surface, 115
Toshiba Medical Systems, 102
Total body mapping, 309
TPS implant, 32
Tracking system, 528
Tract-based studies, 417

Tractography, 404, 408, 409, 411, 412, 416, 418, 548
Trained classification, 38
Training, 216, 395
  data, 36, 298
  sample, 280
  set, 296
Transducer, 11–13, 82, 83, 91, 97, 99, 331
Transfer
  function, 383, 384
  syntax, 435
Transform, 523
  affine, 286
  axis-symmetric, 286
  B-spline, 286
  diffeomorphic, 139
  rigid, 286
  similarity, 286
Transformation model, 132
  affine, 133
  non-rigid, 134
  projective, 134
  rigid, 133
Transitivity
  error, 509
  property, 502
Translation, 394
Translational research, 563
Transmission, 2
  ratio, 11
Transmitter coil, 63
Transparency, 227, 381, 384
  modulation, 382
Transparent shading, 22
Transporter-enriched nuclei, 61
Transport theory, 382
Treatment
  follow-up, 102
  planning, 22
TREC, 483, 506
TRECgenomics, 483
Triangle
  mesh, 387
  surface, 385
Triangular inequality, 206
Triangulated mesh, 280
Triangulation, 22, 290, 380
Trigonometric basis function, 135

Trilinear interpolation, 382, 383, 390
Tungsten, 6
T1-weighted MRI, 64, 229
T2-weighted MRI, 64, 66, 229, 273
Two-level rendering, 383
Two-tensor
  method, 410
  model, 417
Two-way beam, 100

UCLA, 56, 534
UID, 450, 481
UK, 550
Ultra high-resolution MRI, 75
Umbilical point, 185
Umbrella, 387
  operator, 386
  region, 386
Under-segmentation, 32, 498, 504
Uniformity, 204
Unimodal registration, 21
United Kingdom, 528
United States, 57, 81, 308, 331, 333, 335, 352, 362, 397, 459, 487, 499, 532, 534, 536–538, 541, 542, 550, 554
University of Iowa, 509
Unix, 449
Unsharp masking, 556
Unsupervised
  classification, 38
  clustering, 231
Untrained classification, 38
U of U, 532, 534
UPF, 539
Upper airway, 275
US, 547
Usability, 552, 563
USB, 437
USC, 465
Use case, 450
User-driven model, 552
User interaction, 102, 552

Validation, 145, 412, 464
  protocol, 132
Valmet software, 506
Value
  domain, 19

range, 14, 21, 25
Vanderbilt database, 508
Variance, 204
Vascular
    catheter, 562
    disease, 554
    stenosis, 555
    structure, 389
    tree, 389
VBM, 147
VD, 504
Vector
    field, 407, 542
    space, 206
    visualization, 524
Vectorized language, 536
Velocity vector, 242
Vena cava, 87
Vendor-driven model, 552
Venn diagram, 504
Ventricular hypertrophy, 246
Verification database, 325
Vertex
    computation, 385
    connectivity, 290
Vessel, 554
    architecture, 331
    centerline, 390
    diameter, 390, 391
    lumen, 554
    tree segmentation, 538
Vesselness, 185
    filter, 393
Video endoscopy, 395, 397
Viewing plane, 24
Virtual
    camera, 396
    colonoscopy, 397, 547
    endoscopy, 395, 396
Visage imaging, 542
Visiana Ltd, 37
Visible
    Human, 24
    light image, 442
    woman, 273
VisTrails, 534
Visual
    appearance, 280
    cortex, 66, 190

feature, 484
programming, 531
query, 482
smoothing, 387
word, 475
Visualization, 2, 12, 22, 36, 101, 290,
        379, 408, 524, 526, 527, 532, 533,
        535–539
    mode, 13
    technique, 389
VL, 536
VME, 528
VOI, 432
VOI LUT, 440
VOLCANO, 511
Volume
    cell, 380
    growing, 32
    mesh, 542
    rendering, 22, 24, 101, 274, 380, 382,
        389, 398, 532, 562
    segmentation, 532
    visualization, 524, 541
Volume-to-volume registration, 294
Volumetric
    imaging, 92, 101
    overlap, 500
VolView, 524, 541
Voxel, 8, 22
    neighborhood, 411
Voxel-based approach, 374
Voxel-based registration, 141
Voxel-based similarity, 141
Voxel-based studies, 417
Voxel–Man, 24
VPE, 531, 533, 534
VR, 429, 542
VSD, 87
VSG, 526
VTK, 523, 527–530, 533–535, 539
vtkINRIA3D, 529

Wall
    motion, 89
    thickening, 89
Walsh function, 166, 167
Washington University, 56, 532
Water, 28
    parting, 34

Water-based gel, 12
Watershed
  from markers, 124
  lines, 34, 122
  over-segmentation, 123
  segmentation, 524
  transform, 33, 122, 542
Wavelet, 72, 166, 191, 298
  basis function, 135
  coefficient, 337
  transform, 25, 199
W3C, 461
Web-based interface, 485
Web service, 534, 552
White matter, 31, 200, 228–230, 234, 273, 413
Whole-body MRI, 45, 55
Williams index, 505
Windowing, 556
Windows, 449
Worklist server, 529
WS, 252, 262
WSDL, 450
WxWidgets, 539

XA, 432
XA/XRF, 440
XIP
  Platform, 532
  Reference Host, 532
XML, 450, 452
X-ray, 4
  absorption, 20
  beam, 558
  fluoroscopy, 97
  image, 233
  mammography, 335
  photon, 557
XRF, 432

Young's modulus, 138
YUV, 315

Zernike feature, 209
Zernike moment, 204, 208
Zero crossing, 189, 190
ZIB, 542
Zooming, 394

Printed by Books on Demand, Germany